I0065829

Principles of Paleoecology

Principles of Paleoecology

Editor: Anne Offit

RCALLISTO REFERENCE

www.callistoreference.com

Callisto Reference,
118-35 Queens Blvd., Suite 400,
Forest Hills, NY 11375, USA

Visit us on the World Wide Web at:
www.callistoreference.com

© Callisto Reference, 2017

This book contains information obtained from authentic and highly regarded sources. Copyright for all individual chapters remain with the respective authors as indicated. All chapters are published with permission under the Creative Commons Attribution License or equivalent. A wide variety of references are listed. Permission and sources are indicated; for detailed attributions, please refer to the permissions page and list of contributors. Reasonable efforts have been made to publish reliable data and information, but the authors, editors and publisher cannot assume any responsibility for the validity of all materials or the consequences of their use.

ISBN: 978-1-63239-836-9 (Hardback)

The publisher's policy is to use permanent paper from mills that operate a sustainable forestry policy. Furthermore, the publisher ensures that the text paper and cover boards used have met acceptable environmental accreditation standards.

Trademark Notice: Registered trademark of products or corporate names are used only for explanation and identification without intent to infringe.

Printed in the United States of America.

Cataloging-in-publication Data

Principles of paleoecology / edited by Anne Offit.
   p. cm.
Includes bibliographical references and index.
ISBN 978-1-63239-836-9
   1. Paleoecology. 2. Paleobiology. 3. Ecology. I. Offit, Anne.
QE720 .P75 2017
560.45--dc23

Table of Contents

Preface

This book includes some of the vital pieces of work being conducted across the world, on various topics related to paleoecology. It strives to provide a fair idea about this discipline and to help develop a better understanding of the latest advances within this field. Paleoecology refers to the study of fossils, sub-fossils, fossil organisms and their remains to examine the past ecosystem. The main aim of paleoecology is to understand the life cycle, environmental conditions, living interactions and deaths of organisms, in order to reconstruct natural environment. This book brings forth some of the most innovative concepts and elucidates the unexplored aspects of this field. For all readers who are interested in this subject, the case studies included in this text will serve as an excellent guide to develop a comprehensive understanding. It will serve as a valuable source of reference for graduate and post graduate students.

Every book is initially just a concept; it takes months of research and hard work to give it the final shape in which the readers receive it. In its early stages, this book also went through rigorous reviewing. The notable contributions made by experts from across the globe were first molded into patterned chapters and then arranged in a sensibly sequential manner to bring out the best results.

It has been my immense pleasure to be a part of this project and to contribute my years of learning in such a meaningful form. I would like to take this opportunity to thank all the people who have been associated with the completion of this book at any step.

Editor

Compound-Specific $\delta^{15}N$ Amino Acid Measurements in Littoral Mussels in the California Upwelling Ecosystem: A New Approach to Generating Baseline $\delta^{15}N$ Isoscapes for Coastal Ecosystems

Natasha L. Vokhshoori*, Matthew D. McCarthy*

Ocean Sciences Department, University of California Santa Cruz, Santa Cruz, California, United States of America

Abstract

We explored $\delta^{15}N$ compound-specific amino acid isotope data (CSI-AA) in filter-feeding intertidal mussels (*Mytilus californianus*) as a new approach to construct integrated isoscapes of coastal primary production. We examined spatial $\delta^{15}N$ gradients in the California Upwelling Ecosystem (CUE), determining bulk $\delta^{15}N$ values of mussel tissue from 28 sites between Port Orford, Oregon and La Jolla, California, and applying CSI-AA at selected sites to decouple trophic effects from isotopic values at the base of the food web. Bulk $\delta^{15}N$ values showed a strong linear trend with latitude, increasing from North to South (from ~7‰ to ~12‰, $R^2 = 0.759$). In contrast, CSI-AA trophic position estimates showed no correlation with latitude. The $\delta^{15}N$ trend is therefore most consistent with a baseline $\delta^{15}N$ gradient, likely due to the mixing of two source waters: low $\delta^{15}N$ nitrate from the southward flowing surface California Current, and the northward transport of the California Undercurrent (CUC), with ^{15}N-enriched nitrate. This interpretation is strongly supported by a similar linear gradient in $\delta^{15}N$ values of phenylalanine ($\delta^{15}N_{Phe}$), the best AA proxy for baseline $\delta^{15}N$ values. We hypothesize $\delta^{15}N_{Phe}$ values in intertidal mussels can approximate annual integrated $\delta^{15}N$ values of coastal phytoplankton primary production. We therefore used $\delta^{15}N_{Phe}$ values to generate the first compound-specific nitrogen isoscape for the coastal Northeast Pacific, which indicates a remarkably linear gradient in coastal primary production $\delta^{15}N$ values. We propose that $\delta^{15}N_{Phe}$ isoscapes derived from filter feeders can directly characterize baseline $\delta^{15}N$ values across major biochemical provinces, with potential applications for understanding migratory and feeding patterns of top predators, monitoring effects of climate change, and study of paleo-archives.

Editor: Arga Chandrashekar Anil, CSIR- National institute of oceanography, India

Funding: Sources of funding that have supported this work were awards received from Friends of Long Marine Lab (http://seymourcenter.ucsc.edu/about-us/people/board/; http://seymourcenter.ucsc.edu/wp-content/uploads/2013/10/SREA-2013-2014-guidelines.pdf) and from the Myers Oceanographic Trust. Funding for Open Access provided by the University of California, Santa Cruz, Open Access Fund. The funders had no role in study design, data collection and analysis, decision to publish, or preparation of the manuscript.

Competing Interests: The authors have declared that no competing interests exist.

* E-mail: nvokhsho@ucsc.edu (NLV); mccarthy@pmc.ucsc.edu (MDM)

Introduction

Isotope spatial gradients, or *Isoscapes*, are maps of systematic isotope variation and provide important biogeochemical information. Isoscapes are becoming increasingly important tools to characterize major biogeochemical zones and gradients in the ocean, and have been also used in ecological studies to help constrain animal migration and fish stock patterns (e.g., [1,2]). Isoscapes of nitrogen (N) stable isotope values ($\delta^{15}N$) can be particularly informative, because such measurements have the potential to identify major ocean transitions between eutrophic/mesotrophic and oligotrophic regions, the balance of fundamental N cycle processes (e.g., N fixation vs. denitrification), and also basic ecological and food web relationships across major habitat zones. For example, water-column denitrification has a very large isotope effect (ε) of 25–30‰ [3] which greatly increases the $\delta^{15}N$ value of all organisms in areas where this process is important. However, this could be rapidly changing in many ocean regions, linked to oceanographic climate events associated with a shifting

climate (e.g., [4] Detailed isoscapes can ultimately provide a link between biogeochemical process and larger food webs, a key for understanding marine ecosystems. This is especially critical at a time when both natural and anthropogenic perturbations may be rapidly shifting fundamental biogeochemical processes (e.g., [5]), and potentially entire food web structures [6,7].

However, the information potential inherent in $\delta^{15}N$ values also presents significant challenges for interpretation of bulk $\delta^{15}N$ values of organic matter. First, isoscapes are typically constructed from measurements in secondary or higher consumers. This approach provides a temporally integrated measurement; however, by definition, it also results in measured $\delta^{15}N$ values being offset from "baseline" $\delta^{15}N$ values of primary production, since the ^{15}N content of a consumer increases substantially with each trophic transfer. An average trophic enrichment factor (TEF) of ~3.4‰ is often assumed [8–10], however it has been shown that the TEF values in fact vary substantially: not only between species, but also depending on tissue type, life stage, growth rate, and a host of other factors [10–12]. Further, for many oceanographic applications, such as understanding shifting gradients in primary production or N cycle processes, it is really the "baseline" $\delta^{15}N$ value that is of primary interest (i.e., the $\delta^{15}N$ value of primary production or N sources at the base of food webs). Because the bulk isotope value in a consumer is the combined signal of the baseline value *and* subsequent trophic effects, it is extremely difficult to isolate either factor with confidence.

Compound-specific isotope analysis of amino acids (CSI-AA) is a rapidly evolving technique that can address many inherent issues with bulk isotope data. For $\delta^{15}N$ values, a seminal study by McClelland & Montoya (2002) demonstrated strong differential ^{15}N enrichment of different groups of amino acids (AA) with trophic transfer. One group of AA has strongly elevated $\delta^{15}N$ values with each trophic transfer ($\sim$4–8‰), and are now termed the "Trophic AAs." A second group of AA, now termed the "Source AAs," in contrast has relatively constant $\delta^{15}N$ values with trophic transfer, and so largely preserves $\delta^{15}N$ values from the base of the food web. This pattern of AA differential enrichment has now been verified across a wide range of photoautotrophs and primary consumers [13,14] and also in higher trophic organisms [15–19].

Most $\delta^{15}N$ CSI-AA studies to date have focused on nitrogen isotopic values of two main AAs: phenylalanine (Phe), as the best indicator of baseline $\delta^{15}N$ value, and glutamic acid (Glu), as the best indicator for relative trophic transfer. The relative predictability of ^{15}N offsets between Glu and Phe with trophic transfer has also led to an explicit equation now used widely to calculate CSI-AA based trophic position (*see methods*). Based on these findings, CSI-AA patterns ($\delta^{15}N_{AA}$) have now been used to not only estimate trophic position (TP) [16,20,21] and to trace source or microbial re-working of organic N sources [4,19,22–24], and animal movement across broad ocean basins [19,25]. Taken together, CSI-AA work to date strongly suggests that if $\delta^{15}N_{AA}$ is applied in appropriate heterotrophic organisms, the source AA should be able to indicate baseline $\delta^{15}N$ isoscapes, decoupled from influence of trophic transfer.

The California mussel (*Mytilus californianus*) is a sessile resident of intertidal zones, which continuously filters particulate organic matter (POM). As such, mussels correspond closely to an ideal "baseline indicator" organism (i.e., a long-lived primary consumer; [10]). Because mussels temporally integrate filtered POM into their tissues and shells over annual to decadal timescales, they have been widely used as both sentinel organisms for marine pollutants (e.g., Mussel Watch Project: http://ccma.nos.noaa.gov/about/coast/nsandt/musselwatch.aspx), as well as to attempt reconstruction of ocean water composition and conditions [6–10,26–29]. In contrast, many other organism types have important drawbacks for constructing representative isoscapes. For example, highly mobile top predators may rapidly transit distinct biogeochemical zones (e.g. [8–15,18,19,30]), and thus attenuate isotopic variability. In contrast, short-lived organisms (such as zooplankton) can be assumed to not move widely, but because of relatively fast growth rates and rapid N turnover times may not integrate variation, but rather are subject to short temporal isotopic changes in the environment (e.g. [16]). Because of their sessile nature, cosmopolitan distribution, and continuous integration of water column food sources, mussels have major advantages as a potential basis for coastal isoscapes.

Here we examined $\delta^{15}N_{AA}$ patterns in California mussels (*Mytilus californianus*) across 10 degrees of latitude in the coastal zone of the California Upwelling Ecosystem (CUE). The CUE is part of the greater California Current System (CCS), and is a highly dynamic region where we would anticipate not only large potential variation in baseline $\delta^{15}N$ values, but also the potential

for rapid future change linked to a warming climate [26,28,29]. Our overall goal was to explore whether $\delta^{15}N$ values of source AA, and in particular Phe ($\delta^{15}N_{Phe}$), may can serve as a new, direct proxy for constructing isoscapes of integrated $\delta^{15}N$ values of primary production within highly dynamic coastal regions. We compared bulk $\delta^{15}N$ and $\delta^{15}N_{AA}$ patterns in mussels to first test dependence of bulk isotopic variability on baseline $\delta^{15}N$ values, using CSI-AA to constrain variations in TP. We also compared mussel $\delta^{15}N$ values with literature values in more offshore sample types (zooplankton and POM), to examine if our results may also apply to the larger CCS. Our results indicate that source AA values in mussels are likely represent a direct record of variation in baseline $\delta^{15}N$ values, and suggest that in the CA coast region isoscapes based on $\delta^{15}N_{Phe}$ closely follow variations in nitrate $\delta^{15}N$ values.

Methods

Sample Collection and Preparation

Ethics Statement. California mussels (*Mytilus californianus*) analyzed for this study were collected from 28 different sites between Coos Bay, Oregon and San Diego, California (Table 1), under a permit provided by the California Department of Fish and Wildlife.

Mussels were collected in the winter (Dec – Feb) of 2009–2010. Sites were chosen to be approximately evenly distributed along the CA coastline, with $\sim$80 km geographic separation between each sampling site. Our main goal here was to sample mussels from a wide geographic range across the CCS, although for observing finer scale local or regional variations, a finer-scale sampling strategy would like be required. Typically 5 individual mussels were collected from each site, all between 30–40 mm maximum shell length, which were immediately placed on dry ice until further preparation. The adductor muscle of each individual was dissected for analysis. This tissue was selected because isotopic values in muscle tissue have shown relatively long turnover times; based on past growth data, mussels of this size would be expected to integrate approximately annual variability in suspended food source isotopic values for each location sampled [31]. The dissected adductor tissue was carefully separated from other tissue types, rinsed with deionized water, refrozen, and then freeze-dried for 48 hrs. Lipids were removed following the methods of Dobush et al. (1985) [32], using petroleum ether in a Dionex Accelerated Solvent Extractor (Bannockburn, IL). Finally, in preparation for CSI-AA, composite samples were made from a subset of 13 collection sites (Fig.1a). For each location chosen for CSI-AA (based on the bulk $\delta^{15}N$ record), 1 ± 0.05 mg of lyophilized tissue was weighed and combined for each individual mussel ($n = 5$). Further CSI-AA preparation proceeded as described below.

Bulk Stable Nitrogen Analysis

Stable nitrogen isotope analyses were conducted using standard protocols in the Stable Isotope Lab at the University of California, Santa Cruz (UCSC-SIL). Briefly, homogenized muscle tissue of each individual was weighed into tin capsules and combusted. Isotope values determined on a Carlo Erba 1108 elemental analyzer (Lakewood, NJ) coupled to a Thermo Finnigan Delta Plus XP isotope ratio mass spectrometer (San Jose, CA) (EA-IRMS). Analytical error associated with this measurement typically $<\pm0.15$‰ based on sample replicates. Stable isotopes are reported using standard delta (δ) notation in parts per thousand (‰):$\delta^{15}N = [(R_{sample}/R_{standard}) - 1] \times 1,000$, where R is the ratio of heavy to light isotope, R_{sample} is from the sample, and the $R_{standard}$ is atmospheric N_2 (air) for carbon, as provided by

Table 1. Collection sites and bulk $\delta^{15}N$ values for *Mytilus californianus*.

Site	Identifier	Habitat Type	Latitude	Longitude	n	$\delta^{15}N$	SD
Humbug Mtn./Port Orford, OR	HMPO	Rocky	42°43'N	124°28'W	6	7.8	0.2
Meyer's Creek Beach, OR	MCPR	Rocky	42°18'N	124°25'W	5	7.4	0.3
Pelican State Beach, CA	PSB	Rocky	42°00'N	124°12'W	6	8.3	0.2
Lagoon Creek, CA	LC	Rocky	41°36'N	124°06'W	5	8.3	0.4
Humboldt Lagoon	HL	Rocky	41°13'N	124°06'W	5	8.6	0.2
Luffenholtz Beach, CA	LB	Rocky	41°02'N	124°07'W	6	8.2	0.3
Point Cabrillo Lighthouse, CA	PCL	Rocky	39°21'N	123°48'W	5	8.4	0.3
Schooner Gulch	SG	Rocky	38°52'N	123°39'W	5	9.2	0.4
Stillwater Cove Marine	SWC	Rocky	38°32'N	123°17'W	5	9.8	0.2
Bodega Bay, CA	BB	Rocky	38°19'N	123°04'W	5	11.1	0.5
Pacifica, CA	PAC	Rocky	37°39'N	122°29'W	5	10.0	0.1
Half Moon Bay, CA	HMB	Jetty	37°29'N	122°27'W	5	9.3	0.3
Davenport, CA	DAV	Rocky	37°00'N	122°10'W	5	10.0	0.3
Santa Cruz, CA	SC	Rocky	36°56'N	122°03'W	6	10.9	0.2
Moss Landing, CA	ML	Jetty	36°48'N	121°46'W	4	10.2	0.2
Asilomar, CA	ASI	Rocky	36°38'N	121°56'W	5	9.4	0.3
Rocky Point, CA	RP	Rocky	36°24'N	121°54'W	6	10.0	0.4
Mill Creek, CA	MC	Rocky	35°58'N	121°29'W	5	9.5	0.3
Morro Bay, CA	MB	Rocky	35°22'N	120°51'W	5	10.0	0.2
Gaviota, CA	GAV	Rocky	34°28'N	120°13'W	4	10.4	0.1
Santa Barbara, CA	SB	Rocky	34°23'N	119°42'W	5	11.4	0.2
Ventura, CA	VEN	Jetty	34°16'N	119°17'W	3	10.7	0.1
Malibu, CA	MAL	Rocky	34°01'N	118°45'W	6	10.1	0.2
Topanga, CA	TOP	Rocky	34°02'N	118°34'W	4	10.7	0.4
Venice Beach, CA	VB	Jetty	33°58'N	118°28'W	4	11.3	0.4
San Clemente, CA	SCL	Pier	33°25'N	117°37'W	5	10.3	0.3
Oceanside, CA	OCE	Jetty	33°13'N	117°23'W	5	10.8	0.3
La Jolla, CA	LAJ	Rocky	32°50'N	117°16'W	5	11.5	0.3

Reported $\delta^{15}N$ values represent averages for all individuals collected from each location (4 to 6 individuals, 35–45 cm size range; *see methods*). "Identifier" indicates the abbreviation used in text for specific sites; "habitat type" indicates if site was natural rocks or an artificial structure; n = number of individuals collected; SD is the standard deviation for the $\delta^{15}N$ values measured from all individual mussels sampled from a given site.

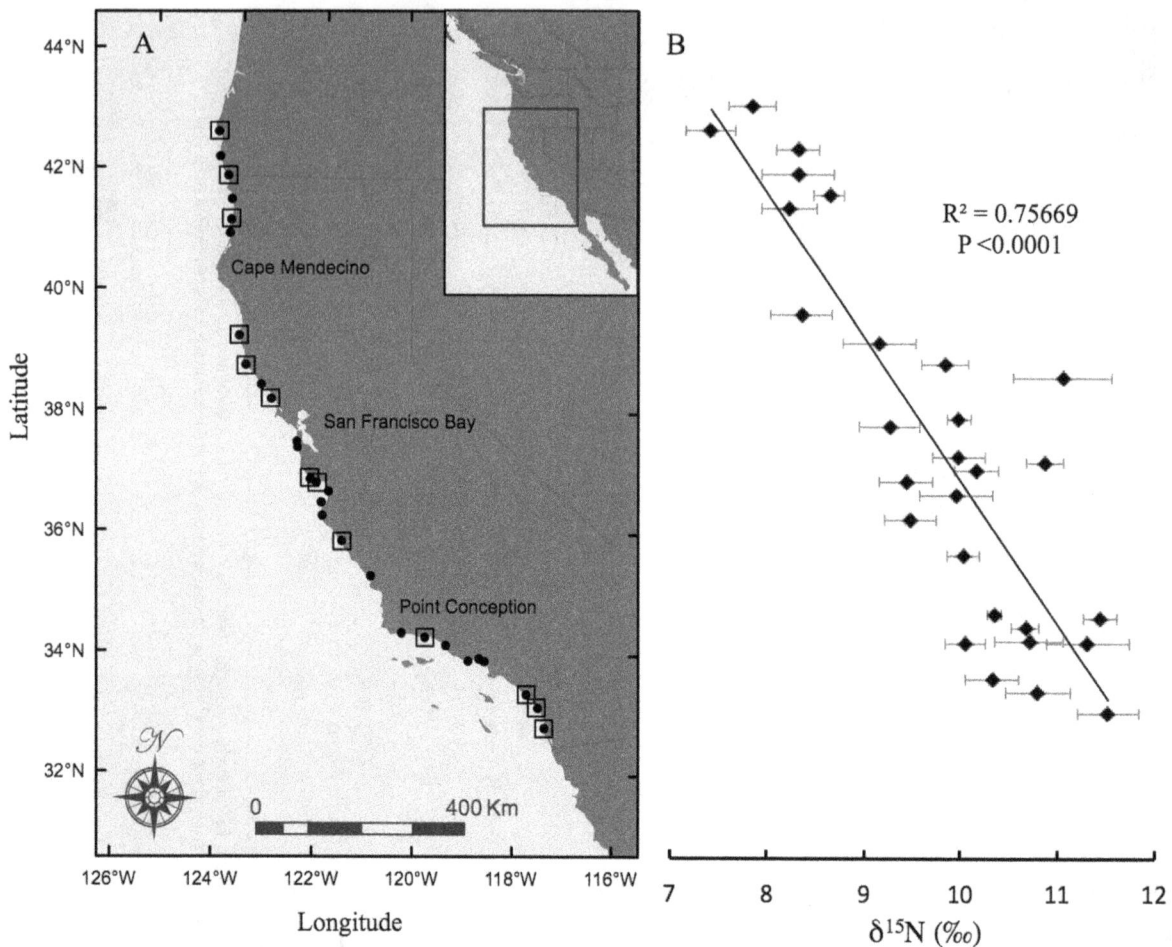

Figure 1. Collection sites and bulk δ¹⁵N values. Site-specific bulk $\delta^{15}N$ values of *Mytilus californianus* as a function of latitude, in context of a map of collection sites on the California coast. (A) Filled circles indicate all sampling sites, and map locations correspond directly to bulk analysis values in panel B; open squares represent sites chosen for compound-specific isotope analysis. (B) Filled diamonds indicate average $\delta^{15}N$ for five individuals sampled from each site; error bars indicate ±1SD. Regression line and statistics indicate strong linear relationship of $\delta^{15}N$ values versus latitude.

pulses of calibrated CO_2 reference gas. For details on correction calculations and normalization to international standards refer to UCSC-SIL website (http://es.ucsc.edu/~silab/index.php).

Compound-specific amino acid $\delta^{15}N$ analysis

Amino acid $\delta^{15}N$ values were measured as Trifluoroacetyl isopropyl ester (TFA-IP) AA derivatives, following protocols described in detail elsewhere (e.g.,. Briefly, samples were hydrolyzed (6 N HCl, 20 hr at 110°C) under nitrogen, and TFA derivatives subsequently prepared from free AA using a modified version of the protocol described by Silfer (Silfer et al. 1991): isopropyl esters were made with a 1:5 mixture of Acetyl Chloride (AcCl):2-propanol (110°C, 60 minutes), and then acylated using a 1:3 mixture of Dichloromethane:Trifluroacetyl acetate (DCM:TFAA) (100°C, 15 minutes). Derivatized AAs were dissolved in DCM to a final ratio of approximately 4 mg of original tissue to 250 µl DCM.

After derivatization, samples were analyzed by a Varian gas chromatograph coupled to a Finnegan Delta-Plus isotope ratio mass spectrometer (GC-IRMS). AAs were separated using a 50 m, 0.32 ID Hewlett Packard Ultra-1 column with 1 µm film thickness. Under our analytical conditions, $\delta^{15}N$ values could be

reproducibly measured for alanine (Ala), aspartic acid + asparagine (Asp), glutamic acid + glutamine (Glu), leucine (Leu), isoleucine (Ile), proline (Pro), valine (Val), glycine (Gly), lysine (Lys), serine (Ser), phenylalanine (Phe), threonine (Thr), and tyrosine (Tyr) (Fig. S4). Most AAs were measured with a standard error of <1.0‰ (based on $n = 4$ injections), and the average mean deviations for individual AA $\delta^{15}N$ measurements across all tissue sample replicates was 0.5‰.

Amino Acid Categories and Trophic Position Calculations

In all results and discussion, measured AA are grouped into one of three categories: "Trophic" vs. "Source" (*after* Popp et al. 2007 [15]) and Thr alone is designated as a "metabolic" AA (*after* Germain et al. 2013 [18]). The measured Trophic AA (with large expected enrichment in ¹⁵N with trophic transfer) were Glu, Asp, Ala, Ile, Leu, Pro and Val. The measured Source AA (with expected little to no change in $\delta^{15}N$ at higher trophic levels) were Phe, Gly, Ser, Lys and Tyr. The $\delta^{15}N$ values of Thr exhibit an apparent inverses isotopic fractionation with trophic transfer, however are also highly variable with organism type [18], so this AA is considered outside the basic Trophic vs. Source division.

Based on this framework, we used unweighted averages of AA groupings, as well as specific TP estimates, to analyze our data. Average Trophic and Source $\delta^{15}N$ values were calculated:

$$Source\ AA\ \delta^{15}N = Average\ \delta^{15}N[Phe,Gly,Ser,Lys,Tyr]\quad(1)$$

$$Trophic\ AA\ \delta^{15}N = Average\ \delta^{15}N \\ [Glu,Asp,Ala,Ile,LeuPro,Val]\quad(2)$$

For explicit TP calculations, we used the "canonical" AA's (Glu and Phe) to calculate TP of mussels in the CUE, after Chikraishi et al. 2009:

$$TP_{Glu-Phe} = \frac{(\delta^{15}N_{Glu} - \delta^{15}N_{Phe}) - 3.4}{7.6} + 1 \quad(3)$$

where, $\delta^{15}N_{Glu} - \delta^{15}N_{Phe}$ are measured values, +3.4 is the assumed isotopic difference between the Glu and Phe in primary producers (also referred to as the β value), and +7.6 is the assumed ^{15}N enrichment in Glu relative to Phe with each trophic transfer from food source to consumer, also called the Δ value [13].

Statistical analyses and calculations

Statistical analyses (e.g., Hierarchical cluster analysis and Analysis of Covariance) were conducted using the JMP statistical software package (SAS Inc., Version 10). We used Arc-GIS Spatial Analyst (version 10.1) to produce visual isoscapes of the CUE. Our first model is based on the line for $\delta^{15}N_{Phe}$ values vs. latitude $(y = -0.3328x + 20.053, R^2 = 0.63592)$. Our second model (Fig. S1) is based on one-dimension of $\delta^{15}N$ values along the latitudinal extent of sampling area and interpolates between data points of known $\delta^{15}N$ values and to 100km offshore.

Results

Bulk $\delta^{15}N$ values

Bulk $\delta^{15}N$ values in the adductor muscle of *Mytilus californianus* ranged from 7.4‰ to 11.5‰ (Table 1). Bulk $\delta^{15}N$ values were measured on tissue from multiple individual mussels (4–6, but typically 5) collected from each site to gauge intra-site variability in individuals. Standard deviations on average $\delta^{15}N$ values for individual mussels from the same sites ranged from 0.1 to 0.5‰ (Table 1). The average standard deviation for all intra-site comparisons, across all locations, was 0.3‰. This value is close to EA instrument error (~0.2 ‰), and therefore indicates an extremely small degree of variation in individual mussel $\delta^{15}N$ values within specific sites, implying instead strong homogeneity of $\delta^{15}N$ values for mussel populations. When plotted as a function of latitude, the average bulk $\delta^{15}N$ values for mussels from each site have a strong linear trend (Fig. 1B). The average bulk $\delta^{15}N$ values increase by 0.41‰ per degree of latitude from north to south $(R^2 = 0.755$ and $P < 0.0001)$. We note that all sampling sites were located in exposed waters, and variable habitat type (i.e. rocky, jetty, etc.; see table 1) did not appear to be a major factor in the overall latitudinal trend in isotopic value.

Amino Acid $\delta^{15}N$ values

Of the 28 sites measured for bulk analysis, 13 were chosen for CSI-AA (Fig 1A). Samples for CSI-AA were chosen first to obtain

relatively even geographic spacing, with specific locations within geographic regions then selected to capture maximum offsets in the north to south bulk $\delta^{15}N$ trend (Fig. 1B). Measured $\delta^{15}N$ values for individual AAs ranged from $-1.0‰$ to $16.0‰$ (Table 2). In all samples, Thr was distinct, with the lowest $\delta^{15}N$ values, the Source AA group always had intermediate values, and the Trophic AAs always had the highest $\delta^{15}N$ values (Table 2; Fig. S2). Over the entire data set, the range of the averaged Source AA $\delta^{15}N$ (*see methods*) was 5.1 to 10.3‰ (SD = 1.3‰, $n = 69$) and the averaged Trophic AA values are 10.3 to 15.9‰ (SD = 0.4‰, $n = 91$). While precision for individual AA $\delta^{15}N$ measurements varied (Table 2), it was typically $< 1‰$, with the average analytical standard error across all AA we measured at all sites as 0.8‰ $(n = 160)$.

We focused on Glu and Phe $\delta^{15}N$ values as the best proxies for Trophic and Source AA groups, respectively, as has been indicated by a number of recent papers [19,33–36]. Glu and Phe $\delta^{15}N$ values both correlated significantly with average values for Trophic and Source AA groups respectively (Phe vs. average Source AA, $R^2 = 0.782$; $P = 0.0006$; Glu versus average Trophic AA's, $R^2 = 0.546$, $P = 0.0049$), confirming the validity of this approach (see also [25]). Both Phe and Glu $\delta^{15}N$ values also tracked changes in bulk $\delta^{15}N$ with latitude (Fig. 2). The $\delta^{15}N$ values of Phe and bulk adductor muscle had a strong and significant linear relationship with latitude $(P = 0.0028$ and $P = 0.0011$, respectively). In contrast, there was more variability in the Glu data. The relationship of Glu vs. bulk $\delta^{15}N$ was not significant at 95% confidence $(P > 0.05)$, however a Fit Model run of Analysis of Covariance shows that $\delta^{15}N$ values of bulk, Phe and Glu all share a common slope (effects test, $P = 0.0050$); in other words, the slope of $\delta^{15}N$ change with latitude for Glu and Phe are not significantly different from the slope of bulk $\delta^{15}N$ change with latitude.

Mussel Trophic Position and Trophic Enrichment Factors

The TP of mussels calculated using Eq. 3 ranged from 1.0 to 1.8 with an average TP of 1.4±0.3 (Table 2). TP had no correlation with latitude $(P = 0.706)$, indicating that despite local variability, mussels' suspended POM food sources had similar average TP in all CA coastal regions. Across all mussel samples analyzed with CSI-AA, the average $\delta^{15}N_{Glu} - \delta^{15}N_{Phe}$ offset was 6.5‰. Prior work indicates that these mussels feed primarily on microalgae [37–39], coupled with data for $\delta^{15}N_{Glu} - \delta^{15}N_{Phe}$ offsets in phytoplankton and marine macroalgae [13,22], this average offset would indicate an average TEF$_{Glu-Phe}$ for *Mytilus californianus* of 3.1‰ (Fig. S3).

Discussion

This study investigated if $\delta^{15}N$ CSI-AA values measured in mussel "bio-archives" can represent a new approach to understanding baseline $\delta^{15}N$ patterns in dynamic coastal regions. We hypothesize that the potential for $\delta^{15}N_{AA}$ values to decouple trophic shifts from baseline $\delta^{15}N$ values may, for the first time, allow construction of isoscapes specifically for temporally integrated $\delta^{15}N$ values for primary production, based tissue samples from heterotrophic organisms. Mussels were chosen for this study because they are sessile, filter-feeding organisms with tissue turnover rates integrating suspended POM food sources on monthly to annual timescales. Other studies have previously CSI-AA or bulk $\delta^{15}N$ patterns in primary producers, zooplankton [31], or mobile top predators, to trace oceanographic processes [15]. However, as noted above, in many heterotrophic organisms multiple variables might complicate inferences about baseline isotopic signals. For plankton these include variability caused by

Table 2. Compound specific amino acid $\delta^{15}N$ values for *Mytilus californianus*.

Site	n	Bulk δ15N	Average Trophic	Average Source	Trophic Position	Glu-Phe	Trophic Glu	±	Asp	±	Ala	±	Ile	±	Leu	±	Val	±	Pro	±	Source Gly	±	Ser	±	Lys	±	Tyr	±	Phe	±	Metabolic Thr	±
HMPO	4	7.8	10.3	5.1	1.3	5.9	12.1	1.3	10.2	0.3	9.8	0.3	9.4	0.6	11.6	0.8	7.6	0.8	11.4	0.4	2.9	0.7	6.0	0.7	nd	nd	nd	nd	6.3	1.4	-2.2	0.3
PSB	4	8.3	11.8	6.3	1.3	5.8	12.0	0.7	10.1	0.4	13.1	0.1	11.7	0.4	12.7	0.2	9.4	0.8	13.2	0.3	6.0	1.0	6.7	1.0	nd	nd	nd	nd	6.2	0.4	-2.1	0.8
HL	4	8.6	12.3	7.6	1.3	5.6	12.7	0.1	10.5	0.2	12.4	0.4	12.3	0.5	13.5	0.2	10.3	1.0	14.0	0.2	7.1	1.2	7.0	1.2	11.0	0.7	6.0	0.5	7.2	0.6	-0.7	0.2
PCL	4	8.4	13.4	5.4	1.6	8.1	14.3	0.1	11.6	0.3	14.4	0.2	13.2	0.7	14.7	0.3	11.0	1.0	14.6	0.5	5.6	0.6	5.6	0.6	3.9	2.2	5.5	0.5	6.2	0.7	-3.1	0.3
SG	4	9.2	12.5	5.8	1.2	5.0	10.5	0.6	11.7	0.2	14.4	0.3	12.3	0.9	13.4	0.1	13.2	0.9	11.6	0.5	6.8	0.1	7.6	0.1	nd	nd	3.2	1.0	5.5	0.3	-0.1	0.3
BB	4	11.1	15.9	8.2	1.8	9.5	17.4	0.8	14.3	0.4	18.1	0.5	16.0	0.5	16.7	0.4	16.5	0.6	12.4	0.5	9.3	1.0	8.6	1.0	8.6	0.6	6.6	0.8	7.9	0.4	-0.9	0.6
DAV	4	10.0	14.0	7.5	1.6	8.0	16.2	0.3	14.0	0.2	12.8	0.6	12.0	0.9	16.4	0.5	10.8	0.9	15.6	0.3	7.4	0.6	9.6	0.6	6.7	0.9	5.8	1.9	8.2	0.8	-0.5	0.5
SC	4	10.9	12.3	5.1	1.7	8.4	15.3	1.1	13.5	0.1	12.1	0.5	12.0	0.7	14.7	0.3	6.0	1.3	12.6	0.5	5.3	1.0	8.4	1.0	nd	nd	nd	nd	6.9	1.0	0.4	0.3
MC	4	10.0	13.3	8.0	1.3	5.4	13.7	0.2	11.8	0.2	13.5	0.7	13.4	0.8	14.4	0.2	11.3	0.2	14.6	0.3	8.3	0.3	7.4	0.3	nd	nd	nd	nd	8.3	0.5	-1.1	0.2
SB	4	11.4	15.1	9.0	1.6	8.3	16.3	0.1	14.0	0.1	16.5	0.2	14.6	0.2	15.7	0.1	15.5	0.6	13.2	0.3	10.0	0.3	9.9	0.3	9.8	0.2	7.4	0.7	8.0	0.4	1.5	0.3
SCL	4	10.3	12.7	8.4	1.2	4.7	13.4	0.1	12.1	0.1	12.9	0.1	11.9	0.1	12.9	0.1	12.7	1.0	13.1	0.1	9.2	0.3	9.2	0.3	9.0	0.6	5.7	1.2	8.7	0.4	2.6	0.2
OCE	4	10.8	13.2	10.3	1.0	3.3	13.9	0.2	12.3	0.1	13.3	0.2	13.0	0.4	13.7	0.2	12.4	0.2	13.8	0.2	11.1	0.2	9.8	0.2	12.1	0.8	7.6	0.8	10.7	0.4	3.2	0.1
LAJ	4	11.5	14.3	7.7	1.4	6.6	15.4	0.3	13.4	0.2	15.1	0.2	13.8	0.2	15.2	0.2	14.6	0.8	12.4	0.5	8.8	0.8	9.8	0.2	4.9	1.4	6.3	1.6	8.8	0.7	-0.4	0.3

$\delta^{15}N$ AA for individual amino acids from 13 individual collection sites, ± the analytical standard deviation from replicate injections (*see methods*). Site abbreviations, bulk $\delta^{15}N$ values, and "n" refer to data for specific sites selected for CSI-AA (Table 1). Trophic, Source and metabolic categories, amino acid abbreviations, and calculated averages for Trophic and Source AA's and Trophic Position are as defined in text

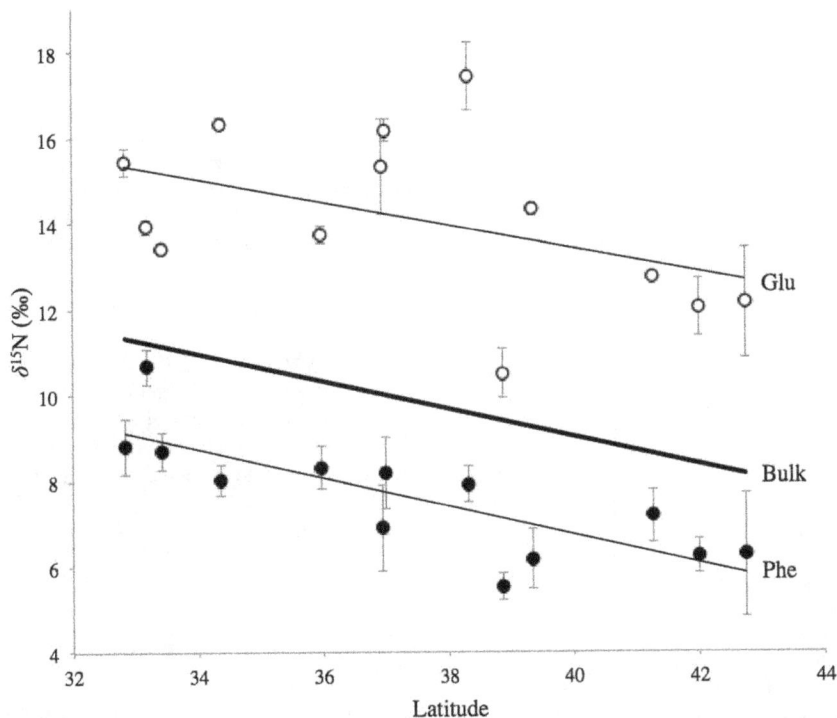

Figure 2. Latitudinal trends in Glu and Phe $\delta^{15}N$ values compared with bulk $\delta^{15}N$. Bold solid line indicates regression for bulk $\delta^{15}N$ values for all sites sampled (as in Fig. 1), thin solid lines indicate linear regressions for Glu (open circles) and Phe (filled circles) ($\delta^{15}N_{Glu} = -0.270x+24.211$, $\delta^{15}N_{Bulk} = -0.313x+21.527$, and $\delta^{15}N_{Phe} = -0.329x+19.927$). ANCOVA analysis indicates that all three share a common slope within error (effects test, $P<0.0050$). Error bars for Glu and Phe indicate analytical standard deviation for CSI-AA performed on a composite sample for all individual mussels from each site (*see methods*).

shorter biochemical turnover rates, coupled with seasonal change in nutrient availability, light intensity and temperature fluctuate [16], while in higher trophic level animals factors such as migration or a mixed diet may dilute the desired signal in question [30]. Overall, we expect that sessile filter feeders such as mussels are likely to be are among the best organisms for baseline source records in systems where they occur.

$\delta^{15}N$ latitudinal gradient in the California Upwelling Ecosystem

We hypothesize that the strong $\delta^{15}N$ gradients with latitude are driven by the mixing of two NO_3^- source waters, coupled by upwelling in the CUE. In this region, northern low-^{15}N water is brought south by surface flow of the main CCS [3,40]. At the same time, southern source of elevated ^{15}N water originating from the zone of denitrification in the ETNP [33,34,36,41,42] is brought north via the California Undercurrent (CUC). The source area for the CUC (approximately south of the tip of Baja CA peninsula) is one of the major persistent oxygen minimum zones (OMZ) in the world ocean [3,40] accounting for 35–45% of global pelagic denitrification [36,41–43]. Bacterial denitrification in the low-oxygen water columns has a very large positive fractionation factor ($\varepsilon \sim 20$–30‰; e.g., [3,34]), and therefore imprints a distinct signal in surrounding waters, such that subsurface NO_3^- values for the southern CUC near the tip of Baja can approach +14‰ [36,43–45]. The CUC then moves northward, with its flow attenuating as it progresses along the CA margin. The core of the CUC is near 150 m, directly in source-depth regions for upwelled waters [34,43]. The main isotopic endmembers for inorganic nitrogen along our study region are therefore open Pacific nitrate

from CCS ($\sim$ 5‰, e.g. Sigman et al., 2009), mixed with the ^{15}N enriched nitrate being carried northward by the CUC ($\sim$9–10 ‰; [44–46]), and brought to the surface locally via upwelling. We note, however, that while all previous literature clearly indicates an expected change in baseline $\delta^{15}N$ with latitude, it cannot indicate exact $\delta^{15}N$ endmembers for the CA coast region we sampled. This is because of the relative paucity of direct nitrate ^{15}N measurements, and the inherent temporal and geographical variation in these measurements, even within similar regions (e.g., [43,46–48]).

A complimentary forcing for $\delta^{15}N$ trends could therefore also be variability of upwelling intensity with latitude. The North American west coast is commonly described in terms of three distinct upwelling regions, characterized by differences in overall annual upwelling intensity: Baja California (21–30°N), continental US (30–48°N) and British Columbia and Alaska (48–60°N) [46,47]. While winds generally increase northward, annual wind intensity is most consistent year-round south of Pt. Conception, strongest seasonally along central CA coast, and generally weakest north of Cape Blanco [39,46–48]. Our study site does not cross all three of these main regions, however it seems possible that the transition between the southern and central zones of upwelling intensity could influence the overall latitudinal trend. Overall, while expected variation in average upwelling intensity is consistent with our observations of latitudinal $\delta^{15}N$ change, wind forcing alone cannot not explain the clear linear decrease in $\delta^{15}N$ values with increasing latitude.

Local denitrification is another process that could also contribute to regional $\delta^{15}N$ baseline values. Water column denitrification in the CCS has been documented in borderland basins (such as in the Santa Barbara Basin, SBB; [47]), where water column exchange with the ocean is blocked by basin sills, causing basin water to become O_2 deficient [34,44,47,49]. It has also been shown to occur in areas along the Oregon coast, due to advection of oxygen-poor water masses onto continental shelves [50]. However, if local denitrification were a main factor driving relative $\delta^{15}N$ values, we would predict far more localized $\delta^{15}N$ variability. Therefore, while this cannot be ruled out as contributing to $\delta^{15}N$ values for specific locations, it seems highly unlikely as the major forcing for such a regular gradient. Finally, both water temperature and sampling season might be considered as additional factors. As noted above, all mussels were collected in the winter season of 2009–2010. While it is possible that mussel metabolism may change throughout the year (high vs. low feeding seasons), the specific tissue analyzed (adductor muscle, *see methods*) and mussel size class were specifically selected to isotopically integrate over an approximate yearly time frame. This assumption is supported by preliminary data of samples collected in both summer and winter season of the same year for selected sampling sites, for which no significant effect on the observed latitudinal trend was observed [39,45,51]. Water temperature is also a general function of latitude in the CCS at all times of year. Change in water temperature might affect the isotopic gradient either directly via changing mussel metabolism, or indirectly as a proxy for upwelling strength. However, in contrast to the mussel N isoscape, the major temperature changes along the CA coast are not linear, but rather shift more strongly at the boundary of the Southern CA Bight, with temperatures generally much warmer south of Pt. Conception, and consistently much cold temperatures (due to stronger upwelling) in central and northern CA.

Overall, the strong latitudinal $\delta^{15}N$ trend recorded in mussel tissues seems most consistent with the endmember mixing outlined above, consistent with both modeling and prior discrete sampling. For example, $\delta^{15}N$ values of sediment traps and sediment cores contrasted between central CA vs. the Southern CA bight have indicated $\delta^{15}N$ values are generally more enriched in the Southern CA bight vs. Northern CA, consistent with our measurements [34,44,49,52,53]. In addition, basin-scale modeling of $\delta^{15}N$ variation [3,45,51,54] also predict a south to north trend of decreasing baseline $\delta^{15}N$ values, driven by the ETNP denitrification endmember. Our study therefore represents perhaps the strongest confirmation to date of both model predictions, and also prior discrete-location sampling results. However, no prior sample set has ever tested CCS latitudinal $\delta^{15}N$ variation at such high resolution, based on an archive coupling unambiguous source location with approximately annual signal integration. In particular, the striking linear trend in our data is a novel, and also perhaps a surprising finding. This indicates that the diminution in ^{15}N-enriched nitrate supply via the CUC (or relative mixing with southerly CCS) is remarkably regular in the CUE: across the 10 degrees of latitude that we sampled, $\delta^{15}N$ change was remarkably consistent (0.41‰±0.04 per degree). We suggest that the ability to capture this regional trend at such high precision is linked to the integrative property of filter feeding consumers, as well as the longer-turnover tissue that we sampled. Overall, we propose that Fig. 1b indicates the integrated approximately annual gradient in ^{15}N values in coastal CUE waters with latitude. If correct, this also suggests that surveys of costal mussel $\delta^{15}N$ values might constitute a powerful new tool for constraining physical mixing and circulation models, since they would show the effective mixing of two source waters in great detail.

While satellite data has documented changing global ocean surface chlorophyll concentrations, leading to predictions of declining primary production in the world's oceans due to increased stratification associated with warming [49,52,53,55], the effects of a warming climate on CCS biogeochemistry remain unclear. Some studies have proposed that some CCS zones are already showing an opposite trend of increasing productivity, linked to increased nutrient supply [3,22,28,54]. If the ocean nitrate endmember were to increase over time, this should result in gradual decrease in $\delta^{15}N$ values in the CUE, and potentially also a change in the slope of the clear latitudinal gradient we observe. At the same time, increasing stratification in CCS waters is also proposed as one main consequence of warming, and this has already been documented [13,49,56]. If this decreased the effective supply of CUC water and associated nitrate, it could also lead to lower $\delta^{15}N$ values. However, the potential effects of natural climatic perturbations (e.g. El Nino Southern Oscillation – ENSO, and Pacific Decadal Oscillation – PDO cycles) are currently very difficult to decouple from longer term trends. Given that our current understanding of physical and biological responses of the CUE to a changing climate remains poor. Repeated sampling of mussels could provide a time and geographically integrated record of baseline isotopic change in this system, revealing longer term trends in the regional $\delta^{15}N$ gradient, due to either natural fluctuations or climate change, at near annual resolution.

Do coastal mussel $\delta^{15}N$ data also reflect broader California Current $\delta^{15}N$ values?

An important question is to what degree $\delta^{15}N$ data derived from mussels may reflect isotopic values of broader coastal waters, as opposed to only littoral sources and process. This is likely to be a function of relative time scale: the time frame over which mussels integrate $\delta^{15}N$ of primary production, vs. the mixing time scale for littoral water with more seaward coastal water masses. If water mixing is relatively rapid vs. sampled tissue isotopic turnover, then it is possible mussels would reflect isotopic values within the broader CUE, and possibly into shoreward extent of the CCS. In contrast, if upwelling and nutrient utilization in the littoral zones are rapid and strongly localized, then littoral mussel $\delta^{15}N$ values could be mostly decoupled from values in more offshore coastal waters. To definitively address this question, an extensive sampling program would likely be required, comparing offshore/onshore POM isotope values with those recorded in mussel tissues.

However, for the well-studied Monterey Bay region, past work offers extensive data sets for both coastal and offshore $\delta^{15}N$ values in both organisms and detrital OM. We therefore compiled $\delta^{15}N$ values for a range of sample types from the Monterey Bay region, and compared these with $\delta^{15}N$ values for mussels at our sampling sites in or near the bay (Fig. 3). Specifically, we compared average $\delta^{15}N$ values for 2 herbivorous and 2 carnivorous zooplankton species [22,45,57] and also OM in sediment traps (450 m) and surface sediment (950 m) samples[13,15,16,24,34,35,39]. Because many of these samples are not primary consumers, $\delta^{15}N$ change due to trophic transfer must be taken into account for any direct comparison. We therefore assumed that mussels and herbivorous zooplankton, as primary consumers would feed at the same TP as mussels (TP = 2), requiring no correction. For carnivorous zooplankton (secondary consumers) we assumed one additional trophic transfer (TP = 3), and therefore adjusted reported $\delta^{15}N$ values by 3.4‰ (the most broadly accepted average bulk TEF value). For sediment trap and surface sediment samples, we used recent results from Monterey Bay long-term sediment trap records, which have indicated an average trophic position (determined by CSI-AA) of 1.6 (Sherwood and McCarthy, *unpublished data*; also similar to TP for POM at Station ALOHA,[20,33,45]). We therefore adjusted both trap and surface sediment values for 1.5 TP (+1.7‰).

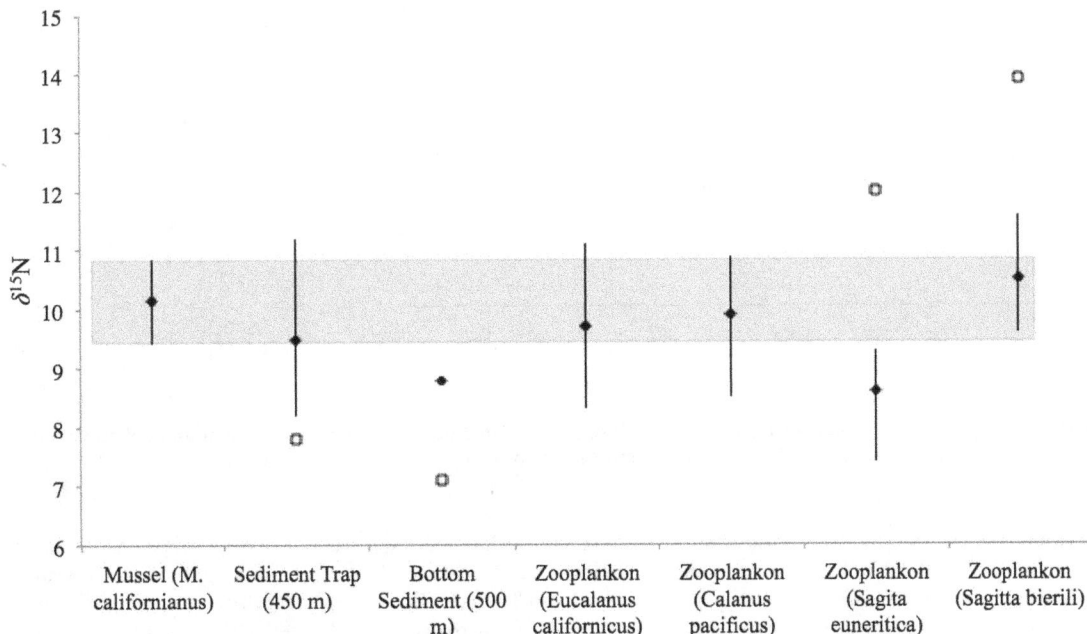

Figure 3. $\delta^{15}N$ values for littoral mussels compared with offshore sample types in Monterey Bay region. $\delta^{15}N$ values for *Mytilus californianus* (this study) compared with literature values for sediment trap and bottom sediments (Altabet et al. 1999), primary consumer zooplankton (*Eucalanus californicus* and *Calanus pacificus*) and secondary consumer zooplankton (*Sagitta euneritica* and *Sagitta bierii*) (Rau et al. 2003) from Monterey Bay. For literature sample types diamonds indicate values corrected for estimated trophic position, open squares are the reported average literature value. Shaded band indicates range of our measured $\delta^{15}N$ values for mussels from three Monterey Bay sampling sites (SC, ML, ASI); for literature sample types error bars indicate main range for reported values.

While we acknowledge that this approach can provide only a very general initial comparison, the results are nevertheless quite encouraging (Fig.3). For most sample types, the Monterey Bay adjusted values fall directly within the $\delta^{15}N$ range for local mussels. This suggests that littoral mussels may in fact reflect $\delta^{15}N$ values more broadly for local coastal waters. Given the high wind stress and mixing characteristic of the Central and Northern CA coasts, this may not be surprising. However, clearly this represents only a preliminary comparison, and rests on a range of assumptions that remain to be fully tested (e.g. that time of sampling is relatively unimportant, or that bulk TEF values are accurately estimated). To fully explore the potential of littoral mussels as integrators of coastal isoscapes, we suggest a synoptic sampling program comparing offshore/onshore mussel isotopic values will be required.

Mussel trophic position

CSI-AA provides a unique opportunity to decouple the effects of trophic transfer from $\delta^{15}N$ values at the base of the food web. Based on differential enrichment behavior of the Trophic vs. Source AAs introduced above [13,15,16,21,24,35,39,52], CSI-AA allows for a direct assessment of the role for TP variation may play in bulk $\delta^{15}N$ value trends. Our calculations of mussel TP for an extended population along the entire CA coast represents, to our knowledge, the first wide-ranging CSI-AA survey of any filter feeding mollusk population in nature. The lack of any trend in TP with latitude (Fig. 4) indicates that mussels spanning the entire CA coast feed at a very similar TP, therefore likely on similar food sources, independent of location. The consistency of TP is interesting, given the previously documented localized variation in $\delta^{13}C$ values for mussels from these same locations (Vokhshoori et al. in press). The similarity of TP from all locations therefore

supports the conclusion that $\delta^{13}C$ variation is primarily driven by changes in baseline $\delta^{13}C$ values. Since TP does not change with latitude also strongly supports our basic hypothesis that the overall $\delta^{15}N$ trend with latitude (Fig. 1b) is also driven by north to south variation in baseline $\delta^{15}N$ values, most likely linked to nitrate $\delta^{15}N$ values.

However, the exact TP values (average TP = 1.4 ± 0.2; Fig. 4a) calculated using the standard Glu-Phe approach (*see methods*, Eq. 1), also are lower than would be expected. Mussels predominately feed on POM derived from primary production [58,59], so as primary consumers the mussel TP values should be at least 2. There are at least two possible explanations for the lower TP indicated by CSI-AA. One relates to non-algal food sources that might contribute to mussel AA, for example detrital POM (Vokhshoori et al. in press) or non-algal primary production sources. For example, some primary producers can have different baseline CSI-AA patterns from marine microalgae (e.g., sea-grasses; [2,13,15,37,39]), such that substantial contributions from non-microalgal sources would change the calculated TP calculated using the standard equation (*see Text S1*). However, $\delta^{13}C_{AA}$ source fingerprinting applied to these same mussels has verified a dominantly marine microalgal diet [38,39], in agreement with ecological expectations diet (e.g.,[3,33]). This suggests that if different source $\delta^{13}C_{AA}$ patterns do account for the offset, it is more likely due to variations among different microalgal groups. Given the relatively limited current $\delta^{13}C_{AA}$ data on different marine algal lineages, this is possible. An alternate possibly, is that the change in Glu $\delta^{15}N$ values with trophic transfer in mussels may be smaller than the TEF factor now most commonly applied (*and assumed in Equation 1; see methods*). Accumulating evidence now suggests that CSI-AA trophic enrichment factors may be specific for different groups of organisms [18,34,36,40–42,60], however TEF values offsets have so far been documented only higher TP

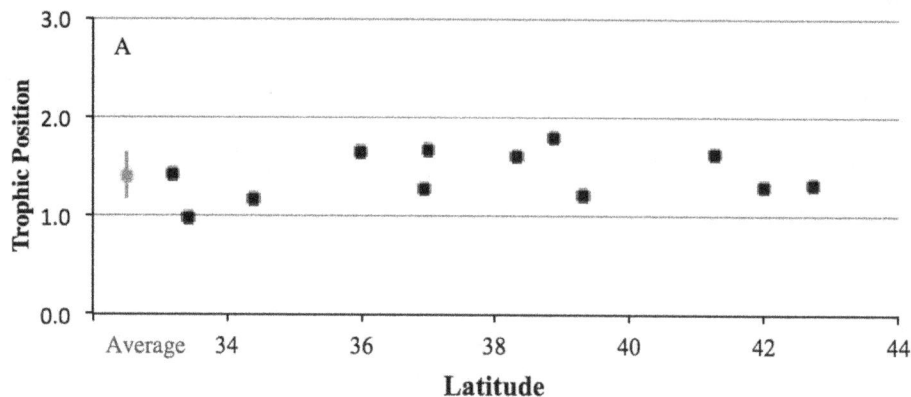

Figure 4. CSI-AA based trophic position for mussels from California Coast. Calculated trophic position (TP) for composites of mussels from the sites selected for CSI-AA (filled squares) and average TP ± 1SD for all mussels sampled along CA coast (grey square)

predators. Ultimately, distinguishing between these possibilities is beyond scope of our current data, however we provide a broader explanation of the underlying issues in Text S1.

Overall, however, it is important to stress that apparently low TP values for mussels do not bear in any significant way on our main observations. Specifically, the constant TP with latitude indicated by CSI-AA data is independent of exact TP estimates. However, TP data does suggest that controlled feeding experiments with filter-feeding mollusks, together with a more extensive survey of variation in the δ^{15}N-AA in different algal types, will be needed to clearly interpret TP values derived from mussel tissue or shells. Such work might also be important for future potential development of CSI-AA patterns in archeological mussel shells as potential paleoceanographic bioarchives.

AA-CSIA: new tool to reconstruct primary production δ^{15}N isoscapes

As noted above, all literature to date has indicated that Phe δ^{15}N values are closely linked to δ^{15}N values at the base of the food web, such that $\delta^{15}N_{Phe}$ values in a heterotroph can be used to estimate δ^{15}N values of average primary production sources [3,5,15,16,34–36,43,44]. Given that these mussels feed almost uniquely on microalgae [37–39,46–48,59] we therefore hypothesize that mussel $\delta^{15}N_{Phe}$ should represent a temporally integrated value for δ^{15}N of coastal phytoplankton production. Sampling $\delta^{15}N_{Phe}$ in mussel populations along a coastline should therefore yield, for the first time, a way to construct an integrated isoscape of baseline δ^{15}N values.

A direct comparison between bulk δ^{15}N and $\delta^{15}N_{Phe}$ values is one way to test this idea (Fig. 5). If we assume average complete NO_3^- utilization for this region (at least over ~annual time frames mussels integrate; e.g., [49,52]), then the slope of the regression for bulk δ^{15}N values (Fig. 1b) should also represent the *gradient* in NO_3^- δ^{15}N values along the CA coast. In this case, the "baseline" δ^{15}N values should also be essentially equivalent with NO_3^- δ^{15}N values. However, the bulk δ^{15}N relationship of course cannot directly represent baseline values, due trophic transfer enrichment factors, as well as tissue-specific offsets. In Fig. 5 we therefore

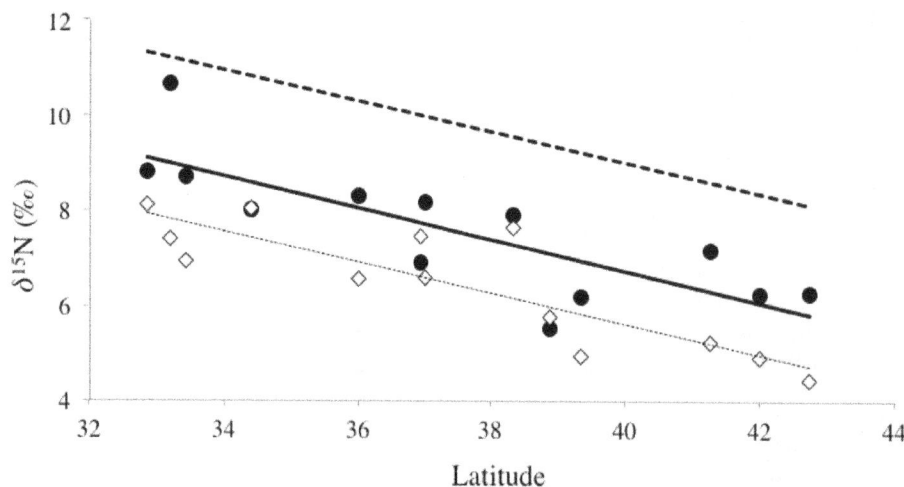

Figure 5. Two approaches for the estimation of baseline CUE δ^{15}N values from mussel isotopic data. The CSI-AA approach, based on average values for $\delta^{15}N_{Phe}$ (filled circles, solid regression line) predicts average baseline δ^{15}N values most consistent with expected NO_3^- δ^{15}N gradients along the CA coast. An alternate approach is based on measured bulk δ^{15}N values, adjusted for an assumed trophic position (open diamonds, dashed regression line). This approach cannot take into account either TEF or tissue-specific fractionations, and returns lower than expected values in most locations. The regression for measured bulk δ^{15}N values in adductor muscle tissue (heavy dashed line) is provided for reference.

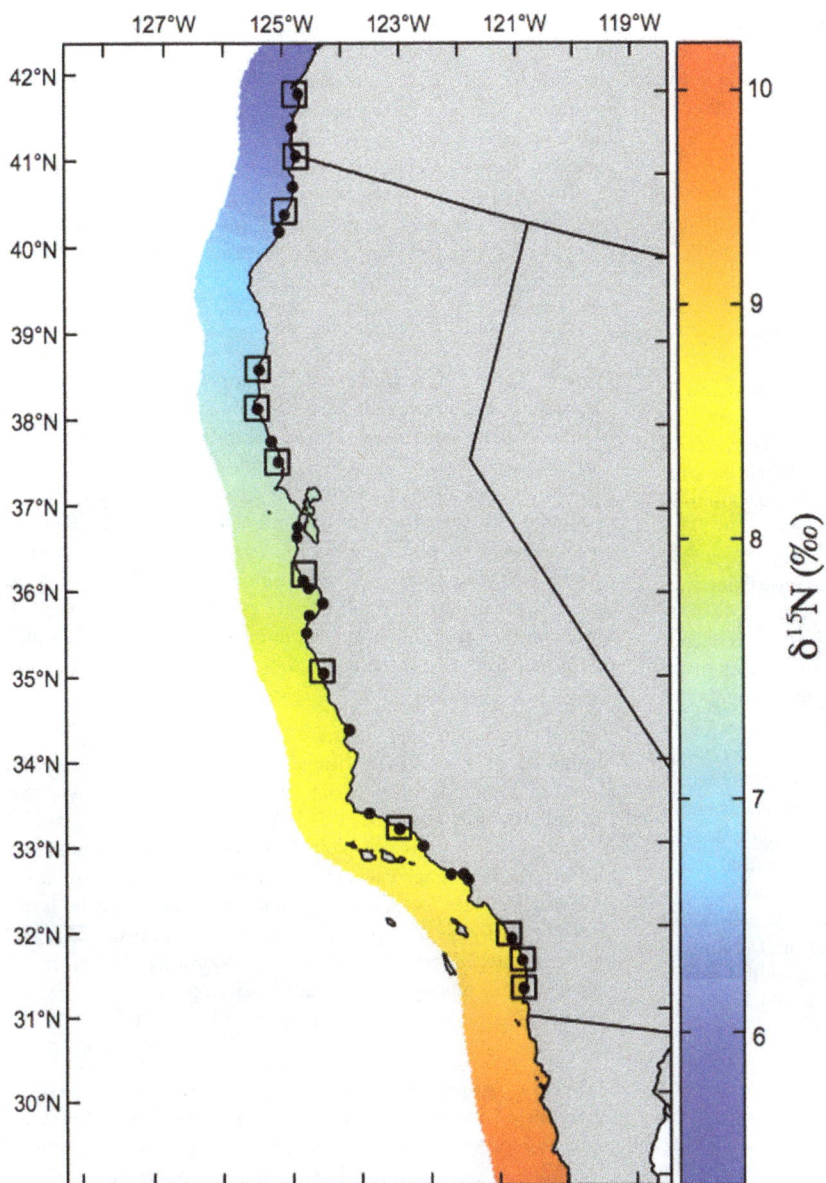

Figure. 6. $\delta^{15}N_{Phe}$ isoscape of the California Upwelling Ecosystem. Color gradient bar indicates $\delta^{15}N$ values. Isoscape is based on the linear relationship of $\delta^{15}N_{Phe}$ vs. latitude ($R^2 = 0.635$, $P = 0.0011$). Squares represent sampling sites chosen for compound-specific isotope analysis; small black dots show all sampling sites for reference.

compare two possible approaches for estimating the baseline $\delta^{15}N$ values from measured tissue data. The first relies only on bulk $\delta^{15}N$ results, and assumes a standard bulk TEF value of 3.4‰, for mussels feeding at TP = 2.0. This approach predicts baseline $\delta^{15}N$ values are far more reasonable vs. the measured bulk tissue values (Fig. 5; open diamonds). However, the results also appear to underestimate $\delta^{15}N$ of NO_3^- along the CA coast (~4.5 to 8‰ north to south), when compared with previous literature data [3,22]. This may not be surprising, since by definition this approach cannot take into tissue-specific $\delta^{15}N$ offsets, nor the actual TEF values for mussels. In contrast, our proposed CSI-AA approach uses $\delta^{15}N_{Phe}$ values as a direct proxy baseline $\delta^{15}N$. The fact that the slope of $\delta^{15}N_{Phe}$ vs. latitude is identical within error to bulk $\delta^{15}N$ (*see results*) strongly supports the idea of a constant offset from primary production. The $\delta^{15}N_{Phe}$ regression-derived baseline values are universally heavier than the bulk- $\delta^{15}N$ derived data

discussed above, and are a closer match for reported NO_3^- values in northern vs. southern CA waters.

We therefore hypothesize that $\delta^{15}N_{Phe}$ is ultimately more accurate representation of baseline $\delta^{15}N$, because it requires no assumptions about TEF values in any specific organism. However, in order to derive precise baseline $\delta^{15}N$ predictions based on $\delta^{15}N_{Phe}$, it will also be necessary to have robust calibrations for the offset between $\delta^{15}N_{Phe}$ and average algal $\delta^{15}N$ values. The close match we observe between the $\delta^{15}N_{Phe}$ regression values and expected NO_3^- $\delta^{15}N$ values of this region (Fig.5; [13,49]) strongly suggests this offset cannot be very large in this system. This would agree with recent analysis of $\delta^{15}N_{AA}$ patterns measured in a range of phytoplankton species [22,45]. However, other work has indicated larger offsets in some macroalgal and also micro-algal species tested in feeding experiments [13,39]. An alternate approach for the future also could be to derive more broad-based corrections, based on $\delta^{15}N$ values of multiple Source AA. Together with indications regarding the importance of representative β values for TP calculations, this further underscores the need for future work aimed at a more robust understanding of $\delta^{15}N_{AA}$ patterns across representative algal sources.

Baseline $\delta^{15}N$ isoscapes from CSI-AA data

Taken together, these results suggest that CSI-AA has the potential, for the first time, to allow direct reconstruction of $\delta^{15}N$ isoscapes of primary production, based on $\delta^{15}N_{Phe}$ values measured in consumers. A baseline $\delta^{15}N$ isoscape for the CUE (Fig. 6), derived on our $\delta^{15}N_{Phe}$ values, represents to our knowledge the first such application. While the general *trend* of decreasing $\delta^{15}N$ values with latitude is similar to broad trends predicted in regional or basin-scale models [45], we suggest that the new potential for CSI-AA to directly produce baseline $\delta^{15}N$ isoscapes represents a major advance. Further, in initial data for selected resampling has shown that specific site-to-site offsets in bulk $\delta^{15}N$ values have so far been highly reproducible [39]. This suggests that, while our initial CUE baseline isoscape is clearly based on relatively few locations, specific geographic variations are may also be meaningful (Fig. S1). While further sampling will be required to verify this conclusion, if mussel-derived $\delta^{15}N$ values can indicate reproducible, fine scale geographic variation in baseline $\delta^{15}N$ values, then our results suggest the potential to create highly detailed spatial maps of isotopic baselines, even in complex coastal environments.

Overall, a CSI-AA approach for constructing baseline isoscapes could have broad importance in both modern and paleoceanographic studies. While CSI-AA based isoscapes could also be generated from other consumers, we suggest that for coastal zones mussels may be a particularly useful bioarchive. The combination of ubiquitous occurrence in many coastal regions, relatively long-term integration of microalgal isotopic signatures, and unambiguous source locations together would provide strong confidence in geographic patterns. We suggest that the degree to which mussel-derived isoscapes also reflect more offshore coastal waters will be an important topic for future work. If our mussel data indicates broad similarity to near-shore coastal isotopic data as a general result (Fig. 3), then sampling within largely existing shore-based programs might rapidly produce detailed, annualized, baseline isoscapes for the entire CCS. Such data could be invaluable in understanding the changing environmental factors driving spatial variability within the CCS; for example the effects of ENSO and PDO cycle effects on baseline $\delta^{15}N$ isoscape gradients, and also provide a more clear understanding of isotopic baselines needed to evaluate possible long-term trends linked to climate change. We also note that isoscapes constructed using CSI-AA from mussels also would not necessarily be limited to coastlines. Mussels frequently attach to the base of fixed moorings located offshore (e.g. http://www.mbari.org/oasis/index.html), and therefore might be used to examine temporal change in isoscapes in many offshore instrumented locations. Finally, our results also suggest potential for paleoceanographic reconstructions. Mussel shells are often the major species found in archeological middens widely distributed from Baja California to Alaska along the US west coast. If source AA $\delta^{15}N$ patterns were also well preserved in archeological shell, this could potentially extend the reconstruction of coastal baseline isoscapes back through much of the Holocene.

Supporting Information

Figure S1 Alternate $\delta^{15}N$ Isoscape approach. Alternate $\delta^{15}N$ Isocape of the California Upwelling Ecosystem, showing $\delta^{15}N$ gradients between sampling stations. As in text Fig. 6 in main text, color gradient indicates $\delta^{15}N$ values. However, this isoscape interpolates between $\delta^{15}N$ at each specific site. While CSI-AA data coverage was not large in this study, preliminary re-sampling has indicated offsets are reproducible. While clearly additional sampling would be required to verify such variation, this approach directly suggests the potential for high resolution isoscapes that capture finer scale regional patterns. Given the ubiquity of mussels along the CA coast, as well as relative ease of sampling, such high resolution coastal isoscapes of baseline $\delta^{15}N$ might be readily constructed.

Figure S2 $\delta^{15}N_{AA}$ patterns in the California Mussel (*Mytilus californianus*). $\delta^{15}N$ amino acid signatures of Mytilus californianus from 13 sampling sites selected for CSI-AA(values based on n = 4 analytical replicate injections). Absolute $\delta^{15}N$ values normalized to the $\delta^{15}N_{Phe}$, so that patterns can be compared. Measured amino acids are categorized into Trophic, Source, and Metabolic (M), based on relative changes with trophic transfer (see main text). Site and amino acid abbreviations are as defined in main text. Overall $\delta^{15}N_{AA}$ patterns conform closely to those expected from other heterotrophic organisms, with Trophic AA enriched in ^{15}N vs. Source AA, and Thr strongly depleted in ^{15}N.

Figure S3 Low CSI-AA based Mussel Trophic Position Results. Relationships between measured $\delta^{15}N_{Glu-Phe}$ values vs. expectations for standard CSI-AA TP equations. Measured $\delta^{15}N_{Glu-Phe}$ of mussels are plotted vs. latitude (filled diamonds). Shaded bar on average value represents $\pm$ 1SD for entire data set. Assumed β values for primary producers are indicated by lower dotted line, ($\beta_{Glu-Phe}$, 3.4 per mil). Commonly assumed TEF$_{Glu-Phe}$ for a single trophic transfer for a primary consumer (7.6 per mil) is represented by upper dashed line. Arrow represents $\Delta_{Glu-Phe}$ the theoretical isotopic enrichment from a TP1 to a TP2.

Figure S4 Representative chromatogram of a GC-IRMS analysis of amino acids. Mussel amino acid gas chromatogram. A representative gas chromatogram of derivatized individual amino acids from *Mytilus californianus*. Abbreviations: Ala, alanine; Gly, Glycine; Thr, threonine; Ser, serine; Val, valine; Leu, leucine; Ile, isoleucine; Nor, Norleucine (internal standard); Pro, proline; Asp, aspartic acid, Met, Methionine; Glu, glutamic acid; Phe, phenylalanine; Lys, Lysine.

Acknowledgments

We thank N. Quintana-Krupinsky, F. Batista, E. Gier and D. Andreasen for helping to collect, prepare, and/or process samples, and J. Felis for his support on map construction.

Author Contributions

Conceived and designed the experiments: NLV MDM. Performed the experiments: NLV. Analyzed the data: NLV MDM. Contributed reagents/materials/analysis tools: MDM. Wrote the paper: NLV MDM.

References

1. Graham BS, Koch PL, Newsome SD, McMahon KW, Aurioles D (2009) Using Isoscapes to Trace the Movements and Foraging Behavior of Top Predators in Oceanic Ecosystems Dordrecht: Springer Netherlands. pp. 299–318. doi:10.1007/978-90-481-3354-3_14.

2. McMahon KW, Ling Hamady L, Thorrold SR (2013) A review of ecogeochemistry approaches to estimating movements of marine animals. Limnol Oceangr 58: 697–714. doi:10.4319/lo.2013.58.2.0697.

3. Sigman DM, Kash KL, Casciotti KL (2009) Ocean process tracers: nitrogen isotopes in the ocean. Encyclopedia of ocean science, 2nd edn Elsevier, Amsterdam.

4. Liu Z, Altabet MA, Herbert TD (2008) Plio-Pleistocene denitrification in the eastern tropical North Pacific: Intensification at 2.1 Ma. Geochem Geophys Geosyst 9: n/a–n/a. doi:10.1029/2008GC002044.

5. Sherwood OA, Guilderson TP, Batista FC, Schiff JT, McCarthy MD (2013) Increasing subtropical North Pacific Ocean. Nature: 1–26.

6. Jackson JBC (2001) Historical Overfishing and the Recent Collapse of Coastal Ecosystems. Science 293: 629–637. doi:10.1126/science.1059199.

7. Ware DM, Thomson RE (2005) Bottom-up ecosystem trophic dynamics determine fish production in the Northeast Pacific. Science 308: 1280–1284.

8. DeNiro MJ, Epstein S (1978) Influence of diet on the distribution of carbon isotopes in animals. GEOCHIMICA ET COSMOCHIMICA ACTA 42: 495–506.

9. Vander Zanden M, Rasmussen JB (2001) Variation in δ15N and δ13C trophic fractionation: implications for aquatic food web studies. Limnol Oceangr 46: 2061–2066.

10. Post DM (2002) Using stable isotopes to estimate trophic position: models, methods, and assumptions. Ecology 83: 703–718.

11. McCutchan JH, Lewis WM, Kendall C, McGrath CC (2003) Variation in trophic shift for stable isotope ratios of carbon, nitrogen, and sulfur. Oikos 102: 378–390.

12. Vanderklift MA, Ponsard S (2003) Sources of variation in consumer-diet ?15N enrichment: a meta-analysis. Oecologia 136: 169–182. doi:10.1007/s00442-003-1270-z.

13. Chikaraishi Y, Ogawa NO, Kashiyama Y, Takano Y, Suga H, et al. (2009) Determination of aquatic food-web structure based on compound-specific nitrogen isotopic composition of amino acids. Limnology and Oceanography: Methods 7: 740–750.

14. Chikaraishi Y, Kashiyama Y, Ogawa NO, Kitazato H, Ohkouchi N (2007) Metabolic control of nitrogen isotope composition of amino acids in macroalgae and gastropods: implications for aquatic food web studies. Mar Ecol Prog Ser 342: 85–90.

15. Popp BN, Graham BS, Olson RJ, Hannides C, Lott MJ, et al. (2007) Insight into the Trophic Ecology of Yellowfin Tuna, Thunnus albacares, from Compound-Specific Nitrogen Isotope Analysis of Proteinaceous Amino Acids. Terrestrial Ecology 1: 173–190.

16. Hannides CC, Popp BN, Landry MR, Graham BS (2009) Quantification of zooplankton trophic position in the North Pacific Subtropical Gyre using stable nitrogen isotopes. Limnol Oceangr 54: 50.

17. Dale JJ, Wallsgrove NJ, Popp BN, Holland KN (2011) Nursery habitat use and foraging ecology of the brown stingray Dasyatis lata determined from stomach contents, bulk and amino acid stable isotopes. Mar Ecol Prog Ser 433: 221–236. doi:10.3354/meps09171.

18. Germain LR, Koch PL, Harvey J, McCarthy MD (2013) Nitrogen isotope fractionation in amino acids from harbor seals: implications for compound-specific trophic position calculations. Mar Ecol Prog Ser 482: 265–277.

19. Ruiz-Cooley RI, Ballance LT, McCarthy MD (2013) Range Expansion of the Jumbo Squid in the NE Pacific: δ15N Decrypts Multiple Origins, Migration and Habitat Use. PLoS ONE 8: e59651. doi:10.1371/journal.pone.0059651.g003.

20. Hannides CCS, Popp BN, Anela Choy C, Drazen JC (2013) Midwater zooplankton and suspended particle dynamics in the North Pacific Subtropical Gyre: A stable isotope perspective. Limnol Oceangr 58: 1931–1946. doi:10.4319/lo.2013.58.6.1931.

21. Choy CA, Davison PC, Drazen JC, Flynn A, Gier EJ, et al. (2012) Global Trophic Position Comparison of Two Dominant Mesopelagic Fish Families (Myctophidae, Stomiidae) Using Amino Acid Nitrogen Isotopic Analyses. PLoS ONE 7: e50133. doi:10.1371/journal.pone.0050133.s005.

22. McCarthy MD, Lehman J, Kudela R (2013) Compound-specific amino acid Î15N patterns in marine algae: Tracer potential for cyanobacterial vs. eukaryotic organic nitrogen sources in the ocean. GEOCHIMICA ET COSMOCHIMICA ACTA 103: 104–120. doi:10.1016/j.gca.2012.10.037.

23. Sigman DM, DiFiore PJ, Hain MP, Deutsch C, Karl DM (2009) Sinking organic matter spreads the nitrogen isotope signal of pelagic denitrification in the North Pacific. 36: L08605. doi:10.1029/2008GL035784.

24. McCarthy MD, Benner R, Lee C, Fogel ML (2007) Amino acid nitrogen isotopic fractionation patterns as indicators of heterotrophy in plankton, particulate, and dissolved organic matter. GEOCHIMICA ET COSMOCHIMICA ACTA 71: 4727–4744. doi:10.1016/j.gca.2007.06.061.

25. Seminoff JA, Benson SR, Arthur KE, Eguchi T, Dutton PH, et al. (2012) Stable Isotope Tracking of Endangered Sea Turtles: Validation with Satellite Telemetry and δ15N Analysis of Amino Acids. PLoS ONE 7: e37403. doi:10.1371/journal.pone.0037403.s002.

26. Kahru M, Mitchell BG (2000) Influence of the 1997-98 El Niño on the surfacechlorophyll in. Geophys Res Lett 27: 2937–2940. Available: http://onlinelibrary.wiley.com/store/10.1029/2000GL011486/asset/grl13695.pdf?v = 1&t = hj1w7gpw&s = 1ed5d0f26e9b4f694f3c377a2f099c51154a2858.

27. Ford HL, Schellenberg SA, Becker BJ, Deutschman DL, Dyck KA, et al. (2010) Evaluating the skeletal chemistry of Mytilus californianusas a temperature proxy: Effects of microenvironment and ontogeny. Paleoceanography 25: PA1203. doi:10.1029/2008PA001677.

28. Rykaczewski RR, Dunne JP (2010) Enhanced nutrient supply to the California Current Ecosystem with global warming and increased stratification in an earth system model. Geophys Res Lett 37: n/a–n/a. doi:10.1029/2010GL045019.

29. Rykaczewski RR, Checkley DM (2008) Influence of ocean winds on the pelagic ecosystem in upwelling regions. Proceedings of the National Academy of Sciences 105: 1965–1970.

30. Olson RJ, Popp BN, Graham BS, López-Ibarra GA, Galván-Magaña F, et al. (2010) Food-web inferences of stable isotope spatial patterns in copepods and yellowfin tuna in the pelagic eastern Pacific Ocean. Progress in Oceanography 86: 124–138. doi:10.1016/j.pocean.2010.04.026.

31. Raikow DF, Hamilton SK (2001) Bivalve diets in a midwestern U.S. stream: A stable isotope enrichment study. Limnol Oceangr 46: 514–522.

32. Dobush GR, Ankney CD, Krementz DG (1985) The effect of apparatus, extraction time, and solvent type on lipid extractions of snow geese. Canadian Journal of Zoology 63: 1917–1920.

33. Asmus RM, Asmus H (1991) Mussel beds: limiting or promoting phytoplankton? Journal of Experimental Marine Biology and Ecology 148: 215–232.

34. Altabet MA, Pilskaln C, Thunell R, Pride C, Sigman D, et al. (1999) The nitrogen isotope biogeochemistry of sinking particles from the margin of the Eastern North Pacific. Deep-Sea Research Part I 46: 655–679.

35. Décima M, Landry MR, Popp BN (2013) Environmental perturbation effects on baseline δ15N values and zooplankton trophic flexibility in the southern California Current Ecosystem. Limnol Oceangr 58: 624–634. doi:10.4319/lo.2013.58.2.0624.

36. Voss M, Dippner JW, Montoya JP (2001) Nitrogen isotope patterns in the oxygen-deficient waters of the Eastern Tropical North Pacific Ocean. Deep-Sea Research Part I 48: 1905–1921.

37. Larsen T, Ventura M, Andersen N, O'Brien DM, Piatkowski U, et al. (2013) Tracing Carbon Sources through Aquatic and Terrestrial Food Webs Using Amino Acid Stable Isotope Fingerprinting. PLoS ONE 8: e73441. doi:10.1371/journal.pone.0073441.s010.

38. Vokhshoori NL, Larsen T, McCarthy MD (2014) Reconstructing δ13C isoscapes of phytoplankton production in a coastal upwelling system with amino acid isotope values of littoral mussels. Mar Ecol Prog Ser. doi:10.3354/meps10746.

39. Vokhshoori N (2013) Using compound-specific amino acid isotope analysis in littoral mussels to generate baseline δ15N and δ13C isoscapes for the California Upwelling Ecosystem. Master's Thesis. Santa Cruz: University of California, Santa Cruz. 125 pp.

40. Gruber N, Sarmiento JL (1997) Global patterns of marine nitrogen fixation and denitrification. Global Biogeochem Cycles 11: 235–266.

41. Cline JD, Richards FA (1972) Oxygen deficient conditions and nitrate reduction in the eastern tropical North Pacific Ocean. Limnol Oceangr: 885–900.

42. Codispoti LA, Richards FA (1976) An analysis of the horizontal regime of denitrification in the eastern tropical North Pacific. Limnol Oceanogr 21: 379–388.

43. White AE, Foster RA, Benitez-Nelson CR, Masqué P, Verdeny E, et al. (2013) Nitrogen fixation in the Gulf of California and the Eastern Tropical North Pacific. Progress in Oceanography 109: 1–17. Available: http://www.sciencedirect.com/science/article/pii/S0079661112001115.

44. Sigman DM, Granger J, DiFiore PJ, Lehmann MM, Ho R, et al. (2005) Coupled nitrogen and oxygen isotope measurements of nitrate along the eastern North Pacific margin. Global Biogeochem Cycles 19: n/a–n/a. doi:10.1029/2005GB002458.

45. Somes CJ, Schmittner A, Altabet MA (2010) Nitrogen isotope simulations show the importance of atmospheric iron deposition for nitrogen fixation across the Pacific Ocean. Geophys Res Lett 37.

46. Schwing FB, O'Farrell M, Steger JM, Baltz K (1996) Coastal Upwelling indices west coast of North America. NOAA Tech Rep NMFS SWFSC 231: 144.

47. Sigman DM, Robinson R, Knapp AN, van Geen A, McCorkle DC, et al. (2003) Distinguishing between water column and sedimentary denitrification in the Santa Barbara Basin using the stable isotopes of nitrate. Geochem Geophys Geosyst 4: n/a–n/a. doi:10.1029/2002GC000384.

48. Checkley DM Jr, Barth JA (2009) Patterns and processes in the California Current System. Progress in Oceanography 83: 49–64. doi:10.1016/j.pocean.2009.07.028.

49. Kienast SS, Calvert SE, Pedersen TF (2002) Nitrogen isotope and productivity variations along the northeast Pacific margin over the last 120 kyr: Surface and subsurface paleoceanography. Paleoceanography 17: 7-1-7-17. doi:10.1029/2001PA000650.

50. Grantham BA, Chan F, Nielsen KJ, Fox DS, Barth JA, et al. (2004) Upwelling-driven nearshore hypoxia signals ecosystem and oceanographic changes in the northeast Pacific. Nature 429: 749–754.

51. Brandes JA, Devol AH, Deutsch C (2007) New Developments in the Marine Nitrogen Cycle. Chem Rev 107: 577–589. doi:10.1021/cr050377t.

52. Waser N, Harrison PJ, Nielsen B, Calvert SE, Turpin DH (2010) Nitrogen isotope fractionation during the uptake and assimilation of nitrate, nitrite, ammonium, and urea by a marine diatom. Limnol Oceangr 43: 215–224. Available: http://www.jstor.org.oca.ucsc.edu/stable/pdfplus/2839209.pdf?acceptTC = true&acceptTC = true&jpdConfirm = true.

53. Behrenfeld MJ, O'Malley RT, Siegel DA, McClain CR, Sarmiento JL, et al. (2006) Climate-driven trends in contemporary ocean productivity. Nature 444: 752–755. doi:10.1038/nature05317.

54. Aksnes DL, Ohman MD (2009) Multi-decadal shoaling of the euphotic zone in the southern sector of the California Current System. Limnol Oceangr 54: 1272.

55. Polovina JJ, Howell EA, Abecassis M (2008) Ocean's least productive waters are expanding. Geophys Res Lett 35: L03618. doi:10.1029/2007GL031745.

56. McKinnell SM, Dagg MJ (2010) Marine ecosystems of the North Pacific Ocean, 2003-2008. PICES Special Publication 4: 393.

57. Rau GH, Ohman MD, Pierrot-Bults A (2003) Linking nitrogen dynamics to climate variability off central California: a 51 year record based on 15N/14N in CalCOFI zooplankton. Deep Sea Research Part II: Topical Studies in Oceanography 50: 2431–2447. doi:10.1016/S0967-0645(03)00128-0.

58. Duggins DO, Simenstad CA, ESTES JA (1989) Magnification of secondary production by kelp detritus in coastal marine ecosystems. Science 245: 170–173.

59. Bracken ME, Menge BA, Foley MM, Sorte CJ, Lubchenco J, et al. (2012) Mussel selectivity for high-quality food drives carbon inputs into open-coast intertidal ecosystems. Mar Ecol Prog Ser 459: 53–62. doi:10.3354/meps09764.

60. Lorrain A, Graham B, Ménard F, Popp B, Bouillon S, et al. (2009) Nitrogen and carbon isotope values of individual amino acids: a tool to study foraging ecology of penguins in the Southern Ocean. Mar Ecol Prog Ser 391: 293–306. doi:10.3354/meps08215.

2

Detecting Human Presence at the Border of the Northeastern Italian Pre-Alps. ^{14}C Dating at Rio Secco Cave as Expression of the First Gravettian and the Late Mousterian in the Northern Adriatic Region

Sahra Talamo[1]*, Marco Peresani[2], Matteo Romandini[2], Rossella Duches[2,5], Camille Jéquier[2], Nicola Nannini[2], Andreas Pastoors[3], Andrea Picin[3,4,6], Manuel Vaquero[4,6], Gerd-Christian Weniger[3], Jean-Jacques Hublin[1]

1 Department of Human Evolution, Max Planck Institute for Evolutionary Anthropology, Leipzig, Germany, **2** Universitá di Ferrara, Dipartimento di Studi Umanistici, Ferrara, Italy, **3** Neanderthal Museum, Mettmann, Germany, **4** Institut Català de Paleoecologia Humana i Evolució Social (IPHES), Tarragona, Spain, **5** Museo delle Scienze, Trento, Italy, **6** Universitat Rovira I Virgili, Area de Prehistòria, Tarragona, Spain

Abstract

In the northern Adriatic regions, which include the Venetian region and the Dalmatian coast, late Neanderthal settlements are recorded in few sites and even more ephemeral are remains of the Mid-Upper Palaeolithic occupations. A contribution to reconstruct the human presence during this time range has been produced from a recently investigated cave, Rio Secco, located in the northern Adriatic region at the foot of the Carnic Pre-Alps. Chronometric data make Rio Secco a key site in the context of recording occupation by late Neanderthals and regarding the diffusion of the Mid-Upper Palaeolithic culture in a particular district at the border of the alpine region. As for the Gravettian, its diffusion in Italy is a subject of on-going research and the aim of this paper is to provide new information on the timing of this process in Italy. In the southern end of the Peninsula the first occupation dates to around 28,000 ^{14}C BP, whereas our results on Gravettian layer range from 29,390 to 28,995 ^{14}C years BP. At the present state of knowledge, the emergence of the Gravettian in eastern Italy is contemporaneous with several sites in Central Europe and the chronological dates support the hypothesis that the Swabian Gravettian probably dispersed from eastern Austria.

Editor: David Caramelli, University of Florence, Italy

Funding: The authors have no support or funding to report.

Competing Interests: The authors have declared that no competing interests exist.

* E-mail: sahra.talamo@eva.mpg.de

Introduction

Numerous sites throughout the Italian Peninsula and the western Balkans document key events between the late Middle Palaeolithic and the Mid-Upper Palaeolithic. Focusing on the northern Adriatic Sea rim which includes the Venetian region and the Dalmatian coast, the millennia preceding the demise of Neanderthals are recorded in very few sites which displayed data of variable relevance [1–3]. Settlements were logistically structured in accordance with the vertical displacement of economic activities at mountain districts sheltered sites were repeatedly used to accomplish different types of complex tasks or were inhabited for short-term occupations, as it has been suggested from the fractionation of stone tool production sequences [3]. Flint provisioning and lithic economy was therefore fully organized and reveal how human land-use varied accordingly to the geographical location and function of the sites [3].

Even scarcer in this area is the archaeological evidence of the Mid-Upper Palaeolithic, a period better known along the Tyrrhenian Sea and the southern Adriatic coasts, where evidence of intense Gravettian occupation can be found [4].

One of the most debated issue is whether the Gravettian developed from a local Aurignacian [5–8] or results from immigration or cultural diffusion processes through various corridors between European regions [4,9–11]. This paper will not enter into this broader issue, instead it will deal with the Northern Italian evidence and the role of two possible passage-ways, one from the west (France) and one from the east (Balkan region) [9,12–14].

The earliest Italian Gravettian groups is documented around 28,000 ^{14}C BP in Paglicci Cave in the southern end of the Peninsula [15–17], and the majority of the sites, adjacent to the two opposite Italian coasts, are recorded at 26,000–24,000 ^{14}C BP (Figure 1) [12].

All along the Tyrrhenian coast, the lithic assemblages described at Mochi rockshelter and La Cala Cave suggest an influence from the French Gravettian [9,12,18]. In contrast, the conspicuous Gravettian evidence from Paglicci Cave, in the South Adriatic area, shows discrepancy with the Tyrrhenian belt from a technological point of view. This indication suggest signatures of cultural influence from a possible eastern route starting from the Carpathian basin [9] and crossing the trans-Adriatic region, when

Figure 1. The earliest Mid-Upper Palaeolithic sites in Italy. Rio Secco is marked in red, the sea level is at −80 m (Base map from NASA http://www2.jpl.nasa.gov/srtm/world.htm) [20].

the sea level at that time was estimated at about 80 m lower than present day [19,20] (Figure 1). Nevertheless, evidence across the Adriatic coast is still too scanty, mostly due to ephemeral field researches, aside a reduced number of sites; e.g. Broion Rock-shelter and Fonte delle Mattinate and the above mentioned Paglicci Cave [11,21–23].

Moreover, Paglicci is not the key site to understand the issue of the local development of the Italian Gravettian because Aurigna-cian and Early Gravettian assemblages show an abrupt change with neither transitional nor formative characters [11,14].

As it is shown the Gravettian settlement of Italy is spatially sparse; in this context the recently investigated cave of Rio Secco, located in the northern Adriatic region, provides evidence on the late Mousterian and the earliest Gravettian, due to a set of new radiometric dates on bone and charcoal samples. Considering its geographic setting between the upper Adriatic Plain and the Pre-Alps, Rio Secco Cave holds a strategic position to investigate the mobility pattern of the Palaeolithic hunter-gatherers across the natural corridor between the Italian Peninsula and the Carpathian Basin.

Rio Secco Cave Consideration

The Site of Rio Secco Cave

Rio Secco Cave is situated in the northeastern portion of the Italian Peninsula, near the village of Clauzetto (Pordenone), at 580 m asl on the Pradis Plateau in the eastern part of the Carnic Pre-Alps. The Pradis Plateau comprise an area of 6 sq km, enclosed on three sides by mountains peaking from 1,148 m to 1,369 m and to the south by the foothills, facing the present-day Friulian Plain (Figure 2). Rio Secco Cave is a large sheltered cave opening on the left slope of a stream gorge at about 20 m above the present day stream bed. Facing south, the shelter has a wide and flat roof derived from the collapse of large slabs of the stratified limestone. The sheltered area is enclosed from the outside by a ridge of large boulders. The cave opens in the middle of the wall and continues as a gallery for 12 m until the sediments completely fill it up. In the outer area the fill forms a slope-waste deposit thickening along the present day drip line where the boulders define the original extension of a vast roof.

The presence of Palaeolithic settlements at Rio Secco Cave was detected in 2002 after a test-pit [24] and an archaeological excavation has been carried out at the site since 2010.

Figure 2. Southern view on the Pradis plateau from the Mount Rossa edge (1,309 m). The position of Rio Secco Cave in the gorge is marked. Background, the alluvial plain crossed from the Tagliamento River at the center.

Stratigraphy

The cave is filled with an ensemble of sedimentary bodies of differing volume, shape, composition and origin, grouped into four macro-stratigraphic units and separated by erosional and sedi-mentary discontinuities [25]. From the top, the macro-units are 1, which originated during historical times, BR1, BR2 and BIO1 (Figure 3).

Macro-unit BR1 includes layer 4 and an anthropic horizon containing Gravettian flint artifacts, layer 6. The most relevant features are angular to subangular stones, with fragments of karst limestone pavement that originated from the collapse of the vault. Layer 6, with organic matter and micro-charcoal has been exposed at the entrance of the cave shelter, approximately 20 cm below the top of BR2: it is thin, planar, discontinuous, and contains rare bones and lithics (Figure 3).

Macro-unit BR2 is a massive open-work stone-supported breccia made of angular boulders and randomly deposited stones. It lies in the external zone but ends 1 m behind the drip line in the SE zone of the cavity, where it seals the layer 5 top. Large patches have been reworked by marmots, as demonstrated by bones, an articulated skeleton found within the tunnels, several burrows and dens.

The sedimentary body below BR2 is composed of stones and loamy fine fraction and is labeled BIO1 due to the intense bioturbation caused by the activity of marmots, responsible for mixing, displacing portions of anthropic sediment, and scattering Mousterian flint implements, bones and charcoals. At the top of this macro-unit, one finds layer 5 top, a brown level of variable thickness with archaeological content. Due to its variable thickness, layer 5 top has been locally divided in two arbitrary cuts, I and II. Below, the loamy, dark yellowish-brown layer 7 has been found only in some squares under the cave vault and not in the external zone, where it is cut by the burrows. The upper boundary with layer 5 top is marked by an increasing frequency of bones and lithics, some of which also bear signatures of accidental heating. Sandwiched between the two anthropic horizons layers 7 and 8, layer 5 is made of stones and loamy fine fraction with dispersed bones and lithic implements frequently affected from post-depositional alteration. Layer 8 continues in the inner cavity and is best described as 10 cm thick, stony, with dark brown loamy fine fraction, frequent tiny charcoals, small and burnt bones. Layer 8 lies over layer 9, possibly a fifth macro-unit, made of stones and yellowish brown sandy-loam, with no charcoal or other finds.

Cultural Sequence

The archaeological contents of BR1 and BIO1 include numerous lithic artifacts ascribed to the Middle Palaeolithic (layers 5 top, 7, 5, 8) and Upper Palaeolithic (layer 6 and correlated arbitrary cuts 4c and 4d) and a few bone retouchers [25]. The Mousterian assemblages are characterized by the use of Levallois and discoid technologies (Figure 4). Layer 8 has yielded scrapers of variable type and size and flakes with patterns typical of Levallois technology. Layer 5 has produced evidence of the use of Levallois technology as well, represented by recurrent unipolar flakes and centripetal flakes and cores, of discoid technology represented from core-edge removal flakes and pseudo-Levallois points and retouched tools, mostly scrapers. Layer 7 has produced flakes and a few tiny scrapers. In layer 5 top lithic items are varied: Levallois and discoid flakes, short blades and short bladelet cores. The Upper Palaeolithic of layer 6 consists of a handful of pieces technologically characterized by blade/bladelet production. The tools are three burins on truncation made on blades and on rejuvenation blades (Figure 4). One of them shows remarkable negatives of several burin spalls, of which one was refitted and for this reason it should be interpreted as a bladelet core. In addition, there are two end scrapers produced on cortical flakes, one of which is thick and large. Among the projectile pieces, we count one backed bi-truncated bladelet, one possibly unfinished backed point and one undeterminable backed fragment.

Evidence for the use of fire has been found in layers 8 and 7 by tiny dispersed charcoals, burnt bones and heat-affected flints. In layer 6 two hearths have been brought to light, even if partially affected from post-depositional disturbances, labeled as US6_SI and US6_SII. The former is an agglomeration of charcoals mostly disaggregated around a large piece of charred wood (Figure 5). This hearth has been cut by illegal excavations in the back of the cave. Traces of ash are lacking, but there is a thin reddish horizon below the level of charcoals. The hearth US6_SII is a small agglomeration of charcoal largely disturbed by several interlaced burrows.

Faunal Remains

Every stratigraphic unit contained animal bone remains. The colonization of the cave fill by marmots is clearly documented by diagnostic signatures observed in BR1 and BR2, such as dens, chambers and articulated skeletons. There are fewer faunal remains in the Gravettian layers in comparison with the Mousterian.

The archaeozoological analysis, still in progress, reveals among the ungulates the presence of caprids (*Capra ibex* and *Rupicapra rupicapra*) and remains of *Bos*/*Bison* (*Bison priscus*/*Bos primigenius*). Traces of human modification on the bones include cut-marks on shafts of caprids, partly combusted, and on a marmot clavicle. One partially burned epiphysis of the scapula of *Castor fiber* has been found associated to the hearth US6_SI.

In the Mousterian sequence, carnivores (brown bears, cave bears, mustelids and canids) predominate over the ungulates, which rather than caprids (chamois and ibex) or bovids, consist more of cervids such as red deer, roe deer, elk and wild boar (Peresani et al., in press). Bones are mostly fragmented, due to post-depositional processes as well as human and carnivore activity. Human interest in ungulates is evidenced by cut marks on red deer. Also the remains of Ursus spelaeus and Ursus sp. from layers 7 and 5 top show traces of butchering, skinning and deliberate fracturing of long bones [26].

This faunal association with cervids and, in particular, deer, elk, roe deer and wild boar is indicative of forest vegetation and marsh environment somewhere in the Pradis Plateau. The presence of

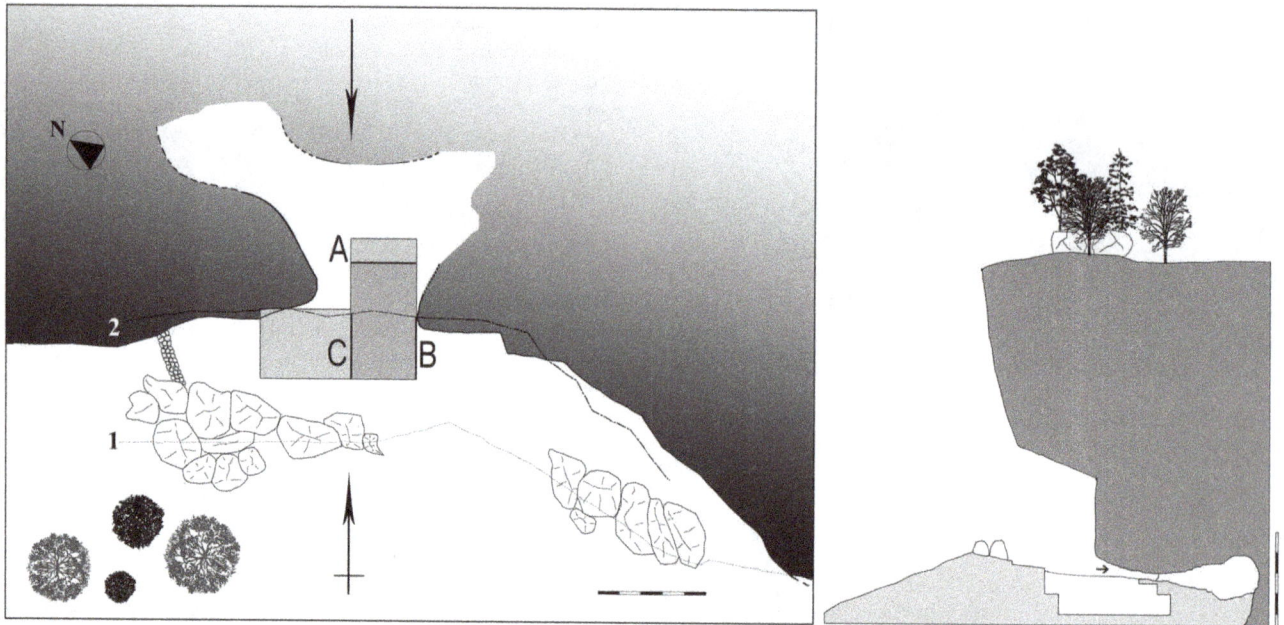

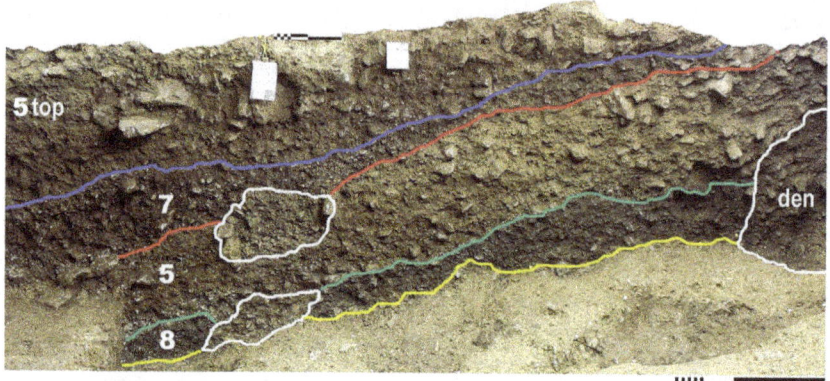

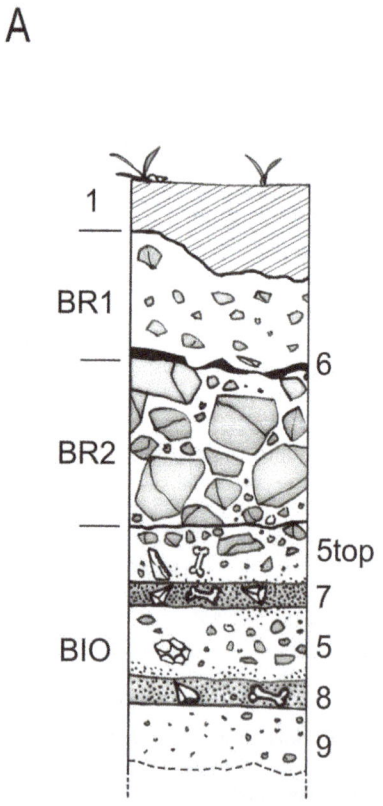

Figure 3. Sketch map and section of the site. Position of the excavated area and the stratigraphic exposures: A – section showing portions of layer 6 embedded in macro-unit BR1; B - section showing the Mousterian layers from 5 top to 8; C – the main sagittal section exposed in 2010 with the reworked sediment sealing the Mousterian sequence from 5 top to 8 (after [25]).

bovids and caprids suggest the existence of patchy woodland compatible with the mountain context. Cave bears were well adapted to this kind of environment, and used the cavities for hibernation, as suggested from the faunal assemblage recovered during the last field-campaigns.

Materials and Methods

Ethics Statement

All necessary permits were obtained from the Archaeological Superintendence of the Friuli-Venezia Giulia for the described study, which complied with all relevant regulations. The identification numbers of the specimens analyzed are: GRS13SP57-89, GRS13SP57-138, GRS13SP57-153, GRS13SP57-125, GRS13SP57-37, GRS13SP57-11, GRS13SP57-18, GRS13SP57-46, GRS13SP57-2, GRS13SP57-4.

Repository information: the specimen is temporary housed at the University of Ferrara, in the Section of Prehistory and Anthropology, Corso Ercole I d'Este Ferrara, Italy, with the permission of the Archaeological Superintendence of the Friuli-Venezia Giulia.

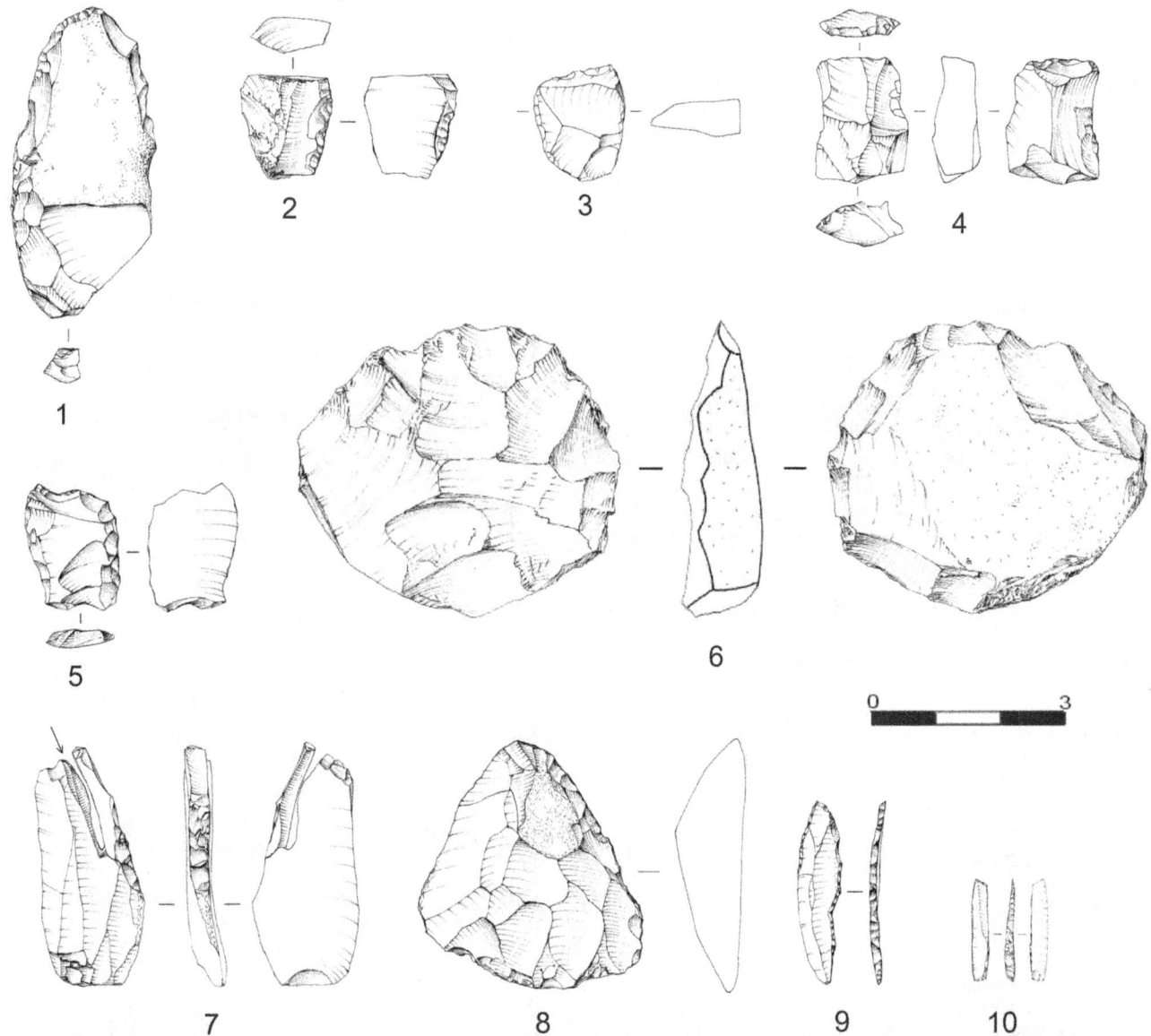

Figure 4. Lithic implements from Mousterian layers 8, 7, 5 and 5 top. Mousterian layers 8 (1), 5 (6), 7 (2, 3) and 5 top (4, 5). Scraper (1), scraper shortened by distal truncation and thinned on the dorsal face (2), core-edge removal flake from discoid core (3), bladelet core (4), double scraper shortened by proximal truncation (5), Levallois centripetal core (6). Gravettian implements: burin with refitted burin spall (7), end-scraper on large retouched flake (8), possibly unfinished backed point (9), double truncated backed bladelet (10). Drawn by S. Muratori.

Figure 5. US6_SI hearth brought to light at the entrance of the cave.

Samples Selection and Radiocarbon Pretreatment

We selected 10 well preserved thick cortical bone fragments with and without cut marks from each layer. Four bones from layer 7 (three of them with cut marks), four bones from layer 5 (three with cut marks) and two charcoal samples from the hearth SI of layer 6.

Bone collagen was extracted at the Department of Human Evolution, Max Planck Institute for Evolutionary Anthropology (MPI-EVA), Leipzig, Germany, using the ultrafiltration method described in Talamo and Richards [27]. The outer surface of the bone samples was first cleaned by a shot blaster and then 500 mg of bone was taken. The samples were then decalcified in 0.5 M HCl at room temperature until no CO_2 effervescence was observed, usually for about 4 hours. 0.1 M NaOH was added for 30 minutes to remove humics. The NaOH step was followed by a final 0.5 M HCl step for 15 minutes. The resulting solid was gelatinized following Longin (1971) at pH 3 in a heater block at 75°C for 20 h. The gelatin was then filtered in an Eeze-Filter™ (Elkay Laboratory Products (UK) Ltd.) to remove small (<80 μm) particles. The gelatin was then ultrafiltered with Sartorius "Vivaspin 15" 30 KDa ultrafilters [28]. Prior to use, the filter was cleaned to remove carbon containing humectants [29]. The samples were lyophilized for 48 hours.

The collagen extract was weighed into pre-cleaned tin capsules for quality control of the material. Stable isotopic analysis was evaluated using a ThermiFinnigan Flash EA coupled to a Delta V isotope ratio mass spectrometer.

The two charcoal samples were sent directly to the Klaus-Tschira-AMS facility of the Curt-Engelhorn Centre in Mannheim, Germany, where they were pretreated with the ABA method [30].

Results and Discussion

^{14}C Results

At Rio Secco Cave the C:N ratio of all the samples are 3.2 which is fully within the acceptable range (between 2.9 and 3.6), and all of them displayed a high collagen yield, mostly ranging between 2.4 to 8.2%, substantially higher than 1% of weight for the standard limit [31,32] (Table 1).

Once these criteria were evaluated, between 3 and 5 mg of the collagen samples were sent to the Mannheim AMS laboratory (Lab code: MAMS), where they were graphitized and dated [30].

The radiocarbon results are listed in table 1. All dates were corrected for a residual preparation background (generally<

Table 1. Radiocarbon dates of Rio Secco Cave.

MPI Lab nr.	Square	U.S.	Material	%Coll	δ^{13}C	δ^{15}N	%C	%N	C:N	AMS Lab nr.	^{14}C Age	1σ Err
S-EVA 26233	I 13	6_SI 2	Charcoal	ABA pretreatment						MAMS-15906	28,995	135
S-EVA 26234	I 13e	6_SI 4	Charcoal	ABA pretreatment						MAMS-15907	29,390	135
S-EVA 25353	I14 b	5 top	Bone with cut marks	5.7	−19.6	2.6	43.7	16.1	3.2	MAMS-15230	44,100	660
S-EVA 25355	G14III	5 top I	Bone with cut marks	3.5	−19.5	4.1	40.1	14.7	3.2	MAMS-15231	45,695	790
S-EVA 25356	H14IV	5 top II	Bone	2.4	−20.6	0.8	37.6	13.8	3.2	MAMS-15232	43,210	600
S-EVA 25357	I14II	5 top I	Bone with cut marks	7.6	−22.3	2.1	42.4	15.5	3.2	MAMS-15233	45,740	800
S-EVA 25359	H14h	7	Bone	5.2	−22.2	1.0	42.4	15.6	3.2	MAMS-15235	46,320	1430
S-EVA 25361	H13IV	7	Bone with cut marks	8.2	−21.8	2.4	44.3	16.2	3.2	MAMS-15236	>49,000	
S-EVA 25362	H13IV	7	Bone with cut marks	4.1	−21.8	1.2	41.6	15.3	3.2	MAMS-15237	44,560	1150
S-EVA 25363	H14g	7	Bone with cut marks	5.5	−21.6	1.6	43.9	15.9	3.2	MAMS-15238	44,770	1180

Isotopic values, C:N ratios, amount of collagen extracted (%Coll) refer to the >30 kDa fraction. The results of AMS radiocarbon dating of 10 samples from Rio Secco Cave of layer (U.S.). 6 (Gravettian cultural sequence), layers (U.S.). 5 top and 7 (Mousterian cultural sequence). δ^{13}C values are reported relative to the vPDB standard and δ^{15}N values are reported relative to the AIR standard.

0,0025 F ^{14}C, equivalent to ca. >48,000 ^{14}C years BP) estimated from pretreated ^{14}C free bone samples, kindly provided by the ORAU and pretreated in the same way as the archaeological samples.

The uncalibrated radiocarbon dates of late Mousterian (layer 7) range from >49,000 to 44,560 ^{14}C years BP. The four dates of layer 5 range from 45,740 to 43,210 ^{14}C years BP. The uppermost layer (layer 6), which was identified as a Gravettian layer, ranges from 29,390 to 28,995 ^{14}C years BP. There is no discrepancy between the results obtained on bones with cut marks and without cut marks.

Comparison with Previous ^{14}C AMS Results

A series of radiocarbon dates were previously obtained from layers 8, 5 and 6 [24,25] (Table 2). The two dates in Layer 8 show a strong discrepancy in the results, in fact the ultrafiltered bone gives an age older than 48,000 ^{14}C BP but the charcoal result, pretreated with ABOX-SC, displayed an age of 42,000±900 ^{14}C BP. The main argument for this difference has to be found, as described above, in the stratigraphic entities of the layer, in fact it contains frequent tiny charcoals of undetermined conifer, small bones and burnt bones. Moreover deformations, removal and various crossings by marmots and other minor bioturbations affect this layer. In addition, a test-pit opened during the last field campaign (summer 2013) had detected no archaeological traces at 1,5 meters underneath this layer, thus excluding possible pollution from older deposits. For this reason we considered the youngest date (OxA-25359 ^{14}C Age 42,000±900) as an outlier.

Layer 5 has produced an age that is too young compared with our new results (LTL-429A, ^{14}C Age 37,790±360) [24]. The sample was selected from the test pit investigated in 2002 and at that time it was not possible to recognize bioturbation produced by marmots. This result is not included in the Bayesian model, discussed below.

Two other charcoal samples in layer 6 were dated at Poznan AMS laboratory pretreated using the ABOX-SC method; these results are consistent with our new results. We incorporate them in the Bayesian model for the distribution of ages.

Discussion of Chronology

The radiocarbon dates we produced were calibrated using OxCal 4.2 [33] and IntCal13 [34], (Table 3). The Bayesian model, which was built using the stratigraphic information, includes a sequence of 3 sequential phases, the two Mousterian Layers 7 and 5 top and the Gravettian layer 6 (Figure 6).

The agreement indices were applied to show how the unmodelled calibrated distribution agrees with the distribution after Bayesian modelling. The agreement index is expected to be above 60% when the dates concord with the stratigraphy. The

t-type outlier analysis, performed to detect problematic samples, with prior probabilities set at 0.05, was also incorporated within the Bayesian model [35].

A start calibrated boundary for the lower part of the sequence (Layer 7) at Rio Secco Cave cannot be defined. What we can determine is that the lower level of Layer 7 is older than 49,000 ^{14}C BP. With this new determination we can confirm the former date obtained on ultrafiltered bone at ORAU (OxA-25336, ^{14}C Age >48,000) in the layer below layer 7 (layer 8) [25] and the only other date (OxA-25359 ^{14}C Age 42,000±900) in layer 8 is confirmed to be an outlier. OxCal finds a good agreement index (A_overall = 71.9%), between the full set of finite radiocarbon dates and the stratigraphic information; the results of the outlier detection method confirm ideal posterior probability for all the samples.

The upper boundary of layer 7, calculated by OxCal, ranges from 49,120 to 47,940 cal BP (68.2%); the layer 5 top ranges from 47,940 to 45,840 cal BP (68.2%) (Table 3).

uAround the Alpine regions Neanderthal sites with comparable age ranges are rare [3,36]. In northern Italy four sites can be considered; Fumane Cave, San Bernardino Cave unit II, Broion Cave and Generosa Cave [3]. In Slovenia on the Šebreljska Plateau the Divje Babe I Cave is contemplated [37] and moving east, in the Drava basin, a layer chronologically consistent with Rio Secco, is found at Vindija Cave [38,39].

The charcoal samples, from the archaeological horizon US 6_SI located between layers BR2 and BR1 range from 33,480 to 30,020 cal BP (68.2%) (Table 3). These ranges clearly place the upper part of Rio Secco in the early Gravettian period and confirm its archaeological assessment.

It should be noted that the charcoal samples dated at Mannheim yielded consistent age with the previous radiometric dates obtained at Poznan for the same horizon.

Here it is useful to remember that strong progress has been achieved in the last decade on the radiocarbon method. Calibration is now possible back to 50,000 cal BP [34,40] and claims of fundamental limitations are not justified [41]. Moreover, samples selection and specific pretreatment procedures to remove modern contaminations have been significantly improved [27,42–44].

An accurate sample selection, more specialized pretreatment protocols, the control of isotopic values of bone collagen, in case the samples pretreated were bones and the requirement of several dated samples per layer are fundamental criteria that should be considered in order to establish the radiocarbon chronology of the archaeological sites.

Normally the risk of underestimating the true age of the samples is higher when the samples are at the limit of the radiocarbon method. However the chronological reassessment of Geißenklösterle, Abri Pataud, Fumane Cave and Mochi rockshelter sites [45–

Table 2. Previous radiometric dates of Rio Secco Cave obtained in 2002.

Context	Nature	Lab. Ref.	^{14}C age BP ±1σErr	Cal. BP 1σ
6, sq.J11, n.3	Charcoal	Poz-41207	27,080±230	31,240–30950
6, sq.J11, n.4	Charcoal	Poz-41208	28,300±260	32,600–31740
5, GRSI	Bone	LTL429A	37,790±360	42,360–41850
8, sq.H11IV, n.17	Charcoal	OxA-25359	42,000±900	46,220–44560
8, sq.H12IV, n.12	Bone	OxA-25336	>48,000	Infinite

Calibrated ages at 1σ error, using OxCal 4.2 [33] and IntCal13 [34].

Table 3. Calibrate boundaries of Rio Secco Cave.

Rio Secco	Modelled Cal BP			
	from	to	from	to
Indice	68.20%		95.40%	
A_model 71.3				
A_overall 71.9				
End Gravettian Layer 6	31,230	30,020	31,500	27,270
Transition Sterile Macro-Unit BR2/**Start Gravettian** Layer 6	35,230	33,480	39,330	33,220
Transition End Mousterian Layer 5 top/Sterile Macro-Unit BR2	47,860	45,840	48,540	43,570
Transition Mousterian Layer 7/5 top	49,120	47,940	49,650	47,460
Start Mousterian Layer 7	50,070	48,670	51,020	47,830

Calibrate boundaries provided by OxCal 4.2 [33] using the international calibration curve IntCal 13 [34].

48] demonstrated that this problem might occur also between 30,000–20,000 ^{14}C years BP.

Bearing in mind this fundamental issue, Rio Secco Cave layer 6 shows the newest radiometric assessment of the Italian late Mid-Upper Palaeolithic. Moreover the comparison with the single dates of layer 23 in Paglicci Cave, permits to ascribe Rio Secco as the oldest Early Gravettian site in Italy.

At this stage of our investigation, the backed pieces and the burins introduced and reduced on site are an expression of short term occupations by hunter gatherers equipped with previously retouched tools made of high quality flints collected outside the Carnic Pre-Alps [25]. However, further investigation is required.

The appearance of the early Gravettian in Europe predates the last phases of the Aurignacian [49]. Although some similarities have been detected with the Ahmarian assemblages of the Near East [49], a local development of the Gravettian technological innovations from the Aurignacian substrate was suggested at Geißenklösterle in layer AH II [50] and at Abri Pataud in layer 6 [51]. Generally speaking the Gravettian might be interpreted as a macro techno-complex characterized by different and synchronic geographic variants [52]. To the north of the Alps, the key Swabian Gravettian facies include the lithic assemblages of the sites Geißenklösterle layer AHI, Hohle Fels layer II, Sirgenstein layer II, Brillenhöhle, Weinberghöhlen and Willendorf II layer 5, which are comparable with the Rio Secco age range, (Table S1) [53].

In central Europe between northern Austria, Moravia and southern Poland one finds a second early Gravettian techno-complex, named the Pavlovian, [54,55]. It is represented at the key sites of Dolní Vstonice II, Pavlov I, Předmostí I and Krems [49,56], which are contemporaneous with Rio Secco layer 6 (Table S1) [57]. This cultural entity differs from the Swabian Gravettian due to the presence in the toolkit of geometric microliths, micro-burins and Pavlovian points [53].

Furthermore, in the Italian Peninsula local developments of the Gravettian have not been recorded so far [11] and the similarities documented in the lithic assemblages of level 23 of Paglicci Cave and Kostienki 8/II [16,58] draw attention to the broader Gravettian diffusion from central Europe.

Current evidences make us inclined on the cultural diffusion hypothesis, and the Rio Secco site provides new insight on the two natural corridors used to reach the Italian Peninsula, the Adriatic southern coast from Croatia [9,20] and the bridge to the north-east from the Carpathian regions. Further researches on the raw materials provenance will shade light on the exploitation of

southern or eastern Alpine outcrops determining the foraging radius of these earliest Gravettian groups.

Conclusion

At the junction between the North Adriatic Plain and the eastern Alps, the chronometric refinement of a new site, Rio Secco Cave, contributes to enhance the investigation of the prehistoric human occupation during the mid-Late Pleistocene. Although not completely explored, Rio Secco Cave fills an important chrono-logical gap and preserves an archive of potential interest for understanding the study of the late Neanderthals, the dispersal of Mid-Upper Palaeolithic populations and the diffusion of the Gravettian culture. Nevertheless, the new set of dates does not cover the millennia of the Middle-Upper Palaeolithic transition in the second half of MIS3, a period chronometrically secured from key-sequences in neighboring regions [45]. Before claiming human ecological or economic factors leading to this dearth of evidence, more data are required from the study of the stratigraphic sequence. The detection of possible stops in the gradation processes of the cave deposit, which may have produced alterations, consolidations, weathering or, alternatively erosions, could explain the complete removal of traces of Aurignacian occupations.

The continued implementation of the project with fieldwork and laboratory studies will provide new elements necessary to better understand the settlements in this area, previously considered so marginal in comparison with the North Adriatic Plain, extending towards the south. At the present stage of research, the Gravettian archaeological record at Rio Secco Cave is scarce compared with the Mousterian one, due to the thinning of layer 6 and its partial reworking produced by illegal excavations in the inner cavity. We cannot exclude that the rockfall that occurred after the late Middle Palaeolithic induced the Gravettian foragers to place their settlement under the present-day rockshelter just in front of the cave entrance. Nevertheless, the few flint artifacts give economic hints of potential interest. The ^{14}C results show that the excavated archaeological horizon Layer 6 belongs to the early Gravettian time period. At the present state of knowledge, with our new ^{14}C dates, the emergence of the early Gravettian in eastern Italy is contemporaneous with the Swabian Gravettian and the Pavlovian.

The broad expansion of Swabian Gravettian and Pavlovian techno-complexes is explained by high mobility patterns of the hunter-gatherers with transport of exogenous raw materials up to

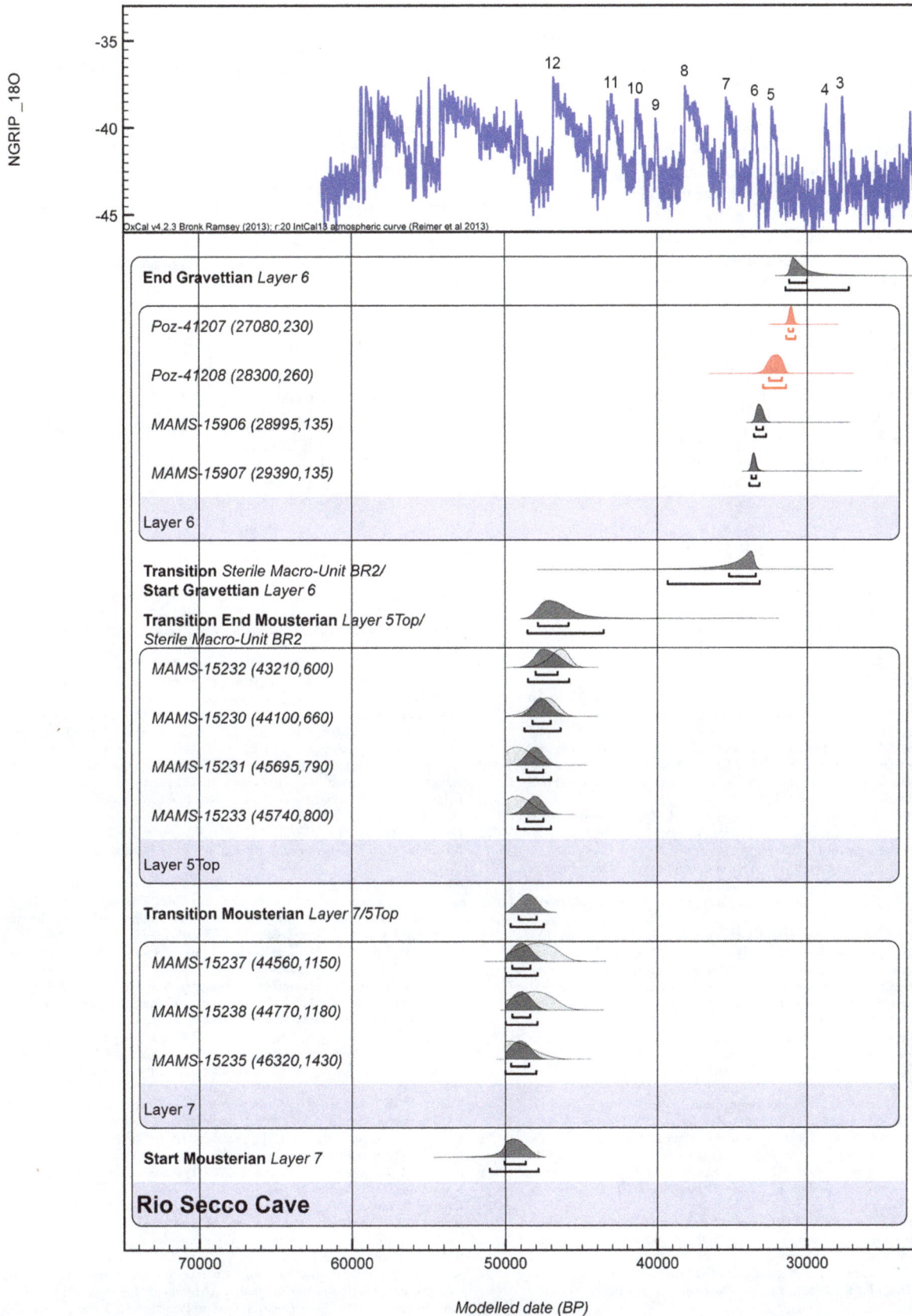

Figure 6. Calibrated ages and boundaries. Calibrated ages and boundaries calculated using OxCal 4.2 [33] and IntCal13 [34]. Rio Secco ages are in black and the previous radiometric results from Poznan (Lab. code Poz-) are in red. The results are linked with the (NGRIP) δ18O climate record.

200 km [50,59–61]. In this scenario the mechanism of the culturally mediated migration might have facilitated the diffusion of the Gravettian innovations and their assimilations in the technical behaviors in the neighboring regions.

Although the absence of diagnostic lithic tools at Rio Secco Cave layer 6 doesn't allow a correlation of the lithic assemblages with the central European techno-complexes, the radiometric dates support the hypothesis of dispersal of the Swabian Gravettian probably from the eastern Austria (Figure 1). In the neighborhood of Rio Secco Cave there are several gorges originating from a combination of tectonic uplift, karstic processes and run-off erosion. Along these gorges, several shelters and caves were formed in the walls and many others at the base of rock walls. Only a few of them (Verdi Caves and Clusantin Cave) have been explored for the presence of Pleistocene fills and have yielded Mousterian and late Epigravettian evidence for human frequentation [62,63]. This situation suggests that the absence of the Mid-Upper Palaeolithic in the eastern Alps may reflect a lack of the archaeological investigation rather than a gap in prehistoric human presence. So far Rio Secco Cave yields new insights for the presence of the last Neanderthals and the spread of Gravettian populations into the junction between the plain and the alpine regions.

References

1. Turk I, Kavur B (1997) Survey and description of Palaeolithic tools, fireplaces and hearths. In: Turk I, editor. Mousterian « bone flute » and other finds from Divje Babe I cave site in Slovenia. Ljubljana: Založba ZRC: Opera Intituti Archaeologici Sloveniae. pp. 119–156.
2. Karavanić I, Janković I (2006) The Middle and Early Upper Paleolithic in Croatia. Opuscula Archaeologica 30: 21–54.
3. Peresani M (2011) The End of the Middle Paleolithic in the Italian Alps: An Overview of Neanderthal Land Use, Subsistence and Technology. In: Conard NJ, editor. pp. 249–259.
4. Mussi M (2002) Earliest Italy. An Overview of the Italian Paleolithic and Mesolithic; Jochim MA, editor. New York: Kluwer Academic.
5. Klima B (1959) Zur Problematik des Aurignacien und Gravettin in Mittel-Europa. Archaeologica Austriaca 26: 35–51.
6. Flas D (2000–2001) Étude de la continuité entre le Lincombien-Ranisien-Jerzmanowicien et le Gravettien aux pointes pédonculées septentrional. Préhistoire européenne 16–17: 163–189.
7. Moreau L (2012) Le Gravettien ancien d'Europe centrale revisité: mise au point et perspectives. L'anthropologie 116: 609–638.
8. Valoch K (1981) Beitrag zur Kenntnis des Pavlovien. Archeologické Rozhledy 33: 279–298.
9. Gambassini P (2007) Traits essentiels du Gravettien en Italie. Paléo19: 105–108.
10. Svoboda J (2007) On Modern Human Penetration to Northern Eurasia: the Multiple Advances Hypothesis. In: Mellars P, Boyle K, Bar-Yosef O, Stringer C, editors. Rethinking the human revolution New behavioural and biological perspectives on the origin and dispersal of modern humans. Cambridge: McDonald Institute Monographs. pp. 329–339.
11. Borgia V, Ranaldo F, Ronchitelli AT, Wierer U (2011) What differences in production and use of Aurignacian and early Gravettian Lithic assemblages? The case of Grotta Paglicci (Rignano Garganico, Foggia, Southern Italy). In: Goutas N, Klaric L, Pesesse D, Guillermin P, editors. À la recherche des identités gravettiennes: actualités, questionnements et perspectives Actes de la table ronde sur le Gravettien en France et dans les pays limitrophes, Aix-en-Provence, 6–8 octobre 2008. Paris: Société préhistorique française. pp. 161–174.
12. Mussi M (2000) Heading south: the gravettian colonisation of Italy. In: Roebroeks W, Mussi M, Svoboda J, Fennema K, editors. Hunters of the Golden Age The Mid Upper Palaeolithic of Eurasia 30,000–20,000 BP: University of Leiden. pp. 355–374.
13. Floss H, Kieselbach P (2004) The Danube Corridor after 29,000 BP – New results on raw material procurement patterns in the Gravettian of southwestern Germany. Mitteilungen der Gesellschaft für Urgeschichte 13: 61–78.
14. Wierer U (2013) Variability and standardization: The early Gravettian lithic complex of Grotta Paglicci, Southern Italy. Quaternary International 288: 215–238.
15. Palma di Cesnola A (2004) Le industrie degli strati 24–22. In: Grenzi C, editor. Paglicci L'Aurignaziano e il Gravettiano antico. Foggia: Palma di Cesnola, A. pp. 111–207.
16. Palma di Cesnola A (2006) L'Aurignacien et le Gravettien ancien de la grotte Paglicci au Mont Gargano Aurignacian and early Gravettian of Paglicci cave – Mount Gargano. L'anthropologie 110: 355–370.
17. Boscato P (2007) Faunes gravettiennes à grands mammifères de l'Italie du Sud: Grotta della Cala (Salerno) et Grotta Paglicci (Foggia). Paléo 19: 109–114.
18. Palma di Cesnola A (2001) Le paléolithique supérieur en Italie; Millon J, editor. Grenoble: J. Millon.
19. Waelbroeck C, Labeyrie L, Michel E, Duplessy JC, McManus JF, et al. (2002) Sea-level and deep water temperature changes derived from benthic foraminifera isotopic records. Quaternary Science Reviews 21: 295–305.
20. Antonioli F (2012) Sea level change in Western-Central Mediterranean since 300 Kyr: comparing global sea level curves with observed data. Alpine and Mediterranean Quaternary 25: 15–23.
21. Giaccio B, Rolfo MF, Bozzato S, Galadini F, Messina P, et al. (2004) La risposta ambientale ed umana alle oscillazioni climatiche sub-orbitali dell'OIS 3: Evidenze geoarcheologiche dalla piana di colfiorito (appennino centrale). Il Quaternario 17: 231–247.
22. De Stefani, Gurioli M F, Ziggiotti S (2005) Il Paleolitico superiore del Riparo del Broion nei Colli Berici (Vicenza). Rivista di Scienze Preistoriche Suppl. 1: 93–107.
23. Silvestrini M, Peresani M, Muratori S (2005) Frequentazioni antropiche allo spartiacque appenninico nella fase antica del Paleolitico superiore: il sito di Fonte delle Mattinate (Altopiano di Colfiorito). Riunione Scientifica Istituto Italiano di Preistoria e Protostoria Atti XXXVIII: 69–79.
24. Peresani M, Gurioli F (2007) The Rio Secco cave, a new final Middle Paleolithic site in North-Eastern Italy. Eurasian Prehistory 5: 85–94.
25. Peresani M, Romandini M, Duches R, Jéquier C, Nannini N, et al. (2014) New evidence for the Neanderthal demise and earliest Gravettian occurrences at Rio Secco Cave, Italy. Journal of Field Archaeology In Press.
26. Romandini M, Nannini N, Tagliacozzo A, Peresani M (2013) Hunting bear during the Late Mousterian. Evidence from the North of Italy. European Association for the Study of Human Evolution ESHE Proceedings, 3rd Meeting: 189.
27. Talamo S, Richards M (2011) A comparison of bone pretreatment methods for AMS dating of samples >30, 000 BP. Radiocarbon 53: 443–449.
28. Brown TA, Nelson DE, Vogel JS, Southon JR (1988) Improved Collagen Extraction by modified Longin method. Radiocarbon 30: 171–177.
29. Brock F, Bronk Ramsey C, Higham T (2007) Quality assurance of ultrafiltered bone dating. Radiocarbon 49: 187–192.
30. Kromer B, Lindauer S, Synal H-A, Wacker L (2013) MAMS – A new AMS facility at the Curt-Engelhorn-Centre for Achaeometry, Mannheim, Germany. Nuclear Instruments and Methods in Physics Research Section B: Beam Interactions with Materials and Atoms 294: 11–13.
31. Ambrose SH (1990) Preparation and Characterization of Bone and Tooth Collagen for Isotopic Analysis. Journal of Archaeological Science 17: 431–451.
32. van Klinken GJ (1999) Bone Collagen Quality Indicators for Palaeodietary and Radiocarbon Measurements. Journal of Archaeological Science 26: 687–695.
33. Bronk Ramsey C, Sharen L (2013) Recent and Planned Developments of the Program OxCal. Radiocarbon 55: 720–730.
34. Reimer PJ, Bard E, Bayliss A, Beck JW, Blackwell PG, et al. (2013) IntCal13 and Marine13 Radiocarbon Age Calibration Curves 0–50,000 Years cal BP. Radiocarbon 55: 1869–1887.

Acknowledgments

The Grotta Rio Secco research project has been promoted by the Administration of the Clauzetto Municipality and coordinated by the University of Ferrara in collaboration with the Friuli Venezia Giulia Region and with permission of the Archaeological Superintendence of the Friuli-Venezia Giulia. M. Peresani structured the research project of Rio Secco and directs the fieldwork; M. Romandini coordinates the fieldwork. The Rio Secco project is also co-designed by the Neanderthal Museum (A. Pastoors & G.C. Weniger) and the Universitat Rovira y Virgili at Tarragona (M. Vaquero). A. Picin is supported by the Fuhlrott Research Fellowship of the Neanderthal Museum Foundation. Logistic assistance was furnished by the Clauzetto Public Administration. S.T. and J-J. H. wishes to thank Lysann Rädisch and Sven Steinbrenner for the laboratory work.

Author Contributions

Conceived and designed the experiments: ST. Performed the experiments: ST. Analyzed the data: ST. Contributed reagents/materials/analysis tools: MP MR RD CJ NN A. Picin MV A. Pastoors GCW. Wrote the paper: ST MP JJH MR A. Picin RD CJ NN.

35. Bronk Ramsey C (2009) Dealing with outliers and offsets in radiocarbon dating. Radiocarbon 51: 1023–1045.

36. Tozzi C (1994) Il Paleolitico inferiore e medio del Friuli. Riunione Scientifica Istituto Italiano Preistoria e Protostoria Atti XXIX: 19–36.

37. Blackwell BAB, Yu ESK, Skinner AR, Turk I, Blickstein JIB, et al. (2007) ESR Dating at Divje babe I, Slovenia. DIVJE BABE I Upper Pleistocene Palaeolithic site in Slovenia Part I: Geology and Palaeontology: Opera Instituti Archaeologici Sloveniae.

38. Krings M, Capelli C, Tschentscher F, Geisert H, Meyer S, et al. (2000) A view of Neandertal genetic diversity. Nature Genetics 26: 144–146.

39. Serre D, Langaney A, Chech M, Teschler-Nicola M, Paunovic M, et al. (2004) No Evidence of Neandertal mtDNA Contribution to Early Modern Humans. PLoS Biology 2: 0313–0317.

40. Reimer PJ, Baillie MGL, Bard E, Bayliss A, Beck JW, et al. (2009) IntCal09 and Marine09 Radiocarbon Age Calibration Curves, 0–50,000 Years cal BP. Radiocarbon 51: 1111–1150.

41. Talamo S, Hughen KA, Kromer B, Reimer PJ (2012) Debates over Palaeolithic chronology – the reliability of ^{14}C is confirmed. Journal of Archaeological Science 39: 2464–2467.

42. Fiedel SJ, Southon JR, Taylor RE, Kuzmin YV, Street M, et al. (2013) Assessment of Interlaboratory Pretreatment Protocols by Radiocarbon Dating an Elk Bone Found Below Laacher See Tephra at Miesenheim IV (Rhineland, Germany). Radiocarbon 55: 1443–1453.

43. Brock F, Geoghegan V, Thomas B, Jurkschat K, Higham TFG (2013) Analysis of Bone "Collagen" Extraction Products for Radiocarbon Dating. Radiocarbon 55: 445–463.

44. Haesaerts P, Damblon F, Nigst P, Hublin J-J (2013) ABA and ABOx Radiocarbon Cross-Dating on Charcoal from Middle Pleniglacial Loess Deposits in Austria, Moravia, and Western Ukraine. Radiocarbon 55: 641–647.

45. Higham T, Brock F, Peresani M, Broglio A, Wood R, et al. (2009) Problems with radiocarbon dating the Middle to Upper Palaeolithic transition in Italy. Quaternary Science Reviews 28: 1257–1267.

46. Higham T, Jacobi R, Basell L, Bronk Ramsey C, Chiotti L, et al. (2011) Precision dating of the Palaeolithic: A new radiocarbon chronology for the Abri Pataud (France), a key Aurignacian sequence. Journal of Human Evolution 61: 549–563.

47. Douka K, Grimaldi S, Boschian G, del Lucchese A, Higham TFG (2012) A new chronostratigraphic framework for the Upper Palaeolithic of Riparo Mochi (Italy). Journal of Human Evolution 62: 286–299.

48. Higham T, Basell L, Jacobi R, Wood R, Bronk Ramsey C, et al. (2012) Testing models for the beginnings of the Aurignacian and the advent of figurative art and music: The radiocarbon chronology of Geißenklösterle. Journal of Human Evolution62: 664–676.

49. Svoboda JA (2007) The Gravettian on the Middle Danube. Paléo 19: 203–220.

50. Moreau L (2009) Geißenklösterle. Das Gravettien der Schwäbischen Alb im europäischen Kontext. Kerns Verlag. Tübingen.

51. Pesesse D (2010) Quelques repères pour mieux comprendre l'émergence du Gravettien en France. Bulletin de la Société Préhistorique Française 107: 465–487.

52. Klaric L, Guillermin P, Aubry T (2009) Des armatures variées et des modes de productions variables. Reflexions à partir de quelques examples issus du Gravettien d'Europe occidentales (France, Portugal, Allemagne). Gallia Préhistorie 51: 113–154.

53. Moreau L (2010) Geißenklösterle. The Swabian Gravettian in its European context. Quartär 57: 79–93.

54. Einwogerer T, Handel M, Neugebauerer-Maresch C, Simon U, Steier P, et al. (2009) [14]C Dating of the Upper Palaeolithic Site at Krems-Wachtberg, Austria. Radiocarbon 51: 847–865.

55. Beresford-Jones D, Taylor S, Paine C, Pryor A, Svoboda J, et al. (2011) Rapid climate change in the Upper Palaeolithic: the record of charcoal conifer rings from the Gravettian site of Dolní Věstonice, Czech Republic. Quaternary Science Reviews 30: 1948–1964.

56. Jöris O, Neugebauer-Maresch C, Weninger B, Street M (2010) The Radiocarbon Chronology of the Aurignacian to Mid-Upper Palaeolithic Transition along the Upper and Middle Danube. In: Neugebauer-Maresch C, Owen LR, editors. New Aspects of the Central and Eastern European Upper Palaeolithic – methods, chronology, technology and subsistence. Wien: Österreichische Akademie der Wissenschaften. pp. 101–137.

57. Djindjian F, Kozłowski JK, Otte M (1999) Le Paléolithique supérieur en Europe; Colin, editor. Paris.

58. Sinitsyn A (2007) Variabilité du Gravettien de Kostienki (Bassin moyen du Don) et des territoires associés. Paléo 19: 179–200.

59. Svoboda J (2000) Hunting in Central Europe at the End of the Last Glacial, La chasse dans la Préhistoire. Hunting in Prehistory. Actes du Colloque international de Treignes 3–7 Octobre 1990. Liege: ERAUL. pp. 233–236.

60. Oliva M (2000) Some thoughts on Pavlovian adaptations and their alternatives. In: Roebroeks W, Mussi M, Svoboda J, Fennema K, editors. Hunters of the golden age: The Mid Upper Palaeolithic of Eurasia 30,000–20,000 BP. Leiden: University of Leiden. pp. 219–229.

61. Digan M (2008) New technological and economic data from La Vigne-Brun (unit KL19), Loire: a contribution to the identification of early Gravettian lithic technological expertise. Quartär 55: 115–125.

62. Bartolomei G, Broglio A, Palma di Cesnola A (1977) Chronostratigraphie et écologie de l'Epigravettien en Italie. In: Sonneville-Bordes Dd, editor. La fin des temps glaciaires en Europe-Chronostratigraphie et écologie des cultures du Paléolithique final. Paris: Colloques Internationaux du C.N.R.S. 297–234.

63. Peresani M, Duches R, Miolo R, Romandini M, Ziggiotti S (2011) Small specialized hunting sites and their role in Epigravettian subsistence strategies. A case study in Northern Italy. In: Bon F, Costamagno S, Valdeyron N, editors. Hunting Camps in Prehistory Current Archaeological Approaches, Proceedings of the International Symposium, May 13–15 2009. University Toulouse II Le Mirail, Palethnology. pp. 251–266.

3

Quantifying Regional Vegetation Cover Variability in North China during the Holocene: Implications for Climate Feedback

Guo Liu, Yi Yin, Hongyan Liu*, Qian Hao

College of Urban and Environmental Sciences, Peking University, Beijing, China

Abstract

Validating model simulations of vegetation-climate feedback needs information not only on changes in past vegetation types as reconstructed by palynologists, but also on other proxies such as vegetation cover. We present here a quantitative regional vegetation cover reconstruction for North China during the Holocene. The reconstruction was based on 15 high-quality lake sediment profiles selected from 55 published sites in North China, along with their modern remote sensing vegetation index. We used the surface soil pollen percentage to build three pollen-vegetation cover transfer models, and used lake surface sediment pollen data to validate their accuracy. Our results showed that vegetation cover in North China increased slightly before its maximum at 6.5 cal ka BP and has since declined significantly. The vegetation decline since 6.5 cal ka BP has likely induced a regional albedo change and aerosol increase. Further comparison with paleoclimate and paleovegetation dynamics in South China reproduced the regional cooling effect of vegetation cover decline in North China modelled in previous work. Our discussion demonstrates that, instead of reconstructing vegetation type from a single site, reconstructing quantitative regional vegetation cover could offer a broader understanding of regional vegetation-climate feedback.

Editor: Dorian Q. Fuller, University College London, United Kingdom

Funding: This study is granted by National Natural Science Foundation of China (NSFC, numbers 41071124, 31021001 and J1103406). The funders had no role in study design, data collection and analysis, decision to publish, or preparation of the manuscript.

Competing Interests: The authors have declared that no competing interests exist.

* E-mail: lhy@urban.pku.edu.cn

Introduction

The need for reducing uncertainty in global climate change predictions has highlighted the importance of integrating model evaluations, on-site and remote sensing monitoring, paleoecological investigation, and small–scale manipulative experiments [1], [2]. Paleoecological data, with their unique advantage in addressing earth system processes at large temporal scales and under extreme conditions, could offer unique insights in this respect [2], [3]. The application of paleoecological data in reconstructions, however, requires the conversion from geological proxies to time-series of reconstructed variables [4].

Vegetation type can be reliably interpreted from pollen records through palynological methods [5], [6], [7], but is hard to quantify and therefore difficult to compare with model simulations or remote sensing monitoring. Modern remote sensing techniques, on the other hand, have allowed the evaluation of vegetation through numerical characteristics that offer important information for model simulations [8], [9]. Using similar palynological methods, with remote sensing data as a modern analogue, vegetation in the past could be reconstructed in a different manner, avoiding a discrete and discontinuous vegetation type reconstruction while providing a detailed description of past vegetation that is comparable with model output. Although it remains uncommon, this approach to vegetation reconstruction has been successfully exploited in some former studies [10], [11], [12].

While paleoenvironmental reconstruction could provide evidence of ecological and climatological processes at large temporal scales, distinguishing variations due to regional trends from those due to local heterogeneity is difficult and thus interpretations are often restricted to being local-scale and case-specific when based on one single site [4], [13]. By combining data from multiple sites, vegetation reconstruction at a regional scale could provide a broader insight into regional scale earth system process: this has already been achieved in many data-rich regions such as Europe [6] and North America [14]. Published late-Quaternary pollen records from northern Asia have increased rapidly in number over the past several decades [15], [16], [17]; in North China, however, reliable pollen data sites with relatively high resolution remained scarce [10] until the collection of higher quality data in recent years (Tab. S1). In this study, we selected 15 sediment profiles with relatively high resolution and reliability from 55 published sites in North China (Table S1), and reconstructed the regional vegetation cover changes during the Holocene using a remote sensing vegetation index (NDVI, normalized deference vegetation index) as a modern analogue. This quantitative regional reconstruction provided an opportunity to examine millennial-scale regional vegetation-climate interaction, and supported the hypothesis of a cooling effect of vegetation decline [18], [19] [20] through comparison with previous model simulations [1].

Data and Methods

1. Pollen dataset

One major obstacle to past vegetation reconstruction is the deficiency of modern lake sediment analogues. Lake sediment profiles are widely used due to their ability to preserve signals of past environmental change; without these data, reconstructions are restricted to surface soil as the modern analogue for lake sediment profiles [21], [22], [23], with much uncertainty remaining with respect to the relationship between surface soil pollen and lake sediment pollen [24], [25]. In this study, both surface soil pollen data and lake surface sediment pollen data were utilized to render the reconstruction as reliable as possible: three simple pollen-NDVI transfer models were built from the modern surface soil pollen and modern NDVI, and were validated by the dataset from modern lake surface sediments. Finally, the most accurate transfer model was applied to the sediment profiles.

A total of 461 published surface-soil pollen records from across North China were collated in this study, collected in regions with mean annual precipitation ranging from 100 to 700 mm (Fig. 1). Over 60 pollen taxa were found in these records, of which *Pinus*, *Betula*, *Quercus*, *Artemisia* and Chenopodiaceae appeared most frequently; these taxa have already been identified as indicators of vegetation type in our study area [26], [27]. Redundancy analysis (RDA) with all the major taxa in the soil pollen (Fig. 2a in [35]) showed that these 5 taxa (especially Pinus, Betula and Quercus) could explain most of the geographical variation of the soil pollen dataset; in addition, comparing the R^2 of models constructed by these 5 most common taxa or by 18 common taxa showed that the difference (<0.05) is non-significant. While making use of more information from the pollen data, using more pollen types could at the same time increase the risk of model over-fitting, especially by artificial neural networks. In addition, models will be sensitive to taxa that only occur in few samples, which could introduce bias and uncertainty in the reconstruction. Therefore, the percentages of these five taxa were chosen as dependent variables to establish pollen-NDVI models. Lake surface sediments (upper 5 cm) of 27 perennial lakes with low human disturbance were collected and analysed, to validate the surface-soil pollen-NDVI models constructed by surface soil samples. The model with the highest accuracy was then applied to the selected sediment profiles.

To select the most appropriate data from the 55 available profiles in or near our study area, three criteria were used: 1, the profiles cannot be too far away from the 27 lakes with surface sediment pollen analysis data, to guarantee the effectiveness of the validation; 2, the profiles must cover the last 10 ka, to guarantee the same number of samples for every period; and 3, the profile must have a high temporal resolution, as the trend in vegetation cover largely depends on the time period selected, thus a higher temporal resolution will decrease the uncertainty in period selection. Among all the 55 profiles, 34 were close to the 27 lakes used in this study; among these 34 profiles, 29 had an average resolution higher than 500a; and among these 29 profiles, 15 covered the last 10 ka. Proportions of the five chosen taxa were digitised from these 15 profiles and the corresponding vegetation cover was then reconstructed (Fig. 1). We then resampled the 15 reconstructed vegetation cover time series to a resolution of 200a because different profiles cover different time periods, such that resampling to a finer resolution allowed better use of the data. Finally, the 15 profiles were averaged to yield the regional vegetation cover changes.

2. Modern NDVI distribution patterns

The GIMMS NDVI (normalized difference vegetation index) dataset [28], [29], [30], running from 1982 to 2006 with a spatial resolution of 8 km×8 km, was used to calculate the NDVI of each sampling site. The lakes in our sediment pollen dataset had an average diameter of less than 1 km and the local impact of lakes on albedo and vegetation was thus ignored at the scale of an NDVI pixel. We then calculated the average of the yearly maximum NDVI values from 1982 to 2006 for each site with both surface soil and lake surface sediment samples in our analysis. Distributions of averaged NDVI and mean annual precipitation, as well as vegetation types, are plotted in Fig. 1.

3. Pollen-NDVI transfer models

Three pollen-NDVI models were constructed using pollen records in surface soil and were verified by corresponding records in the lake surface sediment, with the most accurate model being used to reconstruct NDVI from the pollen spectrum in the sediment cores. Models of pollen-NDVI relationships were built using an artificial neural network (ANN), the modern analogue technique (MAT) and linear regression (LR). Then the lake surface sediment pollen and associated NDVI dataset were applied to these models to verify their accuracy.

Firstly, we constructed a model of the relationship between the pollen spectra and NDVI by applying an ANN, which is a very flexible nonlinear method that is able to precisely simulate complex mapping [31]. In our study, the ANN used the percentages of *Pinus, Betula, Quercus, Artemisia* and Chenopodiaceae of each sample as input variables and the corresponding NDVI as the target variable. After experimenting with different parameter matching configurations, we chose three hidden layers with node numbers of 8, 10, 10 and the activation functions *tansig, logsig, logsig* (in Matlab R2009b), respectively.

Next we applied MAT, which has been widely used in vegetation reconstruction with pollen data [32], using the squared chord distance (SCD) as the index of similarity [33], [34]:

$$SCD = \sum_{i=1}^{n} (\sqrt{x_i} - \sqrt{y_i})^2$$

where SCD is the squared chord distance between two multivariate samples X and Y, and x, y are the proportions of species i in samples X and Y. SCD values can range from 0 to n (number of species), with 0 indicating pollen spectra identical to those of the samples being compared. For each fossil record, five samples with the smallest SCD (with the sample itself excluded) were selected from the surface soil pollen dataset and the corresponding NDVI was reconstructed by the weighted-average:

$$NDVI_r = \frac{\sum_{i=1}^{5} NDVI_i / SCD_i}{\sum_{i=1}^{5} 1/SCD_i}$$

where $NDVI_r$ is the reconstructed NDVI, and $NDVI_i$ and SCD_i are the NDVI and squared chord distance of each sample.

The third method of reconstruction applied here was linear regression (LR). It can be assumed that for each taxon existing in a biome, every individual occupies a certain ratio of the area and contributes a corresponding value to the total NDVI; therefore, the relationship between total NDVI and the proportion of each taxon may have a linear component. LR has the significant advantage of being intuitive, with the respective parameters clearly showing the contribution of each taxon.

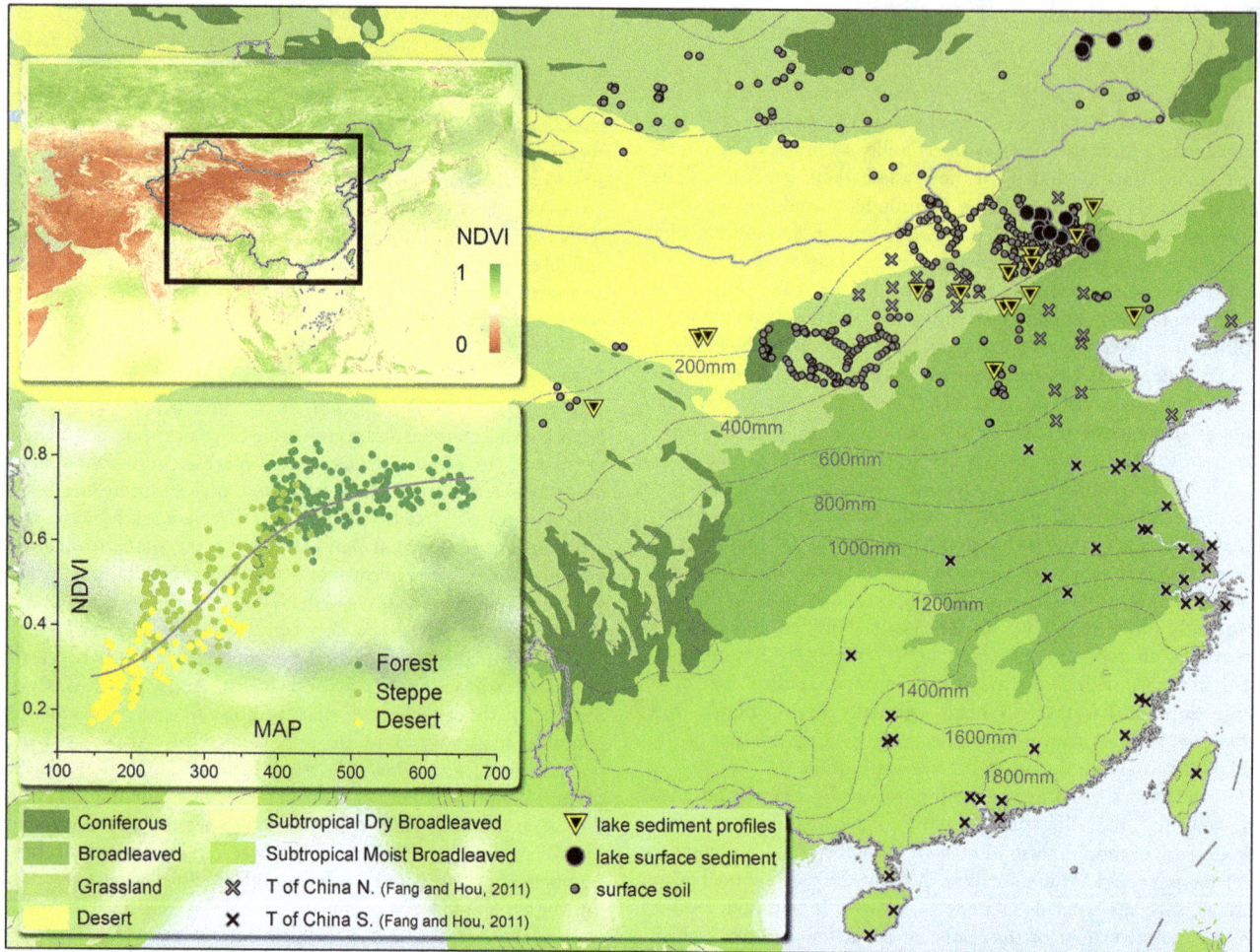

Figure 1. Sample locations and modern NDVI distribution. Modern NDVI data were acquired by averaging data for August from 1982–2006 of the GIMMS dataset. NDVI in South China is homogeneous at a high level, while that of North China varies widely with precipitation. Samples of NDVI-MAP relationships were randomly chosen from grid points with natural vegetation in our study area (North China) and fitted by a logistic curve. Sites with surface soil pollen, lake surface pollen and sediment profiles are distributed around the 400 mm isohyet; some sites with surface soil pollen samples are located in Mongolia, but in the same biome and precipitation regime. *T of China N.* and *T of China S.* indicate the paleotemperature records used in the temperature reconstruction of North China and South China [70], respectively.

These three models of pollen spectra and NDVI, built from our surface-pollen dataset, could have been evaluated by means of the adjusted correlation coefficient and residual error between the estimated NDVI and observed NDVI. A proper approach, however, has to extend its accuracy into the fossil sediment pollen record. Therefore, we used our sediment pollen dataset to test the reliability of extending the models into the sediment pollen record by calculating the correlation coefficient between the simulated NDVI and observed NDVI of lake surface sediment samples (Fig. 2b).

All the above calculations were performed with the Matlab R2009b software.

4. Statistical analysis of the reconstructed NDVI

To maximise accuracy of the results, the 15 sediment profiles were selected to occupy the same time range (11.5 cal. ka BP to present) and to have comparatively high resolution (200 yr or higher). Each profile was resampled to 200 years temporal resolution before reconstruction. The reconstruction thus produced 15 comparable time series ranging from 11.5 cal ka BP to present. The stalagmite δ^{18}O from a former study was sampled

from Dongge cave in the Pacific monsoon region of China as a proxy for precipitation in the monsoon region, and was linearly correlated with the reconstructed vegetation cover. Meanwhile, the coarse sand (>63 μm) percentage from Anguli Nuur Lake in North China, located in the central-eastern region of our study area in the forest-steppe ecotone about 250 km northwest of Beijing, was used as an indicator of local soil coarsening, and was correlated with the reconstructed vegetation cover.

Results

The ANN yielded the highest accuracy in the pollen-NDVI transfer model and was therefore used in our vegetation cover reconstruction. Each of the three models (ANN, MAT and LR) performed well in establishing the relationship between surface soil pollen and NDVI, with adjusted R^2 values of 0.62, 0.49 and 0.41 (P<0.01), respectively. In the validation by lake surface sediment pollen, only ANN and LR continued to produce reliable results, with adjusted R^2 values of 0.56 and 0.47 (p<0.01), respectively. Therefore, only the reconstruction by ANN was used in the following analysis.

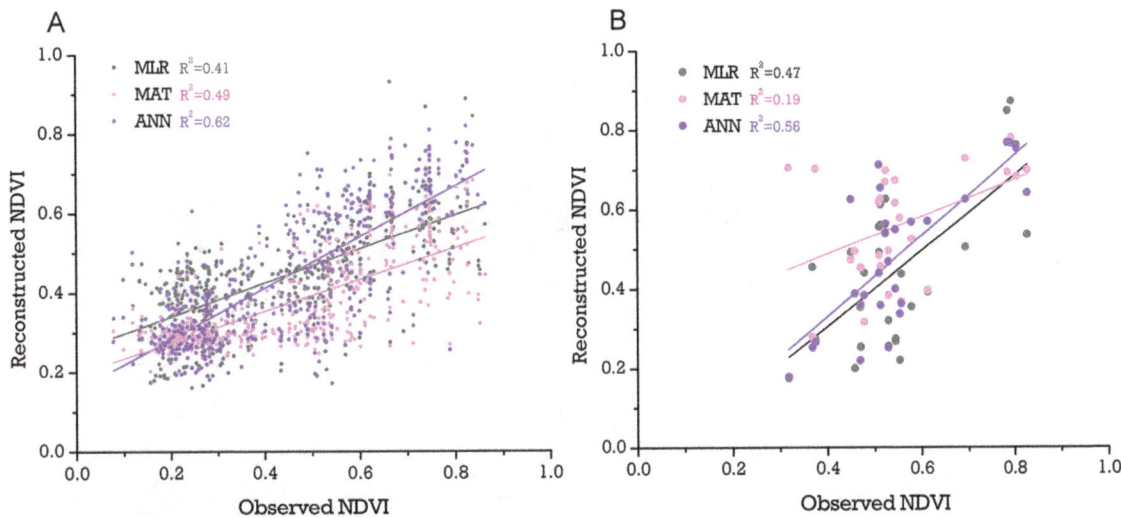

Figure 2. Result of model reconstruction and verification. In subplot A, LR, MAT and ANN all performed well in the model construction phase and passed the 0.01 significance level test; in subplot B, when verified by lake surface sediment pollen data, MAT failed to produce a reliable result. ANN was chosen for the reconstruction. All R^2 data are the adjusted R square.

Reconstruction results show that vegetation cover in North China increased slightly before its maximum at 6.5 cal. ka BP and has since declined significantly (Fig. 3). The Holocene can be divided into 3 periods according to the averaged results from 15 profiles with 200-year resolution, as follows. *Period III* (11.5–6.5 cal. ka BP): most profiles in this period were identified as forest in former studies; vegetation cover fluctuated greatly but showed an overall slight increase before reaching its maximum at 6.5 cal. ka BP. *Period II* (6.5–3 cal. ka BP): a greater number of profiles were identified as steppe or forest steppe in this period, regional vegetation cover declined significantly from 0.57 to approximately 0.46. *Period I* (3 cal. ka BP- present): most profiles were identified as steppe, but the decline in Period II ended and there was even a slight increase in the most recent 1 cal. ka BP. Although 15 profiles are not sufficient to map the distribution of vegetation cover, we calculated and plotted the mean reconstructed vegetation cover of each profile for 10–9, 7–6 and 2–1 cal. ka BP (Fig. 4) to indicate the state of each period. Upper and lower quartiles in each 200-yr period were also calculated, showing the difference within sub-regions with different levels of vegetation cover.

Discussion

1. Vegetation cover changes and their driving forces

The number of profiles adequate for regional vegetation reconstruction is directly related to the indicating area of lake sediment pollen and the vegetation heterogeneity in the studied area. Former studies conducted in small lakes with an assumption of neutral atmospheric conditions in regions dominated by forest generally showed that the source area of sediment pollen is within a radius of 1 km [35], [36], [37], [38]. With a source area of this scale, 15 profiles would be a poor representation of this region; however, in areas with a relatively open landscape, e.g. North China and the North American Central Plain, strong winds prevail through most of the year causing long distance transportation of dusts [39], which carry 20–1000 μm particles including pollen grains [40]. Combining surface soil pollen data and remote sensing data in a similar fashion to this study, research in eastern North America has yielded indicating areas with half-widths of 25–75 km [41]. Using a similar method, we have also calculated

the indicating area of surface soil pollen and lake surface sediment pollen (corresponding to the lake sediment profiles used in reconstruction) in our study area (Yin et al., unpublished manuscript) and identified an average indicating area much larger than that of traditional estimates (half-widths of 20 km for surface soil, and over 100 km for surface lake sediment.

Although we could not directly quantify the heterogeneity of paleo-vegetation cover in our study area from the 15 profiles, we were able to assess whether the variability in these profiles could cause uncertainties in our results, and we indirectly evaluated the uncertainty related to vegetation cover heterogeneity. As our conclusions were based on the trend in vegetation cover, the relative change in vegetation cover was more important than the actual value of reconstructed vegetation cover in our study. Despite the different magnitudes of variability, the upper quartile, the lower quartile and the average of the 15 sequences showed the same trends (Fig. 3), indicating that the difference between profiles has little influence on our conclusion. Thus we believe that the 15 profiles are representative of the geographical range they cover. However, as the 15 profiles did not include sediment from the Loess Plateau, which has markedly different vegetation, the reconstructed vegetation cover only includes the regions of North China outside of the plateau.

Average results from the 15 profiles showed that vegetation cover in North China increased slightly from 11 cal. ka BP and reached its maxim at 6.5 cal. ka BP, before decreasing significantly. Calculation of the upper and lower quartiles in each 200-year period revealed the range in levels of vegetation cover between different regions: accordingly, the overall trends in vegetation cover were mainly attributed to changes in regions with high vegetation cover (mostly forest), while regions with low vegetation cover (mostly grassland and desert) remained relatively stable. Our results coincide well with those of former studies of Holocene vegetation types in North China (Tab. S1): in *Period III*, when vegetation cover fluctuated greatly with a slight overall increase, the vegetation type was mainly forest; in *Period II*, vegetation cover declined significantly and changed from forest-dominated to grassland-dominated at most sites; in *Period I*, the decline of vegetation cover slowed and North China was dominated by grassland.

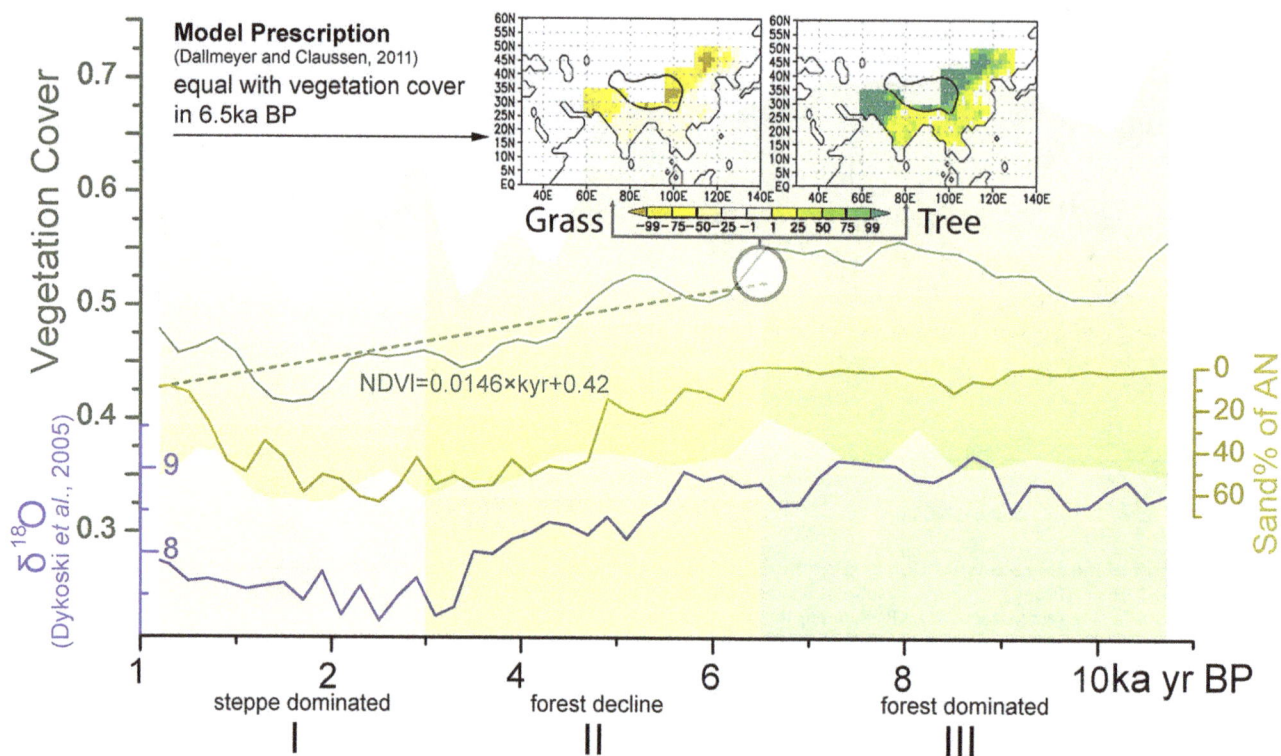

Figure 3. Mean reconstructed vegetation cover of each profile for 1–2, 6–7 and 9–10 ka BP. The circle diameters show the mean values of the reconstruction within 1 ka, while modern NDVI and modern MAP are plotted in the background. Because the heterogeneity between sites is much larger than the variation, the diameters of the circles were plotted according to the relative values compared to the mean value of the entire time series at each site.

Regional reconstruction of vegetation cover provides an opportunity for studying regional driving forces of vegetation cover, and the partial correlation between reconstructed temperature and stalagmite $\delta^{18}O$ has revealed precipitation as the main driving factor of vegetation cover change. Precipitation in the Pacific monsoon regions of China during the Holocene could be indicated by the stalagmite $\delta^{18}O$ proxy from Dongge Cave [42], [43]. Linear regression shows that this proxy is significantly and negatively correlated with the reconstructed vegetation cover in North China ($R^2 = 0.701$, P<0.01), indicating the influence of precipitation on vegetation (a lower $\delta^{18}O$ indicates higher precipitation). However, a potential risk of comparing the stalagmite $\delta^{18}O$ from Dongge Cave with results from our reconstruction is the uncertainty in the $\delta^{18}O$ interpretation. Although former studies have shown that $\delta^{18}O$ of stalagmites in the region of Dongge Cave can be interpreted reliably as an indicator of the overall precipitation at a continental scale [43], [44], it has been argued that $\delta^{18}O$ may also be influenced by temperature [45], [46]. Studies based on modern observations in North China, however, have shown that precipitation dominates vegetation growth while temperature only plays a minor role [47], [48], [49]. Thermal regulation of vegetation has been found in areas with vegetation decline due to warming-induced drought [48], [49], [50], which casts further doubt on the relationship between decreasing temperature and declining vegetation cover through the Holocene [51]. In addition, we conducted a model simulation using the ORCHIDEE dynamic global vegetation model (DGVM) to test the influence of precipitation and temperature on vegetation cover in this region, and have also found that precipitation strongly controls vegetation cover in

North China, while temperature has a relatively weak influence (Liu *et al.*, unpublished data).

2. Possible impacts of vegetation cover change

While vegetation cover is mainly controlled by precipitation, significant change on a regional scale could in return have impacts on regional climate. Among these impacts are the decrease in albedo and the increase of aerosol production. The degradation from forest to grassland or from grassland to desert is accompanied by an increase in albedo [52], [53]; in addition, NDVI, indicating vegetation cover change in our study, has a linear negative relationship with albedo [54], [55], and its reconstruction implies the albedo increase in *Period I*. A more significant mechanism contributing to albedo changes is related to the transformation between vegetation types. In the winter of higher latitudes, grassland is entirely covered by snow whereas forest can penetrate through snow cover and generally remains exposed. Therefore, the transformation from forest to grassland could significantly change the albedo of the land cover [56], [57]. This transformation has taken place in the past 6500 years in regions with high vegetation cover, corresponding to those regions which have been mainly attributed to the overall vegetation cover change owing to their position in the upper quartile of the reconstruction.

The other possible impact of vegetation cover decline is the increase in aerosol production: soil dust plays an important role in the production of aerosol, and could significantly alter the regional radiation balance [58], [59]. Soil coarsening or sandification, the main source of soil dust, could be indirectly inferred from the coarse sand percentage in sediment records. Vegetation cover decline may lead to soil sandification and an increase in wind

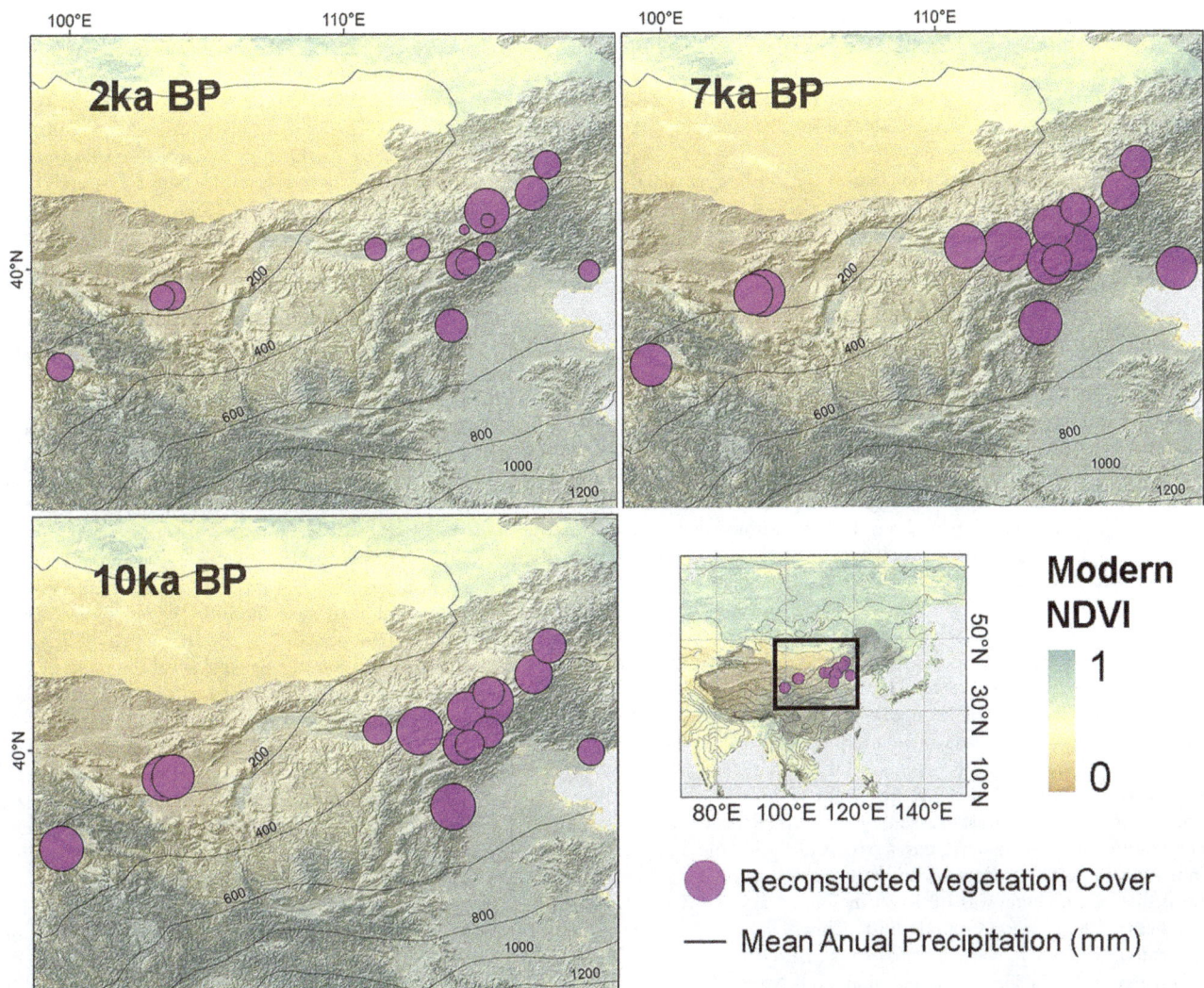

Figure 4. Reconstructed vegetation cover, related factors and prescribed values in previous studies. *Vegetation Cover* shows the reconstructed NDVI averaged from 15 sequences, with linear regression after 6.5 ka yr BP showing a decrease of 0.13. The light green area indicates the upper quartile and lower quartile of the 15 sequences, indicating that areas with higher NDVI experience larger fluctuations and were the major contributors to the vegetation decline in *Period II*. Sand% of AN shows the coarse sand (>63 µm) percentage in sediment cores from Anguli Nuur (inversely scaled). Stalagmite $\delta^{18}O$ of Dongge Cave (as $\delta^{18}O$, values are negative), coarse sand percentage of Anguli Nuur and reconstructed NDVI are significantly correlated with each other. Prescribed values in the model simulation by Dallmeyer and Claussen (2011) coincide with the vegetation cover at 6.5 cal ka BP as reconstructed here; the colour bar shows the percentage of grass/trees in comparison to the modern vegetation distribution.

velocity, both of which can contribute to a higher coarse sand percentage in lake sediments. Changes in wind velocity are attributed to the shift from forest to grassland, with a corresponding reduction in surface roughness [60], [61]. The vegetation type derived from this core has, however, shown the dominance of grassland since 5.0 cal. ka BP [62], while the coarse sand percentage has increased in tandem with the regional vegetation cover decline since then (except for a decrease in the past thousand years that coincides with the recovery of vegetation cover) (Fig. 4). A connection between coarse sand percentage and monsoon intensity has been reported throughout the last glacial maximum, but has mostly been observed at the larger temporal scale of the glacial-interglacial cycle and on plateaus such the Chinese Loess Plateau [63], [64]. The fluctuation of monsoon intensity during the Holocene was, however, much milder; in addition, the location of Anguli Nuur lake within a basin reduces its sensitivity to monsoon-controlled changes in wind velocity. A more likely

explanation is that the soil coarsening or sandification accompanied vegetation decline [65], [66], which can still be observed around Anguli Nuur Lake. Through the increase in dust transferred to the atmosphere, desertification could become a further source of aerosol [58], [59], [67]. The regional average vegetation cover is used in this study instead of that solely reconstructed from the sediment core of Anguli Nuur Lake due to the lower credibility of reconstruction from a single site. Although this coarse sand percentage is mainly a local indicator, the correlation reflects the relationship between regional vegetation cover decline and local soil sandification and implies the consequential increase of dust and aerosol.

Although the carbon emission accompanying the vegetation cover decline could have an impact at larger scales, both of the impacts discussed above suggest a cooling effect on regional climate (Fig. 5). Interactions between the regional vegetation-climate processes make it difficult to evaluate the overall effect of these feedbacks; however, regional comparison of paleoenvironmental reconstructions could provide evidence for the overall response over a long period of time, disregarding the uncertainties in the mechanisms' details. While the vegetation cover has declined significantly in North China since 6.5 cal. ka BP, vegetation in South China has been relatively steady and constantly dominated by forest [68], and its NDVI has remained saturated even to modern times [48]. Comparison between the vegetation records and temperature records in these two regions could thus offer insight into the overall thermal feedback of forest. Moreover, the vegetation-climate interaction in these regions has been discussed through a model simulation in a previous study [1], whose result implied a cooling effect of vegetation cover decline in North China that can be compared to our reconstruction.

As prescribed in the model, approximately 70% forest was added and 60% steppe removed in our study area, while vegetation change was comparably small in South China [1]. This prescription change, according to the modern distribution of NDVI (Fig. 1), is equivalent to a change of 0.1–0.15 in the NDVI of North China. Our result showed that, within the past 10,000 years, the vegetation cover consistent with these prescribed conditions occurred at 6.5 cal. ka BP (0.13 higher that modern times in NDVI, Fig. 3). Modelling results have indicated that North China was 1~1.5 K warmer in the cold/dry season while South China was 0.25~0.5 K warmer in the cold/dry season [1]. These results can then be compared to reconstructed temperatures for this period. Reconstructed temperatures around 6.5 cal. ka BP from single sites have shown a more significant warming, specifically 1.7–2.6 K, 3.0 K and 3.0–4.0 K for North China, and 1.0 K, 1.7 K and 3.0–3.5 K for South China [69], with a similar north-south contrast. Recent synthetic reconstructions of temperature in these two regions agree better with the model simulation, in which the linear trend in the temperature time series shows that North China was 1.8 K warmer and that South China (Southeast China and Central East China in the original publication) was 0.4 K warmer at 6.5 cal ka BP [70].

Situated in the Pacific monsoon region, both North China and South China are controlled by the same monsoon system (referred to as the SE Asian Monsoon in previous studies) [71], [72], and have shared similar climate histories during the Holocene. The differences in their temperature changes could be attributed, at least partially, to the contrasting vegetation cover changes. An opposing view could be that the vegetation cover decline was a result of the temperate change. However, as discussed in *Section 4.1.*, warming generally intensifies the water deficit of vegetation in North China [73], [74]; therefore it is unlikely that the decrease in temperature could have led the decline in vegetation cover of North China. Although agreement between model simulation and

the mechanisms we have suggested cannot rigorously prove the causal-consequence relationship between vegetation cover and temperature, it strongly supports former modelling studies and clearly increase support for the existence of this feedback. While global warming due to greenhouse gas emissions is commonly regarded as a serious issue for the future, this cooling effect might partially compensate the decrease in carbon sinks and thus counteract the effects of global warming. Estimates of the intensity of this feedback are evidently critical for the precision of model simulations.

3. Prospective regional vegetation cover reconstruction

The importance of integrating models and paleoecology data, in both the process of constraining inverse modelling and in the structure adjustment in forward modelling, has been recognized with the development of mechanistic, comprehensive and complex structures in ecological models and with the rapid growth of data in paleoecology [3]. As many have argued before, the assimilation of proxy data into models has become important for the improvement of both palaeovegetation reconstructions and model simulations [75], [3]. As a result, while ecological models have been developed in preparation for the data assimilation application, palaeoclimate and palaeovegetation are being reconstructed in a manner that can be directly utilized by models [76], [77], [78], and intensive integration of paleoecological data and model simulations has recently been achieved [79], [80]. Reliable methodologies and a paradigm for single-site based vegetation type reconstructions has been developed and recognized, but the implications of such reconstructions are, however, often site-specific and locally-restricted, with results that are difficult to quantify. We have shown in this study that regional vegetation cover reconstruction, with the help of newly developed vegetation indices in remote sensing, could offer new insights in earth system models.

As a trial of this approach, we discuss the cooling effect of vegetation cover decline, which has been crucial not only because it might counteract the organic carbon release, but because it could also result in a regional overall cooling [19], [20]. This is in contrast to the long-held view that afforestation, which opposes the vegetation cover decline, can alleviate global warming [81], [82]. Previous observations, however, were conducted during comparatively short periods of time [19], [43], [83], which might be insufficient for a significant change in vegetation cover to take place. The long temporal scales in paleoecological studies are therefore valuable for the evaluation of similar processes. An ideal regional vegetation cover reconstruction would have a spatial resolution that could demonstrate the pattern of vegetation cover distribution, and could provide boundary conditions or driving

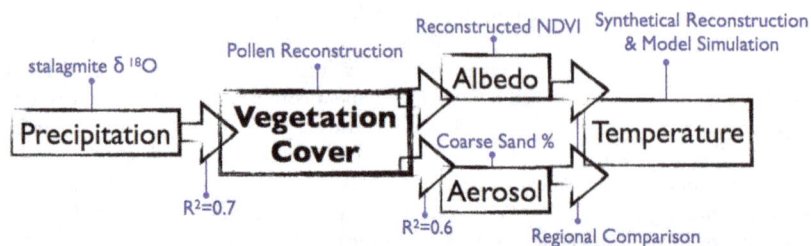

Figure 5. Mechanisms discussed in this study. Black boxes show the mechanisms discussed in the text, and are further explained by the blue annotations or the corresponding R^2 in the linear correlation. While vegetation cover in North China is mainly controlled by precipitation, its decline in the past 6500 years might have led to the changes in both land cover albedo and aerosol production, with a resulting regional cooling effect, as model simulations in previous studies and the regional comparison in this study have shown.

parameters for model simulations. In North China, this could be achieved with the rapid accumulation of paleoecological data; however, at present, in this study the regional spatial pattern of vegetation cover could not be obtained reliably based on 15 profiles (which were selected out of 55 profiles in this region according to the quality control criteria) even if calibration with the modern vegetation cover pattern is taken into account. As a result, regional averaging was a compromise necessary in the comparison between results from the model and those from the reconstruction. Nevertheless, in view of the data-rich enterprise that palaeoecology has become, we here present a method of vegetation reconstruction that yields regional vegetation cover instead of vegetation type, and which is more suitable for addressing regional-scale questions and model-data comparison. Our discussion has demonstrated that this switch in the reconstruction objective could help in validating results from models, as well as in understanding the mechanisms underlying earth system process on a larger scale.

Conclusions

In this study we quantitatively reconstructed the Holocene vegetation cover changes in North China from multiple pollen records, using the modern NDVI as an index. The results shows that:

1. Vegetation cover in North China increased slightly after 11 cal. ka BP and reached its maximum at 6.5 cal ka BP, but has decreased significantly since then. This change was mainly attributed to vegetation dynamics in regions with a high vegetation cover, while regions with low vegetation cover changed relatively little.

2. Vegetation cover decline in North China was mainly controlled by precipitation changes in the Pacific Monsoon region; this vegetation decline could have then induced local soil coarsening and a consequential increase in the production of aerosol.

3. One important implication of this result is its support of a thermal feedback of vegetation cover change as reported in former modelling studies, suggesting that the vegetation cover changes in North China could have had an overall cooling effect on this region during the late Holocene, owing to their alteration of albedo and aerosol production.

4. Combining multiple profiles and using modern vegetation indices as a modern analogue could yield a quantitative reconstruction of regional vegetation cover, which could be used in validating results from models as well as understanding ecological process on a larger scale. This advantage will become even more notable with the rapid accumulation of paleoecological data.

Supporting Information

Table S1 Site description of the 15 selected sediment profiles in North China. This table contains the references and basic information of the 15 profiles used for vegetation reconstruction in this study.

Author Contributions

Conceived and designed the experiments: HL. Performed the experiments: GL YY. Analyzed the data: GL YY QH. Contributed reagents/materials/analysis tools: GL YY QH. Wrote the paper: GL HL.

References

1. Dallmeyer A, Claussen M (2011) The influence of land cover change in the Asian monsoon region on present-day and mid-Holocene climate. Biogeosciences 8: 1499–1519.
2. Reichstein M, Mahecha MD, Ciais P, Seneviratne SI, Blyth EM, et al. (2011) Elk–testing climate–carbon cycle models: a case for pattern–oriented system analysis. iLEAPS Newsl. 11, 14–21.
3. Peng C, Guiot J, Wu H, Jiang H, Luo Y (2011) Integrating models with data in ecology and palaeoecology: advances towards a model–data fusion approach. Ecol. Let. 14, 522–536.
4. Tarasov PE, Guiot J, Cheddadi R, Andreev AA, Bezusko LG, et al. (1999) Climate in northern Eurasia 6000 years ago reconstructed from pollen data. Earth Planet. Sci. Lett. 171, 635–645.
5. Cheddadi R, Lamb H, Guiot J, Van Der Kaars S (1998) Holocene climatic change in Morocco: a quantitative reconstruction from pollen data. Clim. Dyn. 14, 883–890.
6. Parsons R, Prentice IC (1981) Statistical approaches to R-values and the pollen-vegetation relationship. Rev. Palaeobot. Palynol. 32, 127–152.
7. Prentice IC, Guiot J, Huntley B, Jolly D, Cheddadi R (1996) Reconstructing biomes from palaeoecological data: a general method and its application to European pollen data at 0 and 6 ka. Clim. Dyn. 12, 185–194.
8. Sugita S (1994) Pollen representation of vegetation in Quaternary sediments: theory and method in patchy vegetation. J. Ecol. 82, 881–897.
9. Sellers P, Meeson B, Hall F, Asrar G, Murphy R, et al. (1995) Remote sensing of the land surface for studies of global change: Models – algorithms – experiments. Remote Sens. Environ. 51, 3–26.
10. Field CB, Randerson JT, Malmström CM (1995) Global net primary production: combining ecology and remote sensing. Remote Sens. Environ. 51: 74–88.
11. Tarasov P, Williams JW, Andreev A, Nakagawa T, Bezrukova E, et al. (2007) Satellite-and pollen-based quantitative woody cover reconstructions for northern Asia: verification and application to late-Quaternary pollen data. Earth Planet. Sci. Lett. 264, 284–298.
12. Williams J, Shuman B (2008) Obtaining accurate and precise environmental reconstructions from the modern analog technique and North American surface pollen dataset. Quat. Sci. Revi. 27, 669–687.
13. Herzschuh U, Birks HJB, Ni J, Zhao Y, Liu H, et al. (2010) Holocene land-cover changes on the Tibetan Plateau. The Holocene 20, 91–104.
14. Boyle E, Keigwin L (1985) Comparison of Atlantic and Pacific paleochemical records for the last 215,000 years: Changes in deep ocean circulation and chemical inventories. Earth Planet. Sci. Lett. 76, 135–150.
15. Overpeck J, Webb T, Prentice IC (1985) Quantitative interpretation of fossil pollen spectra: dissimilarity coefficients and the method of modern analogies. Quat. Res. 23, 87–108.
16. Velichko AA (1984) Late quaternary environments of the Soviet Union. University of Minnesota Press.
17. Wight Jr HE (1994) Global Climates: since the Last Glacial Maximum. University of Minnesota Press, 169–193.
18. Edwards M, Anderson P, Brubaker L, Ager T, Andreev A, et al. (2001) Pollen-based biomes for Beringia 18,000, 6000 and 0 14C yr BP. J. Biogeogr. 27, 521–554.
19. Davin EL, de Noblet-Ducoudré N (2010) Climatic impact of global-scale deforestation: radiative versus nonradiative processes. J. Clim. 23, 97–112.
20. Bala G, Caldeira K, Wickett M, Phillips T, Lobell D, et al. (2007) Combined climate and carbon-cycle effects of large-scale deforestation. Proc. Natl. Acad. Sci. 104, 6550–6555.
21. Birks HJB (2009) Holocene climate research–progress, paradigms, and problems. Natural climate variability and global warming: a Holocene perspective. Wiley Blackwell, Oxford, 7–57.
22. Lang G (1994) Quartäre Vegetationsgeschichte Europas: Methoden und Ergebnisse. Spektrum Akademischer Verlag. Spektrum Akademischer Verlag.
23. Delcourt HR, Delcourt PA (1991) Quaternary ecology: a paleoecological perspective, first ed. Kluwer Acad. Pub.
24. Wilmshurst JM, McGlone MS (2005) Origin of pollen and spores in surface lake sediments: comparison of modern palynomorph assemblages in moss cushions, surface soils and surface lake sediments. Rev. Palaeobot. Palynol. 136, 1–15.
25. Xu Q, Li Y, Tian F, Cao X, Yang X (2009) Pollen assemblages of tauber traps and surface soil samples in steppe areas of China and their relationships with vegetation and climate. Rev. Palaeob. Palyn. 153, 86–101.
26. Herzschuh U, Tarasov P, Wunnemann B, Hartmann K (2004) Holocene vegetation and climate of the Alashan Plateau, NW China, reconstructed from pollen data. Palaeogeogr., Palaeoclim., Palaeoecol. 211: 1–17.
27. Liu H, Li Y (2009) Pollen Indicators of climate chang and human activities in the semi-arid region. Acta Palaeontol. Sin. 48, 211–221.
28. Tucker C, Pinzon J, Brown M, Molly E (2004) Global inventory modeling and mapping studies (GIMMS) satellite drift corrected and NOAA-16 incorporated

normalized difference vegetation index (NDVI), monthly 1981–2002. University of Maryland.

29. Tucker CJ, Pinzon JE, Brown ME, Slayback DA, Pak EW, et al. (2005) An extended AVHRR 8-km NDVI dataset compatible with MODIS and SPOT vegetation NDVI data. Int. J. Remote Sens. 26, 4485–4498.

30. Pinzon J, Brown ME, Tucker CJ (2005) Satellite time series correction of orbital drift artifacts using empirical mode decomposition. Hilbert-Huang transform: introduction and applications.

31. Chen S, Billings SA (1992) Neural networks for nonlinear dynamic system modelling and identification. Intl. J. Control 56, 319–346.

32. Prell WL (1985) Stability of low-latitude sea-surface temperatures: an evaluation of the CLIMAP reconstruction with emphasis on the positive SST anomalies. Final report. Brown Univ., Providence, RI (USA). Dept. of Geological Sciences.

33. Anderson P, Bartlein P, Brubaker L, Gajewski K, Ritchie J (1989) Modern analogues of late-Quaternary pollen spectra from the western interior of North America. J. Biogeog. 16, 573–596.

34. Overpeck JT, Webb RS, Webb T (1992) Mapping eastern North American vegetation change of the past 18 ka: No-analogs and the future. Geol. 20, 1071–1074.

35. Yin Y, Liu H, Liu G, Hao Q, Wang H (2012) Vegetation responses to mid-Holocene extreme drought events and subsequent long-term drought on the southeastern Inner Mongolian Plateau, China. Agricultural and Forest Meteorology. E-pub ahead of print. doi:10.1016/j.agrformet.2012.10.005. In Press.

36. Gaillard MJ, Sugita S, Bunting MJ, Middleton R, Broström A, et al. (2008) The use of modelling and simulation approach in reconstructing past landscapes from fossil pollen data: a review and results from the POLLANDCAL network. Vegetation History and Archaeobotany. 17, 419–443.

37. Hellman S, Gaillard MJ, Bunting JM, Mazier F (2009) Estimating the relevant source area of pollen in the past cultural landscapes of southern Sweden – a forward modelling approach. Review of Palaeobotany and Palynology. 153, 259–271.

38. Nielsen AB, Sugita S (2005) Estimating relevant source area of pollen for small Danish lakes around AD 1800. The Holocene. 15, 1006–1020.

39. Wishart D (2004) The Great Plains Region, In: Encyclopedia of the Great Plains, Lincoln: University of Nebraska Press, xiii–xviii.

40. Kellogg CA, Griffin DW (2006) Aerobiology and the global transport of desert dust. Trends in ecology & evolution 21, 638–644.

41. Williams JW, Jackson ST (2003) Palynological and AVHRR observations of modern vegetational gradients in eastern North America. The Holocene. 13, 485–497.

42. Yuan D, Cheng H, Edwards RL, Dykoski CA, Kelly MJ, et al. (2004) Timing, duration, and transitions of the last interglacial Asian monsoon. Sci. 304, 575–578.

43. Dykoski CA, Edwards RL, Cheng H, Yuan D, Cai Y, et al. (2005) A high-resolution, absolute-dated Holocene and deglacial Asian monsoon record from Dongge Cave, China. Earth Planet. Sci. Lett. 233, 71–86.

44. Hu C, Henderson GM, Huang J, Xie S, Sun Y, et al. (2008) Quantification of Holocene Asian monsoon rainfall from spatially separated cave records. Earth Planet. Sci. Lett. 266, 221–232.

45. Wang YJ, Cheng H, Edwards RL, An Z, Wu J, et al. (2001) A high-resolution absolute-dated late Pleistocene monsoon record from Hulu Cave, China. Sci. 294, 2345–2348.

46. McDermott F (2004) Palaeo-climate reconstruction from stable isotope variations in speleothems: a review. Quat. Sci. Rev. 23: 901–918.

47. Wu X, Liu H, Ren J, He S, Zhang Y (2009) Water-dominated vegetation activity across biomes in mid-latitudinal eastern China. Geophys. Res. Lett. 36, L04402.

48. Piao S, Mohammat A, Fang J, Cai Q, Feng J (2006) NDVI-based increase in growth of temperate grasslands and its responses to climate changes in China. Glob. Environ. Change 16, 340–348.

49. Zou X, Zhai P, Zhang Q (2005) Variations in droughts over China: 1951–2003. Geophys. Res. Lett. 32: L04707.

50. Li Z, Wang Y, Zhou Q, Wu J, Peng J, et al. (2008) Spatiotemporal variability of land surface moisture based on vegetation and temperature characteristics in Northern Shaanxi Loess Plateau, China. J. Arid Environ. 72: 974–985.

51. Loarie SR, Lobell DB, Asner GP, Field CB (2011) Land-Cover and Surface Water Change Drive Large Albedo Increases in South America. Earth Interact. 15, 1–16.

52. Irvine PJ, Ridgwell A, Lunt DJ (2011) Climatic effects of surface albedo geoengineering. J. Geophys. Res. 116, D24112.

53. Blok D, Schaepman-Strub G, Bartholomeus H, Heijmans MMD, Maximov TC, et al. (2011) The response of Arctic vegetation to the summer climate: relation between shrub cover, NDVI, surface albedo and temperature. Environ. Res. Lett. 6, 035502.

54. Glenn E.P, Huete AR, Nagler PL, Nelson SG (2008) Relationship between remotely-sensed vegetation indices, canopy attributes and plant physiological

processes: What vegetation indices can and cannot tell us about the landscape. Sensors 8, 2136–2160.

55. Betts RA (2000) Offset of the potential carbon sink from boreal forestation by decreases in surface albedo. Nat. 408, 187–190.

56. Lee X, Goulden ML, Hollinger DY, Barr A, Black TA, et al. (2011) Observed increase in local cooling effect of deforestation at higher latitudes. Nat. 479, 384–387.

57. Lau K, Kim K (2007) Cooling of the Atlantic by Saharan dust. Geophy. Res. Lett. 34, L23811.

58. Miller R, Tegen I (1998) Climate response to soil dust aerosols. J. Clim. 11, 3247–3267.

59. Wolfe SA, Nickling WG (1993) The protective role of sparse vegetation in wind erosion. Prog. Phys. Geogr. 17, 50–68.

60. Wieringa J (1986) Roughness-dependent geographical interpolation of surface wind speed averages. Quart. J. R. Meteorol. Soc. 112, 867–889.

61. Yin Y, Liu H, He S, Zhao F, Zhu J, et al. (2011) Patterns of local and regional grain size distribution and their application to Holocene climate reconstruction in semi-arid Inner Mongolia, China. Palaeogeogr., Palaeoclim., Palaeoecol. 307, 168–176.

62. Lu H, An Z (1998) Paleoclimatic significance of grain size of loess-palaeosol deposit in Chinese Loess Plateau. Sci. in China Ser. D: Earth Sci. 41, 626–631.

63. Zhang Z, Zhao M, Lu H, Faiia AM (2003) Lower temperature as the main cause of C4 plant declines during the glacial periods on the Chinese Loess Plateau. Earth Planet. Sci. Lett. 214, 467–481.

64. Glantz MH, Orlovsky N (1983) Desertification: A review of the concept. Desert. Control Bull. 9, 15–22.

65. Frumkin A, Stein M (2004) The Sahara–East Mediterranean dust and climate connection revealed by strontium and uranium isotopes in a Jerusalem speleothem. Earth Planet. Sci. Lett. 217, 451–464.

66. Tegen I, Harrison SP, Kohfeld K, Prentice IC, Coe M, et al. (2002) Impact of vegetation and preferential source areas on global dust aerosol: Results from a model study. J. Geoph. Res. 107, 4576.

67. Zhao Y, Yu Z, Chen F, Zhang J, Yang B (2009) Vegetation response to Holocene climate change in monsoon-influenced region of China. Earth Sci. Rev. 97, 242–256.

68. Herzschuh U (2006) Palaeo-moisture evolution in monsoonal Central Asia during the last 50,000 years. Quat. Sci. Rev. 25, 163–178.

69. Wang S, Gong D (2000) Climate in China during the four special periods in Holocene. Prog. Nat. Sci. 10, 325–332.

70. Fang X, Hou G (2011) Synthetically Reconstructed Holocene Temperature Change in China. Scientia Geographica Sinica 31, 385–393.

71. Gao Y (1962) Some problems on East-Asia monsoon. Science Press, Beijing.

72. Sun P, Yu Z, Liu S, Wei X, Wang J, et al. (2012) Climate change, growing season water deficit and vegetation activity along the north-south transect of Eastern China from 1982 through 2006. Hydrol. Earth Syst. Sci. Discuss 9, 6649–6688.

73. Chuai X, Huang X, Wang W, Bao G (2012) NDVI, temperature and precipitation changes and their relationships with different vegetation types during 1998–2007 in Inner Mongolia, China. Int. J. of Clim.

74. Widmann M, Goosse H, van der Schrier G, Schnur R, Barkmeijer J (2010) Using data assimilation to study extratropical Northern Hemisphere climate over the last millennium. Clim. Past 6, 627–644.

75. Williams M, Richardson A, Reichstein M, Stoy P, Peylin P, et al. (2009) Improving land surface models with FLUXNET data. Biogeosci. 6, 1341–1359.

76. Guiot J, Wu H, Garreta V, Hatté C, Magny M (2009) A few prospective ideas on climate reconstruction: from a statistical single proxy approach towards a multi-proxy and dynamical approach. Clim. Past 5, 571–583.

77. Wu H, Guiot JEL, Peng C, Guo Z (2008) New coupled model used inversely for reconstructing past terrestrial carbon storage from pollen data: validation of model using modern data. Glob. Change Biol. 15, 82–96.

78. Williams JW, Gonzales LM, Kaplan JO (2008) Leaf area index for northern and eastern North America at the Last Glacial Maximum: a data–model comparison. Glob. Ecol. Biogeogr. 17, 122–134.

79. François L, Utescher T, Favre E, Henrot AJ, Warnant P, et al. (2011) Modelling Late Miocene vegetation in Europe: Results of the CARAIB model and comparison with palaeovegetation data. Palaeogeogr., Palaeoclim., Palaeoecol. 304, 359–378.

80. Nakicenovic N, Inaba A, Messner S, Nilsson S, Nishimura Y, et al. (1993) Long-term strategies for mitigating global warming. Energy 18, 401–401.

81. Ornstein L, Aleinov I, Rind D (2009) Irrigated afforestation of the Sahara and Australian Outback to end global warming. Clim. Change 97, 409–437.

82. Rotenberg E, Yakir D (2010) Contribution of semi-arid forests to the climate system. Sci. 327, 451.

83. Juang JY, Katul G, Siqueira M, Stoy P, Novick K (2007) Separating the effects of albedo from eco-physiological changes on surface temperature along a successional chronosequence in the southeastern United States. Geophys. Res. Lett. 34, L21408.

4

Dietary Ecology of Murinae (Muridae, Rodentia): A Geometric Morphometric Approach

Ana Rosa Gómez Cano[1]*, **Manuel Hernández Fernández**[1,2], **M. Ángeles Álvarez-Sierra**[1,2]

1 Departamento de Paleontología, Facultad de Ciencias Geológicas, Universidad Complutense de Madrid, Madrid, Spain, **2** Departamento de Geología Sedimentaria y Cambio Medioambiental, Instituto de Geociencias (UCM, CSIC), Madrid, Spain

Abstract

Murine rodents represent a highly diverse group, which displays great ecological versatility. In the present paper we analyse the relationship between dental morphology, on one hand, using geometric morphometrics based upon the outline of first upper molar and the dietary preference of extant murine genera, on the other. This ecomorphological study of extant murine rodents demonstrates that dietary groups can be distinguished with the use of a quantitative geometric morphometric approach based on first upper molar outline. A discriminant analysis of the geometric morphometric variables of the first upper molars enables us to infer the dietary preferences of extinct murine genera from the Iberian Peninsula. Most of the extinct genera were omnivore; only *Stephanomys* showed a pattern of dental morphology alike that of the herbivore genera.

Editor: Laurent Viriot, Team 'Evo-Devo of Vertebrate Dentition', France

Funding: This research was supported by projects of the Spanish Ministry of Tecnology and Science CGL2006-01773/BTE, CGL-2008-05813-C02-01/BTE, CGL-2010-19116/BOS and CGL-2011-28877/BTE, as well as by a predoctoral contract of the Spanish Ministry of Education in the program of Formacion de Profesorado Universitario to A.R.G.C. During this study, M.H.F. availed of a UCM contract from the Ramón y Cajal Program, belonging to the Spanish Ministry of Education and Science. This research is a contribution by the research group UCM-910607 on Evolution of Cenozoic Mammals and Continental Palaeoenvironments. The funders had no role in study design, data collection and analysis, decision to publish, or preparation of the manuscript.

Competing Interests: The authors have declared that no competing interests exist.

* E-mail: argomez@ucm.es

Introduction

Rodentia is the most speciose group of mammals and the morphology of their dentition is highly morphologically specialized [1]. Addressing different rodent groups in detail, one can observe that the diversity of each group can also be seen in the disparity of their morphological dental features [2]. Furthermore, the morphological characters could be directly related with the ecological features of each taxon such as its habitat or diet [3–5].

Several authors had demonstrated an interesting relationship between dental features and grazing diet in rodents, inferred by classic morphometric methodologies in lateral view (hypsodonty) [6] or in occlusal pattern [7]. Other studies have pointed out the existence of a close relationship between feeding habits and skulls or mandibles morphology using classical and geometric morphometric tools for quantification of shapes [8–12]. To work with fossil material, however, there is a need to establish a methodology based on the study of isolated teeth because these pieces are the most abundant in the fossil sites where rodents are recorded, particularly in the case of the fossil record of the European Miocene. The geometric morphometric methodologies are of great interest because they enable researchers to quantify the variation in shape and size to develop morphospace, which facilitate the ecological and evolutionary inferences [13,8–10,14].

Previous research on geometric morphometrics of rodent molars provides interesting results with regard to describing ecological preferences [7,15,9,16]. The interest of our study is based upon the use of a methodological approach allowing us to analyse the high morphological diversity within extant and extinct murine rodents and to associate it with their feeding habits. Murines are the largest subfamily of muroid rodents [17]. Furthermore, the wide range of habitats and the large amount of studies on extant and fossil taxa [18,7,19,20,17,2] of this group makes it an interesting one for our study.

Materials and Methods

To developed the study one of the authors (ARGC) visited the Murinae collections at the Museum National d'Histoire Naturelle in Paris (France) during August 2010, to take photos of the first upper molars by the authorization of the Christiane Denys curator of the Rodentia section of Mammalian Department and supervised by Dr. Emmanuelle Stoetzel.

Dietary Categories

Based on descriptions available from the literature (Appendix S1 and references therein) we classified each extant genus into one of three dietary categories [21,10]: 1) herbivores (n = 22) if it feeds mostly on plant matter being nearly purely herbivorous; 2) omnivores (n = 40), when it includes both animal-dominated and plant-dominated taxa; or 3) and faunivores (n = 14), with a diet composed primarily of animal matter, being nearly purely faunivorous (Appendix S1).

Although these dietary categories are a simplification of a complex classification of diets, this categorization has been shown to be useful for examining the relationships between morphology and feeding habits [10].

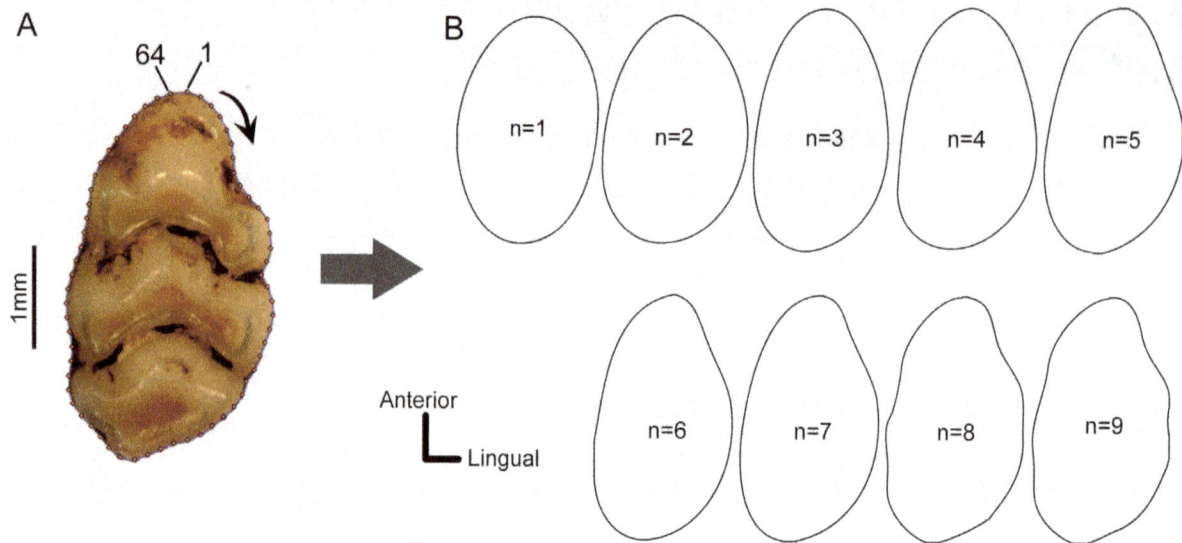

Figure 1. Outline and EFA reconstruction for an increasing number of harmonics. *A*, Outline digitalized from an image of one first upper molar of *Rattus andamanensis* (MNHN 1995 2833) based on 64 points equally distributed using TPSdig2 software version 2.16 [27]; the black arrow indicates the direction of the digitalization process. *B*, Reconstructed outlines for an increasing number of harmonics; n represents the number of harmonics in each outline.

Samples

When ecomorphological studies focus on supraspecific taxa (genera, families, etc.), they can help to reveal the development of diversity in different groups and understand the course of their adaptive evolution [22].

In this study we included material from extant murines from all over the world and extinct Iberoccitanian murine rodents (Appendix S2 and references therein). We chose the first upper molar (M1) based on its highly distinctive features and it is very useful in studies of fossil material [7].

The samples of extant material included in this paper are housed in the Museum National d'Histoire Naturelle (MNHN), Paris. We photographed the first upper molars of all the murine genera available at the MNHN using a Nikon D300s camera fitted with a Nikon AF-S VR 105 mm f/2.8 IF-ED lens. Furthermore, for those extant genera unavailable in the MNHN collections we included scaled photographs from the literature. In general terms, we compiled a database of 232 specimens of right or left first upper molars of extant Murinae (Appendix S2).

Finally, based on both the information on diets available in the bibliography and the specimens available in the collections and literature, we were able to include in our study 76 of the 124 extant genera of Murinae (61.3%) according to the taxonomic revision of [17].

Moreover, in our analysis we included pictures of teeth belonging to the 9 extinct genera of murine (*Anthracomys*, *Castillomys*, *Castromys*, *Huerzelerimys*, *Occitanomys*, *Paraethomys*, *Progonomys*, *Rhagapodemus* and *Stephanomys*; Appendix S2 and references therein), which have been described at the Iberoccitanian (south western Europe) fossil sites across the whole Upper Miocene [23]. As for the extant taxa, we developed another photographic database for the fossil genera. In this case we compiled pictures of first upper molars from the literature. The intensive sampling work at the Iberoccitanian fossil sites and the large amount of detailed studies of these materials over the last decade [24], enabled us to include 169 scaled pictures and drawings, which represent all the fossil genera and most of the species described in our study area (see Appendix S2 for specimen information and references).

Morphological Analysis of the Outline

We chose the outline analysis to describe the morphology of the molars because, besides being effective with regard to describing the location of the tubercles characteristic of the murines molar, it is less sensitive to modifications of the dental pattern occurring with wearing than the landmark analysis [13,25]. Furthermore, whereas individual homologous landmarks are difficult to pinpoint from one molar to another, outline methods have been suggested as be useful tools for the analysis of biological shapes in the absence of homologous landmarks [26].

The molar outline is defined as the two-dimensional projection of the molar viewed from its occlusal side [25]. Following [13] these outlines were digitalized for each tooth as x and y coordinates of sixty-four points equally spaced along the tooth outline with TPSdig2 software version 2.16 [27]. The starting point of each outline was defined at the maximum of curvature in the forepart of the tooth. In order to provide a convenient way to get measurements using right and left molars, left molars were reflected in a mirror and measured as right molars [25].

In order to analyse these *x* and *y* coordinates we applied an Elliptic Fourier Analysis (EFA) [28] to the samples data using EFAwin software [29] which extracts Fourier coefficients from the original outline and normalizes these shape variables (Fig. 1). With this method the complex outline can be described as a sum of trigonometric functions, known as harmonics, of decreasing wavelength. Each harmonic is defined by four Fourier Coefficients (FCs), two for the *x* coordinate (A and B) and two for the *y* coordinate (C and D) [30].

Based in our data (Appendix S3) and in previous studies, which demonstrated that the effect of measurement error for upper molars was limited by considering only Fourier Coefficients up to the ninth harmonic [25,14]. Furthermore, the first harmonic is proportional to the size of each specimen but its four coefficients (A1-D1) are constant due to the standardization [13,31]. Thus, following previous studies we retain nine harmonics, which represents the best compromise between measurement error and information content [25,14]. Thus, we finally we retain 36 FCs from these nine harmonics (i.e. A1–D9), which describes the outline of each specimen.

Since our analysis was performed at the genus level, each set of Fourier coefficients describing the outline for each specimen was averaged per genus (Appendix S4).

Finally, as a support for visual interpretation of shape changes we obtained accurate reconstructions of these average outlines using an inverse Fourier transformation [28,32], which directly provides the Euclidean *xy*-coordinates of the reconstructed outline [33].

Statistical Analysis

In order to evaluate the importance of among-group differentiation relative to within-group variation using the dietary categories as grouping variables we performed a non-parametric multivariate analysis on variance (NPMANOVA, [34]) on the obtained sets of Fourier Coefficients (A1 to D9). Likewise, we evaluated pairwise NPMANOVAs between all pairs of dietary groups by means of a *post-hoc* test (Bonferroni).

Moreover, associated with the analysis of variance, we estimated canonical functions (Canonical Variate Analysis, CVA [35]), which enabled us to plot scores for each dietary group and to visualize the pattern of morphological differentiation in each of them. The CVA produces maximal and second to maximal separation between all groups, and its axes are linear combinations of the original variables. Statistical analyses were performed with PAST v. 2.17 [36].

Additionally, [37] have linked dietary categories with molar size in a group of Primates. We were able to explore this relationship in murine rodents because the EFA calculates the size variable (measured as major axis length) linked to the first FC. Therefore we analysed the differences in size (M1 length) and shape (allometry) by a linear correlation between size and CV1 and CV2 as shape estimators [38]. Furthermore we analysed the differences in size among the feeding habits through ANOVA and *Post-hoc* Tukey tests using SPSS v. 15.

Finally, for the purpose of classifying the murine extinct genera in the dietary categories we assessed a multivariate discriminant analysis on morphological variables (FCs of their M1 and size) employing SPSS v 15.

Results and Discussion

Relation between Shape and Diet

Rather than the high apparently homogeneity of the dentition pattern described for this group [39], we found that shapes defined by the Fourier Coefficients (FC) showed significant differences among the three dietary categories. The FC showed significant differences between dietary categories in the NPMANOVA analysis (F = 6.118; p = 0.0004). The results of the pairwise

comparison of Euclidean distance were significant (p<0.05) for all dietary pairs (table 1).

The two axes of the CVA (Fig. 2) explained 65.8% (CV1) and 34.2% (CV2) respectively of total variation in shape of molar outline. Reconstruction of mean outlines corresponding to theoretical outlines equivalent to the coordinates along the canonical axes showed the differences involved in this differentiation of dietary shape.

The first canonical axis (CV1) showed a morphological gradient in the shape of the outline from tight and elliptical molars, with a prominent cusp (t1) in the anterior edge on the positive side of this axis to wide molars on the negative side of this axis (Fig. 2). It seems that this morphological gradient is related to the differentiation between faunivores and consumers of plant matter (herbivores and omnivores). Whereas faunivore rodents, with more elliptical and tight outlines, were grouped around the highest values of this axis, other murines were placed mostly on the negative part of the axis, presenting rectangular and wide outlines. The presence of broad molars is characteristic for taxa including a plant component in their diets [40]. On the other hand, faunivore genera may show a tendency to reduce occlusal surface in their molars, probably due to the development of robust incisors [10]. Furthermore, the tendency towards buccolingual compression of the M1 shown by faunivore murines agrees with the observed general trend in Carnivora, which present narrowed molars by aligning their cusps anteroposteriorly as a specialization for shearing or slicing [41].

The second canonical axis (CV2) described a variation from more pronounced cusps in an undulated outline on the positive side of the axis towards more symmetrical and rounded outlines on the negative side (Fig. 2). These shapes are congruent with previous works on Murinae rodents, which associated a slender asymmetrical outline with omnivory, which are at the positive boundary of the CV2 axis, and a broader and symmetrical one with herbivory, which is at the negative edge of our CV2 axis [31,42]. This dental feature should be functionally related to a relative increase in occlusal surface and increased grinding efficiency [43], which might be associated with cusp broadening and a progressive increase in enamel surface with tooth wearing. Moreover, the morphological trend of this CV2 represents the variation across the generalist/specialist gradient. This contrast of differentiated outline between generalist and specialist was already highlighted by Renaud et al. [9] in a geometric morphometric study of murine mandibles. The results in that study emphasize the role of ecological diversification in determining the rhythm of morphological evolution, indicating that lower morphological divergences only involve omnivores, instead large morphological divergences involving the specialist taxa, which correspond to diverse feeding behaviours diverging from the omnivorous diet.

Despite the congruence between morphology and feeding habits, the high morphological variability contained within the genera studied as well as the interspecific differences in ecology within each genus are responsible for the existence of overlapping areas among the different dietary categories. Furthermore, the evolutionary history of the genera studied with changes in feeding habits among related lineages, is also likely related to the presence of this overlap. For example, the omnivore *Margaretamys*, which is placed in the overlapping area between the morphospaces defined by faunivore and omnivore murines, is closely related to the faunivore genera *Echiothrix*, *Melasmothrix* and *Tateomys* [44,45]. Likewise, *Micromys* (omnivore) and *Millardia* (herbivore), which also occur in this overlapping area, configure a monophyletic clade with the herbivore *Pogonomys* [44,45].

Table 1. Pairwise comparison of Euclidean distance.

	Herbivore	Omnivore	Faunivore
Herbivore		0.027	0.001
Omnivore	0.027		0.003
Faunivore	0.001	0.003	

P values obtained in the pairwise comparison, p<0.05 indicates significant differences among the dietary groups.

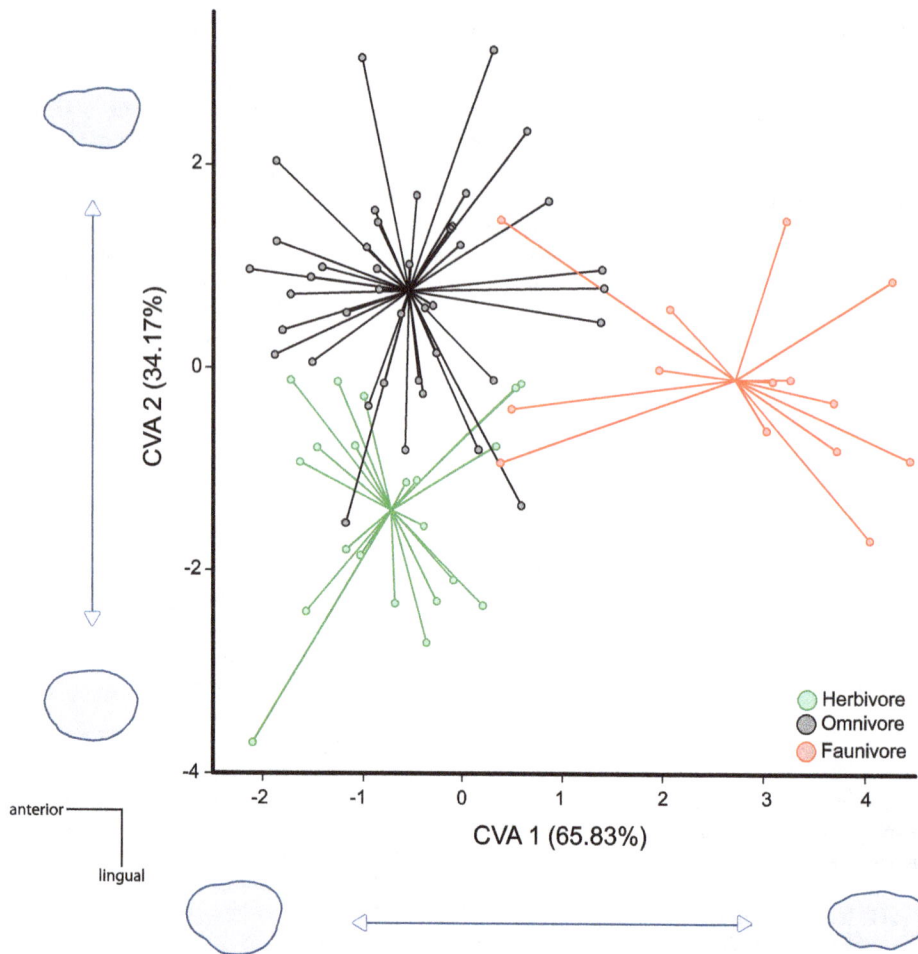

Figure 2. Inter-dietary shape differentiation of the first upper molars. Variation in shapes variation was estimated by the first two axes of a canonical analysis of the EFT Fourier coefficients of the M1 outline. Colours correspond to the different dietary groups. White for herbivores; Grey for omnivores; Black for faunivores. On each axis, shape changes corresponding to the canonical axes are depicted, corresponding to the maximum values of the axes on the plot.

Finally, we did not find a statistically significant relation between size and CV1 (r = 0.023; p = 0.842) nor CV2 (r = 0.028; p = 0.072). Furthermore, there was not significant result in the ANOVA of size and feeding habits in murine rodents. This could be explained because the size factor is part of a more complex pattern linked to the development of the whole dental row and the patterning cascade model of development of the molars, in which are implied the size of other dental elements [21,46] are involved.

Dietary Inference in Extinct Murinae

The discriminant functions for the distinction among different diets in extant murines correctly classified 84.2% of the genera. Percentages of well-classified genera are similarly distributed in each dietary category (table 2). Genera that have been classified into an erroneous dietary group are those included in the overlapping area of the CVA plot (Fig. 2).

Our dietary reconstructions for the extinct murine genera from the Iberoccitanian Upper Miocene indicated an omnivore dietary preference for most of them (table 3), with the exception of *Stephanomys*, which was classified as herbivore. These results, were therefore roughly consistent with previous studies focusing on the evolution of morphological variation in *Occitanomys* [31] or *Progonomys* [43], which also indicated omnivore preferences for

these extinct genera. The reconstructed herbivore diet for *Stephanomys* agrees with the classical assumptions for this genus. Several studies indicate a particular preference of *Stephanomys* for feeding on grass, due to the presence of morphological features in the dental pattern associated with the stephanodonty [31,43,47,40]. This dental pattern is characterised by the presence of longitudinal ridges between molar cusps as well as by broad and symmetrical molars [48,39,49].

When compared to modern faunas from areas under tropical or subtropical climate regimes, such as the ones inferred for the

Table 2. Percentages of well-classified genera in the discriminant analysis.

Diet	Pronosticated group		
	Herbivore	**Omnivore**	**Faunivore**
Herbivore	81.80	18.20	0.00
Omnivore	15.00	82.50	2.50
Faunivore	7.10	14.30	78.60

Table 3. Discriminant analysis results for the 85 genera considered.

Genus	Diet	Probability of belonging to		
		Herbivore	Omnivore	Faunivore
Abditomys	Herbivore	0.979	0.017	0.004
Chiropodomys	Herbivore	0.891	0.108	0.001
Golunda	Herbivore	0.994	0.006	0.000
Hadromys	Herbivore	0.989	0.011	0.000
Haeromys	Herbivore	0.911	0.088	0.002
Hapalomys	Herbivore	1.000	0.000	0.000
Hyomys	Herbivore	0.800	0.200	0.000
Kadarsanomys	Herbivore	0.976	0.018	0.006
Leporillus	Herbivore	0.357	0.434*	0.208
Mallomys	Herbivore	0.849	0.150	0.000
Mastacomys	Herbivore	0.993	0.007	0.001
Melomys	Herbivore	0.957	0.043	0.000
Millardia	Herbivore	0.577	0.423	0.000
Papagonomys	Herbivore	0.990	0.007	0.002
Pelomys	Herbivore	0.364	0.636*	0.000
Phloeomys	Herbivore	0.888	0.109	0.003
Pogonomys	Herbivore	0.911	0.089	0.000
Solomys	Herbivore	0.927	0.066	0.006
Spelaeomys	Herbivore	0.376	0.624*	0.000
Uromys	Herbivore	0.359	0.507*	0.134
Vandeluria	Herbivore	0.522	0.450	0.028
Aethomys	Omnivore	0.050	0.950	0.000
Anisomys	Omnivore	0.641*	0.334	0.025
Apodemus	Omnivore	0.315	0.685	0.000
Apomys	Omnivore	0.017	0.551	0.432
Arvicanthis	Omnivore	0.026	0.973	0.001
Bandicota	Omnivore	0.092	0.906	0.002
Bunomys	Omnivore	0.018	0.982	0.000
Coccymys	Omnivore	0.746*	0.210	0.044
Crateromys	Omnivore	0.040	0.960	0.000
Dasymys	Omnivore	0.918	0.082	0.000
Eropeplus	Omnivore	0.006	0.951	0.043
Grammomys	Omnivore	0.088	0.909	0.003
Hybomys	Omnivore	0.496	0.503	0.001
Hylomyscus	Omnivore	0.455	0.537	0.007
Leggadina	Omnivore	0.013	0.985	0.002
Lemniscomys	Omnivore	0.010	0.989	0.000
Lenomys	Omnivore	0.073	0.927	0.000
Leopoldomys	Omnivore	0.042	0.958	0.000
Lorentzimys	Omnivore	0.016	0.984	0.000
Malacomys	Omnivore	0.305	0.681	0.014
Margaretamys	Omnivore	0.021	0.587	0.392
Mastomys	Omnivore	0.267	0.725	0.008
Maxomys	Omnivore	0.041	0.957	0.001
Micromys	Omnivore	0.403	0.596	0.002
Mus	Omnivore	0.003	0.997	0.000
Nivivemter	Omnivore	0.033	0.104	0.863

Table 3. Cont.

Genus	Diet	Probability of belonging to		
		Herbivore	Omnivore	Faunivore
Notomys	Omnivore	0.078	0.922	0.000
Oenomys	Omnivore	0.298	0.702	0.000
Pitecheir	Omnivore	0.340	0.659	0.000
Praomys	Omnivore	0.006	0.994	0.000
Pseudohydromys	Omnivore	0.000	0.998	0.002
Pseudomys	Omnivore	0.258	0.741	0.001
Rattus	Omnivore	0.024	0.968	0.008
Rhabdomys	Omnivore	0.150	0.839	0.011
Stochomys	Omnivore	0.450	0.544	0.006
Sundamys	Omnivore	0.058	0.909	0.033
Thallomys	Omnivore	0.002	0.967	0.031
Thammomys	Omnivore	0.599	0.400	0.001
Tokudaia	Omnivore	0.228	0.744	0.028
Zelotomys	Omnivore	0.020	0.956	0.024
Zyzomys	Omnivore	0.001	0.999	0.000
Archboldomys	Faunivore	0.000	0.000	1.000
Colomys	Faunivore	0.360	0.578	0.062
Crossomys	Faunivore	0.000	0.000	1.000
Crunomys	Faunivore	0.040	0.819	0.141
Chrotomys	Faunivore	0.000	0.000	0.999
Echiothrix	Faunivore	0.766*	0.149	0.086
Hydromys	Faunivore	0.001	0.001	0.998
Leptomys	Faunivore	0.000	0.001	0.999
Melasmothrix	Faunivore	0.000	0.000	1.000
Parahydromys	Faunivore	0.000	0.000	1.000
Paulamys	Faunivore	0.000	0.000	1.000
Rhynchomys	Faunivore	0.014	0.033	0.953
Sommeromys	Faunivore	0.005	0.066	0.929
Tateomys	Faunivore	0.000	0.000	1.000
Anthracomys		0.000	0.895	0.105
Castillomys		0.225	0.775	0.000
Castromys		0.093	0.906	0.001
Huerzelerimys		0.150	0.850	0.000
Occitanomys		0.173	0.827	0.000
Paraethomys		0.112	0.887	0.000
Progonomys		0.059	0.941	0.000
Rhagapodemus		0.001	0.998	0.000
Stephanomys		0.896	0.104	0.001

The table shows the values of high probability for each genus in the three different dietary categories.
*indicates erroneous classification in the extant genera; Extinct genera are shown in grey.

Iberoccitanian region during the Late Miocene [50,51], the dominance of omnivorous taxa in our results appears to point to subtropical climatic conditions [52], in which herbivore taxa are very scarce. This agrees with the palaeoclimatic data provided for herpetofaunas [53,54] and plants [55–57]. Additionally, this predominance of omnivorous taxa might be also associated with the arrival of murines to southwestern Europe. In general, species capable of exploiting a wide variety of resources tend to be more widespread than the more specialized ones [58,59], and are probably responsible for most of the large dispersals. This is the case for *Progonomys*, the first murine found in Europe [60], as well as for most of the other genera from the Iberoccitanian region [61]. Only *Stephanomys* appears to be the result of strong directional selection towards herbivory [43,62], presumably imposed by aridification in the Iberoccitanian region around 7 ma [53].

Finally, since we included all the species variability of one genus within an average M1 outline for the genus, one can assume the presence of more specialist species within the genera we considered as omnivores. For example, [40] evidenced the diversity of feeding habits of species within *Occitanomys*; *O. adroveri* and *O. sondaari* were specialised in feeding on grass whereas *O. alcalai* was inferred as a non grass feeder. As a consequence of this interspecific variability there can be differences between our results and those from studies focusing on the species level [7,40].

Conclusions

This is the first time to our knowledge, that geometric morphometric comparison of the outlines of the first upper molar in the highly diverse extant murines has enabled the inference of ecological preferences in diet based on dental morphology. In the morphometric space described, all dietary groups were significantly distinguished from each other. Furthermore, based on the data of extant murines, we were able to infer the dietary preferences of nine extinct genera of murine rodents, which have been described at the Iberoccitanian fossil sites from the Upper Miocene.

Finally, Elliptic Fourier Analysis has been shown to constitute an interesting tool for inferring ecological preferences in extinct rodents, which are mostly recorded as isolated teeth in the fossil sites. The results of our study open up possibilities for establishing new comparisons in other mammalian groups, and for making ecological inferences of extinct taxa based upon information referring to their extant relatives. Furthermore, with the increasing development of phylogenetic studies, such inferences would help to map the evolutionary history of feeding habits.

Supporting Information

Appendix S1 Dietary preferences of extant murine genus and references. Diet determination in literature resumed as 1: field data, feeding trials, stomach morphology or bibliographic compilation; 2: stomach content or faecal pellet analyses.

Appendix S2 Collection number and references of the extant and extinct murine rodent used in this work. *indicates the specimens for which we take the photograph in the Musée National d'Histoire Naturelle (Paris); Grey fonts for extinct genera.

Appendix S3 Estimation of the information content in each harmonic based on its amplitude. The amplitudes were cumulated over the total range of harmonics and the information brought by each harmonic was estimated as the percentage of the sum of all harmonic amplitudes. In our case each of the nine first harmonics increased the amount of shape information up to 97% of the total information, meanwhile the subsequent harmonics provided almost no further relevant shape information (Fig. S1).

Appendix S4 Fourier Components and size. Values estimated for each extant and extinct genus considered in this works after EFA.

Acknowledgments

We greatly acknowledge to S. Renaud and P. Chevret (Université Claude Bernard, Lyon) and R. Ledevin (Universität Zürich, Zürich) for their comments during learning of the geometric morphometric methodology. Furthermore, we are very grateful to C. Denys and E. Stoetzel (Musée National d'Histoire Naturelle, Paris) for letting us check out the collections of extant Murinae stored at the Musée National d'Histoire Naturelle (Paris, France). The Editor Laurent Viriot (Ecole normale supérieure de Lyon) and two anonimous referees are acknowledged for their comments on the preliminary version of the manuscript. We would also like to thank all our colleagues and friends from the Paleontology and Paleobiology departments of the UCM and MNCN-CSIC, especially those from the PMMV team (http://www.pmmv.com.es), who contributed with productive discussions on these topics, particularly to Enrique Cantero Hernández and Juan L. Cantalapiedra for their technical support to develop this work during the Somosaguas field work. Furthermore, we wish to show our appreciation to the many paleontologists who conducted fieldwork at the fossil sites studied herein and who published their results on the fossil record of the Iberoccitanian micrommamals from the Middle and Late Miocene, all of which made this research possible. Finally, we wish to thank our Irish translator, Cormac De Brun, who helped us with the English.

Author Contributions

Conceived and designed the experiments: ARGC MHF. Performed the experiments: ARGC. Analyzed the data: ARGC MHF MAS. Wrote the paper: ARGC MHF MAS.

References

1. Hunter JP, Jernvall J (1995) The hypocone as a key innovation in mammalian evolution. Proceedings of the National Academy of Sciences of the United States of America 92: 10718–10722.
2. Fabre P-H, Hautier L, Dimitrov D, Douzery EJ (2012) A glimpse on the pattern of rodent diversification: a phylogenetic approach. BMC Evolutionary Biology 12: 88–88.
3. Auffray J-C, Renaud S, Claude J (2009) Rodent Biodiversity in Changing Environments. Kasetsart Journal, Natural Sciences 43: 83–93.
4. Blois JL, Hadly EA (2009) Mammalian Response to Cenozoic Climatic Change. Annual Review of Earth and Planetary Sciences 37: 181–208.
5. Martin SA (2010) Dental adaptation in murine rodents (Muridae): assessing mechanical predictions. Florida: The Florida State University. 73 p.
6. Williams SH, Kay RF (2001) A comparative test of adaptive explanations for hypsodonty in ungulates and rodents. Journal of Mammalian Evolution 8: 207–229.
7. van Dam JA (1997) The small mammals from the Upper Miocene of the Teruel-Alfambra region (Spain): Paleobiology and Paleoclimatic reconstructions. Geologica Ultraiectina 156: 204 pp.
8. Michaux J, Chevret P, Renaud S (2007) Morphological diversity of Old World rats and mice (Rodentia, Muridae) mandible in relation with phylogeny and

adaptation. Journal of Zoological Systematics and Evolutionary Research 45: 263–279.
9. Renaud S, Chevret P, Michaux J (2007) Morphological vs. molecular evolution: ecology and phylogeny both shape the mandible of rodents. Zoologica Scripta 36: 525–535.
10. Samuels JX (2009) Cranial morphology and dietary habits of rodents. Zoological Journal of the Linnean Society 156: 864–888.
11. Cox PG, Rayfield EJ, Fagan MJ, Herrel A, Pataky TC, et al. (2012) Functional Evolution of the Feeding System in Rodents. PLoS ONE 7: e36299.
12. Hautier L, Lebrun R, Cox PG (2012) Patterns of covariation in the masticatory apparatus of hystricognathous rodents: Implications for evolution and diversification. Journal of Morphology 273: 1319–1337.
13. Renaud S, Michaux J, Jaeger J-J, Auffray J-C (1996) Fourier Analysis Applied to *Stephanomys* (Rodentia, Muridae) Molars: Nonprogressive Evolutionary Pattern in a Gradual Lineage. Paleobiology 22: 255–265.
14. Ledevin R, Michaux JR, Deffontaine V, Henttonen H, Renaud S (2010) Evolutionary history of the bank vole Myodes glareolus: a morphometric perspective. Biological Journal of the Linnean Society 100: 681–694.

15. Swiderski D, Zelditch M, Fink W (2000) Phylogenetic analysis of skull shape evolution in marmotine squirrels using landmarks and thin-plate splines. Hystrix-the Italian Journal of Mammalogy 11: 49–75.
16. McGuire J (2009) Geometric morphometrics of vole (*Microtus californicus*) dentition as a new paleoclimate proxy: Shape change along geographic and climatic clines. Quaternary international: 1–8.
17. Musser GG, carleton MD (2005) Superfamily Muroidea. In: Wilson DE, Reeder DM, editors. Mammal Species of the World: A Taxonomic and Geographic Reference. Baltimore: The Johns Hopkins University Press. 894–1531.
18. van de Weerd A (1976) Rodent faunas of the Mio-Pliocene continental sediments of the Teruel-Alfambra region, Spain. Special publication, 2: 1–217.
19. Michaux J, Reyes A, Catzeflis F (2001) Evolutionary history of the most speciose mammals: Molecular phylogeny of muroid rodents. Molecular Biology and Evolution 18: 2017–2031.
20. Jansa SA, Weksler M (2004) Phylogeny of muroid rodents: relationships within and among major lineages as determined by IRBP gene sequences. Molecular Phylogenetics and Evolution 31: 256–276.
21. Kavanagh KD, Evans AR, Jernvall J (2007) Predicting evolutionary patterns of mammalian teeth from development. Nature 449: 427–432.
22. Miljutin A (2011) Trends of Specialisation in Rodents: The Hamsters, Subfamily Cricetinae (Cricetidae, Rodentia, Mammalia). Acta Zoologica Lituanica 21: 192–206.
23. Gómez Cano AR, Hernández Fernández M, Álvarez-Sierra MA (2011) Biogeographic provincialism in rodent faunas from the Iberoccitanian Region (southwestern Europe) generates severe diachrony within the Mammalian Neogene (MN) biochronologic scale during the Late Miocene. Palaeogeography, Palaeoclimatology, Palaeoecology 307: 193–204.
24. Sesé C (2006) Los roedores y lagomorfos del Neógeno de España. Estudios Geológicos 62: 429–480.
25. Renaud S (1999) Size and shape variability in relation to species differences and climatic gradients in the African rodent *Oenomys*. Journal of Biogeography.
26. Van Bocxlaer B, Schultheiß R (2010) Comparison of morphometric techniques for shapes with few homologous landmarks based on machine-learning approaches to biological discrimination. Paleobiology 36: 497–515.
27. Rohlf FJ (2010) tpsDIG2: Thin Plate Spline Digitizing Landmarks. 2.16 ed. New York: State University of New York at Stony Brook, Stony Brook.
28. Kuhl FP, Giardina CR (1982) Elliptic Fourier features of a closed contour. Computer Graphics and Image Processing 18: 259–278.
29. Ferson S, Rohlf FJ, Koehn RK (1985) Measuring shape variation of two-dimensional outlines. Systematic Zoology 34: 59–68.
30. Deffontaine V, Ledevin R, Fontaine MC, Quéré J-P, Renaud S, et al. (2009) A relict bank vole lineage highlights the biogeographic history of the Pyrenean region in Europe. Molecular Ecology 18: 2489–2502.
31. Renaud S, Michaux J (2004) Parallel evolution in molar outline of murine rodents: the case of the extinct *Malpaisomys insularis* (Eastern Canary Islands). Zoological Journal of the Linnean Society: 555–572.
32. Rohlf F, Archie J (1984) A Comparison of Fourier Methods for the Description of Wing Shape in Mosquitoes (Diptera: Culicidae). Systematic Zoology 33: 302–317.
33. Renaud S, Michaux JR (2003) Adaptive latitudinal trends in the mandible shape of *Apodemus* wood mice. Journal of Biogeography 30: 1617–1628.
34. Anderson MJ (2001) A new method for non-parametric multivariate analysis of variance. Austral Ecology 26: 32–46.
35. Hammer Ø, Harper DAT (2006) Chapter 4 Morphometrics. In: Hammer O, Harper DAT, editors. Paleontological data analysis. Malden, Oxford & Carlton: Blackwell Publishing. 78–148.
36. Hammer Ø, Harper DAT, Ryan PD (2001) PAST: Paleontological statistics software package for education and data analysis. Palaeontologia Electronica 4: 9 pp.
37. Boyer DM, Evans AR, Jernvall J (2010) Evidence of Dietary Differentiation Among Late Paleocene-Early Eocene Plesiadapids (Mammalia, Primates). American Journal of Phisical Anthropology 142: 194–210.
38. Cucchi T, Orth A, Auffray J-C, Renaud S, Fabre L, et al. (2006) A new endemic species of the subgenus Mus (Rodentia, Mammalia) on the Island of Cyprus. Zootaxa 1241: 1–36.
39. Misonne X (1969) African and Indo-Australian Muridae: Musee Royal de L'Afrique Centrale - Tervuren, Belguique Annales Serie N° 8. 1–185 p.
40. Casanovas-Vilar I, Van Dam jA, Moyà-Solà S, Rook L (2011) Late Miocene insular mice from the Tusco-Sardinian palaeobioprovince provide new insights on the palaeoecology of the *Oreopithecus* faunas. Journal of Human Evolution 61: 42–49.
41. Ungar P (2010) Mammal teeth: origin, evolution and diversity: The Johns Hopkins University Press. 304 p.
42. Matthews T, Stynder DD (2011) An analysis of the *Aethomys* (Murinae) community from Langebaanweg (Early Pliocene, South Africa) using geometric morphometrics. Palaeogeography, Palaeoclimatology, Palaeoecology 302: 230–242.
43. Renaud S, Michaux J, Schmidt DN, Aguilar J-P, Mein P, et al. (2005) Morphological evolution, ecological diversification and climate change in rodents. Proceedings of the Royal Society B: Biological Sciences 272: 609–617.
44. Huchon D, Chevret P, Jordan U, Kilpatrick CW, Ranwez V, et al. (2007) Multiple molecular evidences for a living mammalian fossil. Proceedings of the National Academy of Sciences of the United States of America 104: 7495–7499.
45. Collen B, Turvey ST, Waterman C, Meredith HMR, Kuhn TS, et al. (2011) Investing in evolutionary history: implementing a phylogenetic approach for mammal conservation. Philosophical Transactions of the Royal Society B: Biological Sciences 366: 2611–2622.
46. Renvoisé E, Evans AR, Jebrane A, Labruère C, Laffont R, et al. (2009) Evolution of mammal tooth patterns: new insight from a developmental prediction model. Evolution 63: 1327–1340.
47. García-Alix A, Minwer-Barakat R, Martín-Suárez E, Freudenthal M, Martín JM (2008) Late Miocene–Early Pliocene climatic evolution of the Granada Basin (southern Spain) deduced from the paleoecology of the micromammal associations. Palaeogeography, Palaeoclimatology, Palaeoecology 265: 214–225.
48. Schaub S (1938) Tertiäre und Quartäre Murinae. Abhandlungen des Schweizerischen paläontologischen Gesellschaft = Mémoires de la Société paléontologique suisse 61: 1–38.
49. López-Martínez N, Michaux J, Hutterer R (1998) The skull of *Stephanomys* and a review of *Malpaisomys* relationships (Rodentia: Muridae): taxonomic incongruence in murids. Journal of Mammalian Evolution 5: 185–215.
50. Van Dam jA, Weltje GJ (1999) Reconstruction of the Late Miocene climate of Spain using rodent palaeocommunity successions: an application of end-member modelling. Palaeogeography, Palaeoclimatology, Palaeoecology 151: 267–305.
51. Costeur L, Legendre S, Aguilar J-P, Lécuyer C (2007) Marine and continental synchronous climatic records: Towards a revision of the European Mid-Miocene mammalian biochronological framework. Geobios 40: 775–784.
52. Hernández Fernández M (2005) Análisis paleoclimático y paleoecológico de las sucesiones de mamíferos del Plio-Pleistoceno de la Península Ibérica. Madrid: Universidad Complutense de Madrid.
53. Böhme M, Ilg A, Winklhofer M (2008) Late Miocene "washhouse" climate in Europe. Earth and Planetary Science Letters 275: 393–401.
54. Böhme M, Winklhofer M, Ilg A (2011) Miocene precipitation in Europe: Temporal trends and spatial gradients. Palaeogeography, Palaeoclimatology, Palaeoecology 304: 212–218.
55. Sanz de Siria Catalán A (1994) La evolución de las paleofloras en las cuencas cenozoicas catalanas. Acta geológica hispánica 29: 169–190.
56. Carrión JS, Fernández S, Jiménez-Moreno G, Fauquette S, Gil-Romera G, et al. (2009) The historical origins of aridity and vegetation degradation in southeastern Spain. Journal of arid environments 74: 731–736.
57. Carrión JS, Leroy SAG (2010) Iberian floras through time: Land of diversity and survival. Review of Palaeobotany and Palynology 162: 227–230.
58. Vrba ES (1980) Evolution, species and fossils: how does life evolve? South African Journal of Science 76: 61–84.
59. Vrba ES (1987) Ecology in relation to speciation rates: some case histories of Miocene recent mammal clades. Evolutionary Ecology 1: 283–300.
60. Mein P, Martin-Suarez E, Agusti J (2009) *Progonomys* Schaub, 1938 and *Huerzelerimys* gen. nov.(Rodentia); their evolution in Western Europe. Scripta Geologica 103: 41–64.
61. Gómez Cano AR, Cantalapiedra JL, Mesa A, Moreno Bofarull A, Hernández Fernández M (2013) Global climatic changes drive ecological specialization of mammal faunas: Trends in rodent assemblages from the Iberian Plio-Pleistocene. BMC Evolutionary Biology 13.
62. van Zon L (2011) Seasonality and survaival: Competition, Speciation and Extinction in Variable Environments. Utrecht: Utrecht University. 82 p.

5

Representation of Dormant and Active Microbial Dynamics for Ecosystem Modeling

Gangsheng Wang[1,2]*, Melanie A. Mayes[1,2], Lianhong Gu[1,2], Christopher W. Schadt[1,3]

1 Climate Change Science Institute, Oak Ridge National Laboratory, Oak Ridge, Tennessee, United States of America, **2** Environmental Sciences Division, Oak Ridge National Laboratory, Oak Ridge, Tennessee, United States of America, **3** Biosciences Division, Oak Ridge National Laboratory, Oak Ridge, Tennessee, United States of America

Abstract

Dormancy is an essential strategy for microorganisms to cope with environmental stress. However, global ecosystem models typically ignore microbial dormancy, resulting in notable model uncertainties. To facilitate the consideration of dormancy in these large-scale models, we propose a new microbial physiology component that works for a wide range of substrate availabilities. This new model is based on microbial physiological states and the major parameters are the maximum specific growth and maintenance rates of active microbes and the ratio of dormant to active maintenance rates. A major improvement of our model over extant models is that it can explain the low active microbial fractions commonly observed in undisturbed soils. Our new model shows that the exponentially-increasing respiration from substrate-induced respiration experiments can only be used to determine the maximum specific growth rate and initial active microbial biomass, while the respiration data representing both exponentially-increasing and non-exponentially-increasing phases can robustly determine a range of key parameters including the initial total live biomass, initial active fraction, the maximum specific growth and maintenance rates, and the half-saturation constant. Our new model can be incorporated into existing ecosystem models to account for dormancy in microbially-driven processes and to provide improved estimates of microbial activities.

Editor: Jonathan H. Badger, J. Craig Venter Institute, United States of America

Funding: This research was funded by the Laboratory Directed Research and Development (LDRD) Program of the Oak Ridge National Laboratory (ORNL) and by the U.S. Department of Energy Biological and Environmental Research (BER) program. ORNL is managed by UT-Battelle, LLC, for the U.S. Department of Energy under contract DE-AC05-00OR22725. The funders had no role in study design, data collection and analysis, decision to publish, or preparation of the manuscript.

Competing Interests: The authors have declared that no competing interests exist.

* E-mail: wangg@ornl.gov

Introduction

Ecologically-important processes such as soil organic carbon and nutrient cycling largely depend on the active fraction of microbial communities [1]. At any given time in a given environment, microorganisms can be in active, dormant, or dead states [2]. Dormancy is considered an evolutionary strategy designed to maintain the genetic code until conditions improve to allow replication [3]. When environmental conditions are unfavorable for growth, e.g., resource limitation, microbes may enter a reversible state of low to zero metabolic activity to alleviate the loss of biomass and metabolic functions [4,5]. The maintenance coefficient (i.e., maintenance cost of C per unit microbial biomass C per unit time) can be two to three orders of magnitude lower in dormant microbes than in metabolically active microbes [6,7]. Many soils have slow organic matter turnover rates with seasonal changes in substrate supply, temperature, and moisture. The complexity of soils in space and time may result in uneven distributions of multiple potentially limiting resources, leading to significant rates of dormancy even when some resources are abundant. When spatial and temporal complexity is combined with differential resource partitioning among species in a community, high rates of dormancy could be a prominent feature in soil systems. Thus it is essential to understand dormancy in order to more accurately predict how active microorganisms contribute to ecosystem processes such as decomposition and nutrient turnover [1].

A complicating factor in studying microbial dormancy is that no single approach can be easily employe to simultaneously measure individual microbial states (active, dormant or dead), and a combination of different techniques is required. Differential staining is often used to segregate physiological states with direct microscopic counting of bacteria and fungi. 'Life-indicating' stains that require the presence of 'standard' physiological abilities, such as the esterase activity needed for fluorescein diacetate cleavage, may distinguish active from dormant+dead cells [8]. When combined with general-purpose stains, these strains can distinguish dormant cells by difference [9]. Combining membrane-permeant with membrane-impermeant nucleophilic stains (e.g., SYTO-9 and propidium iodide respectively) may distinguish live from dead, but not active from dormant [10,11]. Active microbes may or may not be 'viable' with common culture-based techniques, which complicates classification and measurement of dormancy phenomena [5]. Methods such as direct plating, serial dilution and most probable number (MPN) techniques will not distinguish between active and dormant organisms [12]. Substrate Induced Respiration (SIR) or Substrate Induced Growth Response (SIGR) method [13,14] can distinguish active and dormant communities if growth respiration curves are modeled (using initial exponentially-increasing respiration); however, the technique often needs to be combined with microscopy or chloroform fumigation/extraction in order to obtain total live microbial biomass [15,16].

Despite limitations in distinguishing active, dormant and dead microbial biomass, abundant evidence indicates that the majority of environmental microorganisms in a given community may be dormant under natural conditions [1,17]. Alvarez *et al.* [18] reported that only 3.8–9.7% of the total biomass is active in a Typic Argiudoll soil from the Argentinean Pampa. Khomutova *et al.* [19] showed that the fraction of active microbial biomass ranged from 0.02% to 19.1% in the subkurgan paleosoils of different age and 9.2–24.2% in modern background soils. Microbial biomass measured through SIR or SIGR is thought to reflect only the active portion because the maintenance respiration of dormancy biomass is negligible in the initial exponentially-increasing phase [13,16,20]. Through a mathematical analysis of respiration curves, Van de Werf & Verstraete [21] examined 16 soils and found that 4–49% of the total biomass was in an active state; and the active component in undisturbed natural ecosystems (18.8±8.8%, mean±standard deviation) was about 70% of that in arable agricultural soils (25.7±14.8%). Stenström *et al.* [22] showed that the fraction of active biomass typically varied from 5% to 20% in soils without recent addition of substrates. Lennon & Jones [5] found much lower active fractions in soils (18±15%) than in marine (65±19%) and fresh (54±11%) water environments. From the above studies it seems conservative to extrapolate that the active fraction is very likely below 50% of live microbes under most natural soil conditions.

Microbially-mediated processes have been incorporated into ecosystem models [23–28] although continued development is still required to bring microbial processes into global climate models [29–31]. However, these recent models do not consider physiological state changes and assume that measures of microbial biomass constitute the active biomass. The exclusion of dormancy from the microbially-driven ecosystem processes could result in incorrect estimates of total live microbial biomass, which further leads to deficiencies in model parameterization and predictions of soil organic carbon and nutrients.

Generally, there are two strategies to represent physiological states in microbial-ecology models: one is to explicitly separate the total live biomass into two pools, i.e., active and dormant [4,32]; the other is to directly regard the active fraction (i.e., ratio of active biomass to total live biomass) as a state variable [33,34]. Both of these two approaches predict the total live biomass, active and dormant biomass, and the flux between the active and dormant components. Apparently the introduction of the 'active fraction' as a state variable in the latter approach simplifies the model structure since the adaptive variation of microbial composition might be represented by one single variable (active fraction) [34,35]. However, another state variable indicating the microbial biomass pool size (e.g., total live biomass, active biomass or dormant biomass) is still essential for ecosystem modeling since the carbon and nutrient fluxes are pool-size dependent. For example, if we define active fraction and total microbial biomass as state variables, the active and dormant biomass could be determined by them, and the net flux between active and dormant fractions and other related fluxes could also be computed according to the active and dormant biomass constrained by mass balance. The above-mentioned modeling efforts have shown that adequate representation of dormancy and the transitions between the dormant and active states is crucial for modeling important microbially-mediated ecosystem processes.

Here, we review state-of-the-art microbial dormancy modeling approaches and discuss the rationales of these models with a focus on transformation processes between active and dormant states.

We propose an improved synthetic microbial physiology model based on accepted assumptions and examine the model behavior with theoretical and experimental analyses. In this paper, the 'total microbial biomass' refers to the 'total live microbial biomass' unless otherwise stated. Our objective is to clarify the applicability of existing microbial dormancy models and provide a new theoretical basis for representing microbial activity and dormancy in ecosystem models.

Dormancy In Microbial Models

Transformation between active and dormant states

Although Buerger *et al.* [36] argued that dormant microbial cells could reactivate stochastically and might be independent of environmental cues, environmental factors such as substrate availability are often thought to control the transformation between active and dormant states [5]. Most models (see Appendix S1) distinguish the active biomass pool from the dormant pool and define them as two state variables (B_a and B_d) (Fig. 1). Only active microbes (B_a) can uptake substrate and produce new cells. The connection between the active and dormant states is a reversible process including two directional sub-processes, i.e., dormancy (from active to dormant) and reactivation (or resuscitation, from dormant to active). Losses from active biomass include growth respiration and maintenance (maintenance respiration, mortality, enzyme synthesis, etc.) [23]. Dormant microbes still require energy for maintenance and survival although at a lower metabolic rate [5].

The net transformation rate ($B_{a \to d}^N$) from active to dormant state is the difference between the flux from active to dormant ($B_{a \to d}$) and the flux from dormant to active state ($B_{d \to a}$), i.e., $B_{a \to d}^N = B_{a \to d} - B_{d \to a}$. The models of Hunt [37] (Equation S1-1 in Appendix S1) and Gignoux *et al.* [38] (Equation S1-2 in Appendix S1) directly formulate the net flux ($B_{a \to d}^N$) without explicit components for $B_{a \to d}$ and $B_{d \to a}$. The direction of the net flux depends on the maintenance requirement relative to the substrate availability. If the available substrate is less than the maintenance requirement, there is a positive net flux from active to dormant pool, and vice versa. In addition, Hunt [37] assumed a 'buffer zone' for the change of states: when the maintenance requirement surpasses the substrate supply but the deficit is within a small fraction ($1\% \ \mathrm{d}^{-1}$) of B_a, there is no flux between the two states.

Some models define rates for both dormancy and reactivation. In the model of Ayati [39] (Equation S1-3 in Appendix S1), the

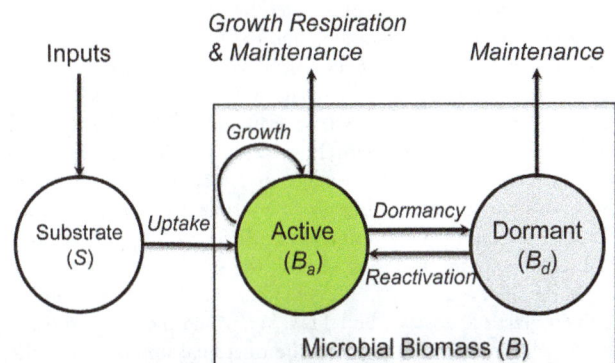

Figure 1. Active and dormant microbial biomass pools in microbial physiology models (modified from Fig. 2 in Lennon & Jones, 2011).

dormant rate ($\gamma_{a \to d}$) increases with declining substrate concentration, and the reactivation ($\gamma_{d \to a}$) only occurs when substrate concentration is higher than the half-saturation constant (K_s). Konopka [32] modified the potential rates for dormancy and reactivation by the relative growth rate (μ/μ_{max}, ratio of true specific growth rate to maximum specific growth rate), i.e, the two rates are multiplied by ($1-\mu/\mu_{max}$) and μ/μ_{max}, respectively (Equation S1-4 in Appendix S1). Similarly, Jones & Lennon [40] postulated two complementary rates ($1-R$ and R) for dormancy and resuscitation (Equation S1-5 in Appendix S1).

Two other models also explicitly formulate the two conversion rates between states but do so using concepts of probability. Bär *et al.* [41] used two complementary factors ($1-J$ and J) to represent the probability for the transition between active and dormant state in addition to an identical potential rate constant for the two processes (Equation S1-6 in Appendix S1). The conceptual model of Locey [42] applies a deterministic dormant rate and a stochastic resuscitation rate (Equation S1-7 in Appendix S1). The potential resuscitation rate is modified by ($1-p$), where p is the probability that a disturbance in the active pool will result in the immigration of one individual from the metacommunity. The probability (J) in Bär *et al.* [41] is explicitly calculated from the environmental cues (e.g., soil moisture), while the cause of the probability (p) in Locey [42] is not elucidated.

Switch function model

In addition to the dormancy and reactivation processes, a key concept in the model developed by Stolpovsky *et al.* [4] is 'switch function' (Equation S1-8 in Appendix S1). The switch function (θ) determines the fraction of active cells taking up dissolved organic carbon (DOC). This function refers to the growth fraction in active biomass (B_a) that consumes substrate and thus is not the same as the active fraction (r) in total biomass (B). Furthermore, the dormancy and reactivation fluxes are set to be proportional to ($1-\theta$) and θ, respectively. θ is formulated by the Fermi-Dirac statistics [4]. Another feature of this model is the consideration of 'depth' of dormancy in reactivation, where the reactivation rate is negatively dependent on the duration of dormancy. The switch function model includes at least 15 model parameters and it is difficult to compute the Gibbs energy change of the oxidation of DOC [4].

The switch function (θ) sets it apart from the conventional Michaelis-Menten (M-M) or Monod kinetics because of its new perspective of thermodynamics. According to the M-M kinetics [43], the substrate saturation level represents the fraction of enzyme-substrate complex (ES) in active enzyme (E_0), where the substrate saturation level is formulated by $S/(K_s + S)$ with S and K_s being the substrate concentration and the half-saturation constant [43]. When the M-M kinetics is applied to describe microbial uptake of substrate, the substrate (or combined with TEA) saturation level (i.e., $\mu(S, TEA)$ in Equation S1-8a of Appendix S1) is a measure of the actively growing fraction in the active microbial community. The switch function is also determined by the saturation levels of substrate and terminal electron acceptor (TEA), i.e., $\mu(S, TEA)$ (Equation S1-8e in Appendix S1). Mathematically the inclusion of both the switch function (θ) and the M-M kinetics (i.e., $\mu(S, TEA)$) might result in double counting of the impact of substrate and TEA. We would recommend using the switch function (θ) to modify the microbial uptake rate if the Gibbs energy change of the oxidation of substrate (ΔG) is tractable and the thermodynamic threshold (G_0) and the steepness of the step function (st) are identifiable.

Physiological state index models

As an alternative to models with two microbial biomass pools (i.e., active and dormant), a further state variable indicating the dormant or active fraction in total biomass has been proposed. Wirtz [44] developed a simple index ($r_d = 0.5-1.0$) representing the dormant microbial biomass as a fraction of the steady-state total biomass (B_{stat}) under the condition of $B_d << B_a$. In case of a net loss of total biomass ($dB/dt<0$), the dormant biomass $B_d = B_{stat} \cdot r_d$; otherwise ($dB/dt>0$), $B_d = B_{stat} \cdot (1-r_d)$. This model has a sudden change of dormant biomass at the transition point (i.e., $dB/dt = 0$) since $r_d > 0.5$.

Different from the dormant index of Wirtz [44], the concept of an active index (i.e., index of physiological state) of soil microbial community has been employed in soil carbon and nutrient cycling models [33,34]. The index of physiological state (r), referring to the activity state, is often defined as the ratio of metabolically active microbial biomass to the total soil microbial biomass [22,33,34].

In the Synthetic Chemostat Model (SCM), the rate of change of the state variable r is described as follows [33,45]:

$$\frac{dr}{dt} = \frac{1}{B}\frac{dB}{dt} \cdot (\phi - r) = \mu \cdot (\phi - r) \tag{1}$$

with

$$\phi = \phi(S) = S^n/(K_r + S^n), \text{or} \phi = S/(K_r + S) \tag{2}$$

where $r = B_a/B$, representing the fraction (hereinafter referred to as 'active fraction') of active biomass in total biomass; μ is the specific growth rate of total biomass; ϕ denotes the saturation level of substrate (S); the simple power ($n = 1$) has been widely used [35] and, in this case ($n = 1$), K_r is called the half-saturation constant.

Blagodatsky & Richter [34] used the expression $\mu(S) = \mu_{max} \cdot \phi(S)$ in their model development. This expression was not derived in the original definition of the specific growth rate (see Equation 3) by Panikov [45] and because its validity cannot be inferred, the concepts will not be addressed here.

According to Panikov's derivation [45], the specific growth rate (μ) follows the general definition [46,47]:

$$\mu = \frac{1}{B}\frac{dB}{dt} \tag{3}$$

Based on Equations 1 and 3, we can derive (see Equation S2-1 in Appendix S2):

$$dB_a/dt = \phi \cdot (dB/dt) \tag{4}$$

$$dB_d/dt = (1 - \phi) \cdot (dB/dt) \tag{5}$$

We find that the model described by Equation 1 is not applicable under low substrate availability, as described below. Generally, the rates of change in biomass pools (B, B_a, and B_d) can be expressed as

$$dB/dt = g^{\pm}(S, B_a) - f^+(S, B_d) \tag{6}$$

$$dB_a/dt = g^\pm(S,B_a) - B_{a \to d}^N \tag{7}$$

$$dB_d/dt = -f^+(S,B_d) + B_{a \to d}^N \tag{8}$$

where $B_{a \to d}^N$ denotes the net dormancy flux; $g^\pm(S, B_a)$ is a function that represents the difference between the substrate uptake and the maintenance requirements of B_a, i.e., the net growth of B_a; and $f^+(S, B_a)$ is a function denoting the maintenance and survival energy costs of B_d. The superscript '$\pm$' in $g^\pm$ indicates the function value of g could be positive at high S or non-positive when the substrate uptake cannot satisfy the maintenance requirements of B_a at low S. The superscript '+' in f^+ implies $f \geq 0$. Note that the function $f^+(S, B_a)$ is not necessarily dependent on S [4].

From Equations 4, 6 and 7, we can obtain

$$B_{a \to d}^N = (1-\phi) \cdot g^\pm(S,B_a) - \phi \cdot f^+(S,B_d) \tag{9}$$

The two terms in the right side of Equation 9 may be regarded as the conversion of B_a to B_d (i.e., $B_{a \to d}$) and the transformation of B_d to B_a (i.e., $B_{d \to a}$), respectively. At high S resulting in $g \geq 0$, Equation 9 may be one of the possible expressions for $B_{a \to d}$ and $B_{d \to a}$. However, at low S leading to $g < 0$ and $B_{a \to d} < 0$, i.e., no active cells become dormant under insufficient substrate, which is inconsistent with the strategy of dormancy for microorganisms when faced with unfavorable environmental conditions [5].

Based on the above analysis, we conclude that the physiological state index model (Equation 1) needs to be improved. In other words, the empirical assumption that the steady state active fraction (r^{ss}) approaches the substrate saturation level (ϕ^{ss}) may not be necessary because this assumption could lead to impractical flux (Equation 9) between dormant and active states under low substrate availability.

A Synthetic Microbial Physiology Model

Based on the aforementioned review and analysis, we have developed a synthetic microbial physiology model component relating to substrate availability. As indicated by Fig. 1, the growth and maintenance functions of active microbes (B_a) are characterized by the maximum specific growth rate (μ_G) and maintenance rate (m_R); whereas the dormant microbes (B_d) cost energy to maintain their basic cellular functions at a much lower specific maintenance rate (denoted by $\beta \cdot m_R$, where $\beta < 1$) [48].

General assumptions

First we define the substrate saturation level (ϕ) as

$$\phi = S/(K_S + S) \tag{10}$$

where the parameter K_s is the half saturation constant for substrate uptake as indicated by the M-M kinetics [43].

Based on the above review of existing dormancy models, the following assumptions are accepted in our new model: (1) the dormancy rate is proportional to the active biomass and the reactivation rate is proportional to the dormant biomass, i.e., $B_{a \to d} \propto B_a$ and $B_{d \to a} \propto B_d$; (2) under very high substrate concentration ($S \gg K_s$), $\phi \to 1$, $B_{a \to d} \to 0$ and $B_{d \to a} \geq 0$; (3) under very low substrate ($S \ll K_s$), $\phi \to 0$, $B_{a \to d} \geq 0$ and $B_{d \to a} \to 0$; (4) based on the assumptions (1–3), we derive that $B_{a \to d} \propto (1-\phi) \cdot B_a$ and

$B_{d \to a} \propto \phi \cdot B_a$; (5) further we assume that the maximum specific maintenance rate for active microbes (m_R with units of h^{-1}) controls both transformation processes since the maintenance energy cost is the key factor regulating the dormancy strategy [5,37,38]. As a result we postulate that

$$B_{a \to d} = (1-\phi) \cdot m_R \cdot B_a \tag{11a}$$

$$B_{d \to a} = \phi \cdot m_R \cdot B_d \tag{11b}$$

Model Description

Equations 11a and 11b only describe the transformation between the active and dormant states. They need to be linked to a microbial growth and maintenance model for depicting microbial dynamics. Our recent work to develop the Microbial-ENzyme-mediated Decomposition (MEND) model [23] suggested that it might be adapted to serve this purpose due to its focus on microbial processes for which we have developed a firm theoretical basis [47,49]. Combining Equations 11a and 11b with the MEND model [23,47], we express the microbial physiology component (see Fig. 1) as a group of differential equations

$$dS/dt = I_s - \frac{1}{Y_G} \cdot \frac{\phi}{\alpha} m_R \cdot B_a \tag{12a}$$

$$dB/dt = d(B_a + B_d)/dt = (\phi/\alpha - 1) \cdot m_R \cdot B_a - (\beta \cdot m_R) \cdot B_d \tag{12b}$$

$$dB_a/dt = (\phi/\alpha - 1) \cdot m_R \cdot B_a - (1-\phi) \cdot m_R \cdot B_a + \phi \cdot m_R \cdot B_d \tag{12c}$$

$$dB_d/dt = -(\beta \cdot m_R) \cdot B_d + (1-\phi) \cdot m_R \cdot B_a - \phi \cdot m_R \cdot B_d \tag{12d}$$

where t is the time scale; ϕ is defined by Equation 10; I_s is the input to substrate pool; Y_G is the true growth yield; m_R denotes the specific maintenance rate at active state (h^{-1}); $\alpha = m_R/(\mu_G + m_R)$ is the ratio of m_R to the sum of maximum specific growth rate (μ_G) and m_R, $\alpha \in (0,0.5)$ since usually $m_R \leq \mu_G$; and β (0–1) is the ratio of dormant maintenance rate to active maintenance rate, i.e., (β m_R) denotes the maximum specific maintenance rate at dormant state.

In summary, there are five parameters (α, β, m_R or μ_G, Y_G, K_s) in the proposed model (hereinafter referred to as the MEND model). From Equations 12b and 12c, we can derive the change rate of active fraction (r) (see Equation S2-2 in Appendix S2)

$$dr/dt = m_R \cdot [(\phi - r) + (\phi/\alpha + \beta - 1) \cdot r \cdot (1-r)] \tag{12e}$$

This equation for r is more complicated than Equation 1 but still practical, given currently available data. Additionally, it implies that r needs not approach ϕ at steady state in our model, whereas $r \equiv \phi$ at steady state is required by the model of Panikov [45].

Steady state analysis

Assuming the input (I_s) is time-invariant, we can obtain the steady state solution to the above new MEND model (see Equations S2-3(a–e) in Appendix S2). Fig. 2 shows the steady

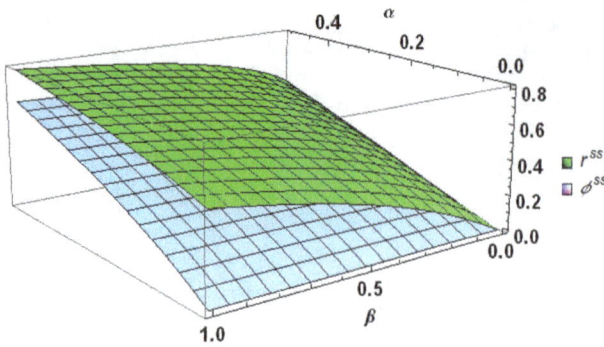

Figure 2. Steady state active fraction (r^{ss}) and substrate saturation level (ϕ^{ss}) as a function of α and β; $\alpha = m_R/(\mu_G+m_R)$, μ_G and m_R (h^{-1}) are maximum specific growth rate and specific maintenance rate for active microbial biomass, respectivly; β denotes the ratio of dormant specific maintenance rate to m_R.

state active fraction (r^{ss}) and substrate saturation level (ϕ^{ss}) as a function of the two physiological indices, i.e., α (0–0.5) and β (0–1). Both r^{ss} and ϕ^{ss} positively depend on α and β and $r^{ss} \geq \phi^{ss}$ for any combinations of α and β. If we consider two extreme values of $\beta\to0$ or $\beta\to1$, the r^{ss} and ϕ^{ss} (see Equations S2-4 and S2-5 in Appendix S2) can be simplified to

$$r^{ss}_{\beta\to0} = \phi^{ss}_{\beta\to0} = \alpha \qquad (13a)$$

$$\begin{cases} r^{ss}_{\beta\to1} = \dfrac{1+\sqrt{1+8\alpha}}{4} \\ \phi^{ss}_{\beta\to1} = \dfrac{4\alpha-1+\sqrt{1+8\alpha}}{3+\sqrt{1+8\alpha}} \end{cases} \qquad (13b)$$

Equation 13 and Fig. 2 indicate that: (1) the steady state active fraction (r^{ss}) is equal to ϕ^{ss} and they are identical to $\alpha = m_R/(\mu_G+m_R)$ only under the condition of $\beta\to0$; (2) the upper bound of r^{ss} is approximately 0.8 at $\alpha\to0.5$ and $\beta\to1$; and (3) with $\alpha\leq0.5$, the maximum r^{ss} is ca. 0.5 if the magnitude of β is around 0.001–0.01 [6]. This threshold value (0.5) of r^{ss} is a reasonable estimate that can explain how the measured active fraction of microbes in undisturbed soils is usually considerably less than the total biomass [5,21,22].

Model simplification under sufficient substrate condition

As mentioned in the Introduction, SIR or SIGR method can distinguish active from dormant composition and the data from these experiments have been widely used to estimate the active microbial biomass and the maximum specific growth rate [13,14]. The simplification of the microbial model under excess substrate has also been employed to estimate maximum specific growth rate (μ_G), active microbial biomass (B_a), and/or total microbial biomass (B) using the SIR or SIGR data [1,13,35]. Here we show the simplification of our model (Equation 12) for conditions appropriate to SIGR or SIR experiments, e.g., the short-term period of exponentially-increasing respiration of active biomass following substrate addition. We will test our reduced and full model with the SIGR data of Colores et al. [13] in the next section.

Under sufficient substrate (i.e., $S \gg K_s$ in Equation 10 thus $\phi\to1$), Equations 12(a–e) can be simplified and integrated for initial conditions, i.e., $S=S_0$, $B=B_0$ and $r=r_0$ at $t=0$ (see

Equations S2-6 and S2-7 in Appendix S2):

$$S(t) = S_0 - \frac{B(t)-B_0}{Y_G(1-\alpha)} \qquad (14a)$$

$$B(t) = B_0 r_0 \cdot e^{(1/\alpha-1)m_R t} + \\ B_0(1-r_0)\cdot[\alpha\cdot e^{(1/\alpha-1)m_R t}+(1-\alpha)\cdot e^{-m_R t}] \qquad (14b)$$

$$r(t) = \frac{[r_0+\alpha(1-r_0)]\cdot e^{(m_R/\alpha)\cdot t}-\alpha(1-r_0)}{[r_0+\alpha(1-r_0)]\cdot e^{(m_R/\alpha)\cdot t}+(1-\alpha)(1-r_0)} \qquad (14c)$$

The CO_2 production rate, $v(t)$, during the exponential growth stage is derived as an explicit function of t (see Equation S2-7d in Appendix S2):

$$v(t) = \frac{dCO_2}{dt} = \\ \frac{B_0(1-Y_G)}{Y_G}\left\{[(m_R/\alpha)\cdot r_0+m_R\cdot(1-r_0)]\cdot e^{(1/\alpha-1)m_R t}-[m_R\cdot(1-r_0)]\cdot e^{-m_R t}\right\} \qquad (14d)$$

The respiration rate, $v(t)$, is associated with two exponential items, i.e., $e^{(1/\alpha-1)m_R t}=e^{\mu_G t}$ and $e^{-m_R t}$. Considering an extreme case that $m_R \ll \mu_G$ (i.e., $\alpha\to0$), Equations 14(b–d) can be further simplified to Equations S2-8(b–d) (see Appendix S2). Equations S2-8b and S2-8c (denoting $B(t)$ and $r(t)$, respectively) are similar to Equations 11 and 10 in Panikov & Sizova [35], respectively. However, Equation S2-8d (for $v(t)$) is different from Equation 13 of Panikov & Sizova [35], where a constant 'A' was added to the exponential. Equation S2-8d is identical to Equation 7 derived for SIGR experiments in Colores et al. [13].

Panikov & Sizova [35] used their Equation 13 to fit respiration rates during the exponentially-increasing (i.e., no substrate limitation) phase (see Fig. 2 in Panikov & Sizova [35] for data and curve fittings). However, these data are based on glucose-induced respiration that includes both basal respiration of native SOC and respiration due to the addition of glucose [13]. The basal respiration rate may be regarded as a constant in certain cases (see Colores et al. [13] and data in Fig. 1 of Blagodatsky et al. [50]). The constant 'A' representing the basal respiration rate was included in Equation 13 of Panikov & Sizova [35] in order to fit the combined respiration from the addition of glucose and basal respiration. However, this constant 'A' cannot be derived from such governing equations as Equations S2-6(a–c) (see Appendix S2) that assume respiration is the sole result of substrate addition. In other words, the equations do not include basal respiration. Certainly, our predicted respiration could include basal respiration as long as (i) a basal respiration rate is added to Eq. 14d ad hoc or (ii) Equations S2-6(a–c), or more commonly Equations 12(a–e), are linked to a soil organic matter (SOM) decomposition model, which can produce decomposed native soil C in addition to the respiration of substrate addition. Because Equation 13 of Panikov & Sizova [35] is not linked to a native C decomposition model, fitting the model to combined native C and substrate respiration data is not appropriate.

Model test I: substrate-induced respiration

In this section, we used the respiration data from ^{14}C-labeled glucose SIGR experiments by Colores et al. [13] to calibrate our

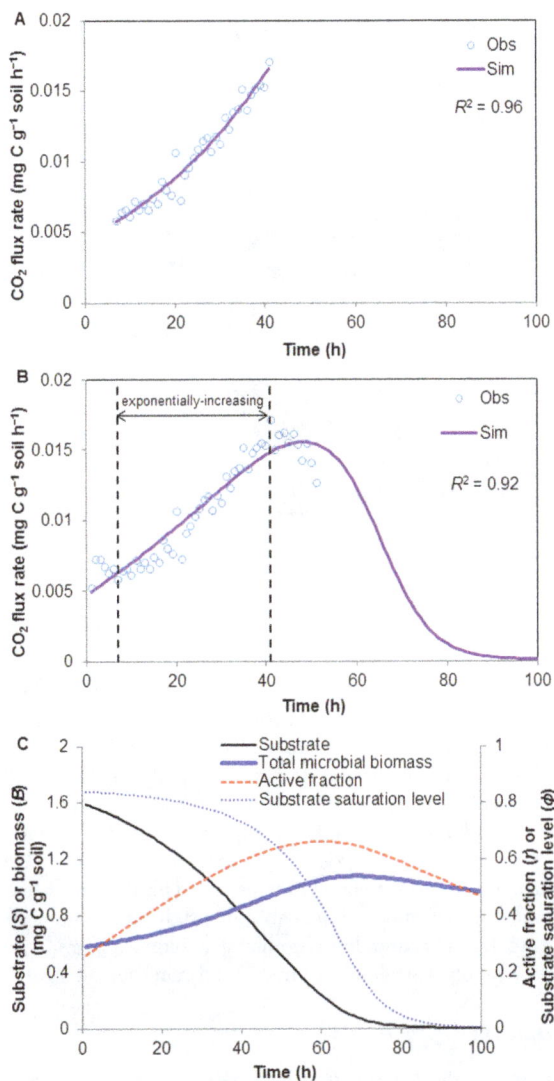

Figure 3. MEND model simulations against the respiration rates due to added ^{14}C-labeled glucose in Colores et al. [13]. (a) Fitting of the respiration rates in the exponentially-increasing phase using Equation 14, 'Obs' and 'Sim' denote observed and simulated data, respectively. (b) Fitting of the respiration rates in both exponentially-increasing and non-exponentially-increasing phases using Equation 12. (c) Simulated substrate (S), total live microbial biomass (B), active fraction (r) and substrate saturation level (ϑ) based on Equation 12.

MEND model. The respiration data only represented the CO_2 production from the added substrate and did not include basal respiration from the native C.

First we employed Equation 14d to fit the respiration rates during the exponentially-increasing stage and the result is shown in Fig. 3a (see original data in Fig. 3 of Colores et al. [13]). The true growth yield (Y_G) was set to 0.5 according to Colores et al. [13]. There are four undetermined parameters (B_0, r_0, μ_G, α) in Equation 14d (with $m_R = \mu_G \cdot \alpha/(1-\alpha)$). We found that only the maximum specific growth rate (μ_G) could be determined with high confidence (coefficient of variation (CV) = 5%) from the exponentially-increasing respiration rates. The CVs of the other three optimized parameters (B_0, r_0, α) were as high as 55–77% (Table 1). However, the initial active microbial biomass ($B_{a0} = B_0 \times r_0$) had a lower uncertainty (CV = 20%) compared to B_0 and r_0. The above

results indicate that the exponentially-increasing respiration rates can only be used to obtain μ_G and B_{a0}.

We then conducted numerical simulations in terms of all data including both exponentially-increasing and non-exponentially-increasing respiration rates (Fig. 3b). The non-exponentially-increasing respiration rates include the lag period before the exponentially-increasing phase and the respiration at longer times after the rates cease to increase exponentially [13]. The latter phase is likely because of the substrate saturation levels (ϕ) become limiting to respiration. We used Equations 12a, 12b, 12e and the corresponding expression for CO_2 flux rate, to allow the substrate saturation level (ϕ) to change with time. Additionally, we used the ranges of μ_G determined above. We used the SCEUA (Shuffled Complex Evolution at University of Arizona) algorithm [51,52] to determine model parameters. The SCEUA is a widely used stochastic optimization algorithm for calibrating hydrological and environmental models [51].

When exponentially-increasing and non-exponentially-increasing data are included together, the CVs of all parameters (B_0, r_0, μ_G, α, K_s, β) are within 25% except β with a high CV of 76% (Table 1). The optimized μ_G values (0.030 ± 0.001 h^{-1}) are almost the same as obtained by Colores et al. [13]. Model estimates of $\alpha = 0.228\pm0.031$ indicate that the maximum specific maintenance rate of active microbes (m_R) is about 30% of μ_G and thus cannot be ignored. The initial active biomass (B_{a0}) is 0.145 ± 0.004 mg C g^{-1} soil (Table 1), which is lower than the values (0.194 ± 0.004 mg C g^{-1} soil) using the SIGR method [13]. This is likely due to the inclusion of maintenance respiration (characterized by m_R, see Equation 14d) in our model even for the exponentially-increasing stage; thus a lower B_{a0} could produce similar CO_2 flux to the case with higher B_{a0} that does not include the contributions from maintenance respiration. Our results also show that the initial active fraction (r_0) is $28.5\pm6.4\%$ and β is 0.025 ± 0.019. The magnitude of β is comparable to the estimation by Anderson & Domsch [6,7]. In addition, the half-saturation constant (K_s) was estimated as 0.275 ± 0.038 mg C g^{-1} soil, which is very close to the values derived from 16 soils by Van de Werf & Verstraete [21]. This K_s value indicates the substrate saturation level (ϕ) is higher than 0.7 before the transition from exponentially-increasing to non-exponentially-increasing phase (Fig. 3c). The changes of substrate (S), total microbial biomass (B) and active fraction (r) with time are also shown in Fig. 3c. In conclusion, the five parameters (B_0, r_0, μ_G, α, K_s) can be effectively determined using both exponentially-increasing and non-exponentially-increasing respiration rates, whereas β may also be determined but with a relatively high uncertainty (CV = 76%) than the other parameters.

Through this experimental analysis, we identified the need for isotopic data to discriminate between basal and substrate-induced respiration. We also discovered that the exponentially-increasing period due to substrate addition can be used to identify only a select set of model parameters (i.e., μ_G and B_{a0}) as also demonstrated by the method of Colores et al. [13]. These parameters, however, can be further applied to longer-term respiration experiments to enable fitting to obtain the remainder of model parameters by using our MEND model. Thus, we have found a new and unique solution to identify different parameters as a function of time, and to effectively use isotopic labeling to yield a specific set of model parameters.

Model test II: intermittent substrate supply

In order to further validate this additional physiological component in the MEND model, we also tested it against a laboratory experimental dataset with intermittent substrate supply [4]. In addition to the substrate, another limiting factor (i.e.,

Table 1. MEND model parameters values used for simulation of respiration rates due to added ^{14}C-labeled glucose in Colores et al. [13].

Parameter	Exponentially-increasing respiration*			All data[†]			Description
	Mean	SD[‡]	CV[§]	Mean	SD	CV	
B_0	0.504	0.279	55%	0.525	0.080	15%	Initial microbial biomass, (mg C g^{-1} soil)
r_0	0.394	0.263	67%	0.285	0.064	23%	Initial active fraction
μ_G	0.027	0.001	5%	0.030	0.001	3%	Maximum specific growth rate (h^{-1})
α	0.185	0.142	77%	0.228	0.031	13%	$m_R/(\mu_G+m_R)$, m_R is maximum specific maintenance rate for active microbes (h^{-1})
K_s	—	—	—	0.275	0.038	14%	Half-saturation constant for substrate (mg C g^{-1} soil)
β	—	—	—	0.025	0.019	76%	Ratio of dormant maintenance rate to m_R
Y_G	0.5	—	—	0.5	—	—	True growth yield, constant
B_{ao}	0.135	0.027	20%	0.145	0.004	3%	Initial active biomass (mg C g^{-1} soil), calculated by $B_0 \times r_0$

*Only the respiration rates during exponentially-increasing phase are used.
[†]All data including both exponentially-increasing and non-exponentially-increasing respiration.
[‡]SD: standard deviation.
[§]CV: Coefficient of variation.

oxygen, O$_2$) was included in this study. For this reason, we also introduced one more parameter (K_o: half saturation constant for O$_2$) to represent the limitation of O$_2$ on the microbial processes sketched in Fig. 1. Similar to substrates, the saturation level of O$_2$ is computed as $O_2/(O_2+K_o)$, where O_2 denotes the concentration of oxygen. The simulated oxygen concentrations by Stolpovsky et al. [4] were used as an input to our model. We used the SCEUA algorithm to determine the six model parameters in addition to the initial value for active fraction (r_0).

A summary of the seven parameters (one of them is r_0) and their fitted values is presented in Table 2. The initial active fraction (r_0) has a median of 0.925 with the 95% confidence interval (CI) of [0.628–1.000]. It means that a high r_0 is required for this experiment, but not necessary to be 1.0 set by Stolpovsky et al. [4]. The model and data are not sensitive to β since its 95% CI covers a wide range from 0.001 to 1. The reason is that the experiment only lasts for a very short time (33 h) so the influence of low metabolic rate at dormant state is insignificant.

Fig. 4 shows that the simulated total biomass (B) and substrate (S) concentrations agree very well with the observations (the coefficients of determination are 0.98 and 0.78 for biomass in Fig. 4a and substrate in Fig. 4b, respectively). Our simulation results indicate that, under limited O$_2$ between 12h and 24 h of

the experiment, the active biomass decreases and the dormant biomass increases. As a result, the active fraction (r) declines from ca. 0.9 to 0.7 (Fig. 4a). For the same period Stolpovsky et al. [4] predicted a decrease of r from 1.0 to ca. 0, which means that all active biomass becomes dormant. Although there were not adequate measurements to confirm either prediction, our predicted changes in the active fraction (r) appear to be more reasonable during such a short experimental time period. This demonstration also shows that our model is capable of producing reasonable change in total, active, and dormant microbial biomass in response to substrate supply as well as an important forcing function (O$_2$).

Conclusions

We show that the physiological state index model (Equation 1) of Panikov [33] can be improved by eliminating the assumption that the steady state active fraction (r^{ss}) approaches the substrate saturation level (ϕ^{ss}). In particular, the model of Panikov [33] indicates that no active cells become dormant under insufficient substrate, which disregards the general nature of the strategy of dormancy in microorganisms when faced with unfavorable environmental conditions [5]. Our analysis also implies that the estimate of respiration rates under sufficient substrate by the

Table 2. MEND model parameter values used for simulation of the experiment described in Fig. 3 of Stolpovsky et al. (2011).

Parameter	Fitted Value*	Initial Range	Description
m_R	0.032, [0.011–0.048]	0.001–0.1	Specific maintenance rate for active biomass (h^{-1})
α	0.099, [0.045–0.181]	0.001–0.50	$m_R/(\mu_G+m_R)$, μ_G is specific growth rate (h^{-1})
K_s	3.110, [1.387–5.652]	0.1–9.0	Half-saturation constant for substrate (mg L^{-1})
Y_G	0.573, [0.463–0.600]	0.2–0.6	Growth yield factor (–)
K_o	0.0008, [0.0007–0.001]	0.005–0.1	Half-saturation constant for dissolved oxygen (mM)
β	0.351, [0.001–1.000]	0.001–1	Ratio of dormant maintenance rate to m_R
r_0	0.925, [0.628–1.000]	0–1	Initial fraction of active biomass (–)

*Medians and 95% confidence intervals of the fitted values from 100 optimization runs, i.e., 100 different random seeds are used for the stochastic optimization algorithm.

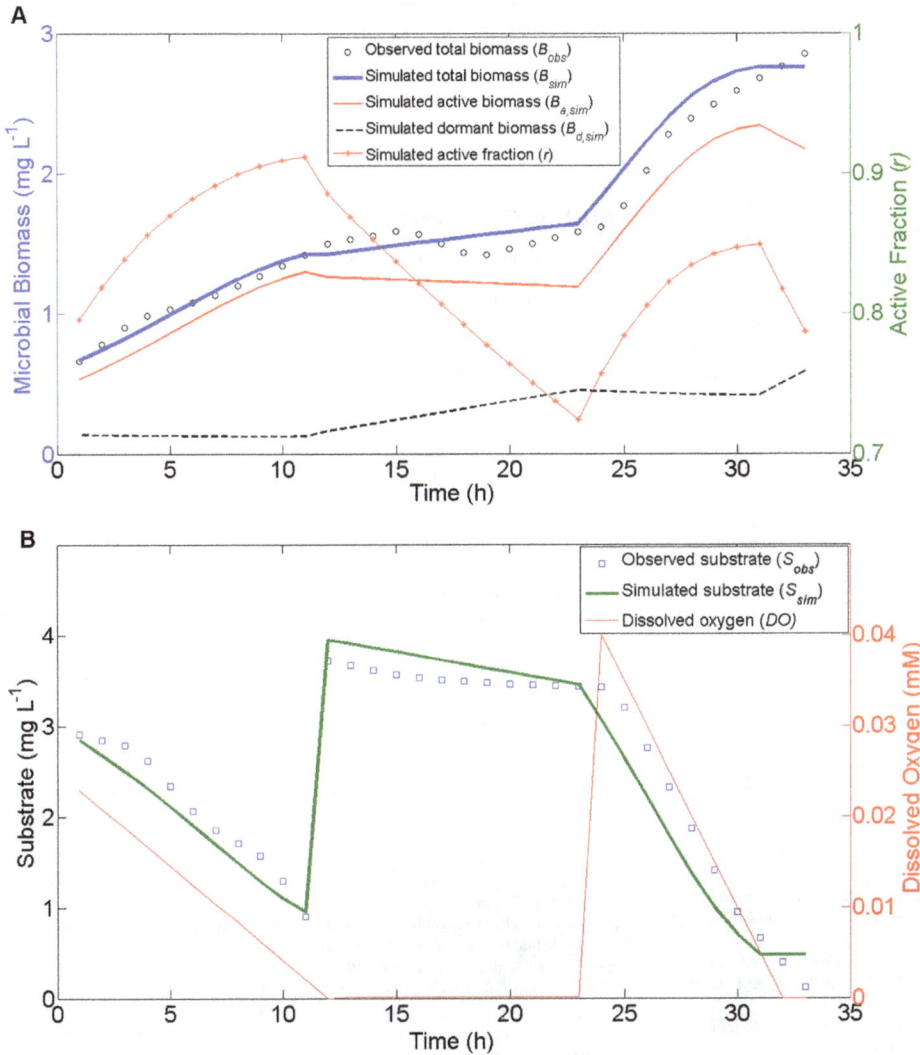

Figure 4. MEND model simulations against the experimental dataset used by Stolpovsky et al. (2011). (a) total live biomass, active and dormant biomass, and active fraction; (b) observed and simulated substrate concentration and prescribed O_2 concentration. There are three manipulations on the substrate and oxygen: (1) at time 0, the substrate (3 mg/L) and O_2 (0.025 mM) are added to the system; (2) after 12 h, the same amount of substrate is injected; (3) at 24 h, additional O_2 (0.04 mM) is injected to the system. The observed concentrations of substrate and total biomass are hourly data interpolated from the original observations in Stolpovsky et al. (2011). We scaled the substrate concentrations (with units of mM in original data) to match the magnitude of biomass concentration in units of mg/L.

physiological state index model is deficient. Pertaining to the switch function model, we argue that the switch function (θ) is also determined by the substrate (or combined with other impact factors) saturation level thus we would recommend using the switch function to modify the microbial uptake rate if the Gibbs energy change of the oxidation of substrate (ΔG) is tractable and the thermodynamic threshold (G_0) and the steepness of the step function (st) are identifiable. Based on the generally accepted assumptions summarized from existing dormancy models, we postulate a synthetic microbial physiology component to account for dormancy. Both the steady state active fraction (r^{ss}) and substrate saturation level (ϕ^{ss}) can be expressed as functions of two physiological indices: α and β. The index $\alpha = m_R/(\mu_G+m_R)$ is composed of μ_G and m_R denoting the maximum specific growth and maintenance rates, respectively, for active microbes. The index β represents the ratio of dormant to active maintenance rates. The value of r^{ss} is no less than ϕ^{ss}, and is equal only under

the condition of $\beta\rightarrow0$, where they are both identical to α. The upper bound of r^{ss} is ca. 0.8 at $\alpha\rightarrow0.5$ and $\beta\rightarrow1$. The maximum r^{ss} is ca. 0.5 if β (≤0.01) following the estimation of Anderson & Domsch [6]. It is evident that r^{ss} could be attenuated further by other limiting factors. The application of the MEND microbial physiology model to an experimental dataset with intermittent substrate supply shows satisfactory model performance (the determination coefficients are 0.98 and 0.78 for microbial biomass and substrate, respectively). The case study on the SIGR dataset indicate that the exponentially-increasing respiration rates can only be used to determine μ_G and B_{a0} (initial active biomass), while the major parameters (B_0, r_0, μ_G, α, K_s) can be effectively determined using both exponentially-increasing and non-exponentially-increasing respiration rates.

In conclusion, the microbial physiology model presented here can be incorporated into existing ecosystem models to account for dormancy in microbially-mediated processes. We have illustrated the impacts of substrate and oxygen availabilities on the physiological states through this study. Other environmental factors, such as soil temperature and soil water potential, could also be introduced into this framework to affect the transformation processes between the two microbial compositions. The changes in the physiological states of microbes could further alter the microbially-driven carbon and nutrient dynamics in ecosystems. Traditional measures of microbial biomass include the entire microbial population, even though dormancy is an important evolutionary strategy for preservation of microbial genetics and function until conditions for growth and replication improve. Parameterizing microbial ecosystem models assuming the entire population is active could therefore lead to significant errors. The approach described here provides a tractable and testable method to include dormancy as a response to external forcing.

Acknowledgments

The authors thank Dr. Sindhu Jagadamma for her helpful comments. Thanks also go to the two anonymous reviewers for their constructive comments.

Author Contributions

Conceived and designed the experiments: GW MAM. Analyzed the data: GW MAM CWS. Wrote the paper: GW MAM LG CWS. Mathematical analysis: GW LG.

References

1. Blagodatsky SA, Heinemeyer O, Richter J (2000) Estimating the active and total soil microbial biomass by kinetic respiration analysis. Biology and Fertility of Soils 32: 73–81.
2. Mason C, Hamer G, Bryers J (1986) The death and lysis of microorganisms in environmental processes. FEMS microbiology letters 39: 373–401.
3. Price PB, Sowers T (2004) Temperature dependence of metabolic rates for microbial growth, maintenance, and survival. Proceedings of the National Academy of Sciences of the United States of America 101: 4631–4636.
4. Stolpovsky K, Martinez-Lavanchy P, Heipieper HJ, Van Cappellen P, Thullner M (2011) Incorporating dormancy in dynamic microbial community models. Ecological Modelling 222: 3092–3102.
5. Lennon JT, Jones SE (2011) Microbial seed banks: the ecological and evolutionary implications of dormancy. Nature Reviews Microbiology 9: 119–130.
6. Anderson TH, Domsch KH (1985) Determination of ecophysiological maintenance carbon requirements of soil microorganisms in a dormant state. Biology and Fertility of Soils 1: 81–89.
7. Anderson TH, Domsch KH (1985) Maintenance carbon requirements of actively-metabolizing microbial populations under in situ conditions. Soil Biology and Biochemistry 17: 197–203.
8. Adam G, Duncan H (2001) Development of a sensitive and rapid method for the measurement of total microbial activity using fluorescein diacetate (FDA) in a range of soils. Soil Biology and Biochemistry 33: 943–951.
9. Jones KH, Senft JA (1985) An improved method to determine cell viability by simultaneous staining with fluorescein diacetate-propidium iodide. Journal of Histochemistry & Cytochemistry 33: 77–79.
10. Boulos L, Prévost M, Barbeau B, Coallier J, Desjardins R (1999) LIVE/DEAD® BacLight™: application of a new rapid staining method for direct enumeration of viable and total bacteria in drinking water. Journal of Microbiological Methods 37: 77–86.
11. Stocks S (2004) Mechanism and use of the commercially available viability stain, BacLight. Cytometry Part A 61: 189–195.
12. Schulz S, Peréz-de-Mora A, Engel M, Munch JC, Schloter M (2010) A comparative study of most probable number (MPN)-PCR vs. real-time-PCR for the measurement of abundance and assessment of diversity of alkB homologous genes in soil. Journal of Microbiological Methods 80: 295–298.
13. Colores GM, Schmidt SK, Fisk MC (1996) Estimating the biomass of microbial functional groups using rates of growth-related soil respiration. Soil Biology and Biochemistry 28: 1569–1577.
14. Anderson JPE, Domsch KH (1978) A physiological method for quantitative measurement of microbial biomass in soils. Soil Biology & Biochemistry 10: 215–221.
15. Jenkinson DS, Powlson DS (1976) The effects of biocidal treatments on metabolism in soil-V: A method for measuring soil biomass. Soil Biology and Biochemistry 8: 209–213.
16. Lodge DJ (1993) Nutrient cycling by fungi in wet tropical forests. In: Isaac S, editor. Aspects of Tropical Mycology: Symposium of the British Mycological Society Cambridge, UK: Cambridge University Press. pp. 37–57.
17. Yarwood S, Brewer E, Yarwood R, Lajtha K, Myrold D (2013) Soil microbe active community composition and capability of responding to litter addition after 12 years of no inputs. Applied and Environmental Microbiology 79: 1385–1392.
18. Alvarez C, Alvarez R, Grigera M, Lavado R (1998) Associations between organic matter fractions and the active soil microbial biomass. Soil Biology and Biochemistry 30: 767–773.
19. Khomutova T, Demkina T, Demkin V (2004) Estimation of the total and active microbial biomasses in buried subkurgan paleosoils of different age. Microbiology 73: 196–201.
20. Orwin KH, Wardle DA, Greenfield LG (2006) Context-dependent changes in the resistance and resilience of soil microbes to an experimental disturbance for three primary plant chronosequences. Oikos 112: 196–208.
21. Van de Werf H, Verstraete W (1987) Estimation of active soil microbial biomass by mathematical analysis of respiration curves: calibration of the test procedure. Soil Biology and Biochemistry 19: 261–265.
22. Stenström J, Svensson K, Johansson M (2001) Reversible transition between active and dormant microbial states in soil. FEMS Microbiology Ecology 36: 93–104.
23. Wang G, Post WM, Mayes MA (2013) Development of microbial-enzyme-mediated decomposition model parameters through steady-state and dynamic analyses. Ecological Applications 23: 255–272.
24. Moorhead DL, Lashermes G, Sinsabaugh RL (2012) A theoretical model of C- and N-acquiring exoenzyme activities, which balances microbial demands during decomposition. Soil Biology and Biochemistry 53: 133–141.
25. Sinsabaugh RL, Manzoni S, Moorhead DL, Richter A (2013) Carbon use efficiency of microbial communities: stoichiometry, methodology and modelling. Ecology Letters 16: 930–939.
26. Schimel JP, Weintraub MN (2003) The implications of exoenzyme activity on microbial carbon and nitrogen limitation in soil: a theoretical model. Soil Biology & Biochemistry 35: 549–563.
27. Lawrence CR, Neff JC, Schimel JP (2009) Does adding microbial mechanisms of decomposition improve soil organic matter models? A comparison of four models using data from a pulsed rewetting experiment. Soil Biology & Biochemistry 41: 1923–1934.
28. Wang G, Post WM, Mayes MA, Frerichs JT, Jagadamma S (2012) Parameter estimation for models of ligninolytic and cellulolytic enzyme kinetics. Soil Biology and Biochemistry 48: 28–38.
29. Todd-Brown KE, Hopkins FM, Kivlin SN, Talbot JM, Allison SD (2012) A framework for representing microbial decomposition in coupled climate models. Biogeochemistry 109: 19–33.
30. Treseder KK, Balser TC, Bradford MA, Brodie EL, Dubinsky EA, et al. (2012) Integrating microbial ecology into ecosystem models: challenges and priorities. Biogeochemistry 109: 7–18.
31. Wieder WR, Bonan GB, Allison SD (2013) Global soil carbon projections are improved by modelling microbial processes. Nature Clim Change 3: 909–912.
32. Konopka A (1999) Theoretical analysis of the starvation response under substrate pulses. Microbial Ecology 38: 321–329.
33. Panikov NS (1996) Mechanistic mathematical models of microbial growth in bioreactors and in natural soils: explanation of complex phenomena. Mathematics and Computers in Simulation 42: 179–186.
34. Blagodatsky S, Richter O (1998) Microbial growth in soil and nitrogen turnover: a theoretical model considering the activity state of microorganisms. Soil Biology and Biochemistry 30: 1743–1755.
35. Panikov NS, Sizova MV (1996) A kinetic method for estimating the biomass of microbial functional groups in soil. Journal of Microbiological Methods 24: 219–230.
36. Buerger S, Spoering A, Gavrish E, Leslin C, Ling L, et al. (2012) Microbial scout hypothesis, stochastic exit from dormancy, and the nature of slow growers. Applied and Environmental Microbiology 78: 3221–3228.
37. Hunt HW (1977) A simulation model for decomposition in grasslands. Ecology 58: 469–484.
38. Gignoux J, House J, Hall D, Masse D, Nacro HB, et al. (2001) Design and test of a generic cohort model of soil organic matter decomposition: the SOMKO model. Global Ecology and Biogeography 10: 639–660.
39. Ayati BP (2012) Microbial dormancy in batch cultures as a function of substrate-dependent mortality. Journal of Theoretical Biology 293: 34–40.

40. Jones SE, Lennon JT (2010) Dormancy contributes to the maintenance of microbial diversity. Proceedings of the National Academy of Sciences 107: 5881–5886.

41. Bär M, Hardenberg J, Meron E, Provenzale A (2002) Modelling the survival of bacteria in drylands: the advantage of being dormant. Proceedings of the Royal Society of London Series B: Biological Sciences 269: 937–942.

42. Locey KJ (2010) Synthesizing traditional biogeography with microbial ecology: the importance of dormancy. Journal of Biogeography 37: 1835–1841.

43. Wang G, Post WM (2013) A note on the reverse Michaelis–Menten kinetics. Soil Biology and Biochemistry 57: 946–949.

44. Wirtz KW (2003) Control of biogeochemical cycling by mobility and metabolic strategies of microbes in the sediments: an integrated model study. FEMS Microbiology Ecology 46: 295–306.

45. Panikov NS (1995) Microbial growth kinetics. London, UK: Chapman & Hall. 378 p.

46. Pirt SJ (1965) Maintenance energy of bateria in growing cultures. Proceedings of the Royal Society of London Series B-Biological Sciences 163: 224–231.

47. Wang G, Post WM (2012) A theoretical reassessment of microbial maintenance and implications for microbial ecology modeling. FEMS Microbiology Ecology 81: 610–617.

48. Hoehler TM, Jorgensen BB (2013) Microbial life under extreme energy limitation. Nat Rev Micro 11: 83–94.

49. Beeftink HH, Van der Heijden RTJM, Heijnen JJ (1990) Maintenance requirements: energy supply from simultaneous endogenous respiration and substrate consumption. FEMS microbiology letters 73: 203–209.

50. Blagodatsky S, Yevdokimov I, Larionova A, Richter J (1998) Microbial growth in soil and nitrogen turnover: model calibration with laboratory data. Soil Biology and Biochemistry 30: 1757–1764.

51. Duan QY, Sorooshian S, Gupta V (1992) Effective and efficient global optimization for conceptual rainfall-runoff models. Water Resources Research 28: 1015–1031.

52. Wang G, Xia J, Chen J (2009) Quantification of effects of climate variations and human activities on runoff by a monthly water balance model: A case study of the Chaobai River basin in northern China. Water Resources Research 45: W00A11.

6

Paleohistology and Lifestyle Inferences of a Dyrosaurid (Archosauria: Crocodylomorpha) from Paraíba Basin (Northeastern Brazil)

Rafael César Lima Pedroso de Andrade[1]*, Juliana Manso Sayão[2]

1 Graduate Student/Programa de Pós-Graduação (CTG), Universidade Federal de Pernambuco, Recife, Brazil, **2** Centro Acadêmico de Vitória, Universidade Federal de Pernambuco, Bela Vista, Vitória de Santo Antão, Pernambuco, Brazil

Abstract

Among the few vertebrates that survived the mass extinction event documented at the Cretaceous–Paleocene boundary are dyrosaurid crocodylomorphs. Surprisingly, there is little information regarding the bone histology of dyrosaurids, despite their relatively common occurrence in the fossil record, and the potential to gain insight about their biology and lifestyle. We provide the first description of the long bone histology of the dyrosaurids. Specimens were collected from the Maria Farinha Formation, in the Paraíba Basin of northeast Brazil. Thin sections of a right femur and left tibia were made. In the left tibia, the cortex consists of lamellar-zonal bone with five lines of arrested growth (LAGs), spaced ~300 μm apart. The tibia contains a small to medium-sized organized vascular network of both simple vascular canals and primary osteons that decrease in density periostially. The femur exhibits a similar histological pattern overall but has double-LAGs, and an EFS layer (the latter is rare in living crocodylians). Secondary osteons occur in the deep cortex near and inside the spongiosa as a result of remodeling in both bones. This tissue pattern is fairly common among slow-growing animals. These specimens were a sub-adult and a senescent. Patterns in the distribution of bone consistent with osteosclerosis suggest that these animals probably hada fast-swimming ecology. Although these results are consistent with the histology in anatomically convergent taxa, it will be necessary to make additional sections from the mid-diaphysis in order to assign their ecology.

Editor: Christof Markus Aegerter, University of Zurich, Switzerland

Funding: This research is supported by Conselho Nacional Pq (Proc. N. 401787/2010-9 grant to J.M.S.) and Coordenação de Aperfeiçoamento de Pessoal de Nível Superior (CAPES fellowship to R.C.L.P.A.). The funders had no role in study design, data collection and analysis, decision to publish, or preparation of the manuscript.

Competing Interests: The authors have declared that no competing interests exist.

* Email: rafael-clpa1@hotmail.com

Introduction

Among the few vertebrates that survived the mass extinction event documented at the Cretaceous–Palaeogene boundary are the dyrosaurids [1], which represent an extinct lineage of Neosuchia. Dyrosaurids are found in transitional marine sediments from the Late Cretaceous to Lower Eocene [2], and they exhibit the primary feature of a long snout [3], [4]. The family was named by Giuseppe de Stefano in 1903 [5] for the genus *Dyrosaurus*, referring to the locality where the holotype was found, Djebel Dyr, Algeria.

In South America, dyrosaurids were previously known only from fragmentary material [6], [7], [8], [9], several cranial elements briefly mentioned in the literature [10], and by the description of an almost complete skull and part of the jaw, ulna, cervical and caudal vertebrae, ribs, dermal scutes and isolated teeth belonging to *Guarinisuchus munizi* [11]. The holotype of *G. munizi* was collected from the Maria Farinha Formation (Paleocene) in the Poty Quarry, which is located close to Recife in northeastern Brazil (Figure 1) [11]. The bones at this site exhibit excellent external preservation in three dimensions, and preserve the internal microstructures as well.

Examination of the microstructure of bones enhance the simple morphological description of the specimens; this approach is a method to obtain certain types of information about the biology of extinct animals, for example, the presence of adaptations to a lifestyle, growth rates, and indications of its ontogenetic stage [12], [13], [14]. The bone structure is composed of mineralized connective tissue (bone tissue) that was produced by hydroxyapatite deposition [15], crystalline calcium phosphate and, in the inner parts, by osteocytes and numerous channels in the form of blood and lymphatic vessels. After death, the organic components, including the cells and blood vessels, decomposed, while the inorganic portion became fossilized, thereby maintaining the bones microstructures and preserving the shape of the decomposed components [16].

Additionally, histological examination has established structural modifications of bone in tetrapod taxa secondarily adapted to life in water, including both fossil and recent taxa [17]. Within this large available data set, the only published thin sections of dyrosaurid bones are of the vertebrae and skull of *Dyrosaurus phosphaticus* [18]. Thus, the paleohistology of dyrosaurids remains largely unexplored. In this study, we present the first histological description of dyrosaurid limb bones, belonging to specimens from the Poty Quarry, Maria Farinha Formation.

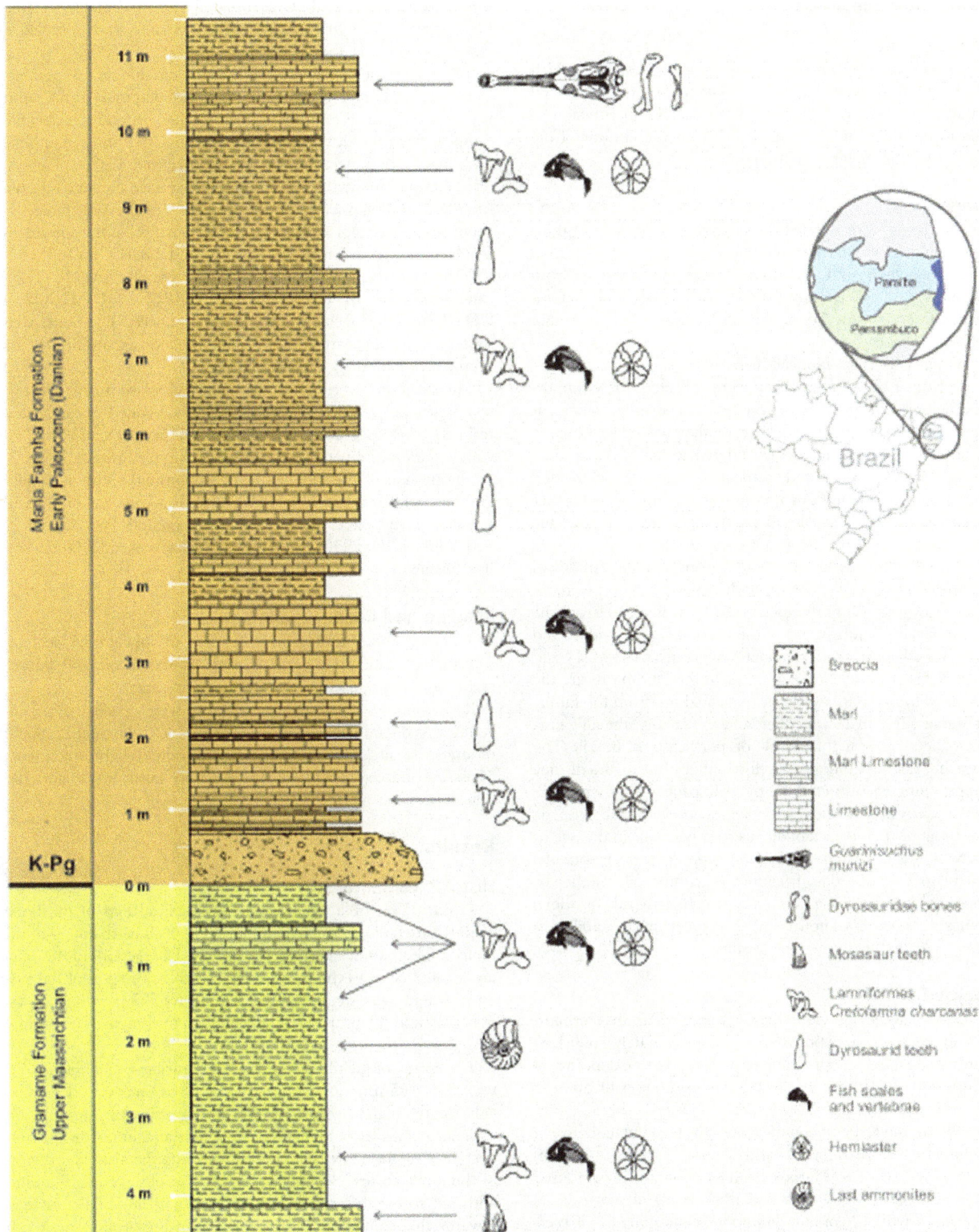

Figure 1. Location of the Paraíba Basin and stratigraphic section of the Maria Farinha Formation at the Poty Quarry, located in the state of Pernambuco, close to the city of Recife, in northeast Brazil (Redrawn from Barbosa et al. 2008 [11]). This section indicates each level of composition (marls, marly-limestones and limestones) and their fossil contents. The position where the dyrosaurid bones were found is indicated above.

Materials and Methods

Ethics Statement

No permits were required for the described study, which complied with all relevant regulations. The author JMS collected these materials with Dr. Antônio Barbosa. No permission is required to use them in research since JMS is the curator of the collection were the material is housed.

Specimens

Two specimens were selected for sampling: the proximal third of a right femur (CAV 0010-V) and the distal portion of a left tibia including the metaphysis (CAV 0011-V). Both specimens are from the paleontological collection of the Centro Acadêmico de Vitória (Universidade Federal de Pernambuco), Vitória de Santo Antão, Pernambuco, Brazil.

The femur and tibia sampled here were found at the type locality of *Guarinisuchus munizi*. They were collected a few months after the holotype, from the same horizon but in rolled blocks. It was not possible to identify whether or not they belong to the same individual, so each element was considered to belong to distinct specimens. They were found associated with an elongated mandible clearly belonging to a dyrosaurid and also near to other isolated appendicular elements. This strongly suggests that the bones sampled here also belong to a dyrosaurid.

The overall crocodyliform record from South American Paleocene deposits is rather slim in both the number of specimens and taxa recovered. Only three crocodyliform lineages cross the K-Pg boundary in South America: the marine Dyrosauridae, and two true crocodylians, the semiaquatic alligatoroids [19–21] and the terrestrial sebecosuchians (e.g. [22]). In the Paraíba Basin, the only crocodyliforms that have been recovered are from the family Dyrosauridae [23]. In fact, neither sebecosuchians nor alligatoroids have been reported from all of northeastern Brazil. The Sebecosuchia have a Gondwanan distribution, but in Brazil they are almost entirely restricted to the Mesozoic [24], with one reported occurrence in the Paleocene deposits of São José de Itaboraí Basin [25]. The lack of other crocodyliforms in the age or proximity of Maria Farinha Formation supports their taxonomic assignment to Dyrosauridae. It cannot be confirmed whether or not they belong to *G. munizi*, for which femoral or tibial autapomorphies are not known, but this is a strong possibility as they were collected from the type locality.

Geological Setting

The Paraíba basin, previously known as the Paraíba-Pernambuco Basin, occupies an exposed area of approximately 7600 km^2 and a submerged area of approximately 31,400 km^2, extending on the continental shelf down to the bathymetric quota of 3000 m [29].

In this basin, the K-Pg boundary represents the contact between the Gramame (Maastrichtian) and Maria Farinha (Danian) Formations (Fig. 1). The Maria Farinha Formation is composed of limestones, marly limestones and thick levels of marls in its lower portion, while dolomitic limestones containing fossil reefs and lagunal reefs characterize its upper portion [30], [31]. This formation exhibits the regressive characteristics of high- to low-energy oscillations.

Slide Preparation

For this analysis, the distal metaphysis of the left tibia and the proximal portion of the right femoral diaphysis were sampled. Typically, the mid-diaphysis is sampled in paleohistological studies, as this region preserves more cortical bone and more of the growth record [26]. Unfortunately, the mid-diaphyses were not preserved in either bone, so the elements were sampled as close as possible to the midshaft.

To prepare the histological slides, a 0.5 cm sample was obtained from each specimen (femur and tibia). The diaphysis of the femur was sampled just distal to the fourth trochanter, and the tibia was sampled from the distal metaphysis (Figure 2). Prior to sampling, both bones had been mechanically prepared with the use of airscribes and manual tools. No external surface damage resulting from preparation was observed in neither of the specimens. To preserve the external morphological information of the specimens, molds in silicon rubber (RTV CAL/N - ULTRALUB QUÍMICA LTDA, São Paulo, Brazil) and resin casts (RESAPOL T-208 catalyzed with BUTANOX M50 - IBEX QUIMICOS E COMPÓSITOS, Recife, Brazil) were produced. The bones were subsequently measured and photographed according to the protocol proposed by Lamm (2013) [27].

Thin sections were produced using standard fossil histology techniques [27], [28]. The samples were embedded in epoxy clear resin RESAPOL T-208 catalysed with BUTANOX M50 and cut with a diamond-tipped blade on a saw (multiple brands). Next, the mounting-side of the sections were wet-ground using a metallographic polishing machine (AROPOL-E, Arotec LTDA) with Arotec abrasive papers of increasing grit size (60/P60, 120/P120, 320/P400, 1200/P2500) until a final thickness of ~30–60 microns was reached.

Imaging and Image Analysis

To observe the histological structures, an optical microscope in transmitted light mode with parallel/crossed nicols and fluorescence filters were used to enhance birefringence.

Representative histological images were taken using an AxioCam digital sight camera (Zeiss Inc., Barcelona, Spain) mounted to an Axio Imager.M2 transmitted light microscope (Zeiss Inc. Barcelona, Spain). Images were taken at 5× and 10× total magnification.

Results

Histological features

Tibia. The tibia preserves only the distal end of the bone, from the metaphysis to the epiphyses (Fig. 2). The diameter of this bone at the point of sampling is 2.94 cm. The sections of this bone are located close to the metaphysis, which is a region of the long bone where periosteal cortices are always thinner than in the midshaft, and the proportion of spongiosa to compact bone is high (Fig. 3A).

The cortex of the tibia, composed of lamellar-zonal bone (LZB), was 1,119.65 μm in width (Fig. 3B). Osteocyte density is consistently high throughout the cortex, especially around the vascular canals. It is higher in the center and inner regions, but decreases somewhat moving periosteally into the avascular regions of the outer cortex. Secondary osteons are observed in the inner and mid cortex, but this is only local remodeling. They increase towards the spongiosa, which shows many resorption rooms. In the spongiosa the secondary osteons occasionally anastomose with each other.

In the primary bone tissue, the vascular network consists of simple vascular canals and primary osteons that vary in their diameter (from 52.55 μm to 176.77 μm). The orientation of the vascular network is predominantly longitudinal and the canals do not show anastomoses. Five lines of arrested growth are visible (LAGs) (Fig. 4B, C; note black arrows). The zones between them are slightly variable in width around the section, but are generally

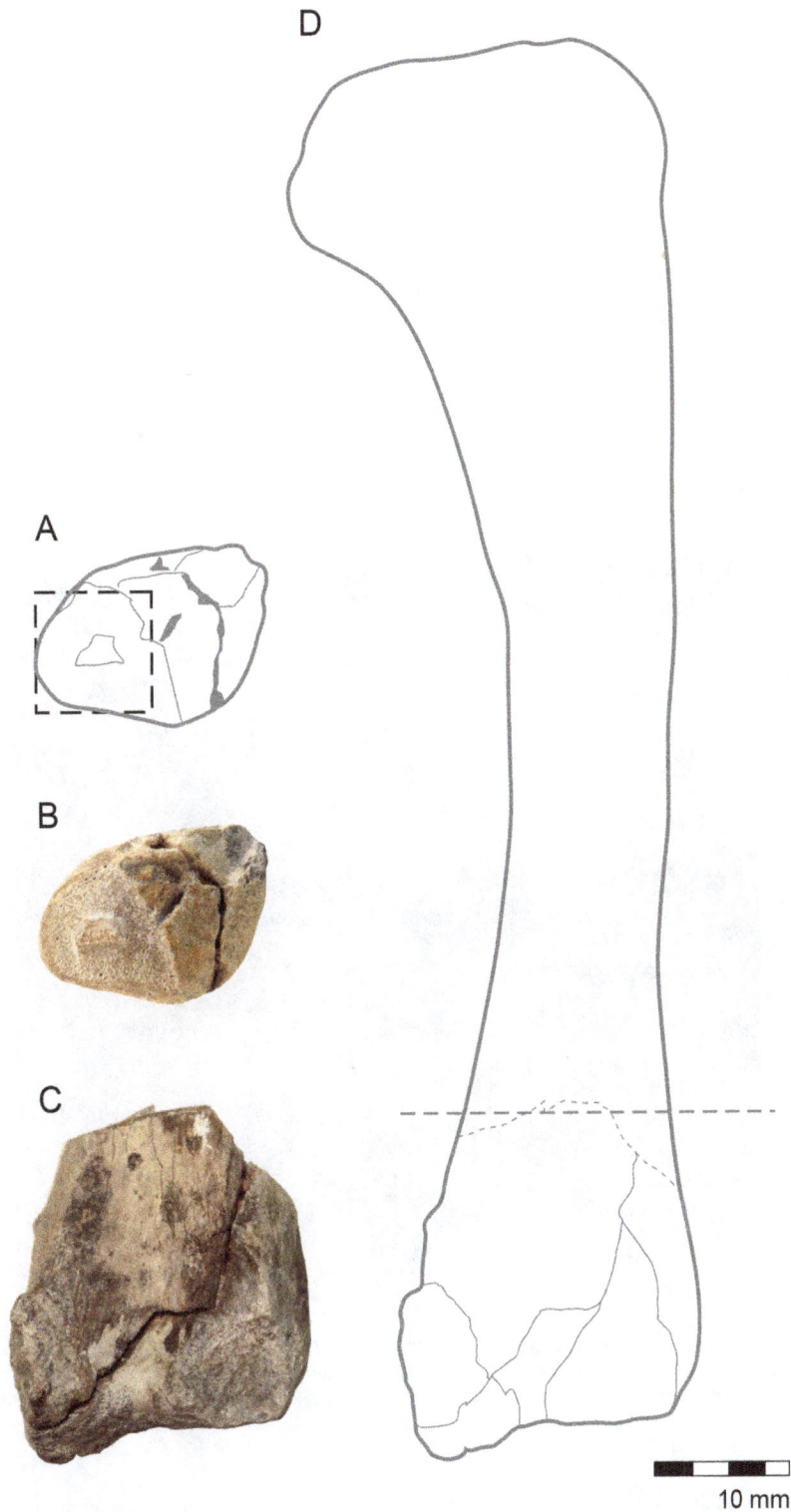

Figure 2. Aspects of external and internal anatomy in the left tibia, CAV 0011-V. (A) The dotted square represents the region where the analysis and images have been made. (B) A view of the section made for sample collection, which corresponds to a distal metaphiso-epiphyseal portion where the cortex is thinner than in the diaphyseal portion. (C) Posterior view of the distal portion of the tibia, all that was preserved in this element. (D) Reconstruction of the complete tibia; the dotted line indicates where the cut was made for the sample collection.

approximately 300 μm (Fig. 4C). Thus, this region of the bone characterizes 5 cycles (annuli-zones) during the life of the animal. However, sampling at any point except the mid-diaphysis can underestimate the true number of LAGs [32], so it is possible that some cycles may have been obscured by remodeling and resorption.

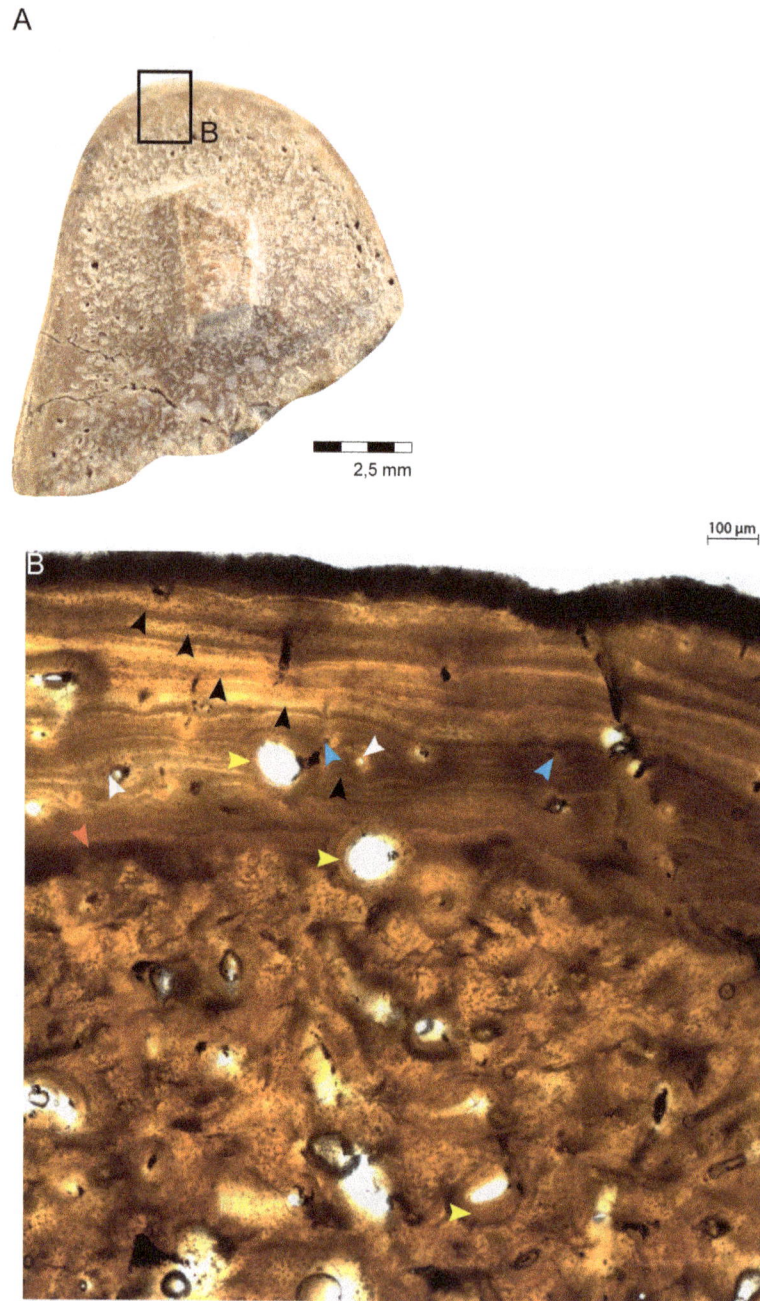

Figure 3. General microstructure anatomy of the tibia CAV 0011-V. (A) View of the complete transect. Black box indicates, respectively, where the related image B was taken. (B) View of the compact periosteal cortex and spongiosa (note the relation of the two tissues). Blue arrows-primary osteons; white arrows- primary osteons; yellow arrows- secondary osteons; black arrows- LAGs and red arrow- transitional area from cortex to remodeled bone.

Femur.   The proximal third of the femur is preserved (Fig. 5B). The diameter of the bone at the point of sampling is 3.10 cm, and the cortex is 1,997.49 μm thick. The sampled section of this bone is located within the diaphysis, closer to the midshaft than the tibia. In this region, the proportion of compact to cancellous bone is always higher than it is in either the metaphysis or the epiphysis (Fig. 5A).

The femur exhibits a lamellar-zonal histological pattern that is similar to that of the tibia, but has some particularities. Most notably, the femur shows an outermost layer of closely-spaced lines, the external fundamental system (EFS), that characterizes

the end of bone growth. The femur also exhibits a different pattern of cortical LAGs in comparison to the tibia. Five lines can be observed, but four of them (two pairs) are double LAGs. One pair is visible in the inner cortex. Moving periosteally, there is a single LAG at midcortex and then another double LAG in the outer cortex. The variation in the zones between LAGs is greater in the femur than in the tibia (Fig. 6B). In the tibia the distance between the LAGs is about 300 μm, whereas in the femur the distance between the closest LAGs are 20 μm and the most distant is 194. In the EFS, there are numerous, closely-spaced LAGs, none of which are very distinct from each other. This prevents an exact

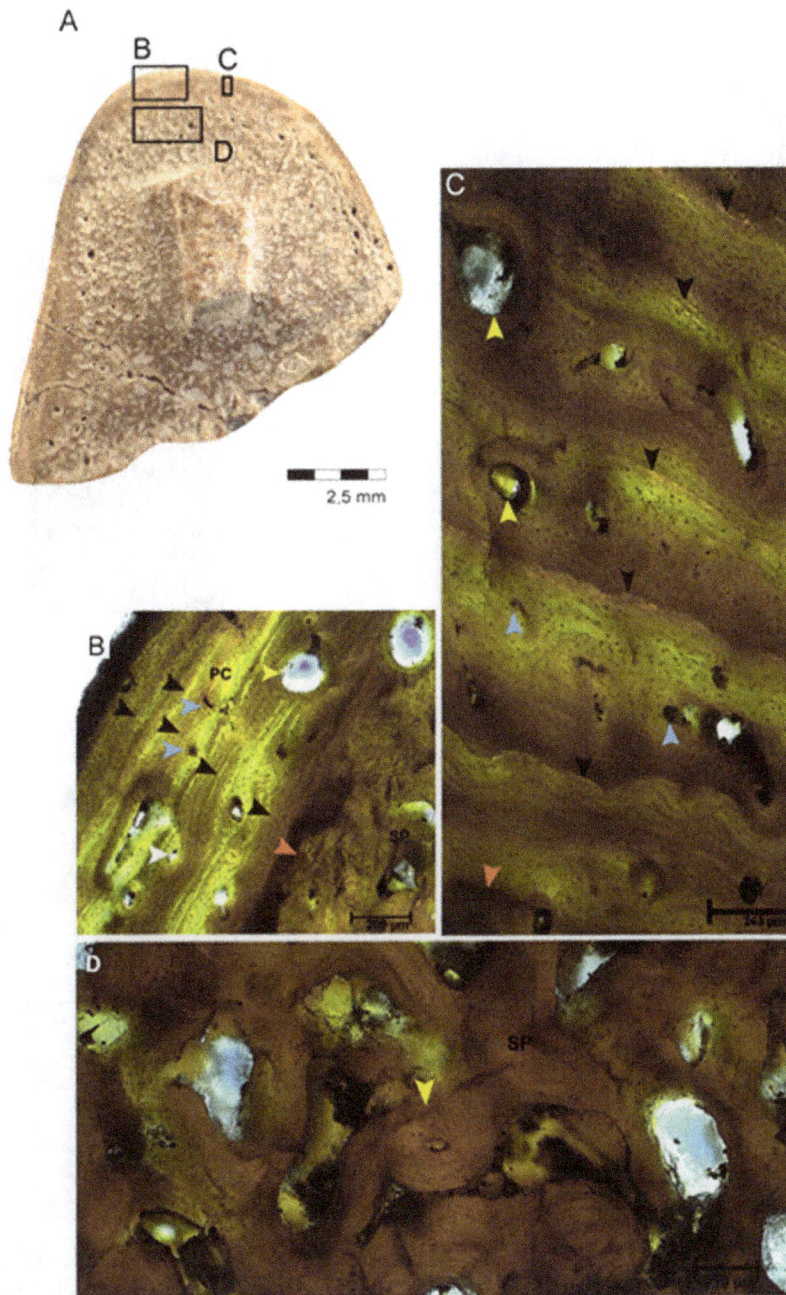

Figure 4. Histological characteristics of the tibia CAV 0011-V. (A) View of the complete transect. Black boxes indicate, respectively, where the related images were taken. (B) View of the cortex exhibiting lamellar-zonal bone tissue with circumferentially organized simple vascular canals (blue arrows), primary osteons (white arrow), secondary osteons (yellow arrow), LAGs (black arrows) and transicional area (red arrow). Periosteal surface to upper left. (C) middle-cortical area showing the five lines of arrested growth (black arrows), secondary osteons (yellow arrows), vascular canals (blue arrows) and contact between the compacta and the spongiosa (red arrow). Periosteal surface to top. The secondary osteons can be found in its inner portion, and the LAGs are indicated by the black arrows. (D) An extended view of the spongiosa where the erosion rooms are visible and infilled by calcite. Rough surfaces of the trabeculae indicate they were formed by resorption. The trabeculae themselves show signs of secondary remodeling, including secondary osteons (yellow arrow). Abbreviations: PC, periosteal compact cortex; SO, secondary osteons; SP, spongiosa.

number from being determined. As in the tibia, the femur is comprised of LZB, and osteocyte density is consistently high throughout the cortex, especially around the vascular canals. It is higher in the center and inner regions, but decreases moving periosteally into the avascular regions of the outer cortex (Fig. 6B).

The vascular canals vary from 22.12 μm to 85.92 μm in diameter. The orientation of the vascular network is predomi-

nantly longitudinal, but unlike the tibia there are some circumferential anastomoses. Secondary osteons are most common in the innermost cortex near the spongiosa, but decrease in density periosteally. The inner and mid cortex show only local remodeling, Secondary osteons diameter ranges, on average, from 150.95 μm to 292 μm. In the spongiosa, the secondary osteons occasionally anastomoses with each other.

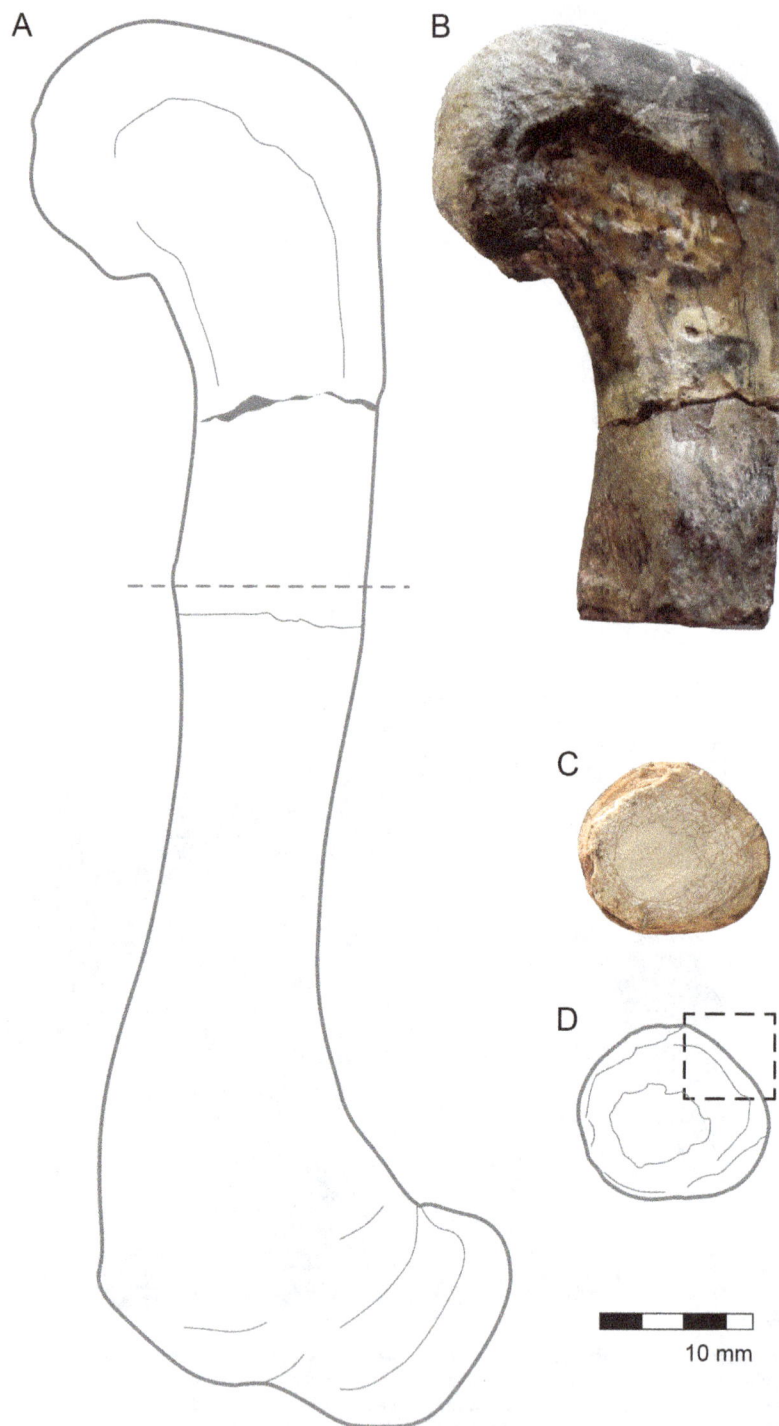

Figure 5. Aspects of the external and internal anatomy in the right femur CAV 0010-V. (A) Drawn reconstruction of the complete femur and the dotted line indicates where the cut was made for the sample collection. (B) Frontal view of the femur proximal portion. (C) A view of the section fabricated for sample collection, which corresponds to a diaphyseal portion where the cortex reaches its region of higher thickness. Note that both the cortex and the spongiosa are of equivalent quantity. (D) The dotted, square, in the draw, represents the region where the analysis and images have been made.

Discussion and Functional Inferences

The morphological specializations of bones in aquatic tetrapods are among the most striking examples of adaptive convergence [33]. Beyond the modification involving the external morphology of their bones, these organisms also share another convergence that is perhaps more striking: the structural specialization exhibited on the inner organization and histological features of their skeletons [34]. These specializations are sufficiently common among the various extant and extinct tetrapod taxa secondarily

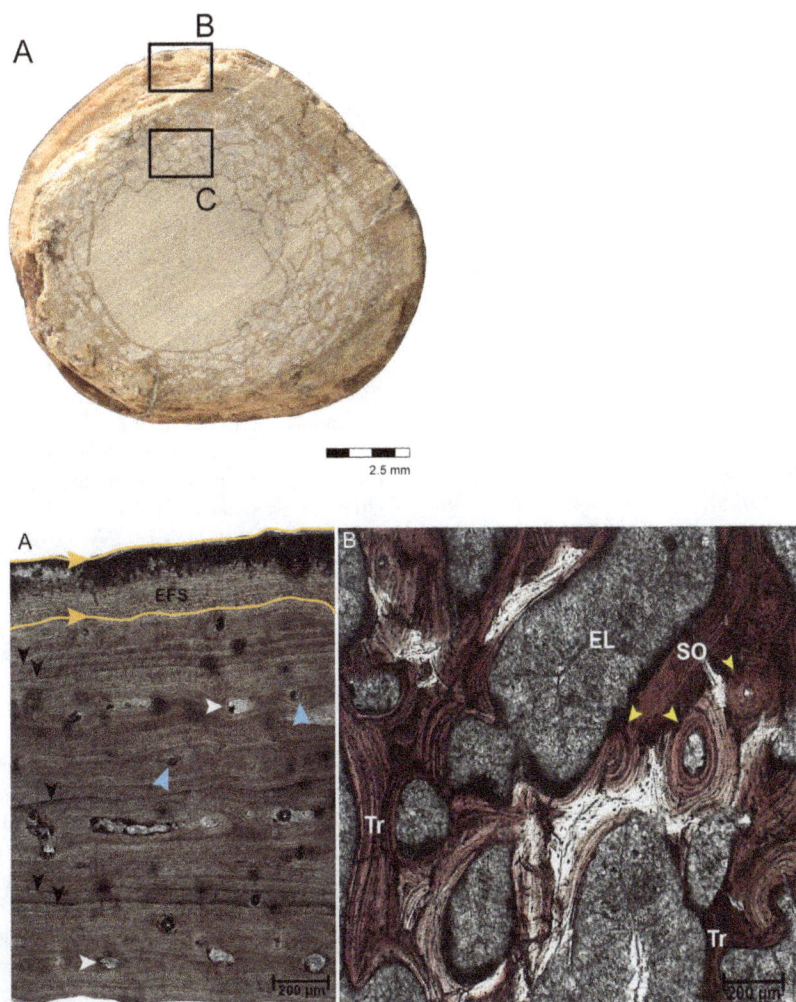

Figure 6. Histological characteristics of the femur, CAV 0010-V. (A) View of the complete transect. Black boxes indicate, respectively, where the related images were taken. (B) A view of the cortex with black arrows indicating one double LAG on the top a single one in the middle and double LAG is also indicating in the bottom and white arrows indicating the primary osteons in this area. The femur exhibits LZB tissue (Lamelar-Zonal Bone) with annuli between the LAGs, along with deposition of the external fundamental system (EFS- delimited by orange lines). (C) Significantly remodeled spongiosa of the femur supported by the presence of secondary osteons and trabeculae. The erosion lacuna is filled with minerals. TR - trabecula; SO - secondary osteons; EL - erosion lacuna.

adapted to an aquatic life that they seem to represent a necessary and unavoidable consequence of this adaptation [35], [36], [17].

The histological pattern (LZB) observed in the femur and tibia consists of primary periosteal deposits composed of bone tissue with simple vascular canals and primary osteons. The presence of cyclic growth marks are conspicuous, partly caused by endogenous physiological rhythms. These rhythms are synchronized and amplified by the seasonal variations in the environment, such as temperature, light, hygrometry, or food availability, indicators of the circannual periodicity of the growth cycles [37], [38], [39]. These indicators are represented by a phase of rapid growth and a subsequent phase of slow growth or temporary growth cessation occurring each year.

The annual periodicity of LAGs has been established empirically in the long bones of living crocodylians. Osteoderms of Nile crocodiles have been shown to be deposited in annual cycles, as well. Juveniles injected with tetracycline were shown to deposit zones during the hot season, whereas annuli corresponded to the cool seasons. Therefore, the presence of one zone and one annulus marks the passage of one year in Nile crocodiles [39]. Annual

LAGs are common in many living and extinct crocodyliforms, for example, in the limbs of the marine Thalattosuchia [17]. Therefore, we infer these cycles are annual in the dyrosaurid elements examined here.

The femur preserves three annual cycles: the two endosteal and periosteal double-LAGs, and a midcortical single LAG. Double LAGs, which appear as two very closely adjacent twin lines. When this pattern is in the sample, the question of two growth cycles arises; however, these twin LAGs may be counted as a single one representing one year [40]. In the EFS, LAGs can be observed but are closely-spaced and difficult to count. The presence of an EFS indicates that this animal was more or less fully grown at time of death, having completed the active growth phase. We cannot definitively conclude that this is the real number of LAGs, because the midshaft was not preserved, but it has been sampled very close. These data provide more complete information when compared to the tibia which was sampled further from the mid-diaphysis. The tibia showed 5 LAGs and no evidence of double-LAGs, indicating that these cycles represent five distinct moments of slow growth (annuli-zone). These analyses were, similarly, used by Woodward et al., (2013) [41] when they sampled American alligator and found evidence of slow growth and EFS in these captive animals. EFS is a rare structure in crocodilians but it has been found in several Triassic pseudosuchians [41]. The EFS, here reported for the first time in the Dyrosauridae, support the hypothesis that determinate growth may be the rule rather than the exception for sauropsids [41].

It is assumed in this report that these crocodyliforms were not juveniles. Based on the extent of the trabecular bone and the limited number of secondary osteons in the cortex, we conclude that the tibia (CAV 0010-V) belonged to a young adult. The femur (CAV 0011-V) was likely a senescent adult based on the observations of the spongiosa/remodeling processes and the presence of secondary osteons in the cortex. The deposition of EFS layer, in the femur, is also a signal of the end of active bone growth and the onset of skeletal senescence [42]. This indicates that this animal had reached its upper growth asymptote at time of the death [42]. The absence of an EFS in the tibia suggests that the upper growth asymptote had not been reached and the specimen is still in active growth [43]. This is supported by the fact that the LAGs are preserved around the section, and that these animals did not have their cortex significantly remodeled compared to a typical cortex from an older individual, for example in the femur. The blood supply, when present, is arranged in the concentric layers of the longitudinal canals with a certain rate of remodeling that is thought to be responsible for the recycling of fixed calcium in response to the general physiologic needs of the organism. This recycling of calcium is revealed by the presence of secondary osteons and erosion lacunae. This basic type of tissue is by no means exclusive to the Thalattosuchia and Dyrosaridae; such tissue is also encountered in numerous extant and extinct poikilothermic tetrapods, including stegocephalians [44], [45] and chelonians [46], [47].

In *Dyrosaurus phospaticus*, the histological structure of the cortex is also characterized by the presence of LAGs [18]. The structure suggests a cyclic pattern of growth similar to the specimens described above. However, the spongiosa of the vertebral center in *Dyrosaurus phospaticus*, compared to an extant taxon, (e.g. *Crocodylus niloticus)*, exhibits more cavities [18]. Additionally, the bone trabeculae of *D. phospaticus* are thinner compared to *C. niloticus*. Limb bones experience different growth and biomechanical patterns compared to vertebrae and skull bones. In limb bones, such observations and comparisons also characterize an adaptation to an aquatic life. Although this lifestyle has been proposed for dyrosaurs, these are the only specimens sampled for histological studies to date [18].

The histological and physiological analysis of dyrosaurids in this study is similar to those of another group of marine crocodiles, the Thalattosuchia. In this clade, the histology is largely similar to that of CAV 0010-V and of *Dyrosaurus phospaticus* [18]. Two types of structural organization of bone occur in aquatic tetrapods: osteoporotic [48] or pachyostotic. Osteoporotic bone has a very porous/spongy inner structure and is characterized by a loss of bone through accelerated resorption rates. An animal with osteoporotic bones would be better suited for faster swimming [48] because the reduction of bone tissue increases the animal's maneuverability in water. This pattern is exemplified by extant cetaceans and marine turtles, which swim swiftly [48], and also occurs in ichthyosaurs [48–50].

Pachyostotic bone involves a local or general increase in skeletal mass and can be caused by three mechanisms: osteosclerosis (inner compaction of bone), pachyostosis (hyperplasy of compact cortices), or a third mechanism that combines the two previous ones, known as pachyosteosclerosis [17], [51]. The resulting increase in skeletal mass is considered to play the functional role of ballast for buoyancy control and hydrostatic regulation of body trim [51,52].

According to Houssaye (2009) [52], true pachyostosis (cortical hyperplasy) is observed in tetrapods secondarily adapted to life in water and those that live in shallow marine environments [51,52]. Osteosclerosis is observed in fully aquatic forms only. Coastal forms (organisms that are poorly adapted to rapid, sustained swimming for anatomical and/or physiological reasons), exhibit both osteosclerotic and pachyostotic tissue [53]. Truly pachyostotic taxa include Tangasaurus and Hovasaurus [52,54], pachypleurosaurs [52], nothosaurs [52,55], and Ophidiomorpha [52,56]. Taxa that increase bone mass by osteosclerosis include *Claudiosaurus*, some derived mosasauroids and *Placodus* [52].

Notably, none of the marine or aquatic crocodylomorphs previously examined (Crocodylus, Thalattoschians, Pholidosauridae and Dyrosauridae seem to display true pachyostosis [52], although osteosclerosis is common in Thalattosuchia [17].

The tibial metaphysis (CAV 0010-V) has very thin compact cortical bone and extensive trabeculae filling most of the bone in cross-section. Although this pattern is observed in osteoporotic bone, it is also typical of long bone metaphyses, so the ecology cannot be inferred from this section (Fig. 4A).

In the proximal diaphysis of the femur (CAV 0011-V), the cortical compacta is much thinner compared to mid-diaphyseal sections of extant crocodylians [41]. Although this region is naturally expected to display a slightly thinner cortex, it is surprisingly thin despite being sampled much closer the midshaft. Additionally, the trabeculae here are extensive and fill most of the medullary cavity, despite being close to the mid-diaphysis. These trabeculae extend throughout the area normally filled by compact bone, leave a clear medullary cavity (Fig. 6A), and were clearly formed by the resorption and remodeling of primary tissues (Figure 6C). This is consistent with osteoporosis, suggesting a fast-swimming ecology for this animal. There is also some lamellar thickening of the trabeculae in the femur, which may result from the normal "finishing off" of trabeculae during remodeling, but is also consistent with osteosclerosis. If this is the case, osteosclerosis is minimal in these dyrosaurids. However, as osteosclerosis is associated with a fully aquatic lifestyle, this is not incompatible with a fast-swimming ecology.

The histological, ecological, and physiological inferences of the dyrosaurids examined in this study are similar to those of another group of marine crocodiles, the Thalattosuchia and to descriptions of other skeletal elements of *Dyrosaurus phospaticus* [18]. Among the thalattosuchians, observations of the femoral microstructure are especially similar to that described for the Teleosauridae, a group that is clearly aquatic based on morphology and histology, but were not as well adapted as the obligatorily marine animals, such as mosasaurs, elasmosaurs, and metriorhynchid thalattosuchians [33], [17]. In general, dyrosaurids have been reconstructed as being shallow, near-shore marine animals utilizing axial swimming typical of extant crocodylians, with perhaps great tail undulatory frequency and more powerful forward thrust generated by expanded muscles of the tail [57]. These hypotheses are consistent with the histological pattern found here. Therefore, we suggest an aquatic marine condition for the dyrosaurids from the Paraíba Basin, but not one with specialized adaptation for fully marine life.

Although these results are consistent with the histology in anatomically convergent taxa, it will be necessary to make additional sections from the mid-diaphysis in order to confirm that these animals exhibit osteoporosis or osteosclerosis and assign their ecology with more confidence [41,48].

Conclusion

The marine habit was already proposed to the Dyrosauridae. It was based mainly in evidences of body modifications, and for the marine paleoenvironment deposits where the remains are found. The microstructural composition with patterns consistent with osteoporosis can enforce that these animals correspond to animals with adaptations tending to a semi-aquatic life and fast swimming.

This is the first information so far of long bone histological analyses in the Dyrosauridae, and with Thalattosuchia constitutes the only paleohistological evidences about the basal neosuchians. Further analysis, specially using the midshaft information, and from others representatives from this family can strengthen these evidences.

Acknowledgments

We thank Dr. Antônio Barbosa (DEGEO-UFPE) for collecting the material used here, Dr. Cristiano A. Chagas (CAV-UFPE) for imaging of the thin sections, and Lucia Helena de S. Eleutério (CAV-UFPE) for helping in the preparation of the histological slides. Renan A. M. Bantim (UFPE) and Matheus Barbosa for helping with figures edition. We would like to thank Sarah Werning (Stony Brook University) and Alexander W. A. Kellner (MN/UFRJ) for the comments that greatly improved this work.

Author Contributions

Wrote the paper: RCLPA JMS.

References

1. Buffetaut E (1990) Vertebrate extinction and survival across the Cretaceous–Tertiary boundary. Tectonophysics 171: 337–345.
2. Brochu CA, Bouaré ML, Sissoko F, Roberts EM, O'Leary MA (2002) A dyrosaurid crocodyliform braincase from Mali. J Paleontol 76: 1060–1071.
3. Owen RD (1849) Notes on remains of fossil reptiles discovered by Prof. Henry Rodgers of Pennsylvania, U.S., in Greensand formations of New Jersey. Q J Geol Soc Lond 5: 380–383.
4. Denton RKJ, Bobie JL, Parris DC (1997) The marine crocodilian Hyposaurus in North America. In Ancient marine reptiles, editors. J. M. Callaway and E. L. Nicholls. London, UK: Academic Press. 375–397.
5. Stefano G (1903) Nuovi rettili degli strati a fosfato della Tunisia. Boll Soc Geol Ital 22: 51–80.
6. Cope ED (1886) A contribution to the vertebrate paleontology of Brazil. Proc. Am Philos Soc 23: 1–20.
7. Argollo J, Buffetaut E, Cappetta H, Fornari M, Herail, etal. (1987) Découverte de Vertébrés aquatiques présumés Paléocènes dans les Andes septentrionales des Bolivie (Rio Suches, synclinorium de Puntina). Geobios 20: 123–127.
8. Buffetaut E (1991) Fossil crocodilians from Tiupampa, (Santa Lucia Formation, Early Paleocene) Bolivia: a preliminary report. Rev Tec YPFB 12: 541–544.
9. Gayet M, Marshall LG, Sempere T (1991) The Mesozoic and Paleocene vertebrates of Bolivia and their stratigraphic context: a review. Rev Tecn YPFB 12: 393–433.
10. Hastings A, Bloch J (2007) New short-snouted Dyrosaurid (Crocodylomorpha) from the Paleocene of Northern Colombia. J Vert Paleontol 27: 87–92.
11. Barbosa JA, Kellner AWA, Viana MSS (2008) New dyrosaurid crocodylomorph and evidences for faunal turnover at the K–P transition in Brazil. Proc R Soc B 275: 1385–1391.
12. Sayão JM (2003). Histovariability in bones of two pterodactyloid pterosaurs from the Santana Formation, Araripe Basin, Brazil: preliminary results. Geol Soc Spec Publ 217: 335–342.
13. Padian K, Horner JR, de Ricqlés A (2004) Growth in small dinosaurs and pterosaurs: the evolution of archosaurian growth strategies. J Vert Paleontol 24: 555–571.
14. Chinsamy A, Codorniú L, Chiappe L (2009) Palaeobiological implications of the bone histology of Pterodaustro guinazui. Anat Rec 292(9): 1462–77.
15. Chinsamy A, Chiappe LM, Dodson P (1995). Growth rings in Mesozoic birds. Nature 368: 196–197.
16. Ricqlés A, Padian K, Horner JR (1998). Growth dynamics of the Hadrosaurid dinosaur Maiasaura peeblesorum. J Vert Paleontol 18: 72A.
17. Hua S, de Buffrénil V (1996) Bone histology as a clue in the interpretation of functional adaptations in the Thalattosuchia (Reptilia, Crocodylia. J Vert Paleontol 16: 703–717.
18. Buffetaut E, de Buffrénil V, de Ricqlés A, Spinar ZV (1982) Remarques anatomiques et paléohistologiques sur Dyrosaurus phosphaticus, crocodilien mesosuchien des Phosphates yprésiens de Tunisie. Annls paléont 68: 327–341.
19. Bartels WS (1984) Osteology and systematic affinities of the horned alligator Ceratosuchus (Reptilia, Crocodilia). J Paleontol 58: 1347–1353.
20. Bona P (2007) Una nueva especie de Eocaiman Simpson (Crocodylia, Alligatoridae) del Paleoceno Inferior de Patagonia. Ameghiniana 44: 435–445.
21. Pinheiro AEP, Fortier DC, Pol D, Campos DA, Bergqvist LP (2012) A new Eocaiman (Alligatoridae, Crocodylia) from Itaboraí Basin, Paleogene of Rio de Janeiro, Brazil. Hist Biol: 1–12. (doi: 10.1080/08912963.2012.705838).
22. Pol D, Leardi JM, Lecuona A, Krause M (2012) Postcranial anatomy of Sebecusicaeorhinus (Crocodyliformes, Sebecidae) from the Eocene of Patagonia. J. Vert. Paleontol 32: 328–354.
23. Silva MC, Barreto AMF, Carvalho IS, Carvalho MSS (2007) Vertebrados e Paleoambientes do Neocretáceo-Daniano da Bacia da Paraíba, Nordeste do Brasil. Estud Geol 17(2): 85–95.
24. Montefeltro FC, Larsson HCE, Langer MC (2011) A New Baurusuchid (Crocodyliformes, Mesoeucrocodylia) from the Late Cretaceous of Brazil and the Phylogeny of Baurusuchidae. PLoS ONE 6(7): e21916.

25. Kellner AWA, Pinheiro AEP, Campos DA (2014) A New Sebecid from the Paleogene of Brazil and the Crocodyliform Radiation after the K–Pg Boundary. PLoS ONE 9(1): e81386.
26. Francillon-Vieillot HJ, Arntzen W, Geraudie J (1990) Age, growth and longevity of sympatric Triturus cristatus, Triturus marmora- tus and their hybrids (Amphibia, Urodela): A ske- letochronological comparison. J. Herpetol. 24: 13–22.
27. Lamm E-T (2013) Bone Histology of Fossil Tetrapods. In: Padian K, Lamm E-T, editors. Preparation and Sectioning of Specimens. University of California Press. 55–160.
28. Chinsamy A, Raath MA (1992) Preparation of fossil bone for histological examination. Palaeontol Afr 29: 39–44.
29. Nascimento-Silva M, Sial NA, Ferreira VP, Neumann VH, Barbosa JA, et al. (2011) Cretaceous-Paleogene transition at the Paraíba Basin, Northeastern, Brazil: Carbon-isotope and mercury subsurface stratigraphies. J S Am Earth Sci. 1–14.
30. Beurlen K (1967a) Estratigrafia da faixa sedimentar costeira Recife-João Pessoa. Bol Geol 16: 43–53.
31. Beurlen K (1967b) Paleontologia da faixa sedimentar costeira Recife-João Pessoa. Bol Geol 16: 73–79.
32. Stein K, Sander M (2009) Histological core drilling: a less destructive method for studying bone histology. In: Brown MA, Kane JF, Parker WG, editors. Methods In Fossil Preparation: Proceedings of the First Annual Fossil Preparation and Collections Symposium. 69–80.
33. Salgado L, Fernández M, Talevi M (2007) Observaciones histológicas en reptiles marinos (Elasmosauridae y Mosasauridae) Del Cretácio Tardío e Patagonia y Antártida. Ameghiniana. 44(3): 1–13.
34. Kaiser HE (1969) Das abdorme in der Evolution. Acta Biotheor, suppl.9, E.J.Brill (publ.). Leiden.
35. Storrs GW (1993) Function and phylogeny in sauropterygians (Diapsida) evolution. Am J Sci 293A: 63–90.
36. Wiffen J, de Buffrénil V, de Ricqlès A, Mazin JM (1995) Ontogenetic evolution of boné structure in Late Cretaceous plesiosauria from New Zeland. Geobios. 28: 625–640.
37. Castanet JFJ, Meunier A, De Ricqlès (1977) L'enregistrement de la croissance cyclique par les tissue osseux chez les vertébrés poikilothermes: donnés comparative et essai de synthese. Bull Biol Fr Bel 3: 183–202.
38. Buffrénil V (1980) Mise em évidence de l'incidence dês conditions de milliu sur La croissance de Crocodylus siamensis (Schneider 1801) et valeurdes marques de croissance squelettiques pour l'evaluation de l'âge individuel. Arch zool Exp Gen 121: 63–76.
39. Hutton JM (1986) Age determination of living Nile crocodiles from the cortical stratification of bone. Copeia 1986: 332–341.
40. Caetano MH, Castanet J (1993) Variability and microevolutionary patterns in Triturus marmoratus from Portugal: age, size, longevity and individual growth. Amphib-Reptilia 14: 117–129.
41. Woodward HN, Horner JR, Farlow JO (2011) Osteohistological evidence for determinate growth in the American Alligator. J Herpetol, 45(3): 339–342. 2011.
42. Lee AH, O'Connor PM (2013) Bone histology confirms determinate growth and small body size in the noasaurid theropod Masiakasaurus knopfleri. J. Vert. Paleontol 33: 4, 865–876.
43. Horner JR, deRicqlès AJ, Padian K (1999) Variation in dinosaur skeletochronology indicators: implications for age assessment and physiology. Paleobiol 25: 295–304.
44. Gross W (1934) Die Typen des mikroskopischen Knochenbauesbei fossilen Stegocepahlen und Reptilien. Z Anat Entwicklungs 203: 731–764.
45. Ricqlés A (1975a) Quelques remarques paléo-histologiques sur Le probléme de La néoténie chez lês Stégocéphales. Colloq interl 1973: 351–363.
46. Enlow DH, Bown SO (1957) A comparative histological study of fossil and recent bone tissues. Part II. Tex J Sci 9: 186–214.
47. Ricqlés A (1976) On bone histology of living and fossil reptiles, with comments on their functional and evolutionary significance. In: d'A Bellairs A, Cox CB,

editors. Morphology and Biology of Reptiles. Linnean Society Symposium 3, Academic Press, London.pp.123–150.

48. Houssaye A, Scheyer TM, Kolb C, Fischer V, Sander PM (2014) A New Look at Ichthyosaur Long Bone Microanatomy and Histology: Implications for Their Adaptation to an Aquatic Life. PLoS ONE 9(4): e95637. doi:10.1371/journal.pone.0095637.

49. Buffrénil VD, Schoevaert D (1988) On how the periosteal bone of the delphinid humerus becomes cancellous: ontogeny of a histological specialization. J Morphol 198: 149–164.

50. Dumont M, Laurin M, Jacques F, Pellé E, Dabin W, et al. (2013) Inner architecture of vertebral centra in terrestrial and aquatic mammals: a twodimensional comparative study. J Morphol 274: 570–584.

51. Ricqlès A, de Buffrénil V (2001) Bone Histology, heterochronies and the return of tetrapods to life in water: where are we? In: Mazin JM, de Buffrénil V, editors. Secondary Adaptation of Tetrapods to Life in Water. Verlag Dr. Friedrich Pfeil, München, Germany. 289–310.

52. Houssaye A (2009) "Pachyostosis" in aquatic amniotes: a review. Integrative Zool 4: 325–340.

53. Buffrénil V, de Mazin JM (1990) Bone histology of the ichthyosaurs: comparative data and functional interpretation. Paleobiol 16: 435–447.

54. Currie PJ (1982) The osteology and relationships of *Tangasaurus mennelli* Haughton (Reptilia, Eosuchia). Ann S Afr Mus 86, 247–65.

55. Rieppel O (1995) The status of Anarosaurus multidentatus von Huene (Reptilia, Sauropterygia), from the lower Anisian of the Lechtaler Alps (Arlberg, Austria); Palaeontol Z 69, 289–99.

56. Bardet N, Houssaye A, Rage JC, Pereda SX (2008) The Cenomanian-Turonian (Late Cretaceous) radiation of marine squamates (Reptilia): The role of the Mediterranean Tethys. B Soc Geol Fr 176, 605–22.

57. Schwarz D, Frey E, Martin T (2006) The postcranial skeleton of the Hyposaurinae (Dyrosauridae; Crocodyliformes). Palaeontology 49: 695–718.

7

The First Ant-Termite Syninclusion in Amber with CT-Scan Analysis of Taphonomy

David Coty[1]*, **Cédric Aria**[2,3], **Romain Garrouste**[1], **Patricia Wils**[4], **Frédéric Legendre**[1], **André Nel**[1]

1 Muséum National d'Histoire Naturelle, Institut de Systématique, Evolution, Biodiversité, ISYEB, UMR 7205 CNRS UPMC EPHE, Paris, France, **2** Department of Natural History-Palaeobiology, Royal Ontario Museum, Toronto, Ontario, Canada, **3** Department of Ecology & Evolutionary Biology, University of Toronto, Toronto, Ontario, Canada, **4** CNRS UMS 2700, Muséum National d'Histoire Naturelle, Paris, France

Abstract

We describe here a co-occurrence (i.e. a syninclusion) of ants and termites in a piece of Mexican amber (Totolapa deposit, Chiapas), whose importance is two-fold. First, this finding suggests at least a middle Miocene antiquity for the modern, though poorly documented, relationship between *Azteca* ants and *Nasutitermes* termites. Second, the presence of a *Neivamyrmex* army ant documents an in situ raiding behaviour of the same age and within the same community, confirmed by the fact that the army ant is holding one of the termite worker between its mandibles and by the presence of a termite with bitten abdomen. In addition, we present how CT-scan imaging can be an efficient tool to describe the topology of resin flows within amber pieces, and to point out the different states of preservation of the embedded insects. This can help achieving a better understanding of taphonomical processes, and tests ethological and ecological hypotheses in such complex syninclusions.

Editor: Judith Korb, University of Freiburg, Germany

Funding: This study is supported by the French ISYEB, UMR 7205 CNRS, UPMC and EPHE, and the Spanish Ministry of Economy and Competitiveness project CGL2011-23948/BTE. The funders had no role in study design, data collection and analysis, decision to publish, or preparation of the manuscript.

Competing Interests: The authors have declared that no competing interests exist.

* Email: coty.david@gmail.com

Introduction

Ants and termites represent ecologically critical organisms in intertropical and subtropical ecosystems, impacting by their abundance, organization and variety of occupied niches the availability of nutrients as well as the composition of soils [1–5]. Although conspicuous and ecologically meaningful, the relationships between these key eusocial insects are sparsely documented. The data gathered so far have reported on ant predatory behaviour over termites [6–9] and/or termite nest (termitaria) occupation by ants [10–13]. The question of the antiquity of these relationships remains untackled, which overlaps with elucidating the age and stability of modern 'hot spots' of biodiversity occupied by these insects. Ants and termites are recorded since the early Cretaceous [14,15], but there is yet no fossil record of interactions between these two taxa. This is despite the fact that ants are very common in the Neogene Neotropical and Eocene Baltic amber [16]. They can be found in syninclusions with numerous other insects of various groups, therefore giving possibilities to track the origin of extant behaviours involving these organisms in the fossil record [17,18]. The term 'syninclusion' is therefore intended here in the sens of Koteja [19], for multiple organic inclusions in the same piece of amber, essential for understanding arthropods paleobehaviours in past environments.

We aim here to describe the first syninclusion of termites and ants (for a review of syninclusions see [18]) in a piece of Mexican amber from the Totolapa deposit, together with an adult Psocodea (Figures 1A–B, S1, S2). The exceptional feature of this syninclusion lies in the fact that a raider ant (*Neivamyrmex*) and inquiline ants (*Azteca*) are entrapped together with *Nasutitermes* termites,

thus ensuring that these genera were present at exactly the same time and shared at least a part of their ecological niche.

Using CT-scan analysis with the purpose to improve access to taphonomically concealed features, we discovered that our amber piece was the result of several different flows and that the preservation of internal organic structures differed between insects. If tomographic analyses have already been widely used for taxonomical studies of insects, reconstructions of their external and internal morphology [20–26], and to illustrate a syninclusion in amber [27], we also use here the CT-scan as a tool to analyze the results of taphonomical processes in an amber syninclusion.

Material, Locality and Method

This amber piece was discovered in a batch of crude amber acquired by one of us (DC) from locals exploiting the Totolapa amber deposit (Salt River Mine). Later, the piece was offered to the Muséum National d'Histoire Naturelle de Paris (specimen MNHN.F.A49933). Totolapa is a village located in the central depression of Chiapas, 70 km south-east of Tuxtla Gutiérrez, the capital of Chiapas State. The Salt River amber mine, exploited since 2007 by Manuel Ramirez and his son Heriberto, is 1 km north of Totolapa, on the banks of the Salt River. The arthropod fauna collected by DC is currently under study. The age of the main Mexican amber locality, Simojovel, is still in debate, between Late Oligocene to Middle Miocene [28–33]. According to a geological map of the Instituto Nacional de Estatistica y Geografia [34] Totolapa amber would be Eocene in age, but a recent geological study of the Totolapa deposit suggests that the material originates from the Early Miocene Mazantic and Balumtum

A

B

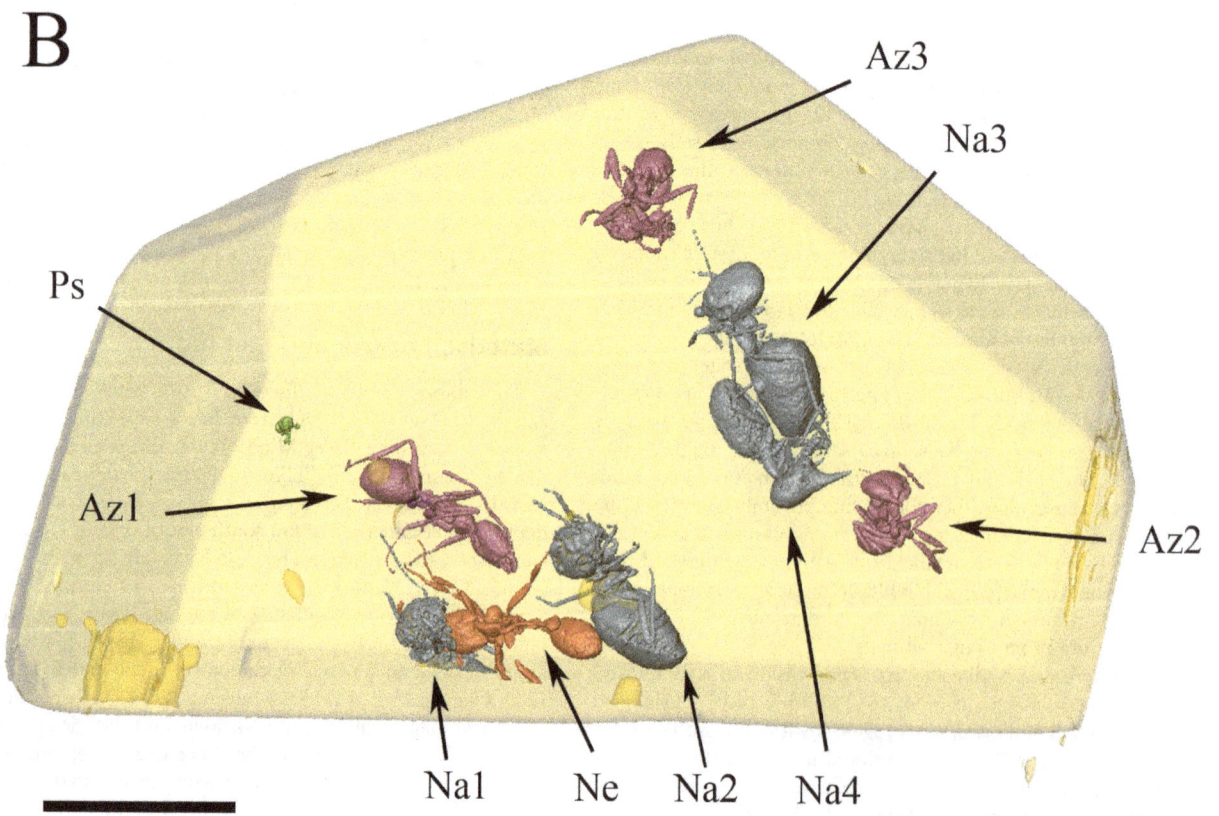

Figure 1. General configuration of the syninclusion. (A) Overview of the amber piece, under optical microscope. scale bar = 3 mm. (B) Three-dimensional replica of the same; colours define taxonomic groups, viz. purple for *Azteca* ants, blue for *Nasutitermes* termites, red for *Neivamyrmex* ant, green for small Psocodea. Labels: Az1, *Azteca* ant nearest to predation scene; Az2 and Az3, two other *Azteca* ants, both trapped in a flow distinct to that of the others inclusions and whose physical density matches that of the *Nasutitermes* soldier; Na1, *Nasutitermes* worker trapped between the *Neivamyrmex* mandibles; Na2, isolated *Nasutitermes* termite closest to predation scene; Na3, *Nasutitermes* worker with damaged gaster; Na4, *Nasutitermes* soldier; Ne, *Neivamyrmex* ant; Ps, Psocodea; Scale bar = 3 mm.

formations on top of Eocene marine facies [35]. As a matter of fact, Lambert et al. [36] suggested after a Carbon13 NMR spectroscopy study made on Baltic, Dominican and Mexican amber, that Simojovel and Totolapa amber came from the same palaeobotanical source; while Dominican amber, even if also closely related, shows more differences with both Mexican deposits. We therefore consider that the age of Totolapa amber is most probably between late Oligocene to middle Miocene and that the producing tree could also be *Hymenaea mexicana* [37] or a *Hymenaea* of undetermined species. Further geological studies are therefore needed to discover the precise age of Totolapa amber.

The original external surface of the amber piece has been removed by polishing; final lustration was done using diatomite powder. The specimens were examined under Nikon SZ10 and Olympus SZX9 stereomicroscopes. Photos were taken with an Olympus E-3 digital camera. Several digital pictures were reconstructed using Helicon Focus software.

X-ray tomography was realised in the AST-RX service (CT scan facility of the MNHN, UMS 2700), using a v|tome|x L240-180 from GE Sensing and Inspection Technologies phoenix|x-ray, with a X-Ray 180 KV/15 W nanofocus transmission tube, as well as a movable detector formed by a 20242 pixels (200 microns pixel). The voxel size of the reconstructed volume is 11.2 µm. 3D reconstructions and movies have been made using AVIZO 7.0 software. Variations in the material density in the amber piece are visible through changes of coloration from black (low density) to white (high density).

No permits were required for the described study.

Systematic palaeontology

Identifications of the specimens were possible at the generic level but not at the specific level, for the reasons indicated below.

Order Isoptera Brullé, 1832; Family Termitidae Latreille, 1802; Subfamily Nasutitermitinae Hare, 1937; Genus *Nasutitermes* Dudley, 1890; *Nasutitermes* species (Figures 1B, 2A–C).

Comments. Typical workers and soldier termites assignable to *Nasutitermes* sp. by the following diagnostic characters: soldier with vestigial mandibles, with points; head capsule rounded, without constriction behind antennae; presence of a glabrous and narrow-tipped conical frontal tube (nasus); pronotum saddle-shaped and proctodeal segment not forming a loop on the right side of abdomen.

In Mexican amber, *Nasutitermes* was hitherto known only from imagos [38], to which our fossil soldier cannot be compared. From the two species known by soldiers in coeval Dominican amber, i.e. *N. electronasutus* Krishna, 1996 and *N. rotundicephalus* Krishna and Grimaldi, 1999, our fossil differs by its bare head as opposed to a head with long setae [39,40]. Finally comparison of a new fossil *Nasutitermes* with the approximately 260 known modern species is highly difficult, given the absence of a reliable key and the uneven reliability of the various descriptions. We therefore do not ascribe our fossil to any particular species, and instead left it as *Nasutitermes* sp.

Order Hymenoptera Linné, 1758; Family Formicidae Latreille, 1809; Subfamily Dolichoderinae Forel, 1878; Genus *Azteca* Forel, 1878; *Azteca* species. (Figures 1A–B, 2A).

Comments. Dolichoderine ant with the following characters: nodiform petiole; unarmed hypostoma and propodeum; developed eyes; vertical first gastral tergite and anterior clypeal margin without a broad median concavity [41].

These *Azteca* ants will be described in a future paper, encompassing all the other *Azteca* present in the David Coty Totolapa amber collection.

Order Hymenoptera Linné, 1758; Family Formicidae Latreille, 1809; Subfamily Ecitoninae Forel, 1893; Genus *Neivamyrmex* Borgmeier, 1940; *Neivamyrmex* sp. (Figures 1A–B, 2A–C).

Comments. Ecitonine ant with the following diagnostic characters: eyes absent or reduced to an ommatidium; promesonotal suture absent or vestigial; antenna 12-segmented; antennal sockets fully exposed; absence of a preapical tooth on inner curvature of mid and hind pretarsal claws.

Only two fossil Ecitoninae, both from Dominican amber, are currently recorded: *Neivamyrmex ectopus* [42] and an undescribed army ant associated with a prey wasp pupa [43]. *Neivamyrmex ectopus* differs from our specimen in having a petiole with a subpetiolar process. Nevertheless, as the cuticle of our fossil specimen is badly preserved, and as we cannot reshape the amber piece (to preserve the syninclusion as a whole) in order to access further taxonomic details, we refrain from ascribing a new species.

General description of the amber piece

Our amber piece is 1.6 cm long, 1.0 cm wide and 1.2 cm high. It contains three *Azteca* ants (specimens Az1, Az2, and Az3 in Figure 1B), one Neivamyrmex ant (specimen Ne in Figure 1B), four *Nasutitermes* termites (Na1, Na2, Na3, and Na4, in Figure 1B) and a Psocoptera (specimen Ps in Figure 1B). The *Neivamyrmex* ant (Ne) holds a minor termite worker (Na1) between its mandibles (Figure 2A). In their vicinity, we can find one of the *Nasutitermes* workers (Na2) an *Azteca* ant (Az1), and a Psocoptera (Ps). Further away are grouped together two contiguous *Nasutitermes* termites, a soldier (Na4) with preserved digestive tube (Figure 2C), and a worker (Na3) with the gaster partly damaged (Figure 2B). The two remaining *Azteca* ants (Az2 and Az3) stand aside from both groups, and one from the other. See Figure S1 for a 3D view of the syninclusion.

CT-scan results

X-ray tomographic analysis revealed that our amber piece is in fact made of eight distinct layers corresponding to different flows, and that the distribution of the insects does not reflect a synchronous event. The layers are delimited by sinuose surfaces whose intersections with the different tomographic slicing are rendered as sinuate lines (variations of matter density visible in the images, see Figures 3, and S2).

The fossil specimens are entrapped in two of the eight visible flows hereby identified (Figures 3, S2). Other flows are devoid of insect inclusions. The first flow with inclusions, herein named the 'predation flow set', contains the *Neivamyrmex* ant holding the minor termite between its mandibles (Ne+Na1), one *Azteca* ant

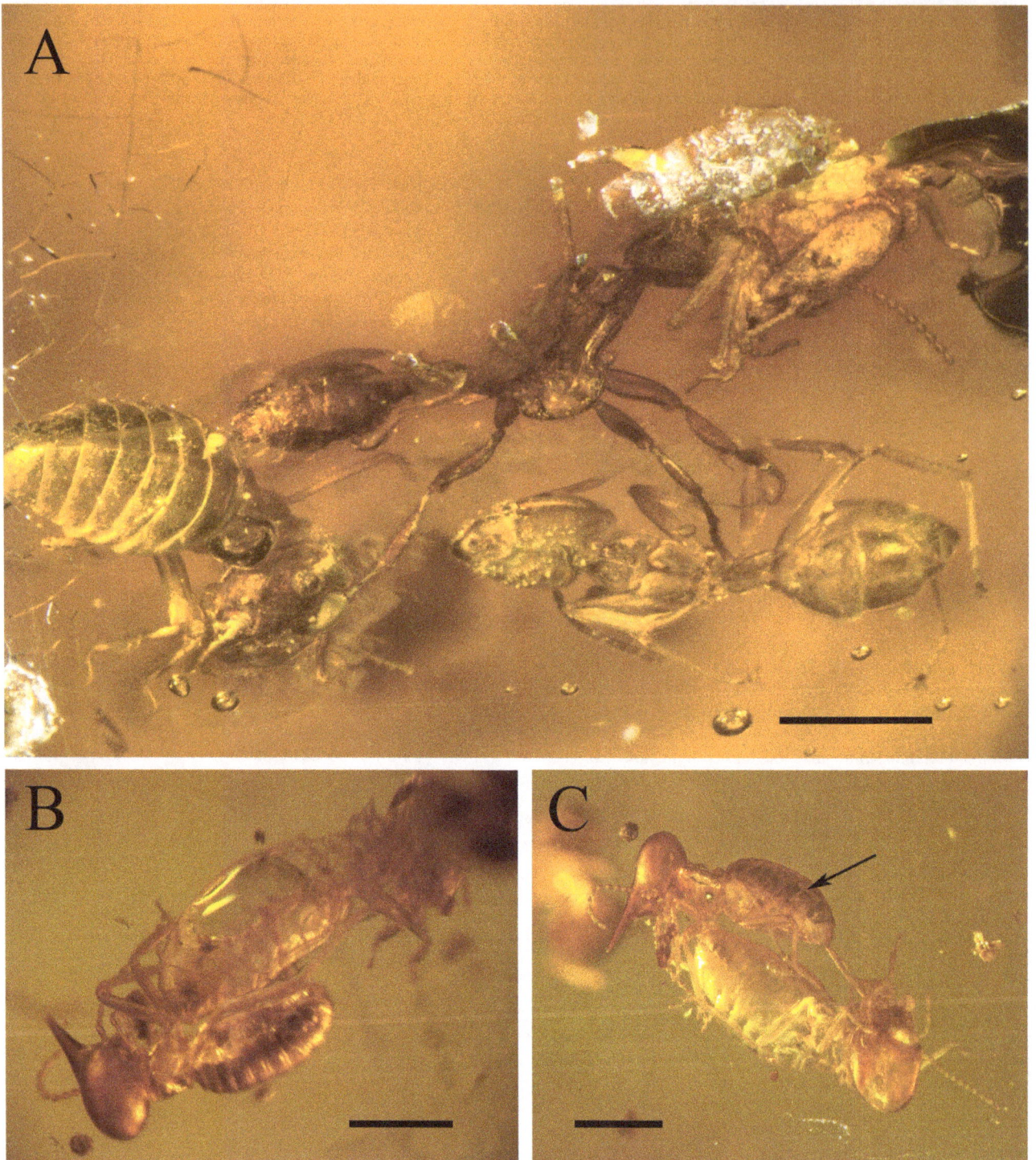

Figure 2. Details of the syninclusion. (A) General side view of the *Neivamyrmex* ant holding a *Nasutitermes* termite (Na1) between its mandibles, under optical microscope, scale bar = 1 mm. (B) detail of damaged gaster of *Nasutitermes* worker (Na3) closely contiguous to a *Nasutitermes* soldier (Na4), scale bar = 1 mm. (C) side view of closely contiguous *Nasutitermes* soldier (Na4) and worker (Na3), black arrow: digestive tube of *Nasutitermes* worker scale bar = 1 mm.

(Az1), one *Nasutitermes* worker (Na2) the Psocoptera (Ps), and the two contiguous *Nasutitermes* termites (Na3+Na4). The second flow, named the '*Azteca* flow set' contains the two isolated *Azteca* specimens (Az2 and Az3).

CT-scan analysis also emphasised variations in the physical density of the specimens. Empty specimens appear in black (low density registered) on the slices obtained with the CT-scan. In our case, the *Neivamyrmex* ant (Ne), the minor termite worker trapped between its mandibles (Na1), one *Nasutitermes* worker (Na2), and one *Azteca* ant (Az1) appear as empty structures inside the amber piece (in black on the slices, see Figures 3, 4C). On the contrary, the *Nasutitermes* soldier (Na4), the *Nasutitermes* worker with the

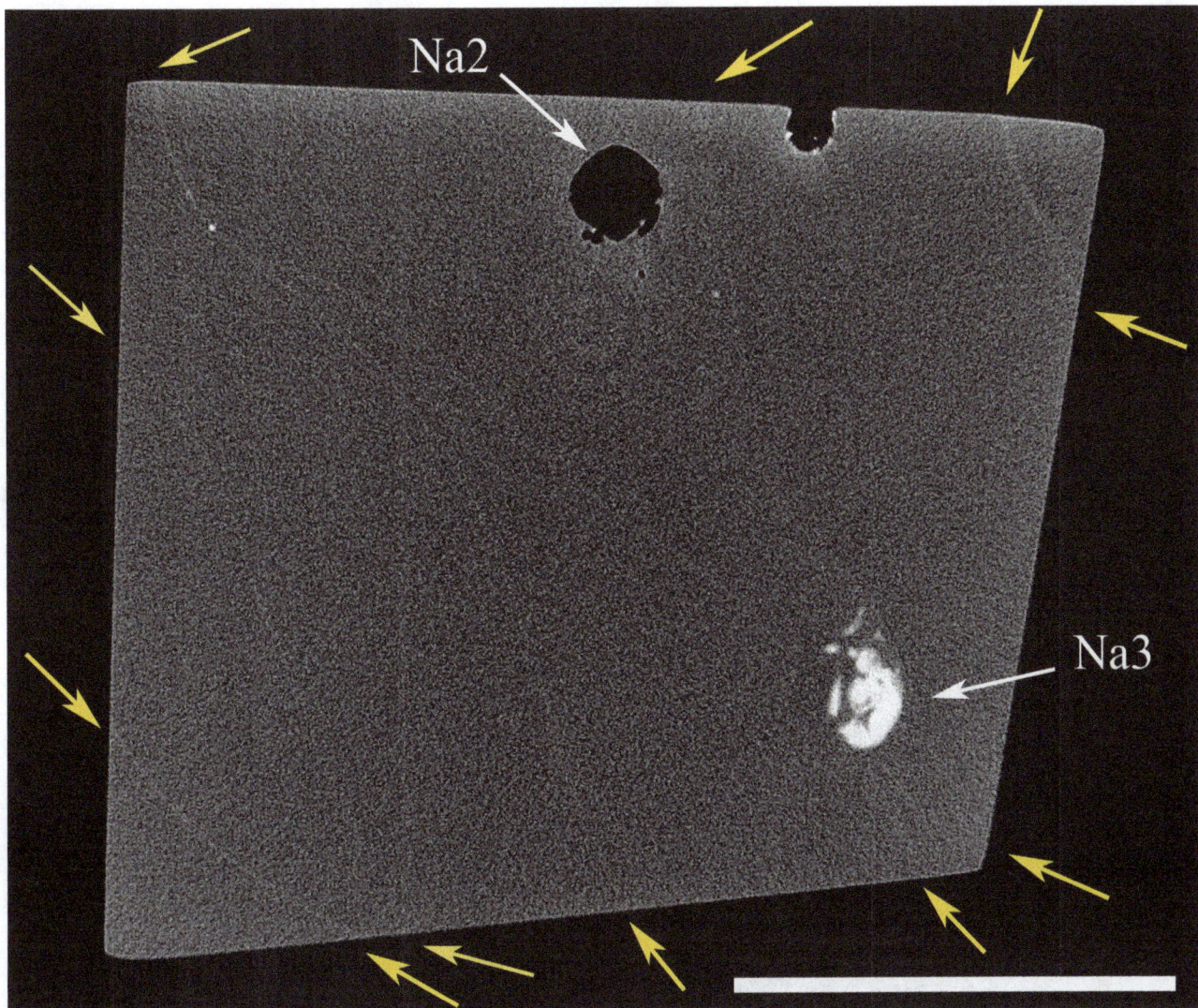

Figure 3. Virtual slicing from CT scan analysis showing flow boundaries. Yellow arrows shows starting and ending points of flows limits. Na2 and Na3 belong to the same flow, strong disparity in density matter between specimens from same taxonomic groups, scale bar = 5 mm.

damaged gaster (Na3) and two *Azteca* ants (Az2 and Az3) appear as full structure, denser than the amber (in white and light gray on the slices, see Figures 3, 4C–F). Thus, two coherent density sets can be distinguished in our piece of amber: the '*Nasutitermes* soldier density set' and the '*Neivamyrmex* density set' (Figures 4A–B).

Discussion

Ants-termites interactions – palaeoecological interpretation of the syninclusion

In the Neotropics, *Nasutitermes* is often involved in relationships with ants, possibly in relation with the fact that it is the termite genus with the highest number of species building conspicuous nests [4,11]. A total of 54 extant ant species have been reported living in the different stages of the *Nasutitermes* nests [11]. Among those, *Azteca* species (*A. chartifex* Forel, 1896 and *A. gnava* Forel, 1906) have been found living at the three categorized termitaria stages (active, decadent, abandoned). The implications of extant *Azteca* in the opportunistic occupation of termitaria have been briefly described elsewhere [10,11], but the nature of interactions is mostly unknown.

In the modern Venezuelan forests, it seems that some ants (including undetermined *Azteca* spp.) occupy the nests of *Nasutitermes corniger* (Motschulsky, 1855) to protect themselves during the flooding events of the wet season [10]. During this temporary association, termites tolerate the predation of ants on their colony, as they themselves take the opportunity to feed on dead ants which constitute a valuable source of nitrogen. Nutrients flows occur both way between termites and ant in such association. Termites may also take benefits of the presence of inquilines ants in their nests to defend their common colony against predators [1,44–46].

Cases of termites occupying parts of an active ant colony are also known [47–50]. Reasons of such cohabitations are generally unknown, although it has been noted that the contacts between ants and termites are rare in such cases, and qualified as neutral. Trager [50] mentioned that this type of association is frequent in the Neotropical region, involving different species of termites and ants.

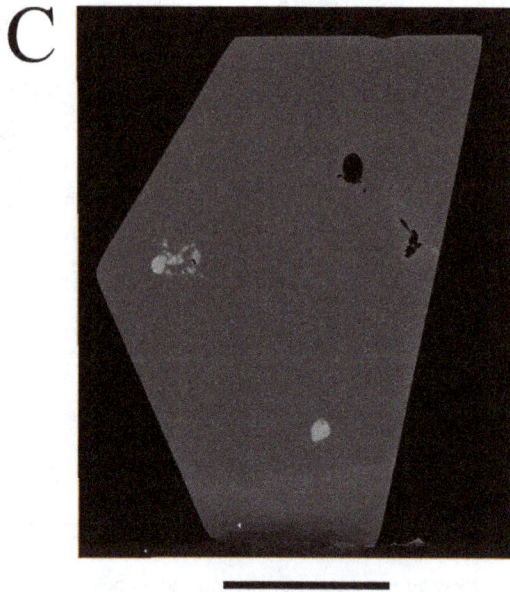

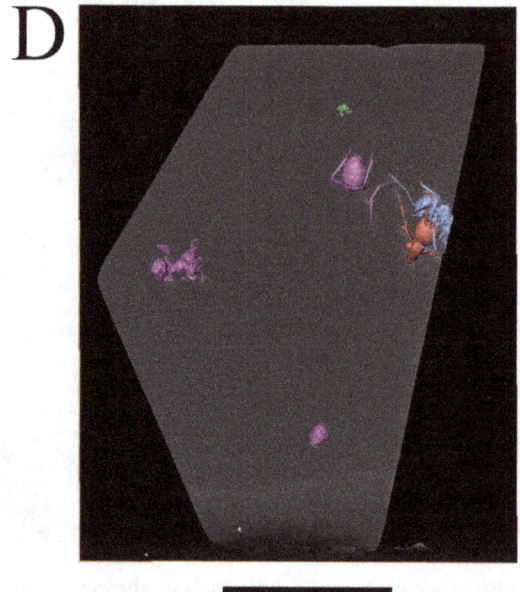

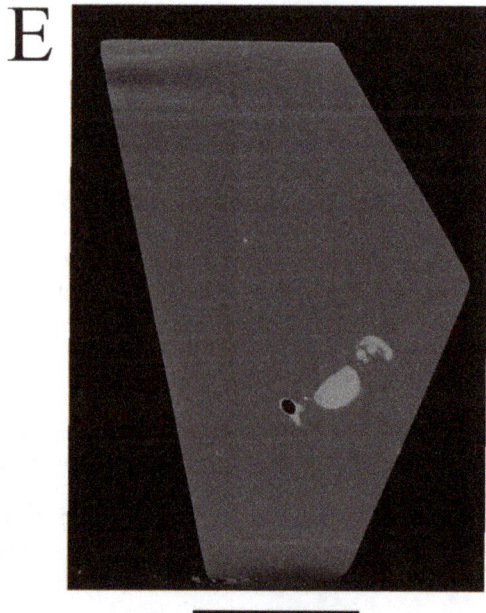

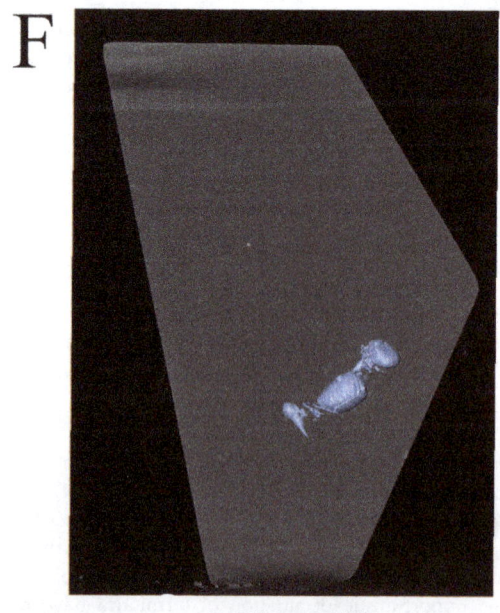

Figure 4. Density differences identified by CT scan. (A–B) Three-dimensional reconstruction artificially coloured representing two different density sets. 'Nasutitermes soldier' set in white, 'Neivamyrmex' set in black, as in CT-scan slices, Scale bar = 5 mm. (C) reslice of the original CT scan slice showing density differences between specimens. High density in white or light gray, low density in black or dark gray (darker than amber). Az2 and Az3 clearly denser than amber, Az1 and Ne empty structures, scale bar = 5 mm. (D) same slice as in B, but with 3D reconstruction of specimens, showing slice position in syninclusion, and specimen correspondences, scale bar = 5 mm. (E) reslice of original CT-scan slices, showing Nasutitermes soldier (Na4) and Nasutitermes worker (Na3), both denser than amber, scale bar = 5 mm. (F) same slice than E, but with 3D reconstruction of specimens, showing positions of specimens, scale bar = 5 mm.

Also called 'The Huns of the insect world' [51], all the modern species of army ants are major predators of both invertebrates and vertebrates [52–54]. They are also known to have a preference for preying on other eusocial insects, and ants in particular [49,54–56]. Cases of predation of army ants on *Nasutitermes* are recorded in the Neotropics [57]. As mentioned by Brady [58], army ants 'never hunt or forage solitarily' but 'dispatch a mass of cooperative, leaderless foragers to locate and overwhelm prey simultaneously'.

By phylogenetic inference, it is possible to say that our Mexican amber *Neivamyrmex* had the same behaviour and biology as its modern relatives (see [59], for inferences from recent fossil). The fact that this *Neivamyrmex* army ant holds a minor termite between its mandibles (Figure 2A) supports this hypothesis. Also, the gaster of the Na3 *Nasutitermes* termite is damaged, clearly showing traces of an ant bite (size and shape of the bite marks visible in Figure 2B). Lastly a phylogenetic inference shows that the *Azteca* ants were not predating the termites but more likely living with them in the same nest, as for their modern relatives. The presence of a *Nasutitermes* soldier (generally confined inside the nest and surging out of it for defence purpose), contiguous to the termite worker that exhibits an ant bite, also enhance the hypothesis (here again for an inference of a modern behaviour, see [59]) that the resin flowed close to a *Nasutitermes* nest.

This beam of evidences suggests that the fossil army ant present in our amber piece was part of a raid, during which the *Nasutitermes* termites and *Azteca* ants, sharing the same nest or interacting in some other degree might have been attacked, as it can typically occur in modern Neotropical settings.

The predation scene could also be the result of a peculiar type of scavenger behaviour, viz. when a predator is embedded while eating a dead insect only partly embedded in resin. However, the tomography shows that there is no discontinuity (a limit between two flows) between these two animals (see below for the study of flows using tomographic slices), which implies that they have been trapped in the same resin flow and thus invalidates this hypothesis.

The stress occurring during the embedment of the living ant together with a termite could have caused it to bite the termite randomly, but in this case, the presence of another termite partly eaten is not explained.

The presence of numerous *Azteca* ants (three specimens) and *Nasutitermes* termites (four specimens) in this small piece of amber, together with a Neivamyrmex ant reduces the possibility that *Azteca* ants and *Nasutitermes* termites have been randomly entrapped in the resin. It suggests that they were defending their common nest against an army ant 'raid' or interacting in some other degree while attacked. A 'raid' of army ants is always a strong perturbation for these eusocial communities, against which they have to defend, which bring many individuals out of their nest, therefore enhancing the probability to have many specimens of these different taxa entrapped together in a resin flow on the tree trunk.

Interpretation of CT-scan results

Structure of the amber. The fact that amber pieces are almost always the result of several flowing event has been pointed out by various authors [60,61]. Although the flow boundaries are partly visible at the surface of the amber piece (Figure 5), they are within the piece only detectable through greenish or brownish translucent surfaces, which appear under some particular view points, and light orientations (Figure 5). The CT-scan analysis therefore represents an informative enhancement allowing the clear mapping of the topology of each flow (Figure S2).

We discard the possibility that those sinuate lines could be cracks that may have occurred during the biostratinomy or the diagenesis of the amber, for two main reasons: 1) cracks never form smooth sinuate surfaces running across all the amber pieces; 2) they never follow the shape of inclusions but often damage them.

The surfaces of the flows appearing in CT-scan slices are always denser (i.e. lighter in the CT scan images) than the surrounding amber (also observed on other CT-scan slices of different amber pieces), suggesting that micro-particles, denser than the resin (not visible under optic microscope), deposited onto the amber surface before the arrival of the succeeding flow, or that a very thin layer of amber is harder at the surface contact between the two flows, as the resin was consolidated by polymerisation through sunlight and wind. The flow surfaces are nonetheless very weakly defined on the CT-scan slices and mostly not visible under optic microscope, possibly suggesting that the delays between flows might have been extremely short to limit the effects of polymerization on the flow surface, and avoid the deposition of dusts and debris on the fresh resin. It would therefore be likely that at least some relative degree of relation could exist between the density level of flow margins (in CT-scan imaging) and the time they were exposed to external elements, but further study on other material will be necessary to confirm such a hypothesis.

Differences of density between specimens. Since the relative density of the insects is unrelated to their taxonomy (*Azteca* ants and *Nasutitermes* termites are present in different density sets), and since there is a clear spatial homogeneity between the two density groups (Figures 4A–B), variations of physical density between the specimens themselves is likely to express taphonomical disparities. This may be related to differences in the preservation of the cuticle and the inner organic structures [61], and differences between the insects before their entombment in the fresh resin, i.e. dead versus living animals, animals with or without filled digestive tubes, etc. As a matter of fact, insects in amber are frequently empty, since most of the internal organic content is anaerobically degraded, as a result of autolysis and the activity of endogenous bacteria [62,63]. On the other hand, the preservation of internal structures (i.e. the digestive tube in the gaster of the *Nasutitermes* soldier, Figure 2C) may be due to a phenomenon of dehydration before the complete embedment in resin, as it has been shown that a pre-entombment dehydration of the insects inhibits the latter degradation process of their body inside the resin [64,65].

Following this logic, it would therefore be possible that the specimens of the '*Nasutitermes* density set' where embedded

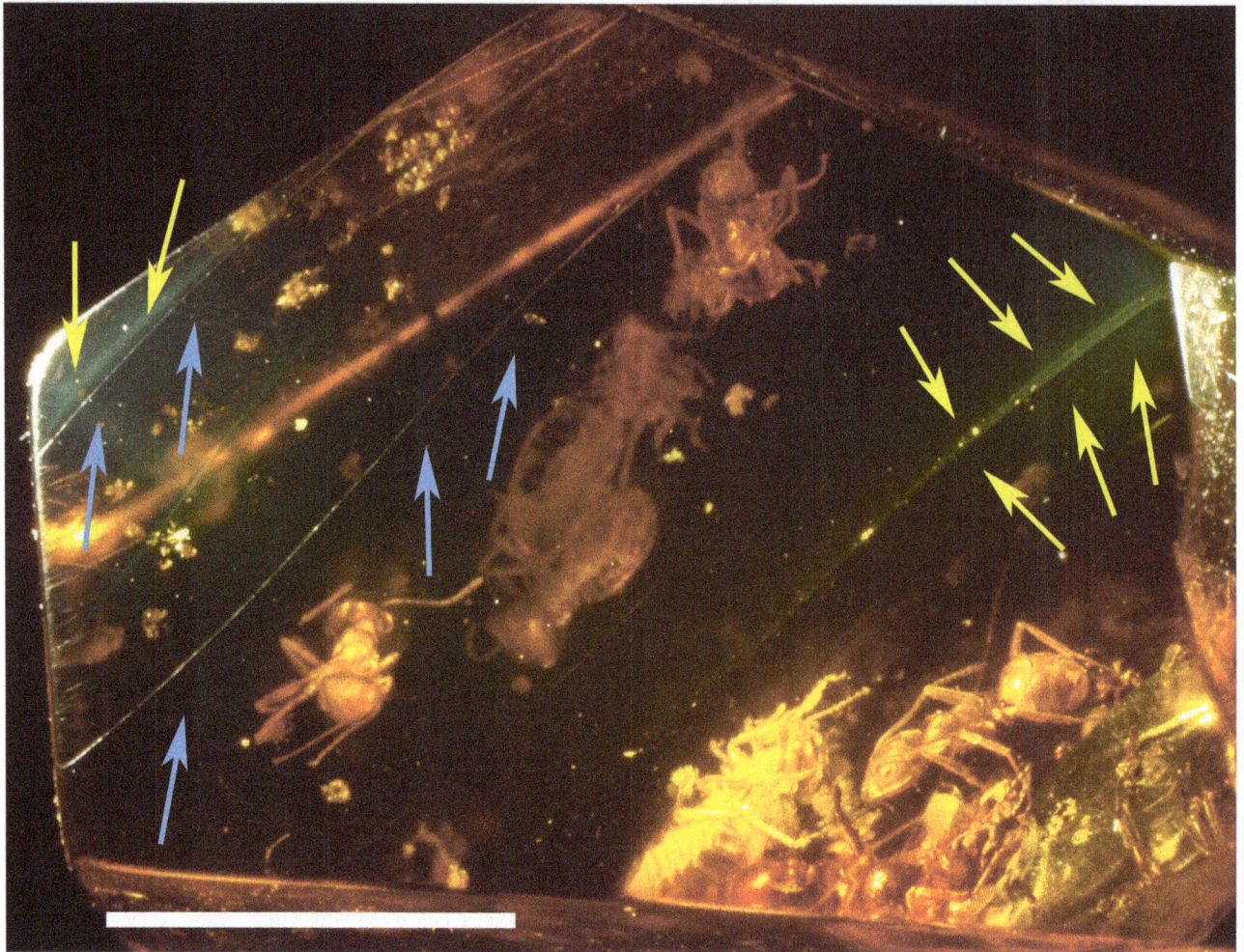

Figure 5. Flow margins visible under optic microscope. Yellow arrows: 'green veils' representing limits between two flows, blue arrows: 'sinuate line' visible on surface of the amber piece, scale bar = 5 mm.

already dead, while the specimens of the '*Neivamyrmex* density set' where embedded alive.

It has to be noted that the presence of a white aureole surrounding the body of inclusions, considered to be a foam of microscopic bubbles by Mierzejewski [66] and Weitschat and Wichard [67] can also help in some cases to reveal which insects where trapped alive or dead in the resin [61], as it is possibly the result of an early diagenesis reaction between fluids, produced by decay and decomposition of labile tissues, with sugar and terpenes in the resin. However, such foams are rare in Mexican and Dominican ambers. In the present case, no foam is visible around our specimens to help us in our analysis.

The two homogenous density groups are not distinguishable under optic microscope. Some preserved internal organs are visible, but the level of preservation of the cuticles is difficult to evaluate. One particularly odd case is the *Nasutitermes* with the damaged gaster (Na3) as this gaster is obviously empty of any kind of internal structures, while CT-scan images show that its whole body is denser than the amber. The fact that the damaged part of the gaster exhibits a bite mark caused by an ant, and is entirely covered by what seems to be the edge of an air bubble (Figure 2B) strongly suggests that this specimen was dead before entombment in fresh resin. In fact, regarding what we mentioned above, if the two closely contiguous *Nasutitermes* (Na3 and Na4) are likely to

have been both dead before entombment in the resin, the similarity of their density levels may not have the same taphonomical origin, and remain to be elucidated for specimen Na3 (*Nasutitermes*).

Biostratinomic processes (the period between the moment when resin is exuded from the tree to the moment when it is buried in sediment) could also be responsible for these differences in matter density, as great differences can occur between flows in term of duration of the flowing event (viscosity), degree of humidity of the air, level and time of exposure to air and to UV, etc. Such variations can therefore create disparities in the taphonomic process between inclusions present in different flows.

If we therefore cross-compare the density distribution between specimens with the different flows in which they were embedded, our amber piece shows that the two density groups almost fit with the distribution of the specimens inside the flows, except for the closely contiguous *Nasutitermes* soldier and the *Nasutitermes* worker with a damaged gaster, as they share the same density level as the two *Azteca* ants that belong to a different flow. Regarding what has been mentioned above, this result can be due to differences between the insects themselves before entombment in the resin.

Conclusion

Our study provides evidence that some degree of relationship between *Azteca* ants and *Nasutitermes* termites might have already existed in Central America during the late Oligocene-middle Miocene period, together with the predation of army ants on other eusocial insects in the same community. However, the condition that led to the apparition of such interactions and their stability through time are still to be elucidated.

We also show here that beside anatomic reconstruction, CT-scan imaging can be used to study the taphonomy of syninclusions by allowing a more exhaustive description of resin flow topology and sequences, as well as a 'cartography' of density patterns of biotic inclusions. The main question to further address is to know in which measure both flow structures of the amber pieces and physical density variations of insect bodies can help to further reconstruct necrolysis, biostratinomic and diagenetic processes that occurred in the amber and its inclusions.

Acknowledgments

We thank the two anonymous referees for their useful comments on the first version of the paper. DC thanks his friends, Manuel Ramirez and his son Heriberto, owners and miners of Salt River Mine, and their family, for their great help in his field research. We also thank Miguel Garcia Sanz (MNHN UMS 2700) for operating CT scan analysis.

Author Contributions

Conceived and designed the experiments: DC CA AN. Performed the experiments: DC RG PW AN. Analyzed the data: DC CA RG PW FL AN. Contributed reagents/materials/analysis tools: DC CA RG PW FL AN. Wrote the paper: DC CA RG PW FL AN.

References

1. Hölldobler B, Wilson EO (1990) The ants. Belknap Harvard University Press, Springer (publ.), Cambridge, Massachusetts, USA: [xiii] + 732.
2. Kaspari M (2001) Taxonomic level, trophic biology, and the regulation of local abundance. Global Ecology & Biogeography 10: 229–244.
3. Lavelle P, Bignell D, Lepage M (1997) Soil function in a changing world: the role of invertebrate ecosystem engineers, European Journal of Soil Biology 33: 159–193.
4. Martius C (1994) Diversity and ecology of termites in Amazonian forests. Pedobiologia 38: 407–428.
5. Bignell DE, Roisin Y, Lo N (2010) Biology of termites: a modern synthesis. Springer publ: 576.
6. Deligne J, Quennedey A, Blum MS (1981) The enemies and defense mechanisms of termites. In Hermann, H. R. (ed.), Social Insects. Volume II, Academic Press, New York: 1–76.
7. Gonçalves TT, Ronaldo R Jr, DeSouza O, Ribeiro SP (2005) Predation and interference competition between ants (Hymenoptera: Formicidae) and arboreal termites (Isoptera: Termitidae). Sociobiology 46: 1–11.
8. Wilson EO (1971) The insect societies. Harvard University Press, Massachusetts: 562.
9. Cornelius ML, Grace JK (1995) Laboratory evaluations of interactions of three ant species with the Formosan subterranean termite (Isoptera: Rhinotermitidae). Sociobiology 26: 291–298.
10. Jaffe K, Ramos SC, Issa S (1995) Trophic interaction between ant and termites that share common nests. Annals of the Entomological Society of America 88: 328–333.
11. Santos PP, Vasconcellos A, Jahyny B, Delabie JHC (2010) Ant fauna (Hymenoptera, Formicidae) associated to arboreal nests of *Nasutitermes* spp. (Isoptera, Termitidae) in a cacao plantation in southeastern Bahia. Revista Brasileira de Entomologia 54: 450–454.
12. Leponce M, Roisin Y, Pasteels JM (1999) Community interactions betweens ants and arboreal-nesting termites in New Guinea coconut plantations. Insectes Sociaux 46: 126–130.
13. Quinet Y, Tekule N, De Biseau JC (2004) Behavioural interactions between *Crematogaster brevispinosa rochai* Forel (Hymenoptera: Formicidae) and two *Nasutitermes* species (Isoptera: Termitidae). Journal of Insect Behavior 18: 1–17.
14. Perrichot V, Lacau S, Néraudeau D, Nel A (2008) Fossil evidence for the early ant evolution. Naturwissenschaften 95: 85–90.
15. Krishna K, Engel MS, Grimaldi DA, Krishna V (2013) Treatise on the Isoptera of the world. Bulletin of American Museum of Natural History 377: 2704.
16. LaPolla JS, Dlussky GM, Perrichot V (2013) Ants and the fossil record. Annual Review of Entomology 58: 609–630.
17. Perkovsky EE, Rasnitsyn AP, Vlaskin AP, Rasnitsyn SP (2012) Contribution to the study of the structure of amber forst communities based on analysis of syninclusions in the Rovno amber (Late Eocene of Ukraine). Paleontological Journal 46: 293–301.
18. Boucot AJ, Poinar GO Jr (2010) Fossil behavior compendium. CRC Press, Taylor & Francis group, Boca Raton, FL, USA: 391.
19. Koteja J (1989) Syninclusions. Wrostek, 8: 7–8.
20. Lak M, Néraudeau D, Nel A, Cloetens P, Perrichot V, et al. (2008) Phase contrast X-Ray synchrotron imaging: opening access to fossil inclusions in opaque amber. Microscopy and Microanalysis 14: 251–259.

21. Grimaldi D, Nguyen T, Ketcham R (2000) Ultra-high-resolution X-ray computed tomography (UHR CT) and the study of fossils in amber. 77–92, in D. Grimaldi (ed.), 2000.
22. Henderickx H, Cnudde V, Masschaele B, Dierick M, Vlassenbroeck J, et al. (2006) Description of a new fossil *Pseudogarypus* (Pseudoscorpiones: Pseudogarypidae) with the use of X-ray micro CT to penetrate opaque amber. Zootaxa 1305: 41–50.
23. Tafforeau P, Boistel R, Boller E, Bravin A, Brunet M, et al. (2006) Applications of X-ray synchrotron microtomography for non-destructive 3D studies of paleontological specimens. Applied Physics A: Materials Science & Processing 83: 195–202.
24. Penney D, Dierick M, Cnudde V, Masschaele B, Vlassenbroeck J, et al. (2007) First fossil Micropholcommatidae (Araneae), imaged in Eocene Paris amber using X-ray computed tomography. Zootaxa 1623: 47–53.
25. Soriano C, Archer M, Azar D, Creaser P, Delclos X, et al. (2010) Synchrotron X-ray imaging of inclusions in amber. Comptes Rendus Palevol 9: 361–368.
26. Sutton MD, Rahman I, Garwood R (2014) Techniques for virtual paleontology (analytical methods in Earth and environmental science). Wiley-Blackwell/John Wiley &Sons: i–viii + 1–200.
27. Penney D, McNeil A, Green DI, Bradley R, Jepson JE, et al. (2012) Ancient Ephemeroptera-Collembola symbiosis predicts contemporary phoretic associations. PLoS ONE 7(10): e47651.
28. Frost SH, Langenheim RL (1974) Cenozoic reef biofacies; Tertiary larger foraminifera and scleractinian corals from Chiapas, Mexico. Northern Illinois University Press, De Kalb, 1–388.
29. Ferrusquía-Villafranca I (2006) The first Paleogene mammal record of Middle America: *Simojovelhyus pocitosense* (Helohyidae, Artiodactyla). Journal of Vertebrate Paleontology 26: 989–1001.
30. Castaneda-Posadas C, Cevallos-Ferriz SRS (2007) *Swietenia* (Meliaceae) flower in Late Oligocene–Early Miocene amber from Simojovel De Allende, Chiapas, Mexico. American Journal of Botany 94: 1821–1827.
31. Solòrzano-Kraemer MM (2007) Systematic, palaeoecology, and palaeobiogeography of the insect fauna from Mexican amber. Palaeontographica (A) 282: 1–133.
32. Vega FJT, Nyborg T, Coutino MA, Solé JM, Hernández-Monzón O (2009) Neogene Crustacea from southeastern Mexico. Bulletin of the Mizunami Fossil Museum 35: 51–69.
33. Perrilliat MC, Vega FJ, Coutino MA (2010) Miocene mollusks from the Simojovel area in Chiapas, southwestern Mexico. Journal of the South American Earth Sciences 30: 111–119.
34. INEGI (1985) Carta Geológica, E15–11 (Tuxtla Gutiérrez), escala 1:250,000. SPP/INEGI, Instituto Nacional de Estadística, Geografía y Informática, Mexico City, Mexico.
35. Durán-Ruiz C, Riquelme F, Coutiño-José M, Carbot-Chanona G, Castaño-Meneses G, et al. (2013) Ants from the Miocene Totolapa amber (Chiapas, México), with the first record of the genus *Forelius* (Hymenoptera, Formicidae). Canadian Journal of Earth Sciences 50: 495–502.
36. Lambert JB, Frye JS, Lee TA Jr, Welch CJ, Poinar GO Jr (1989) Analysis of Mexican amber by Carbon-13 NMR Spectroscopy. Archeological Chemistry 4: 381–388.

37. Poinar GO Jr, Brown AE (2002) *Hymenea mexicana* spp. nov. (Leguminosae: Caesalpinioideae) from Mexican amber indicates Old World connections. Botanical Journal of the Linnean Society 139: 125–132.

38. Krishna K, Engel MS, Grimaldi DA, Krishna V (2013) Treatise on the Isoptera of the world. Bulletin of the American Museum of Natural History 377: 1–2704.

39. Krishna K (1996) New fossil species of termites of the subfamily Nasutitermitinae from Dominican and Mexican amber. American Museum Novitates, 3176: 1–13.

40. Krishna K, Grimaldi DA (2009) Diverse Rhinotermitidae and Termitidae (Isoptera) in Dominican amber. American Museum Novitates 3640: 1–48.

41. Bolton B (1994) Identification guide to the ant genera of the World. Harvard University Press, Cambridge, London, Massachusetts, England: [iv] + 1–222.

42. Wilson EO (1985) Ants of the Dominican amber (Hymenoptera: Formicidae) 2. The first fossil army ants. Psyche 92: 11–16.

43. Poinar GO Jr, Poinar R (1999) The amber forest. A reconstruction of a vanished World. Princeton University Press, Princeton, New Jersey: i–xiii + 1–239.

44. Howse P (1984) Alarm defense and chemical ecology of social insect. In: Lewis, T. (ed.). Insect communication. Academic Press, London: 151–164.

45. Jolivet P (1986) Les fourmis et les plantes, un exemple de coévolution. Boubée Publ., Paris. 255.

46. Highashi S, Ito F (1989) Defense of termitaria by termitophilous ants. Oecologia 80: 145–147.

47. Wheeler WM (1936) Ecological relations of Ponerinae and other ants to termites. Proceedings of the National Academy of Sciences 71: 159–243.

48. Gray B (1974) Associated fauna found in a nests of *Myrmecia* (Hymenoptera, Formicidae). Insectes Sociaux 21: 289–300.

49. Sennepin A (1999) Symbioses entre fourmis et termites: structures et implications. Actes des Colloques des Insectes Sociaux 12: 181–190.

50. Trager JC (1991) A revision of the fire ants *Solenopsis geminata* group (Hymenoptera, Formicidae: Myrmicinae). Journal of the New York Entomological Society 99: 141–198.

51. Wheeler WM (1910) Ants, their structure, development and behavior. Columbia University Biological Series, New York 9: xxvi + 663.

52. Rettenmeyer CW (1963) Behavioral study of army ant. University of Kansas Science Bulletin 44: 281–465.

53. Schneirla TC (1971) Army ants: a study in social organization. W.H. Freeman & Co Ltd, San Francisco: 349.

54. Gotwald WH Jr (1995) Army ants: the biology of social predation. Cornell University Press, Ithaca, New York: 302.

55. LaPolla JS, Mueller UG, Seid M, Cover SP (2002) Predation by the army ant *Neivamyrmex rugulosus* on the fungus-growing ant *Trachymyrmex arizonensis*. Insectes Sociaux 49: 251–256.

56. Le Breton J, Dejean A, Snelling G, Orivel J (2007) Specialized predation on *Wasmannia auropunctata* by the army ant species *Neivamyrmex compressinodis*. Journal of Applied Entomology 131: 740–743.

57. Souza JLP, Moura CAR (2008) Predation of ants and termites by army ants, *Nomamyrmex esenbeckii* (Formicidae, Ecitoninae) in the Brazilian Amazon. Sociobiology 52: 399–402.

58. Brady SG (2003) Evolution of the army ant syndrome: long-term evolutionary stasis of a complex of behavioral and reproductive adaptations. Proceedings of the National Academy of Sciences 100: 6575–6579.

59. Nel A (1997) The probabilistic inference of unknown data in phylogenetic analysis. In: Grandcolas P (ed.). The origin of biodiversity in insects: phylogenetic tests of evolutionary scenarios. Mémoires du Muséum National d'Histoire Naturelle, Paris 173: 305–327.

60. Grimaldi D, Engel MS (2005) Evolution of the Insects. Cambridge University Press, New York/Cambridge, USA, 755 p.

61. Martínez-Delclòs X, Briggs DEG, Peñalver E (2004) Taphonomy of insects in carbonates and amber. Palaeogeography, Palaeoclimatology, Palaeoecology 3225: 1–46.

62. Allison PA, Briggs DEG (1991a) The taphonomy of softbodied animals. In: Donovan S.K. (Ed.), The Processes of Fossilization. Belhaven Press, London: 120–140.

63. Allison PA, Briggs DEG (1991b) Taphonomy of non-mineralized tissues. In: Allison PA, Briggs DEG, (Eds.), Taphonomy: Releasing the data locked in the fossil record. Plenum Press, New York: 25–70.

64. Henwood A (1992a) Insect taphonomy from Tertiary Amber of the Dominican Republic. Ph.D. Thesis, University of Cambridge 166.

65. Henwood A (1992b) Exceptional preservation of dipteran flight muscle and the taphonomy of insects in amber. Palaios 7: 203–212.

66. Mierzejewski P (1978) Electron microscopy study on the milky impurities covering arthropod inclusions in the Baltic amber. Prace Muzeum Zeimi 28: 81–84.

67. Weitschat W, Wichard W (1998) Atlas der Pflanzen und Tiere im Baltischen Bernstein. Pfeil, Munchen: 256.

8

High Diversity in Cretaceous Ichthyosaurs from Europe Prior to Their Extinction

Valentin Fischer[1,2]*, Nathalie Bardet[3], Myette Guiomar[4], Pascal Godefroit[2]

1 Department of Geology, University of Liège, Liège, Belgium, **2** Operational Directory 'Earth and History of Life', Royal Belgian Institute of Natural Sciences, Brussels, Belgium, **3** CNRS UMR 7207, Département Histoire de la Terre, Muséum National d'Histoire Naturelle, Paris, France, **4** Réserve naturelle géologique de Haute Provence, Digne-les-bains, France

Abstract

Background: Ichthyosaurs are reptiles that inhabited the marine realm during most of the Mesozoic. Their Cretaceous representatives have traditionally been considered as the last survivors of a group declining since the Jurassic. Recently, however, an unexpected diversity has been described in Upper Jurassic–Lower Cretaceous deposits, but is widely spread across time and space, giving small clues on the adaptive potential and ecosystem control of the last ichthyosaurs. The famous but little studied English Gault Formation and 'greensands' deposits (the Upper Greensand Formation and the Cambridge Greensand Member of the Lower Chalk Formation) offer an unprecedented opportunity to investigate this topic, containing thousands of ichthyosaur remains spanning the Early–Late Cretaceous boundary.

Methodology/Principal Findings: To assess the diversity of the ichthyosaur assemblage from these sedimentary bodies, we recognized morphotypes within each type of bones. We grouped these morphotypes together, when possible, by using articulated specimens from the same formations and from new localities in the Vocontian Basin (France); a revised taxonomic scheme is proposed. We recognize the following taxa in the 'greensands': the platypterygiines '*Platypterygius*' sp. and *Sisteronia seeleyi* gen. et sp. nov., indeterminate ophthalmosaurines and the rare incertae sedis *Cetarthrosaurus walkeri*. The taxonomic diversity of late Albian ichthyosaurs now matches that of older, well-known intervals such as the Toarcian or the Tithonian. Contrasting tooth shapes and wear patterns suggest that these ichthyosaurs colonized three distinct feeding guilds, despite the presence of numerous plesiosaur taxa.

Conclusion/Significance: Western Europe was a diversity hot-spot for ichthyosaurs a few million years prior to their final extinction. By contrast, the low diversity in Australia and U.S.A. suggests strong geographical disparities in the diversity pattern of Albian–early Cenomanian ichthyosaurs. This provides a whole new context to investigate the extinction of these successful marine reptiles, at the end of the Cenomanian.

Editor: Andrew A. Farke, Raymond M. Alf Museum of Paleontology, United States of America

Funding: VF is an aspirant of the F.R.S.–FNRS. This a PhD thesis grant from the national research fund of Belgium. The funders had no role in study design, data collection and analysis, decision to publish, or preparation of the manuscript.

Competing Interests: The authors have declared that no competing interests exist.

* E-mail: v.fischer@ulg.ac.be

Introduction

Ichthyosauria was a successful clade of marine sauropsids that spanned most of the Mesozoic, from the Olenekian (Early Triassic) to the end of the Cenomanian (Late Cretaceous). When compared to the Triassic and the Jurassic, the Cretaceous record of ichthyosaurs is generally poor [1]. As a result, only minimal attention has been drawn to the Cretaceous representatives of Ichthyosauria in the past. The last in-depth taxonomic reviews of Cretaceous ichthyosaurs are those of McGowan [2], focusing on North American material, and Bardet [3], mainly reviewing Late Cretaceous ichthyosaur occurrences. McGowan [2] merged all valid species within a single genus, *Platypterygius*. Cretaceous ichthyosaurs were then considered as undiversified, despite their worldwide distribution (e.g. [4]). Their extinction, at the Cenomanian–Turonian boundary [3], was therefore considered as inconsequential because the group was already on the decline since the Jurassic [5]. This vision of ichthyosaur evolution has been substantiated by recent reassessments of the abundant Australian and American material, which regarded both these assemblages as monospecific: '*Platypterygius*' *australis* in Australia [6–13] and '*Platypterygius*' *americanus* in U.S.A. [14]. Yet, numerous new forms have recently been described in Canada and western Eurasia, profoundly modifying the traditional view of ichthyosaur's protracted decline in the Cretaceous [1,15–23].

However, these recent findings are widely spread across time (Berriasian–Albian, around 46 Myr) and space (Canada, Argentina, England, Germany, and Russia), and evidence of co-occurring taxa is extremely scarce. Indeed, only three Cretaceous formations have yielded more than one ichthyosaur taxon: the Wabiskaw Member of the Clearwater Formation (early Albian of Canada; two taxa [22,23]), the Loon River Formation (middle Albian of Canada; two taxa [15,16]), and an unnamed formation from the Barremian of Russia (likely two taxa [21]). Therefore, although recent data indicates ichthyosaurs were not a 'dying group' as previously supposed, this new data gives little clues on the ecological diversity and ecosystem control of the Cretaceous ichthyosaurs: were Cretaceous ichthyosaurs a frequent but minor component of marine trophic webs or did they occupy several ecological niches within marine ecosystems as they did in the past (e.g. Early Jurassic Europe [24,25])? Answering this question requires geological formations containing numerous marine tetrapods – a rare resource in the Early Cretaceous strata – but does not necessarily require articulated specimens.

Here, we analyze the diversity of Albian–basal Cenomanian ichthyosaur assemblages of western Europe, by focusing on the Albian Gault Formation (UK), the Albian–Cenomanian Upper Greensand Formation (UK), the basal Cenomanian Cambridge Greensand Member (base of the Lower Chalk Formation, UK), and the Albian part of the Marnes Bleues Formation (France). The abundant material (several thousands specimens in total) from these localities provides precious data on the taxonomic and ecological diversity of some of the last representatives of Ichthyosauria. In order to evaluate this diversity, we (1) thoroughly reassess the taxonomy of the ichthyosaur assemblages from these formations and (2) evaluate the ecological diversity of these taxa by analyzing their tooth shape, tooth wear, and their relative abundances. Then, these western European assemblages are discussed within the worldwide context of ichthyosaur diversity during the Cretaceous by (3) plotting taxonomic richness curves and (4) evaluating geographical disparity of diversity, providing a background for future analyses of their final extinction.

Materials and Methods

Institutional abbreviations

CAMSM: Sedgwick Museum of Earth Sciences, Cambridge University, Cambridge, UK; CM: Carnegie Museum of Natural History, Pittsburg, PA, USA; IRSNB: Royal Belgian Institute of Natural Sciences, Brussels, Belgium; GLAHM: The Hunterian Museum, University of Glasgow, Glasgow, UK; LEICT: New Walk Museum & Art Gallery, Leicester, UK; MJML: Museum of Jurassic Marine Life, Wareham St Martin, UK; NHMUK: Natural History Museum, London, UK; RGHP: Réserve naturelle Géologique de Haute-Provence, Digne-les-bains, France; SSU: Saratov State University, Saratov, Saratov Oblast, Russia.

No permits were required for the described study, which complied with all relevant regulations.

Nomenclatural acts

The electronic edition of this article conforms to the requirements of the amended International Code of Zoological Nomenclature, and hence the new names contained herein are available under that Code from the electronic edition of this article. This published work and the nomenclatural acts it contains have been registered in ZooBank, the online registration system for the ICZN. The ZooBank LSIDs (Life Science Identifiers) can be resolved and the associated information viewed through any standard web browser by appending the LSID to the prefix "http://zoobank.org/". The LSID for this publication is: urn:lsid:zoobank.org:pub:C9E8AE62-3686-4483-8EEB-861B2DCB102C. The electronic edition of this work was published in a journal with an ISSN, and has been archived and is available from the following digital repositories: PubMed Central, LOCKSS, and ORBi.

Assessment of the taxonomic diversity in bone-bed like deposits

Taxonomic diversity. Two bone-bed-like deposits have been investigated during this research: the Upper Greensand Formation and the Cambridge Greensand Member. Their faunal

diversity must be cautiously assessed, because most of the material is disarticulated. In the sections below, we detail the methodology used to evaluate the taxonomic diversity of these remains and the relative abundances of each recognized taxon.

More than one thousand ichthyosaur specimens (without counting the isolated teeth) are held in the Cambridge Greensand Member collections of the CAMSM, IRSNB, GLAHM, LEICT, and NHMUK. Most of them are disarticulated and consist of isolated bones that were either purchased by or donated to these institutions. We accessed and analyzed all these collections; we used a simple, three-step process to assess the taxonomic diversity of these remains. First, we established morphotypes within each series of abundant and usually diagnostic bones (skull roof bones, teeth, humeri, and femora; see Table 1 for a list of the morphotype recognized and Text S5 for a determination key); however, all specimens and all kinds of fragments, including rostra, centra, ribs, gastralia, phalanges, etc. have been investigated. Then, we used articulated specimens from the upper (unreworked) part of the Cambridge Greensand Member and from coeval deposits of the Vocontian Basin (France) to group some of these morphotypes together. Finally, we compared these morphotypes or groups of morphotypes to known taxa in the literature in order to 'translate' these entities into taxa, when possible. However, we refrained from assessing the diversity at the specific level, especially because of the numerous problems related to the species currently referred to as '*Platypterygius*' [18]. Moreover, the taxonomic value of the numerous small morphological variations observed in the sample is difficult to assess. Nevertheless, some bones, such as humeri and femora contain more distinct morphotypes than the number of taxa (genera) recognized, suggesting a higher diversity at a lower taxonomic level, probably reflecting the specific level. On the other hand, some of these morphotypes contain only a few specimens, so intraspecific variation should also be considered as a possible explanation for the high number of humeral and femoral morphotypes. Indeed, slight inter-adult and ontogenetic variability of humeral distal facets has been recognized in the platypterygiine ophthalmosaurid '*P.*' *australis* [26,27].

All the specimens from these deposits cannot be determined, because isolated elements from the rostrum, mandible and axial skeleton are not diagnostic and because of the presence of small, probably juvenile specimens lacking distinguishing features, in addition to damaged specimens. In total, only 124 specimens of the Cambridge Greensand Member (without counting teeth and the three femur morphotypes belonging to Ophthalmosauridae indet. which are described in Text S6) have been assigned to one of the five infrafamilial taxa that we could recognize. Whatever these taxa might be, the Cambridge Greensand Member provides one of the largest samples of a Cretaceous ichthyosaur assemblage, worldwide.

Relative abundances. We counted all diagnosable isolated bones and articulated specimens to estimate the relative abundance of each taxon in the Cambridge Greensand Member. Articulated specimens were counted only once in the total count. Despite their diagnostic features, we did not consider teeth as reliable bones for abundance counts because reptiles shed their teeth; therefore, the relative abundance of tooth morphotypes partly reflects ethological habits and/or physiological features, polluting the signal.

Ecological diversity. We used absolute tooth size, tooth shape, and tooth wear qualitatively to assess the ecological diversity of the ichthyosaurs from the Cambridge Greensand Member and the Marnes Bleues Formation. Intrinsic properties of teeth (size, shape) give an idea of the optimal range of preys that could be processed (e.g. [25,28]), whereas wear gives indications

Table 1. Bone morphotypes recognized here and their assignation.

Bone	Morphotype	Assignation
Basioccipital	BM1	'Platypterygius' sp.
Basioccipital	BM2	Sisteronia seeleyi
Basioccipital	BM3	Acamptonectes sp.
Tooth	TM1	'Platypterygius' sp.
Tooth	TM2	Sisteronia seeleyi
Tooth	TM3	Ophthalmosaurinae indet.
Humerus	HM1	'Platypterygius' sp.
Humerus	HM2	Sisteronia seeleyi
Humerus	HM3	Ophthalmosaurinae indet.
Humerus	HM4	'Platypterygius' sp.
Femur	FM1	'Platypterygius' sp.
Femur	FM2	Ophthalmosauridae indet.
Femur	FM3	Ophthalmosauridae indet.
Femur	FM4	Ophthalmosauridae indet.
Femur	FM5	Cetarthrosaurus walkeri

The morphotype belong to *Cetarthrosaurus walkeri* is placed within the "Femur" category, as suggested by Seeley [106]. In the text, however, we opted for a more conservative position, considering this morphotype as a propodial, because of its unusual morphology.

on the actual use of teeth by a single individual (e.g. [29,30]). A more detailed and quantitative analysis, encompassing numerous craniodental features of Jurassic and Cretaceous taxa is currently in preparation and will be published elsewhere.

Diversity curves

The temporal evolution of two variables is analysed here: the taxonomic diversity at the specific and the generic levels. Both are simple counts of the parvipelvian taxonomic richness for each time interval (the stage level), from the Hettangian (Early Jurassic) to the Turonian (Late Cretaceous). The dataset compiled is available in Text S7. Stages characterize periods of Earth's history with supposed rather constant climate, ocean dynamics, etc., but sometimes greatly differ in duration. Stage duration influences the number of specimens and thus the biodiversity. Rarefaction methods (e.g. [31]) cannot be employed here because numerous stages of Cretaceous record a very small number of specimens and should therefore be omitted from the analysis using this method. We divided the largest stages (Aptian and Albian) into their usual substages (lower and upper Aptian; lower, middle, and upper Albian), based on ammonite stratigraphy [32–38], rather than using temporal bins. The lower Aptian encompasses the ammonite zones from the *oglanlensis* Zone to the *furcata* Zone; the upper Aptian from *subdonosocostatum* Zone to the *Jacobi* Zone; the lower Albian from the *schrammeni/tardefurcata* Zone to the *mammlilatum/auritiformis* Zone; the middle Albian to the *dentatus* Zone to the *lautus* Zone; the upper Albian from the *cristatum* Zone to the *dispar/briacensis* Zone. Using numerical ages from Kuhnt & Moullade [39], Ogg et al. [40], Scott [35] and Gradstein et al. [41], time bins for the stages/substages from the Hettangian to the Turonian have a mean duration 5.06 My, but the standard deviation remains quite high (±2.25 My). At any rate, these durations should not be considered too strictly as the error margin for many stage boundaries can reach ±1 My, and the numerical age for the

substages of the Aptian and Albian are extrapolations based on the calculations of sedimentations rates between dated horizons [35,39]. Nevertheless, this permits to recover stage durations that are comparable. Moreover, this method of splitting the Aptian and the Albian is also useful for better understanding of the extinction of ichthyosaurs by providing a more precise evolution of ichthyosaur diversity near their extinction. But this approach does not mitigate other biases, such as collecting or environmental biases. Corrections exist for some of these factors [42–46] but this would move the results away from the ichthyosaur fossil record itself, an approach we are reluctant to undertake. This has the advantage of being intuitive and plotting 'raw' values, which are directly related to the fossil record itself and how we interpret it.

The specific and generic curves are simple counts of the taxa that we (or the scientific community) recognize as valid for each time bin and the stratigraphic range of each taxa is based on oldest and youngest unambiguous fossil evidences, thus regardless of any phylogenetic ghost lineages. Lazarus ranges are, however, taken into account: for example, if taxon A occurs during the early Hauterivian and the late Aptian, then we consider taxon A as a valid Barremian and early Aptian taxon as well. The problematic genus *Platypterygius* was considered as a single taxon in the generic curves, grouping all species currently referred to it. The generic and specific diversity curves for the Jurassic are added to provide a point of comparison.

Geological setting

The specimens that we have examined are classified by country, and then by formation. Geographic, stratigraphic (encompassing bio- and lithostratigraphic data) and paleoecological data (focusing on the vertebrate content) are given for each formation, when available. These data were taken from the literature and from collaborative investigations and/or personal field observations.

Gault Formation, UK. The Gault is a marl formation occurring in several basins of England, occurring in the East Midland Shelf, the Bedforshire 'Straits', the Wessex Basin, the Wealden Basin, the Vectian Basin [47]; i.e. the whole eastern, southeastern and southern margins of England. The 'Gault' is also recognized as a facies in adjacent basins; for instance, it possibly occurs in the French Paris Basin [33,48,49]. The data presented below is restricted to the Gault Formation, cropping out in the UK, notably in Folkestone (Figure 1).

The Gault Formation encompasses most of the Albian, and passes laterally to the Cambridge Greensand Member/Upper Greensand Formation towards the east [32,47]. In the Cambridgeshire area, the Gault Formation is middle to late Albian in age, whereas its base extends up to the early Albian (*Tardefurcata* Zone) in the Wealden Basin [32,47]. The fossil-rich locality of Folkestone lies within the Wealden Basin. Most of the Aptian–early Cenomanian English ichthyosaurs fossils studied here were collected during the 19th century as 'coprolites' and subsequently acquired by museums [50]; accordingly, there is no precise stratigraphic data linked to these specimens.

The studied specimens from this formation are from the NHMUK collection (19 specimens; see Text S1). Note that the few Gault Formation ichthyosaurs held at CAMSM appear to be lost; we have been unable to locate them in Sedgwick Museum or in the 'stores' at Cambridge University.

Upper Greensand Formation, UK. The Upper Greensand Formation is a glauconitic sandstone reworked from the Gault Formation [47,51]. The Upper Greensand Formation is distinct from the Cambridge Greensand Member. Both these deposits rework the Gault Formation, but they mostly occur in different basins (part of the Vectian and Wealden basins and part of the

Figure 1. General location of the most important late Early Cretaceous ichthyosaur-bearing localities of England: Cambridge and Folkestone.

Bedforshire 'Straits' for the Upper Greensand Formation VS Southern and transitional Provinces for the Cambridge Greensand Member). When the two deposits co-exist (the Bedforshire 'Straits'/Transitional Province, i.e. the Cambridgeshire area, Figure 1), they are separated by an unconformity with the time-gap of slightly variable duration (Hopson, pers. com. to V.F. June 2012). The onset of the Upper Greensand Formation appears diachronic; its total stratigraphic range is lower Albian to lower Cenomanian [47], whereas the Cambridge Greensand Member is strictly early Cenomanian in age [32].

Because both the Upper Greensand Formation and Cambridge Greensand Member can occur together and all specimens were collected without precise stratigraphic data, it is possible that some specimens were listed as belonging to the wrong 'greensand' deposit in the collection database. Text S2 lists all ichthyosaur specimens from the Upper Greensand Formation.

Cambridge Greensand member, UK. The Cambridge Greensand Member is a glauconitic and phosphatic sandstone forming the basal part of the Lower Chalk Formation in the Bedforshire 'Straits' area/Transitional Zone (i.e. central England, East Anglia Massif) [47,51,52]. Hopson et al. [52] revised the stratigraphy of the English Upper Cretaceous. The 'Lower Chalk' of previous authors is called the Grey Chalk Subgroup, containing two formations in the central England zone: the West Melbury Marly Chalk Formation at the base, overlapped by the Zig Zag Chalk Formation. The Grey Chalk Group is strictly

Cenomanian in age (*Mantelliceras mantelli* to *Calycoceras guerangeri* zones; [52]). The Cambridge Greensand Member constitutes the base of the West Melbury Marly Chalk Formation. Glauconitic chalk (the Glauconitic Chalk Member) lies over the Cambridge Greensand Member or the Upper Greensand Formation in some places [52]. Some important articulated specimens (e.g. CAMSM B58257_67, holotype of *Sisteronia seeleyi*) were deposited in this member, as testified by their mode of preservation.

The Cambridge Greensand Member was deposited during the early Cenomanian [53], but reworks the top of the Gault Formation [47,51,54]. The reworked fossils are phosphatized and late Albian in age ([51] and references therein). However, the uppermost part of this deposit contains unreworked, non-phosphatized early Cenomanian specimens embedded in a glauconitic chalk, possibly at the boundary or within the overlying Glauconitic Marl Member ([52,55]; V.F. & N.B., pers. obs., contra Unwin [56]). This permits one to differentiate both assemblages, if needed. Martill & Unwin ([51] and references therein) indicated that the reworked specimen are not older than the *Calihoplites auritus* Subzone, and were therefore probably contemporaneous (i.e. 'Vraconian', see [35,57]) with the large *Platypterygius hercynicus* of northwestern France (MHNH 2010.4; [18]). Microfossil evidence suggests that the time break between the reworked specimens from the Gault Formation and the 'in-place' early Cenomanian ones is probably small [53], although the base of this member is diachronous – as could be expected from such a transgressive/erosive deposit – becoming younger eastwards [58].

The Cambridge Greensand Member ichthyosaur material consists of several thousands specimens – mostly isolated teeth – and has never been reassessed thoroughly since Seeley's catalogue, published in 1869 [50]. Specimens are housed in the CAMSM, GLAHM, IRSNB, LEICT, and NHMUK collections; see Text S3.

The Marnes Bleues Formation, France. The Marnes Bleues Formation was deposited during the Aptian and Albian in the Vocontian Basin [59]. The Vocontian Basin or Vocontian Trough was a deep, highly subsident Mesozoic basin located at the northwestern border of the Tethys, now southeastern France (Figure 2). It represents the deepest structural unit of the Dauphinois Basin, the Vercors carbonate platform representing its shallow part [60]. All southeastern France Albian ichthyosaur remains known so far were found in the Marnes Bleues Formation.

The Marnes Bleues Formation is a monotonous succession of grey marls with a significant lateral variation in thickness and local unconformities ([61,62]; V.F. & M. G., pers. obs.). Several local sandstone and limestone beds interrupt the sequence (e.g. [59,63]; V.F. & M. G., pers. obs.). Cephalopods are rare in this formation, and the age of the horizon of some specimens is only loosely constrained. In the Sisteron locality, two unconformities disturb the sequence: the upper Aptian lies on the truncated middle Aptian, and the last few meters of lower Albian (or the middle Albian) lie on the truncated upper Aptian via a 20 cm-thick glauconitic sandstone layer [59]; Figure 3). The specimens RGHP SI 1, RGHP SI 2, and RGHP SI 3 were found 2, 8, and 25 meters above the Aptian–Albian discordance, respectively, and are late early to middle Albian in age (Figure 3). In the Prads locality, the upper part of the Marnes Bleues Formation crops out, but a Quaternary terrace reworking sandstone clasts of the Oligocene Grès d'Annot Formation truncates the top of the Marnes Bleues Formation. The specimen RGHP PR 1 was found 6.5 m below the base of the Quaternary terrace and is late Albian in age [64] (Figure 4). Text S4 lists all ichthyosaurs from the Marnes Bleues Formation studied in the present paper.

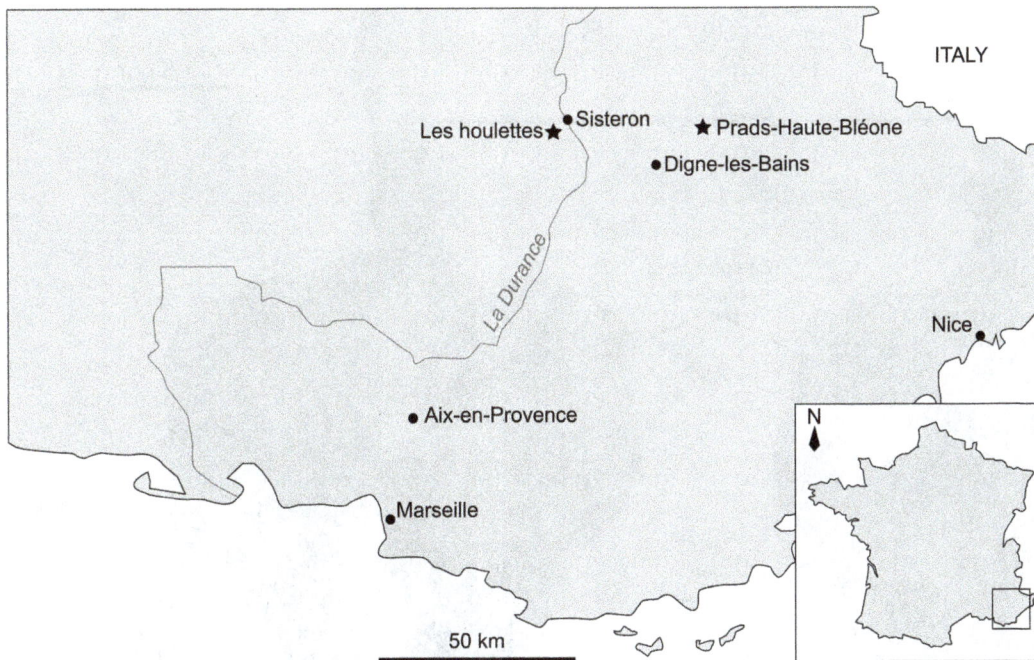

Figure 2. Location of the most important late Early Cretaceous ichthyosaur-bearing localities of the Vocontian Basin in Southeastern France. Stars indicate fossil-localities and plain circles indicate major cities.

Results

Systematic Paleontology

The asterisk (*) next to referred specimens indicates articulated specimens, others are isolated elements.

Ichthyosauria Blainville, 1835 [65]

Ophthalmosauridae Baur 1887 [66]

Platypterygiinae Arkhangelsky 2001 [67] sensu Fischer et al. [20]

Sisteronia seeleyi gen. et sp. nov. urn:lsid:zoobank.org:act:1-B87EED5-6C16-49EE-ADC2-67FEB04819F0

Figures 5, 6, 7

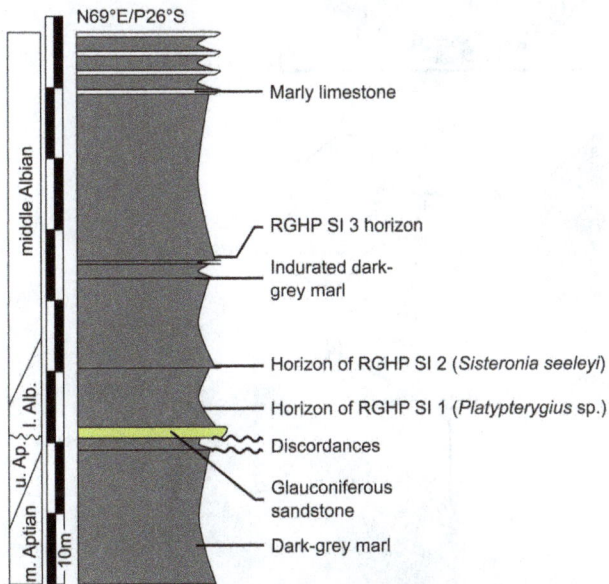

Figure 3. Stratigraphic log of Les Houlettes locality, Sisteron, Alpes de Haute-Provence, France. The position of the stratigraphic boundaries is taken from Bréhéret [59] and personal fieldwork by V.F. and M.G. Abbreviations: Alb, Albian; Ap, Aptian; m., middle; l., lower; u., upper.

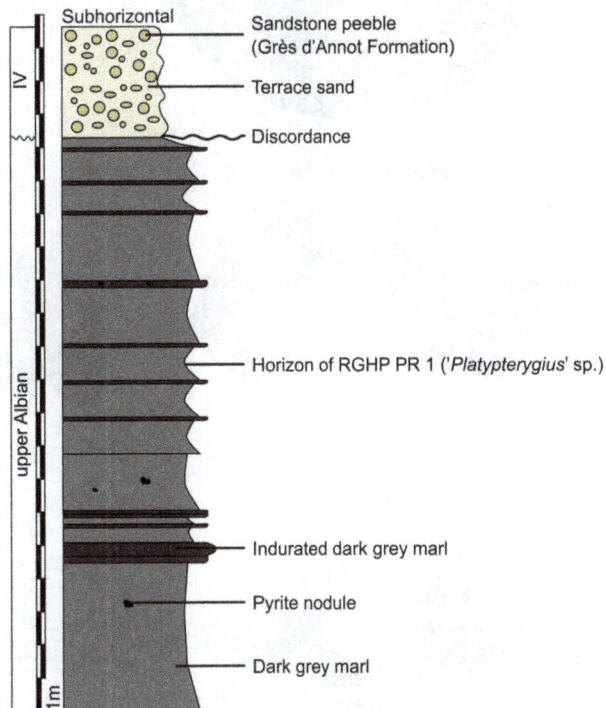

Figure 4. Stratigraphic log at RGHP PR 1's discovery site, Prads-Haute-Bléonne, Alpes de Haute-Provence, France.

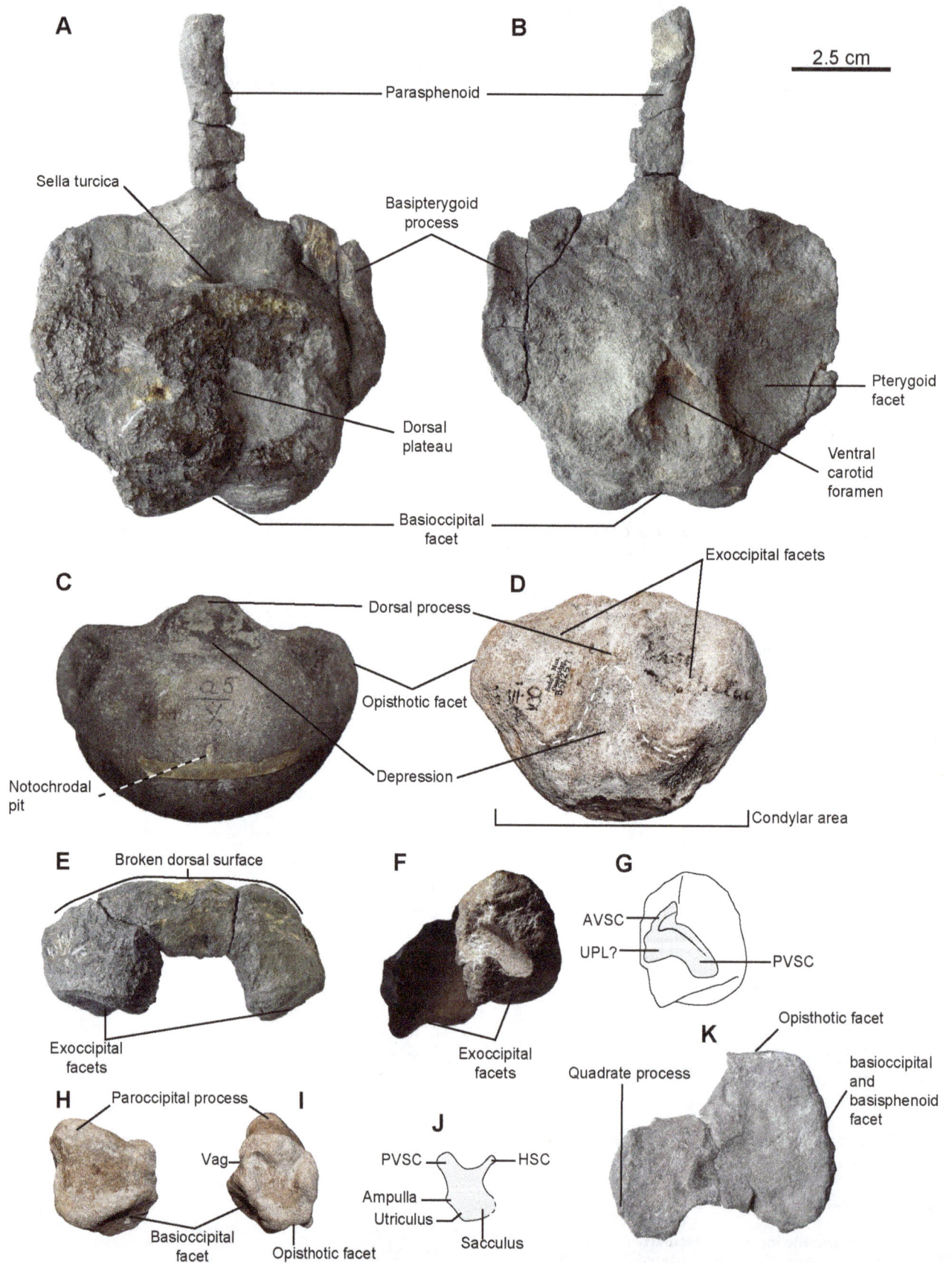

Figure 5. *Sisteronia seeleyi*, basicranium. **A, B: basisphenoid (RGHP SI 2) in dorsal (A) and ventral (B) views.** C: basioccipital (CAMSM B57943) in posterior view. D: holotype basioccipital (CAMSM B58257_67) in dorsal view. E–G: supraoccipital (RGHP SI 2) in posterior (E) and

anterolateral (otic) (F, G) views. H–J: left opisthotic (CAMSM B58257_67) in posterior (H) and anterior (otic) (I, J) views. K: left stapes (RGHP SI 2) in posterior view. Note the extremely reduced (nearly absent) extracondylar area of the basioccipital, a platypterygiine synapomorphy, and the dorsal process posterior to a triangular depression (delineated by the thick dotted line) on the basioccipital, an autapomorphy of *Sisteronia seeleyi*. Abbreviations: AVSC: impression of the anterior vertical semicircular canal of the otic labyrinth; HSC: impression of the horizontal semicircular canal of the otic labyrinth; PVSC: impression of the posterior vertical semicircular canal of the otic labyrinth; UPL: impression of the utricular portion of the otic labyrinth; Vag: vagus foramen.

1889 *Ichthyosaurus campylodon* Lydekker [68]: 19 (NHMUK R16)

1889 *I. campylodon / Ophthalmosaurus?* Lydekker [68]: 20 (NHMUK 44159)

1889 *I. campylodon / Ophthalmosaurus?* Lydekker [68]: 20 (NHMUK 44159a)

2003 Ichthyosauria indet. McGowan & Motani [69]: 27: Figure 37

Holotype. CAMSM B58257_67, an incomplete specimen, including partial basicranium, scapula, humerus, and 5 centra from unreworked (chalky) part of the Cambridge Greensand member (early Cenomanian, Late Cretaceous). The basioccipital is fully ossified and the humerus lacks a rugose texture on its shaft, suggesting a mature specimen [70].

Referred material from the Cambridge Greensand. CAMSM B57943 (basioccipital); CAMSM B57945 (basioccipital); CAMSM B57948 (basioccipital); CAMSM B57950 (basioccipital); CAMSM B57947 (basioccipital); CAMSM B57941 (basioccipital); CAMSM B57951 (basioccipital); CAMSM B57946 (basioccipital); CAMSM B57956 (basioccipital); CAMSM B57954 (basioccipital); CAMSM B58314 (basioccipital); CAMSM TN1727 partim (basioccipital); CAMSM TN1735 partim (6 basioccipitals); CAMSM TN1739 partim (basioccipital); CAMSM TN1751 partim (6 basioccipitals); CAMSM TN1753 partim (basioccipital); IRSNB GS54 (basioccipital); IRSNB GS61 (basioccipital); LEICT G107.1991 (basioccipital); NHMUK 44159 (basioccipital); NHMUK 44159a (basioccipital); CAMSM B57908 (opisthotic); CAMSM B58077_78 (2 opisthotics); CAMSM TN1753 partim (opisthotic); NHMUK R2348 (opisthotic); IRSNB GS10 (opisthotic); CAMSM B58091 (tooth); CAMSM B58092 (tooth); CAMSM TN1716 partim (numerous teeth); CAMSM TN1778 partim (numerous teeth); CAMSM TN1779 partim (numerous teeth); CAMSM B58390 (tooth); NHMUK R1923 (tooth); IRSNB GS23 (tooth); IRSNB GS24 (tooth); IRSNB GS55 to GS58 (teeth); CAMSM TN1755 partim (humerus); CAMSM TN1757 partim (humerus).

Referred material from other deposits. NHMUK R16 partim (teeth, Gault Formation); NHMUK R17 partim (teeth, Gault Formation); NHMUK R2890 partim (opisthotic, Gault Formation); NHMUK 47232 partim (teeth, Gault Formation); RGHP SI 2*, an incomplete skull, containing fragmentary snout and nasals, basioccipital, quadrate, opisthotic, supraoccipital, stapes, teeth from the middle Albian of Sisteron. At least three additional articulated specimens from the middle–late Albian of the Marnes Bleues Formation of the Vocontian Basin are present in the private collection of L. Ebbo [71].

Diagnosis. Platypterygiine ophthalmosaurid characterized by the following autapomorphies: basioccipital with raised process on the floor of foramen magnum; opisthotic with nearly absent paroccipital process (as in juvenile '*P.*' *australis* [26]); tooth with gracile crown and root with rectangular cross-section, the labio-lingual length being usually equal to one half of the anteroposterior length (less conspicuous in anterior- and posterior-most teeth).

Sisteronia seeleyi is also characterized by the following unique combination of features: elongated anterior process of the maxilla, reaching anteriorly the level of the nasal (unlike in *Aegirosaurus* [72]; *Sveltonectes insolitus* [21]); prominent opisthotic facets on basioccipital (shared with *S. insolitus* [21]); expanded sacculus impression on opisthotic (shared with adult '*P.*' *australis* [11] and *A. densus* [20]); anteroposteriorly shortened quadrate condyle (shared with *O. icenicus* [73] and *S. insolitus* [21]); U-shaped supraoccipital (shared

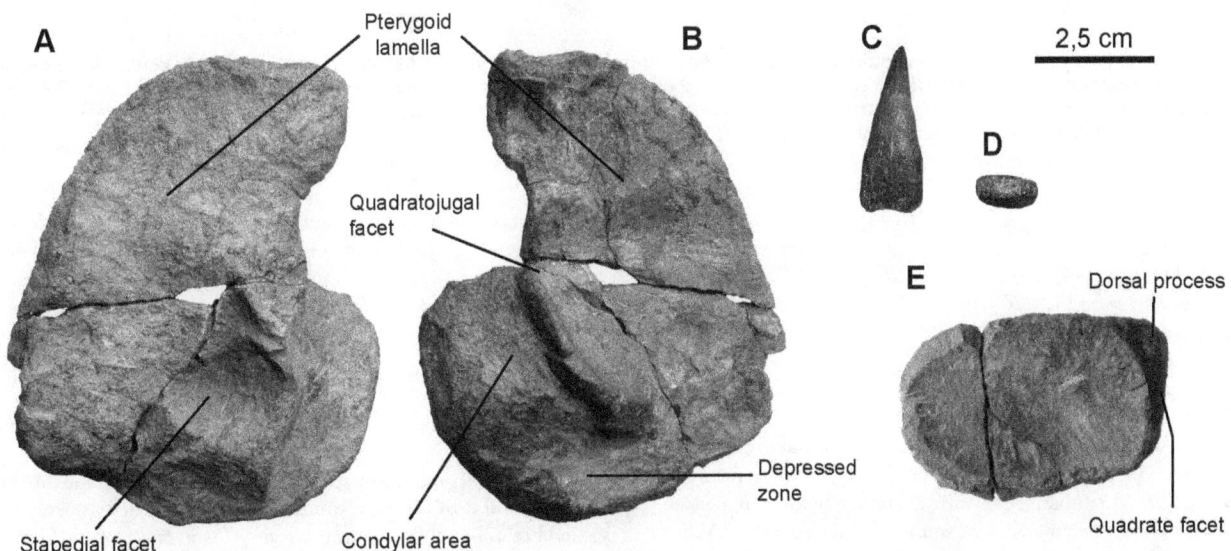

Figure 6. *Sisteronia seeleyi,* **quadrate, tooth and articular.** A, B: right quadrate (RGHP SI 2) in medial (A) and lateral (B) views. C, D: typical mid-rostrum tooth of *Sisteronia seeleyi* (CAMSM TN1779 partim) in labial view (C) and basal (D) views, showing the markedly rectangular cross-section of the root. E: right articular (RGHP SI 2) in lateral view.

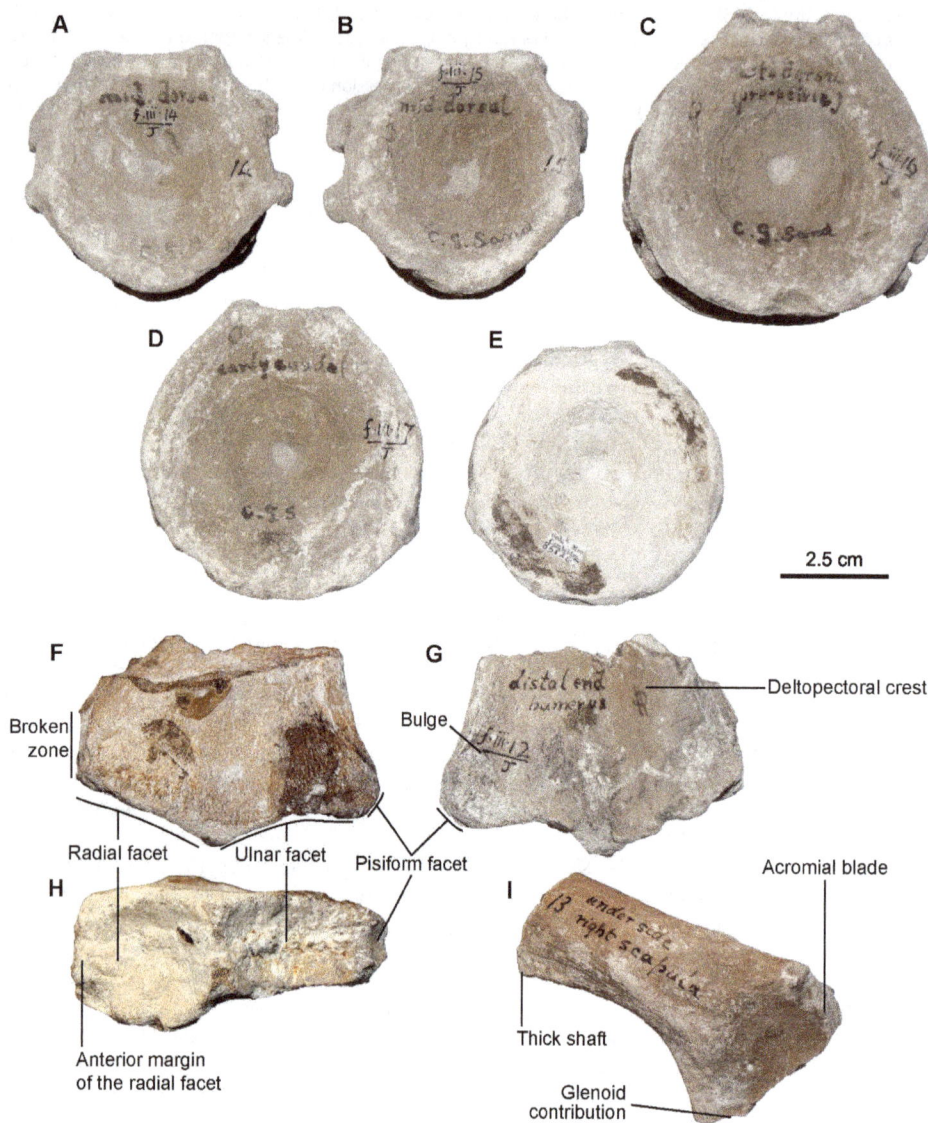

Figure 7. *Sisteronia seeleyi*, **axial and shoulder girdle elements of holotype specimen (CAMSM B58257_67). A–E: centra in anterior view.** A: cervical centrum. B: anterior thoracic centrum. C: posterior thoracic centrum, close to the sacral region. D, E: anterior caudal centra. F–H: left humerus (CAMSM B58257_67) in dorsal (F), ventral (G), and distal (H) views. Note the presence of a facet for a posterior accessory epipodial element, a feature only found in some platypterygiine ichthyosaurs. I: right scapula in anterior view.

with '*P.*' *australis* [11]; '*P.*' *hercynicus* [18,74] and *O. natans* [75]); humerus with a facet for a posterior accessory element (shared with '*P.*' *hercynicus* [74,76]; '*P.*' *americanus* [14]; '*P.*' sp. [16,77]; '*Ophthalmosaurus monocharactus*' [78]).

Stratigraphic range. Early Albian–early Cenomanian (stratum typicum).

Geographic range. Eastern England basins (locus typicus), Vocontian Basin, France.

Note. As mentioned in the 'Referred material from other deposits' section, above, additional articulated specimens from the Albian of the Vocontian Basin are currently held in a private collection. These specimens were studied in the course of V.F.'s PhD thesis [71] and this information is crucial to establish the phylogenetic relationships of *Sisteronia*. Because this material cannot be used for the time being, we refrain from assessing the phylogenetic position of *Sisteronia* in this paper. These data and the phylogenetic placement of *Sisteronia* can be found in V.F.'s thesis

[71]. However, as *Sisteronia* possesses numerous synapomorphies of platypterygiine ophthalmosaurids and lacks the synapomorphies of ophthalmosaurine ophthalmosaurids (see Anatomical Descriptions, below), we confidently place this taxon within Platypterygiinae.

Description

Measurements taken on CAMSM B58257_67 can be found in Table 2.

Basioccipital (morphotype 2, see Systematic Paleontology above for a list of all specimens; Figure 5). The basioccipital is roughly semi-circular in posterior view. As in *Sveltonectes* [21], the basioccipital is wider than high because of the prominence of the bulge-like opisthotic facet, the complete reduction of the extracondylar area ventrally, and the deep exoccipital facets. The extracondylar area is extremely reduced laterally (condyle width = 84.69% of the total width in CAMSM B57943) and

Table 2. Selected measurements on CAMSM B58257_67, holotype of *Sisteronia seeleyi*.

Measurement (mm)	CAMSM B58257_67
Basioccipital height	36.95
Basioccipital width	70.83
Basioccipital length	49.61
humerus distal diameter	68.1
Radial facet length	32.53
Ulnar facet length	31.15
First preserved dorsal centrum height	52.13
First preserved dorsal centrum width	53.03
First preserved dorsal centrum depth	24.24
Last preserved dorsal centrum height	52.65
Last preserved dorsal centrum width	53.34
Last preserved dorsal centrum depth	24.41
First preserved caudal centrum height	63.15
First preserved caudal centrum width	64.17
First preserved caudal centrum depth	23.31
Last preserved caudal centrum height	55.96
Last preserved caudal centrum width	57.26
Last preserved caudal centrum depth	20.99

Measurements are recorded up to the nearest 0.01 mm using a digital caliper.

invisible ventrally in posterior view, a synapomorphy of platypter-ygiine ophthalmosaurids [20]. The condyle is oval and not flattened, and the notochordal pit is located ventral to the central point in most specimens. There is no ventral notch, but the ventral surface is flattened. The stapedial facet is not visible. The exoccipital facets are prominent and bordered medially and posteromedially by a prominent ridge. Both ridges meet medially and form a prominent process dividing the floor of foramen magnum in two in the transverse plane. In dorsal view, this ridge is wave-like and W-shaped. This structure appears ontogenetic, because the smaller basioccipitals have a reduced ridge. The anterior surface is flat and vertical, and the notochordal groove is shallow or absent. Two specimens (CAMSM B57948 and CAMSM B57954) have reduced opisthotic facets, a reduced exoccipital ridge, and deep dorsoventral grooves separating the basisphenoid facet from the opisthotic facet, as in '*P.*' *australis* [11]. They are nevertheless closer to the *Sisteronia* morphotype in general shape and are therefore included in this group.

Opisthotic (CAMSM B57908; CAMSM B58077_78; CAMSM B58257_67* (holotype); CAMSM TN1753 partim; NHMUK R2348; IRSNB GS10; NHMUK R2890 partim; CAMSM 'Saxon Cement works Cambridge 1912'; RGHP SI 2*; Figure 5). The paroccipital process is robust and extremely shortened, unlike that of ophthalmosaurine ichthyosaurs [20,73], and even shorter than in adult '*P.*' *australis* [11] and '*P.*' *hercynicus* [74,76] and resembles that of juvenile '*P.*' *australis* [26]. There is no lateral ridge, unlike in *O. icenicus* and *A. densus* [20,73]. The opisthotic forms two facets medioventrally: a large, rugose, triangular facet facing posteroventrally for the basioccipital and a smaller, roughly triangular facet for the stapes. The stapedial facet is frequently subdivided by a deep anterolateral groove. This deep and narrow groove probably housed the hyomandibular branch of facial (VII) nerve or the glossopharyngeal (IX) nerve [73] and can be extremely complex in some specimens, such as NHMUK

R2890, forming lateral spirals. The otic capsule impression has a deep and elongated impression for the horizontal semicircular canal, a wider and shorter impression for the posterior vertical semicircular canal, and a markedly expanded sacculus, as in adult '*P.*' *australis* [11] and the holotype (adult) specimen of *A. densus* [20].

Stapes (RGHP SI 2*; Figure 5). Both stapes are preserved in RGHP SI 2 but crushed along different planes. The shaft is short and robust unlike in *A. densus* [20]. The opisthotic surface forms a marked angle with the basioccipital/basisphenoid facet. There is no evidence for a hyoid process.

Supraoccipital (RGHP SI 2*; Figure 5). The supraoccipital is U-shaped with a 'squared' opening for the foramen magnum, similar to the condition in '*P.*' *hercynicus* [18,74]. The exoccipital facets are trapezoidal, tapering posteriorly, and are markedly concave. Partial otic impressions are preserved in RGHP SI 2; the impression for the posterior vertical semicircular canal is extremely deep. The utriculus ('utricular portion of labyrinth' of McGowan [79]) impression is a broad semicircular depression that is confluent with the impression for the posterior vertical semicircular canal dorsolaterally. Unlike in '*P.*' *australis* and *A. densus* [11,20], the impression for the anterior vertical semicircular canal is markedly reduced in length and depth and is separated from the rest of the otic impression by a lateral ridge.

Parabasisphenoid (RGHP SI 2*; Figure 5). The basipterygoid process is markedly reduced and forms an elongated bulge on the lateral surface of the basisphenoid. It is even more reduced than in *Sveltonectes*, where it forms a small protruding rod-like process [21], but it may be partly due of the strong diagenetic compaction of this bone in RGPH SI 2. The dorsal plateau appears kidney-shaped, as in *S. insolitus* [21] and unlike those of '*P.*' *australis* (hexagonal [11]), *Brachypterygius* (squared [69]), and *O. icenicus* (rounded [73]). The ventral surface of the basisphenoid bears a wide depression for the medial lamella of the pterygoid. The ventral carotid opening is set in the posterior half of the ventral surface. The posterior surface is divided by a deep median cleft, as in many post-Triassic ichthyosaurs (V.F., pers. obs. on NHMUK and CAMSM material). The parasphenoid is completely fused to the basisphenoid in RGHP SI 2, suggesting a mature age [11], although the ontogenetic significance of this feature has been debated recently [26].

Quadrate (CAMSM B58257_67*; RGHP SI 2*; Figure 6). The quadrate is ear-shaped as in most ophthalmosaurids. The medial surface is flat, and the stapedial articular facet is a deep depression bordered posteriorly and ventrally by a bony ridge. There is no evidence for a marked occipital lamella, unlike in *O. icenicus*, '*P.*' *australis* or *S. insolitus* ([11,21,73], respectively). The lateral surface is smooth and markedly concave. The short condyle is thick along its whole length, and rapidly tapers anteriorly, as in *O. icenicus* and *S. insolitus* [21,73]. The ventral surface of the condyle is concave anteriorly and becomes progressively flat posteriorly. The condyle is separated from the pterygoid lamella by a concave area. Similar quadrates occur in the Cambridge Greensand Member (e.g. CAMSM B57988; CAMSM B57989; NHMUK 35272 [two specimens]; IRSNB GS1; IRSNB GS6; IRSNB GS8), but the lack of clear-cut diagnostic feature prevents confident referral of these isolated bones to *Sisteronia seeleyi*; only the quadrates found in articulation with diagnostic elements are referred to the relevant taxa.

Pterygoid (RGHP SI 2*). A fragmentary pterygoid is preserved in RGHP SI 2. The dorsal lamella has a thick base, and the reception pits for the basipterygoid process are unremarkable, unlike in *A. densus* [20].

Articular (CAMSM B58257_67*; RGHP SI 2*; Figure 6). The left articular is preserved. It appears distinct from that of other

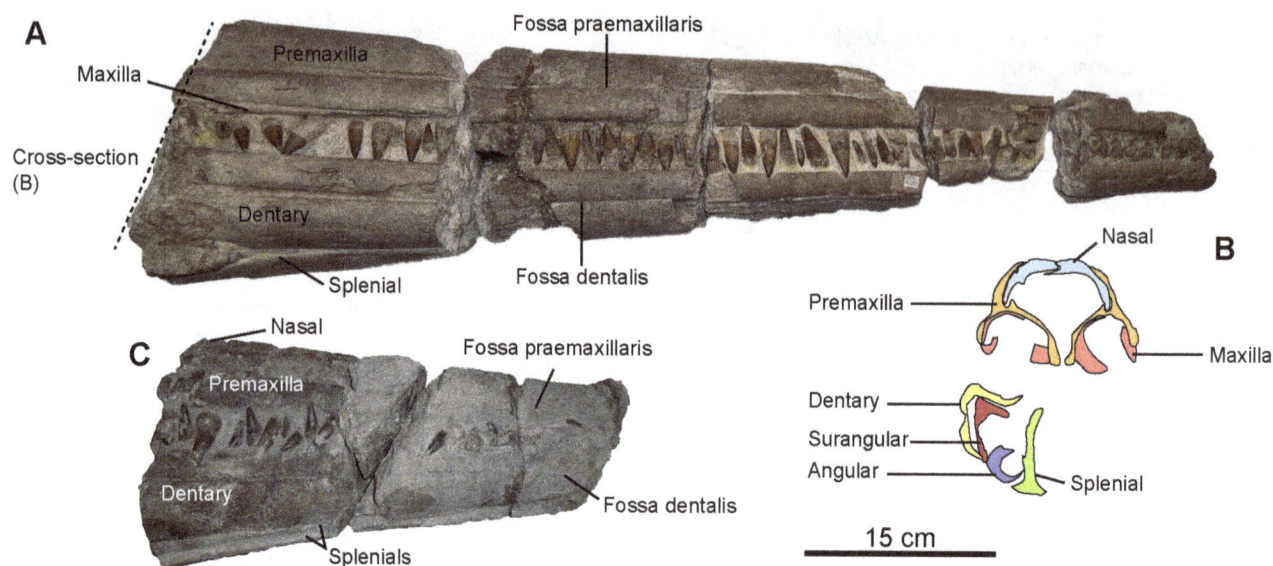

Figure 8. *'Platypterygius'* sp., rostra. A: CAMSM TN283, articulated rostrum in right lateral view. The dashed line indicates the plane and position of the cross-section in B. B: posterior-most cross-section of CAMSM TN283, set posterior to the symphysis. C: RGHP PR 1, articulated rostrum in right lateral view.

ichthyosaurs (e.g. *Ichthyosaurus communis* [79], '*P.*' *australis* [11,26], *O. icenicus* [80]) in being anteroposteriorly elongated (as in *Arthropterygius chrisorum* [81]) and rectangular. It lacks the muscle attachment bulge seen '*P.*' *australis* and *Sveltonectes insolitus* [11,21].

Dentition (morphotype 2; RGPH SI 2*; see Systematic Paleontology above for a complete list of specimens; Figure 6). The teeth are straight generally much smaller than in other coeval taxa; the crown accounts for half of the total height in most teeth. Anterior and median teeth have a slender, straight, a conical crown with well-expressed apicobasal ridges and a markedly laterally compressed, yet quadrangular root. This is not a diagenetic artifact, because a large number of roots have resorption pits that remain perfectly circular and dozens of similar teeth are found in the Gault Formation and Cambridge Greensand Member. Posterior teeth have smaller and more robust crowns, and squarer root cross section. A smooth acellular cementum ring is present, and the root is smooth and lacks a thick layer of cement, unlike in '*Platypterygius*' [82]. It is worth noting that quite similar teeth are found in a juvenile specimen of '*P.*' *australis* (NHMUK unnumbered). This may indicate close relationship between these two taxa and/or potential heterochronial processes related the tooth development.

Centra (CAMSM B58257_67*; Figure 7). A subtle ventral keel occurs on anterior thoracic centra, giving them a pentagonal shape. These centra have prominent diapophyses and parapophyses; horizontal bony ridges follow these apophyses posteriorly. Sacral and anterior caudal centra are weakly amphicœlous and have a circular outline.

Scapula (CAMSM B58257_67*; Figure 7). The medioventral part of the scapula is dorsoventrally compressed and widely expanded anteroposteriorly, to form the articulation area for the coracoid and the glenoid ventrally, and the acromion process anteriorly. Most of the medial part of the proximal surface is missing, so it is impossible to know if the scapular facet and the acromion process were continuous, as in *Ophthalmosaurus icenicus* [83], *Acamptonectes densus* [20], and *Platypterygius americanus* [84], or separated by a deep notch as in *Sveltonectes* [21]. The dorsal surface of the medial part of the scapula is concave, whereas its ventral

surface is flat. The posterior margin of the scapula is markedly curved. Distally, the scapula is thick and rod-like, as in '*P.*' *hercynicus* [74,76] and unlike *O. icenicus* [80,85] and *A. densus* [20].

Humerus (CAMSM B58257_67*; CAMSM TN1757 partim; Figure 7). The anterior surface of the shaft is rounded, whereas the posterior blade is acute and bordered by concave areas, giving the humerus a teardrop shape in cross-section. The deltopectoral crest nearly reaches the distal end of the humerus and merges with the ventral edge of the radial facet. Posterodistally, a bulge is present on the ventral side of the humerus, near the ulnar facet as in *Sveltonectes insolitus* [21] (but a dorsal bulge is also present in *Sveltonectes* [V.F., pers. obs.]). The humerus forms at least three distal facets: a large rounded radial facet, a longer (anteroposterior distance) but thinner (dorsoventral distance) ulnar facet, and a small triangular postaxial accessory facet. This condition has only been reported in some taxa referred to as *Platypterygius* ('*P.*' *hercynicus* [74]; '*P.*' *americanus* [14]; '*P.*' sp. [16,77]). All facets are rugose and concave. The anterodistal extremity of the humerus is damaged. Yet, the anterior edge of the radial facet is preserved, and the shape of the anterior surface of the humerus suggests that a facet for an anterior accessory epipodial element was also present.

Systematic Paleontology

Platypterygius Huene 1922 [86]

'*Platypterygius*' sp.

Figures 8, 9, 10, 11, 12, 13, 14, 15

1869   *Ichthyosaurus platymerus*   Seeley [50]: xvii

1869   *Ichthyosaurus bonneyi*   Seeley [50]: xvii

1889   *Ophthalmosaurus* (?) *cantabrigiensis* Lydekker [68]: 9 (NHMUK 35310)

1889   *Ichthyosaurus campylodon* Lydekker [68]: 17 (NHMUK 47235)

1889   *Ichthyosaurus campylodon* Lydekker [68]: 18 (NHMUK 35254)

1889   *Ichthyosaurus campylodon* Lydekker [68]: 18 (NHMUK 47265)

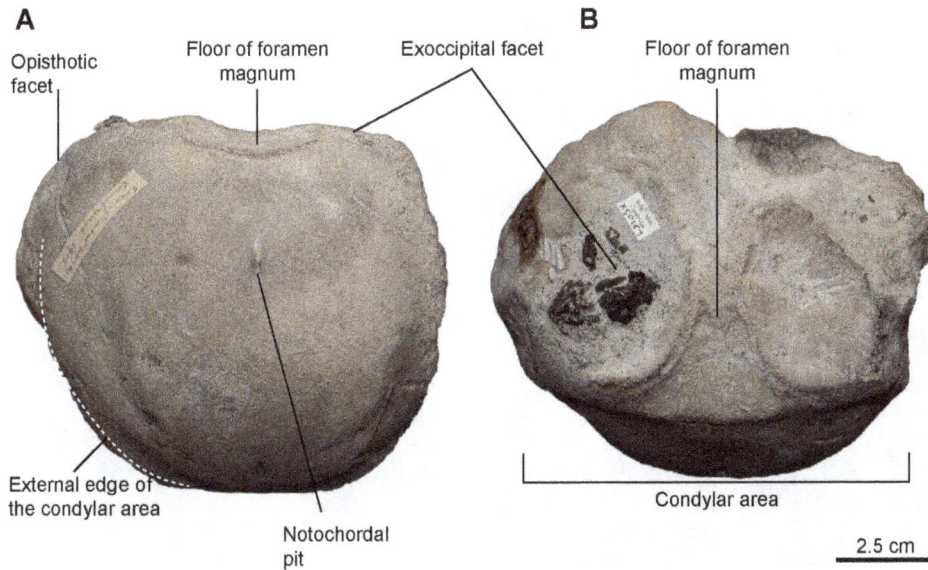

Figure 9. 'Platypterygius' sp., basioccipital (CAMSM X50167) in posterior (A) and dorsal (B) views. Note the extremely reduced extracondylar area, a platypterygiine synapomorphy that appears exaggerated in this taxon.

1889 *Ichthyosaurus campylodon* Lydekker [68]: 18 (NHMUK 30253_4)

1889 *Ichthyosaurus campylodon* Lydekker [68]: 18 (NHMUK 32242)

1889 *Ichthyosaurus campylodon* Lydekker [68]: 18 (NHMUK 35434)

1889 *Ichthyosaurus campylodon* Lydekker [68]: 18 (NHMUK 40358)

1889 *Ichthyosaurus campylodon* Lydekker [68]: 18 (NHMUK 41896)

1889 *Ichthyosaurus campylodon* Lydekker [68]: 18 (NHMUK 32406)

1889 *Ichthyosaurus campylodon* Lydekker [68]: 18 (NHMUK 40095)

1889 *Ichthyosaurus campylodon* Lydekker [68]: 18 (NHMUK 46381)

1889 *Ichthyosaurus campylodon* Lydekker [68]: 18 (NHMUK 47269)

1889 *Ichthyosaurus campylodon* Lydekker [68]: 18 (NHMUK 47235)

1889 *Ichthyosaurus campylodon* Lydekker [68]: 19 (NHMUK R16)

1889 *Ichthyosaurus campylodon* Lydekker [68]: 19 (NHMUK 47270)

1889 *Ichthyosaurus campylodon* Lydekker [68]: 19 (NHMUK 36318)

1889 *Ichthyosaurus campylodon* Lydekker [68]: 19 (NHMUK 36384)

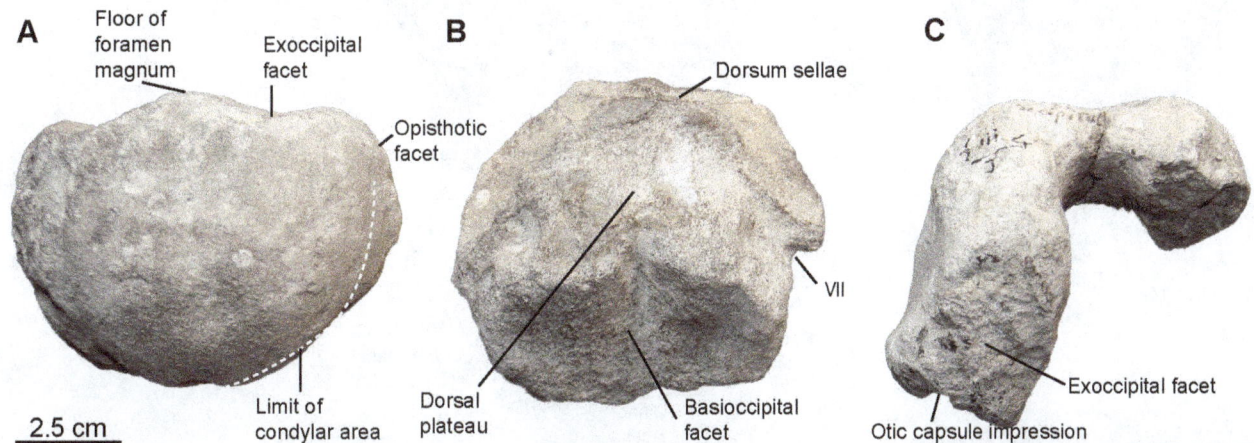

Figure 10. 'Platypterygius' sp., associated basicranium of CAMSM B58250_56. A: basioccipital in posterior view. This basioccipital has a raised floor within the foramen magnum, as in numerous other isolated basioccipitals and 'Platypterygius cf. kiprijanoffi' described by Bardet [71]. B: basisphenoid in dorsal view. C: supraoccipital in posterior view. This specimen also contains a femur (femur morphotype 1). Abbreviations: VII: foramen for the facialis nerve (VII).

Figure 11. 'Platypterygius' sp., teeth (morphotype 1) of medium size. The eight teeth on the left are isolated teeth grouped within the specimen CAMSM B58010 to 58019, and the six teeth on the right are said to have been found associated (specimen CAMSM B76728_45), but their mode of preservation recalls the reworked part of the Cambridge Greensand Member, making it highly unlikely. Note the bulbous and striated root.

1889 *I. campylodon/ Ophthalmosaurus?* Lydekker [68]: 20 (NHMUK 35323)

1960 *Myopterygius campylodon* Delair [87]: 69 (CAMSM B5839_82)

1960 *Myopterygius campylodon* Delair [87]: 70 (NHMUK 40095)

Referred material from the Cambridge Greensand Member. CAMSM TN283* (rostrum and associated 112 teeth); CAMSM B42404_20* (basioccipital, centra); CAMSM

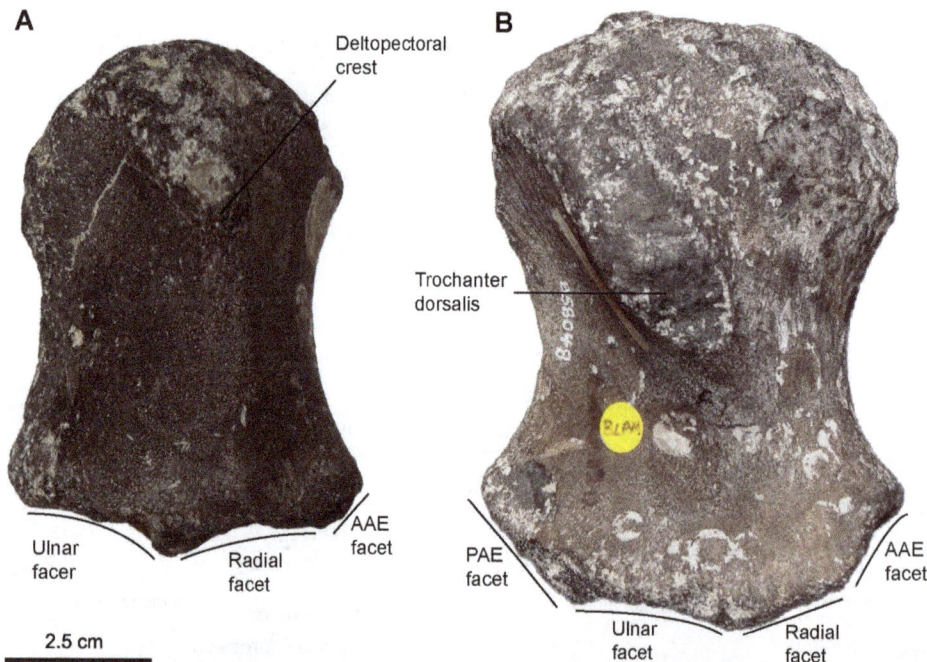

Figure 12. 'Platypterygius' sp., humerus morphotypes. A: Left humerus (morphotype 1) in ventral view (CAMSM TN1757 partim). Note the large radial and ulnar facets set on the same plane. B: Right humerus (morphotype 4) in dorsal view (CAMSM B58048). Note the large four distal facets including one for an anterior and a posterior accessory epipodial element. Abbreviation: AAE, anterior accessory epipodial element; PAE: posterior accessory epipodial element.

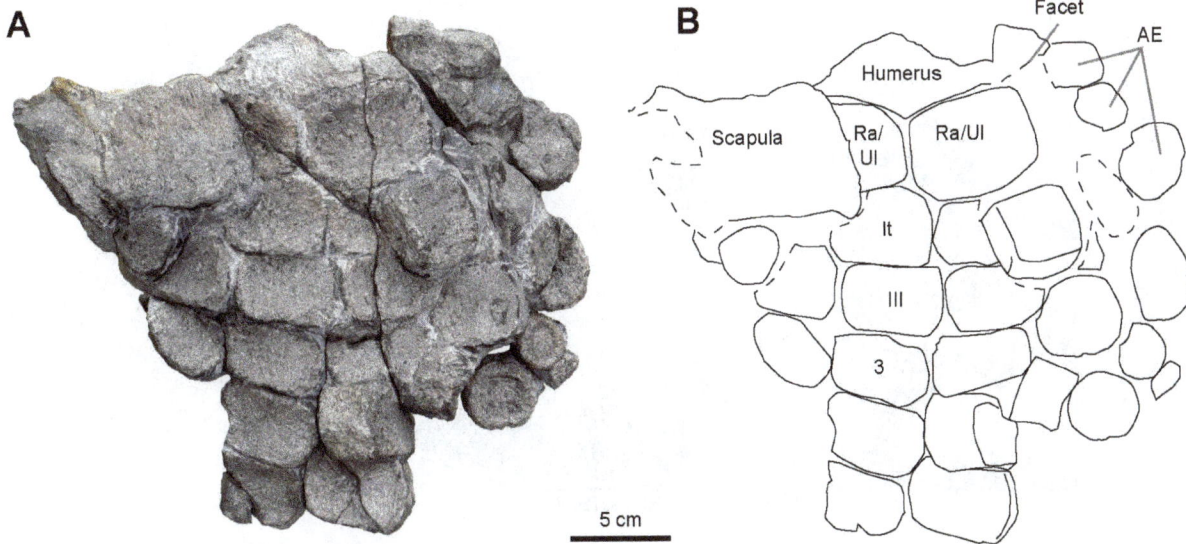

Figure 13. 'Platypterygius' sp., articulated partial forefin (RGHP PR 1), photograph (A) and interpretation (B). The remains are insufficient to characterize which side this forefin is from. Abbreviation: AE: accessory elements; III: carpal 3; It: intermedium; Ra: radius; Ul: ulna; 3: metacarpal 3.

B57939 (basioccipital); CAMSM B57940 (basioccipital); CAMSM B57944 (basioccipital); CAMSM B57959_60* (basioccipital, atlas-axis); CAMSM B58250_56* (quadrate, basioccipital, basisphenoid, supraoccipital, femur); CAMSM B75735 (basioccipital); CAMSM X50161 (basioccipital); CAMSM X50168 (basioccipital); CAMSM X50169 (basioccipital); CAMSM TN1729 partim (basioccipital); CAMSM TN1754 partim (basioccipital); CAMSM TN1755 partim (2 basioccipitals); NHMUK 35323 (basioccipital); several dozens of teeth, including CAMSM B57996_58009, CAMSM B58010 to B58027, CAMSM B58305_13; CAMSM B58379_87, CAMSM B76728_45, CAMSM TN1716 partim, CAMSM TN1778 partim; CAMSM TN1779 partim, NHMUK R625, NHMUK R133b partim, NHMUK R2336 partim (2 teeth), NHMUK 28110 partim, NHMUK 30253 partim, NHMUK 30254 (4 teeth), NHMUK 32406 partim, NHMUK 33242, NHMUK 35254 partim, NHMUK 35432_5, NHMUK 40358, NHMUK 41896, NHMUK 46381, NHMUK 47265_66* (teeth, centra), NHMUK 47269, IRSNB GS21, IRSNB GS25 to GS28, IRSNB GS32 to GS50, IRSNB GS53, IRSNB GS62; CAMSM B97401 partim (humerus morphotype 1); CAMSM B57987 (humerus morphotype 1); CAMSM B58043 (humerus morphotype 4); CAMSM B58048 (humerus morphotype 4); CAMSM B58056 (humerus morphotype 1); CAMSM B58057 (humerus morphotype 1; holotype of *Ichthyosaurus platymerus*); CAMSM B97401 partim (humerus morphotype 1); CAMSM TN1734 partim (humerus morphotype 4); CAMSM TN1751 partim (humerus morphotype 1); CAMSM TN1753 partim (one (humerus morphotype 1and one (humerus morphotype 4); CAMSM TN1757 partim (humerus morphotype 4); NHMUK R2342 partim (two humerus morphotype 4); CAMSM B58058 (femur); CAMSM B58060 (femur); CAMSM B58062 (femur; holotype of *Ichthyosaurus bonneyi*); CAMSM B58063_4 (femur); CAMSM B58361 (femur); CAMSM TN1749 partim (femur); CAMSM TN1748 partim (femur); CAMSM TN1757 partim (2 femora); NHMUK R23412 partim (femur); NHMUK R3510 (femur).

Additional material from other deposits. RGHP SI 1* (basioccipital, centra); RGHP PR 1* (rostrum, teeth, scapula, humerus, forefin); NHMUK 40095 (tooth, Gault Formation); NHMUK 47235* (a dentary and 12 teeth); NHMUK R16 partim (tooth, Gault Formation); NHMUK R2890 (tooth, Gault Formation); NHMUK 36318 (teeth, Gault Formation); NHMUK 36384 (teeth, Gault Formation); NHMUK 47235 (teeth, Gault Formation); NHMUK 47270 (tooth, Gault Formation).

Occurrence. Late Albian of Gault Formation (UK), middle and late Albian of Marnes Bleues (France), earliest Cenomanian of the Cambridge Greensand Member (UK).

Note. This taxon corresponds to most of the material previously referred to as '*Platypterygius*' and '*P. campylodon*' from the Albian–earliest Cenomanian of Europe. *Platypterygius campylodon* was erected on material from the chalk [88,89] and has a complex taxonomic history [2,69,90]; personal observations on the syntypes by V.F. suggest that this material is diagnostic, but appears distinct

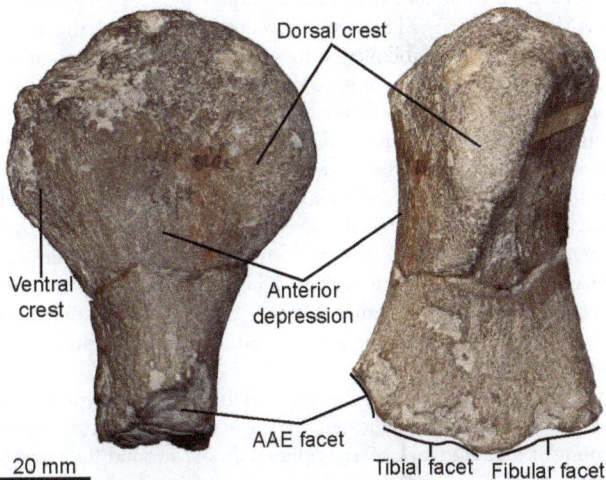

Figure 14. 'Platypterygius' sp., left femur (CAMSM B58058) in anterior (left) and dorsal (right) views. Abbreviation: AAE: anterior accessory epipodial element.

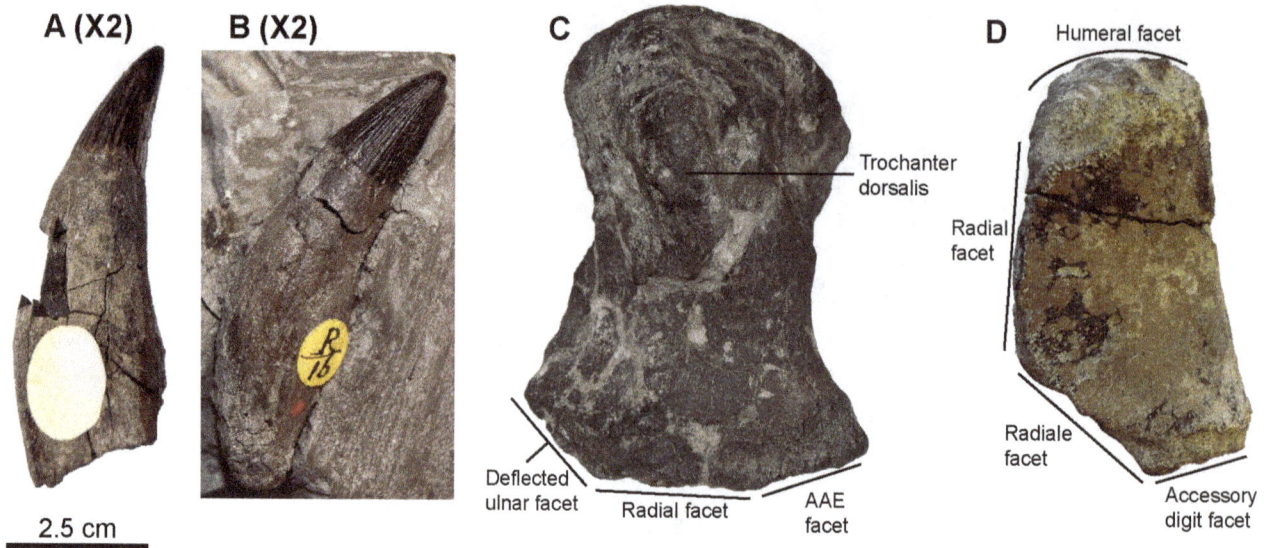

Figure 15. Indeterminate ophthalmosaurine ophthalmosaurids from the Gault Formation and Cambridge Greensand Member. A, B: Teeth (NHMUK R16 partim; magnified two times with respect to other elements); C: right humerus in dorsal view (CAMSM TN1755 partim), note the posterolaterally deflected ulnar facet, an ophthalmosaurine synapomorphy; D: anterior accessory epipodial element in dorsal view (IRSNB GS10). Abbreviation: AAE: anterior accessory epipodial element.

from the abundant material in lower stratigraphic levels (the Gault Formation, the Upper Greensand Formation and the Cambridge Greensand Member). Therefore, pending a thorough reassessment of the species nested within *Platypterygius*, the material outside the chalk cannot be referred to '*P.*' *campylodon* unambiguously. Because phylogenetic and morphological analyses (e.g. [18,20,22]) indicate that *Platypterygius*, as currently defined, is a waste-basket, polyphyletic taxon, it cannot be used unambiguously at the moment either. Therefore, we opt here for a generic rank taxon, '*Platypterygius*' sp., that groups large platypterygiine specimens that share similarities with '*P.*' *hercynicus* and '*P.*' *campylodon* sensu stricto. Detailed analysis of articulated material, such as the specimen described by Bardet [90], could further elucidate the anatomy, relationships and the taxonomic diversity of these large platypterygiine taxa from the Albian–Cenomanian of Eurasia. Because '*Platypterygius*' sp. is based on numerous disarticulated remains, many of which are isolated bones, this taxon will not be counted as a distinct taxon in timebins where '*P.*' *hercynicus* and/or '*P.*' *campylodon* sensu stricto occur.

Description

Premaxilla (CAMSM TN283*; RGHP PR 1*; Figure 8). The premaxilla is elongated and is impossible to distinguish from the dentary in the anterior-most part. Fossa praemaxillaris is shallow and ends anteriorly as a series of deep foramina. A network of very shallow grooves departing from these foramina textures the lateral surface of the very tip of the snout. The dental groove is deep, and the lingual wall is higher than the labial wall. Both these walls are greatly thickened ventrally. An intraosseous channel similar to the Meckelian canal of the dentary is present anteriorly.

Nasal (CAMSM TN283*; RGHP PR 1*; Figure 8). The nasal starts anteriorly as a thin plate covering the internal surface of the premaxilla, before emerging and forming the dorsomedial surface of the rostrum. Unusually, the nasals interlock in a tongue-in-a-groove fashion in CAMSM TN283.

Maxilla (CAMSM TN283*; Figure 8). The maxilla is elongated and low. It emerges at the same level as the nasal in CAMSM TN283, thus differing from Kiprijanoff's '*P.*' *campylodon* material [91], and more posteriorly in RGHP PR 1, where there is no trace of a maxilla even in the posterior-most section of the rostrum. The medial part of the maxilla forms a very thick lingual wall posteriorly.

Basioccipital (morphotype 1; see Systematic Paleontology above for a complete list of specimens; Figures 9, 10). The basioccipital is spherical and usually of large size (except in CAMSM B57944). The condyle is large and markedly rounded and its peripheral edge is slightly flared. The median notochordal pit is teardrop-shaped and is located in the upper half of the condyle. It is sometimes accompanied by a narrow and shallow dorsoventral groove. The extracondylar area is extremely reduced, both ventrally and laterally (condyle width = 87.92% total width in CAMSM X50161). There is no ventral notch, and the extracondylar area is an oblique flat-topped ridge on the lateral surface of the basioccipital. There is no distinctive stapedial facet, and the opisthotic facet is a plateau the barely stands out (if at all) from the extracondylar area, unlike in *Sisteronia seeleyi*. The exoccipital facets are large, oval, slightly concave and lay directly on the body of the basioccipital, unlike in *Sisteronia seeleyi*, where the exoccipital facets are raised. The exoccipital facets are separated medially by a smooth and concave groove forming the base of foramen magnum. This groove is flattened in its middle part and then deepens anteriorly, forming a deep groove housing the notochordal pit anteriorly. The anterior surface is oblique and flat or slightly convex.

This basioccipital morphotype belongs to a platypterygiine ophthalmosaurid, as indicated by the extremely reduced extracondylar area and lack of a peripheral groove around the condyle [20]. Within this clade, only the basioccipital of the genus *Platypterygius* is characterized by a reduction of the opisthotic facets, giving the basioccipital a perfectly circular shape in posterior view [11,90]. In some specimens (e.g. CAMSM B58250_56*), the floor of the foramen magnum is raised and appears very similar to that of 'Platypterygius cf. kiprijanoffi' described by Bardet [90]. Others (e.g. CAMSM X50167) are concave, as in 'P.' hercynicus [76], but not as much as in 'P.' australis [11].

Basisphenoid (CAMSM B58250_56*; Figure 10). The only basisphenoid associated with diagnostic material is incomplete and sheared. The posterior surface is kidney-shaped and slightly concave, with a deep notochordal groove, matching that of the corresponding basioccipital. The ventral carotid foramen is set at the center point. The basipterygoid processes are not preserved.

Supraoccipital (CAMSM B58250_56*; Figure 10). The supraoccipital is markedly U-shaped, as in 'P.' australis [11], 'P.' hercynicus [18,76], and O. natans [83]. The dorsomedial rod is oval in cross-section. Ventrolaterally, the supraoccipital forms an anteroposteriorly-expanded, brick-like exoccipital process. The facet for the exoccipital is flat, rectangular and posteroventrally facing. The anteroventral facet is set at a right angle to the exoccipital facet and bears an impression for the otic capsule, probably the posterior vertical semi-circular canal. This condition differs from 'P.' australis [11], where a T-shaped impression housed the utricle as well. Unlike in 'P.' australis [11], the internal walls of the supraoccipital are smooth and do not bear any foramen.

Dentary (CAMSM TN283*; RGHP PR 1*; Figure 8). The dentary closely resembles the premaxilla, including the shape of the lateral fossa. The lingual wall of the dental groove is also higher than the labial wall.

Rest of the mandible (CAMSM TN283*; RGHP PR 1*; Figure 8). The splenial is the first bone to emerge from the rostrum. It starts anteriorly as a very thin pike of bone, before progressively forming the medial wall of the mandible posteriorly. The angular is long and crescentic in cross-section. It emerges at the level of the symphysis in CAMSM TN283. The surangular is boomerang-shaped in cross-section and emerges≈50 mm after the angular in CAMSM TN283.

Dentition (morphotype 1: see Systematic Paleontology above for a complete list of specimens; Figure 11). The teeth are usually large; the height of the teeth from the middle part of the snout frequently exceeds 5 cm. The crown is conical, straight, robust, and bears numerous deep apicobasal striations. The apex possesses a pitted texture, as described in 'P.' hercynicus [18] and large/adult Aegirosaurus [1]. The angle formed by the crown is wide, usually around 30° (but can reach 37° in some teeth of CAMSM B58010_27). Wide and smooth apicobasal ridges texture the acellular cementum ring. This texture is usually restricted on its apical third, but can cover the whole surface in large teeth. The root is markedly thickened with respect to the acellular cementum ring, and its cross-section is squared. Deep apicobasal ridges occur on the root surface, especially in large teeth. As in all ichthyosaurs, there is a considerable degree of dental variation along the rostrum: anterior teeth are rather smaller, slender, and have a straighter crown whereas posterior teeth are smaller and bulkier, with relatively large recurved crown and short but wide roots with a rounded cross-section.

The squared root in cross-section indicates these teeth belong to a platypterygiine ophthalmosaurid [20]. The general morphology of this tooth morphotype, with bulbous roots, robust crowns and numerous apicobasal ridges on crown, acellular cementum ring and root is typical for the platypterygiine genus 'Platypterygius' (e.g. [18,82,92]; V.F., pers. obs.), commonly found in Albian-Cenomanian sediments of western Europe [18,90,93,94]. Given the complex and nebulous taxonomy of that genus [18], this tooth morphotype is assigned to 'Platypterygius' sp.

Centra (CAMSM B4204_20*; RGHP SI 1*). The height/length ratio is nearly invariable, and close to 2.1. CAMSM B4204_20* contains some of the biggest Cretaceous centra ever reported (up to 240 mm in height).

Scapula (RGHP PR 1*). The scapula is thick proximally, unlike in Sisteronia seeleyi and ophthalmosaurines [20]. The acromial region is not preserved, preventing detailed comparison with other ophthalmosaurids.

Humerus (morphotypes 1 and 4; see Systematic Paleontology above for a complete list of specimens; Figure 12). We refer two distinct humerus morphotypes to 'Platypterygius' sp. The first morphotype contains usually large and stout humeri with thick trochanters, unlike the slender trochanters of Sisteronia seeleyi. In proximal view, this gives the humerus a marked rectangular shape. Both trochanters do not vanish before mid-length. Distally, the humerus possesses two large facets for the radius and the ulna that are parallel to sagittal plane, unlike in coeval ophthalmosaurines (see below). These facets are oval, flattened (unlike S. insolitus [21]), equal in length, and parallel to the sagittal plane (unlike ophthalmosaurines [20]). In some specimens a small and flattened facet for an anterior accessory element occurs at the extremity of an anterodistal process of the humerus. The diminutive size of the facet and the absence of other differences within that morphotype suggest the absence/presence of this facet is variable at the intraspecific level or related to ontogeny, although the possibility that this could represent two distinct species cannot be dismissed.

Humeri belonging to the second 'Platypterygius' sp. morphotype (humerus morphotype 4) have a high, usually short, and markedly oblique trochanter dorsalis (restricted to the proximal half of the humerus), as in some specimens of the ophthalmosaurine morphotype. The deltopectoral crest is high and forms a distal shallow ridge that merge with the ventral edge of the radial facet. Both trochanters are bordered by concave areas and give the proximal surface a concave parallelogram shape. The anterior edge of the humerus is rounded, whereas the posterior edge forms a very acute trailing blade, as in Sisteronia seeleyi. Unusually, this posterior edge is 'trochanter-like', being bordered by concave areas and thickening proximally to form a bulge on posterior end of the glenoid surface. The humerus possesses four distal facets, including two facets for accessory zeugopodial elements: one anteriorly and one posteriorly. Unusually, the posterior accessory facet is large, sometimes larger than the radial facet and faces posterodistally. The anterior accessory facet is the smallest; it is concave, roughly triangular, and faces anterodistally.

The size, stoutness and distal architecture of these humeri correspond to those reported in taxa currently referred to as Platypterygius [14,16,76,77]. The humerus morphotype 1 presents a combination of features (large trochanters; large, flat and oval radial and ulnar facet parallel to the sagittal plane; small to absent anterior accessory facet) that is only found in taxa currently referred to as Platypterygius from the 'middle' Cretaceous of Europe: 'P.' campylodon [91] and 'P.' platydactylus [95], although 'P.' australis possesses many similarities with these forms too [13]. The large four distal facets of the humerus morphotype 4 is a feature only found in some Aptian–Albian taxa currently referred to as Platypterygius as well: 'P.' hercynicus [74,76], and 'P.' sp. from North America [16,77]. Accordingly, we refer both morphotypes to 'Platypterygius' sp., but these morphotypes are likely to represent two distinct species.

Manus (RGHP PR 1; Figure 13). The manus is composed of tightly packed rectangular elements, as is typical for most platypterygiine ophthalmosaurids [20]. The manus architecture appears longipinnate (i.e. with a single digit arising from the intermedium) as in most species referred to as Platypterygius [2,13,14,74,76,84,95,96], Sisteronia (V.F. pers. obs. on uncurated material from southeastern France), and probably *Arthropterygius* [81,97].

Femur (morphotype 1; see Systematic Paleontology above for a complete list of specimens; Figure 14). As in *Sveltonectes insolitus* [21], the dorsal and ventral trochanter of the femur are very high and their morphology matches that of the humeri of ophthalmosaurids, by having a high, plate-like, and oblique dorsal trochanter separated from the slightly thicker ventral trochanter by a flattened area anteriorly. Both trochanters vanish at mid-length. The anterior surface is large and flat, and the posterior edge is rounded, giving the capitulum a rounded triangular shape in proximal or cross-section view. Distally, the femur forms three facets, as in many platypterygiines such as *Maiaspondylus* [22], '*P.*' *americanus* [14], '*P.*' *australis* [13] and '*P.*' *hercynicus* [76]. However, the extra facet is small, triangular and for an anterior accessory element. This condition has only been described in '*P.*' *australis* [13]: the other taxa have an extra facet either for a posterior accessory epipodial element or for the astragalus. The fibular facet is triangular and faces posterodistally. The square-shaped tibial facet is the largest and faces anterodistally.

Out of the several femora morphotypes recognized in the Cambridge Greensand member, only one can be attributed to '*Platypterygius*' sp. with confidence, thanks to an articulated specimen (CAMSM B58250_56) from the upper (chalky) part of the Cambridge Greensand Member. Moreover, similarly large and elongated femora with large trochanters, slightly rounded capitulum and three distal facets are only known in '*P.*' *hercynicus* [76] and '*P.*' *australis* [13].

Systematic Paleontology

Platypterygiinae indet.

1869 *Ichthyosaurus angustidens* Seeley [50]: 3
1869 *Ichthyosaurus bonneyi* Seeley [50] : xvii
1869 *Ichthyosaurus platymerus* Seeley [50] : xvii

Note. As noted by Lydekker [98] and McGowan & Motani [69], Seeley [50] proposed the names *Ichthyosaurus bonneyi, I. doughtyi, I. platymerus* and *I. angustidens* without a formal description or figure, making these taxa nomina nuda. However, we found the holotype specimens for each of these taxa in the CAMSM. Each were placed in a single box and clearly marked as being type specimens. This allows comparison of these taxa with the rest of the Albian record. Given the uncertain future of *Platypterygius* and its species [18], these taxa may therefore have priority over more recent ones, should they be found to belong to the same taxon. Accordingly, these taxa are regarded as nomina inquirenda, even if this. The holotypes of *I. angustidens* (CAMSM B20643, a partial tooth from the Lower Chalk of Hunstanton), *Ichthyosaurus bonneyi* (CAMSM B58062, a femur from the Cambridge Greensand Member), and *I. platymerus* (CAMSM B58057, a humerus from the Cambridge Greensand Member) resemble '*Platypterygius*' sp. However, given the numerous issues inherent to *Platypterygius*,, these species are considered as an indeterminate platypterygiine instead of '*Platypterygius*' sp. for the moment, pending a thorough reassessment of this genus.

Systematic Paleontology

Ophthalmosaurinae Baur 1887 [66] sensu Fischer et al. [20]
Ophthalmosaurinae indet.
Figure 15

1888 *Ophthalmosaurus cantabrigiensis* Lydekker [98]: 310
1889 *Ophthalmosaurus* (?) *cantabrigiensis* Lydekker [68]: 9 (NHMUK 35348)
1889 *Ichthyosaurus campylodon* Lydekker [68]: 19 (NHMUK R16)

2003 *Brachypterygius cantabrigiensis* McGowan & Motani [69]: 34: Figure48

Referred material from the Cambridge Greensand Member. NHMUK 32406 partim (tooth); NHMUK R16 partim (tooth); NHMUK 47268 (5 teeth); CAMSM B58042 (humerus); CAMSM B58045 (humerus); CAMSM B58050 (humerus); CAMSM B58053 (humerus); CAMSM B58055 (humerus); CAMSM TN1727 partim (humerus); CAMSM TN1755 partim (2 humeri); IRSNB GS3 (humerus); LEICT G65.1991 (humerus); NHMUK R2343 (3 humeri); NHMUK R4513 (2 humeri); NHMUK 35348 (humerus); NHMUK 43989 (humerus, holotype of *Brachypterygius cantabrigiensis*); IRSNB GS60 (anterior accessory epipodial element).

Referred material from other deposits. NHMUK R16 partim (teeth, Gault Formation); NHMUK R17 partim (teeth, Gault Formation).

Note. Additionally, Fischer et al. [20] referred eleven basioccipitals, five stapedes and one basisphenoid from the Cambridge Greensand Member to the ophthalmosaurine ophthalmosaurid *Acamptonectes* sp. Fischer et al. [20] misspelled the collection number of a basioccipital referred to as *Acamptonectes* sp.: in their paper, specimen CAMSM B56961 is actually CAMSM B57961. Now that additional ophthalmosaurine ophthalmosaurids have been found in Cretaceous strata of Eurasia [99], the referral of these remains to the Hauterivian genus *Acamptonectes* by Fischer et al. [20] is disputable, even if one basioccipital (CAMSM B57962) and one basisphenoid (NHMUK PV R2341) exhibited autapomorphic features of *Acamptonectes*. Accordingly, we refer all these *Acamptonectes* sp. remains (i.e. CAMSM B57955 [basioccipital], CAMSM B57949 [basioccipital], CAMSM B57942 [basioccipital], CAMSM B57952 [basioccipital], CAMSM B56961 [basioccipital], CAMSM TN1735 partim [basioccipital], CAMSM TN1751 partim [basioccipital], CAMSM TN1753 partim [basioccipital], CAMSM TN1755 partim [basioccipital], GLAHM V.1463 [basioccipital, Newmarket road pits], NHMUK 35301 [basioccipital], CAMSM B58074 [stapes], CAMSM B58075 [stapes], CAMSM B58079 [stapes], CAMSM TN1757 partim [stapes], GLAHM V.1535/1 [stapes], NHMUK R2341 [basisphenoid]) to Ophthalmosaurinae indet. The holotype of *I. cantabrigiensis* (NHMUK 43989) lacks distinguishing features from other ophthalmosaurines; accordingly, this taxon is considered here as nomen dubium.

Description

Dentition (morphotype 3; see Systematic Paleontology above for a complete list of specimens; Figure 15 A, B). The teeth are recurved medially. The crown is conical, textured by light apicobasal ridges, and appears small compared to the apicobasal height of the tooth (19% in NHMUK 47268 partim). The apex is pointed and smooth. Both the acellular cementum ring and the root are smooth (no apicobasal ridges) and their cross-section is rounded. Some teeth have slightly flattened surface of on their roots, but lack the well-defined angles seen in the other tooth morphotypes ('*Platypterygius*' sp. and *Sisteronia seeleyi*).

A squared root section is a synapomorphy of platypterygiine ichthyosaurs [20] (but reversed in *Aegirosaurus* [1,100]). This tooth morphotype does not correspond to *Aegirosaurus* [1], being recurved, having a much smaller crown and a smooth apex. This tooth morphotype is however similar to that of *Ophthalmosaurus icenicus* [73]. Accordingly, we refer the tooth morphotype 3 to Ophthalmosaurinae indet.

Humerus (morphotype 4; see Systematic Paleontology above for a complete list of specimens; Figure 15 C). The humerus is usually small and stout; but larger specimens (such as

CAMSM TN1755 partim) have a more slender shape. The short trochanter dorsalis and the deltopectoral crest are well developed, although the latter may be reduced in some specimens. A similar variability has already been reported in the ophthalmosaurine *A. densus* [20]. The humerus forms three distal facets that are subequal in size. The posterior-most (ulnar) facet is markedly deflected posterolaterally and has a concave margin in dorsal view. The median (radial) facet is the largest and squared or slightly dorsoventrally elongated. The anterior-most (accessory) facet is often slightly deflected anterolaterally.

This humerus morphotype has been interpreted in various ways since Lydekker [68,98]. He considered the three distal facets as indicative of *Ophthalmosaurus*, but the equal size of these three facets in one of these humeri, NHMUK 43989, differed from *O. icenicus*, justifying a new species, *Ophthalmosaurus cantabrigiensis*. Then, McGowan & Motani [69] considered this species to belong to *Brachypterygius*, mainly because it did not resemble *O. icenicus* enough and because they already inferred the presence of *Brachypterygius* in the Cambridge Greensand Member using basicranium evidence. Evidence for a referral of this humerus morphotype to *Brachypterygius* is, however, poor. Indeed, the largest facet on this humerus morphotype (to which the holotype of *O. cantabrigiensis* belongs) is the 'median' facet, a condition never observed in any ichthyosaur whose intermedium contacts the humerus: in these ichthyosaurs, the intermedium facet is less than half the size of the radial or the ulnar facets (*B. extremus* [73,101]; pers. obs. on holotype NHMUK R3177; *Aegirosaurus* [72]; *Maiaspondylus* [15]); a similar interpretation for these morphotype 3 humeri would imply an enormous intermedium, larger than both the radius and the ulna, a condition never seen in Ichthyosauria. Moreover, the radial and ulnar facet are both invariably markedly deflected outwards in the above-mentioned taxa (ibid.), whereas only the ulnar facet is consistently deflected outwards (posteroventrally) in the humerus morphotype 3, as in ophthalmosaurine ichthyosaurs [20]. Kear & Zammit [26] recently casted doubt on the validity of this character by studying two in utero specimens that they referred to the platypterygiine taxon '*Platypterygius*' *australis*, which presumably exhibited the same morphology. However, it is clear that the ossification of the humeri that they figure is far from complete ([26]:Figure 2); thus their shape cannot be assessed unambiguously; moreover, adults representatives of this taxon do not exhibit this peculiar morphology [13]. The degree of deflection of the anterior facet forms a wide spectrum in humerus morphotype 3 (ophthalmosaurine), within which only some (usually small) specimens such as NHMUK 43989 (holotype of *O. cantabrigiensis*), CAMSM B58055, and CAMSM TN1727 partim have a slightly anterolaterally deflected anterior facet. This is likely a juvenile condition that disappears with ontogeny, as in '*P.*' *australis* [26]. Moreover, some specimens of adult ophthalmosaurines also show a slightly deflected anterior facet (e.g. GLAHM 132855, holotype of *A. densus*; LEICT G1.2001.016, *Ophthalmosaurus* sp.; GLAHM V1070, *Ophthalmosaurus icenicus* [20]; V.F., pers. obs. on GLAHM, NHMUK, MJML, and CAMSM material). Similarly, the relative size of the anterior facet in ophthalmosaurines also forms a wide spectrum (e.g. [80,83]; V.F., pers. obs. on GLAHM, NHMUK, MJML, and CAMSM material) within which the holotype of *O. cantabrigiensis* falls satisfactorily. Therefore, we consider the evidence for a referral of this morphotype to *Brachypterygius* as unfounded, and that its morphology falls within the known spectrum for ophthalmosaurines ophthalmosaurids and lacks autapomorphies in the current state of our knowledge. Accordingly, we refer this morphotype to Ophthalmosaurinae indet.

Epipodium (IRSNB GS60; Figure 15 D). IRSNB GS60 is an anterior accessory epipodial element of a forefin. It is elongated proximodistally. This element bears facets for humerus, radius, radiale, and the first autopodial element of the anterior accessory digit. The radial facet is the largest and the humeral and radiale facet are large and equal in size. The humeral and radial facets form a 90° angle. The anterior surface is saddle-shaped rather than convex or flat and its overall shape is not crescent-like. The dorsal half is much thicker than the ventral half.

Accessory epipodial elements are frequent in ophthalmosaurids, but they greatly differ in shape (compare [2,21,77,102]). IRSNB GS60 appears strikingly similar to that of many large specimens of *Ophthalmosaurus icenicus* (V.F., pers. obs. on GLAHM, NHMUK, MJML, and CAMSM material). The lack of a crescentic shape differs from the anterior accessory epipodial element of *Sveltonectes insolitus* and the pisiform of '*P.*' *americanus* [14] and the combination of a proximodistal elongation+a large humeral facet+three additional facets differs from all other platypterygiine ophthalmosaurs for which the epipodium is known [13,16,77,102]. We interpret IRSNB GS60 as an ophthalmosaurinae anterior accessory epipodial element because that morphology has only been found in *O. icenicus* and in the poorly known but probably closely related '*Paraophthalmosaurus*' [103] (V.F. pers. obs. on holotype in SSU) and '*Yasykovia*' [104] so far. Both of these are considered as junior synonyms of *Ophthalmosaurus* by Maisch & Matzke [105] and McGowan & Motani [69].

Systematic Paleontology

Ophthalmosauridae indet.

1869 *Ichthyosaurus doughtyi* Seeley [50] : xvii

Note. The holotype of *I. doughtyi* (CAMSM B58044, from the Cambridge Greensand Member) is a partial humerus, belonging to a juvenile ichthyosaur. The presence of a preaxial accessory facet allows assignment to Ophthalmosauridae, but this specimen lacks diagnostic features. It is therefore referred to Ophthalmosauridae indet. and *Ichthyosaurus doughtyi* is regarded here as a nomen dubium. Several other propodials cannot be assigned more precisely than Ophthalmosauridae indet. These morphotypes are described in Text S6.

Ichthyosauria insertae sedis

Cetarthrosaurus walkeri Seeley, 1873 [106] (Seeley, 1869 [50])

Figure 16

Holotype. CAMSM B58069, a propodial from the Cambridge Greensand Member (Lower Chalk Formation), early Cenomanian, but phosphatized and reworked from the top (late Albian) of the Gault Formation.

Referred material. CAMSM X50170, from the same age and locality as the holotype.

Emended diagnosis. *Cetarthrosaurus walkeri* possesses the following autapomorphies within Ichthyosauria: propodial with hemispherical capitulum disconnected from dorsal and ventral trochanters; elongated and slender shaft (axial length/mid-shaft width ratio = 2.93 in holotype and 3.00 in referred specimen); sheet-like ventral trochanter parallel to the long axis.

Additionally, among Ichthyosauria, the combination of a three-faceted propodial, including a small facet for a preaxial accessory element and a distally-facing ulnar/fibular facet is only shared by: one femur of *Stenopterygius* quadriscissus [86], humerus and femora of some specimens of '*Platypterygius*' sp. from England (this work), humerus and femora of '*P.*' *australis* [13]; humerus of *Caypullisaurus* [107]; an unnamed taxon from Canada [23].

Occurrence. Late Albian of the Gault Formation reworked in the Cambridge Greensand Member. No evidence for presence

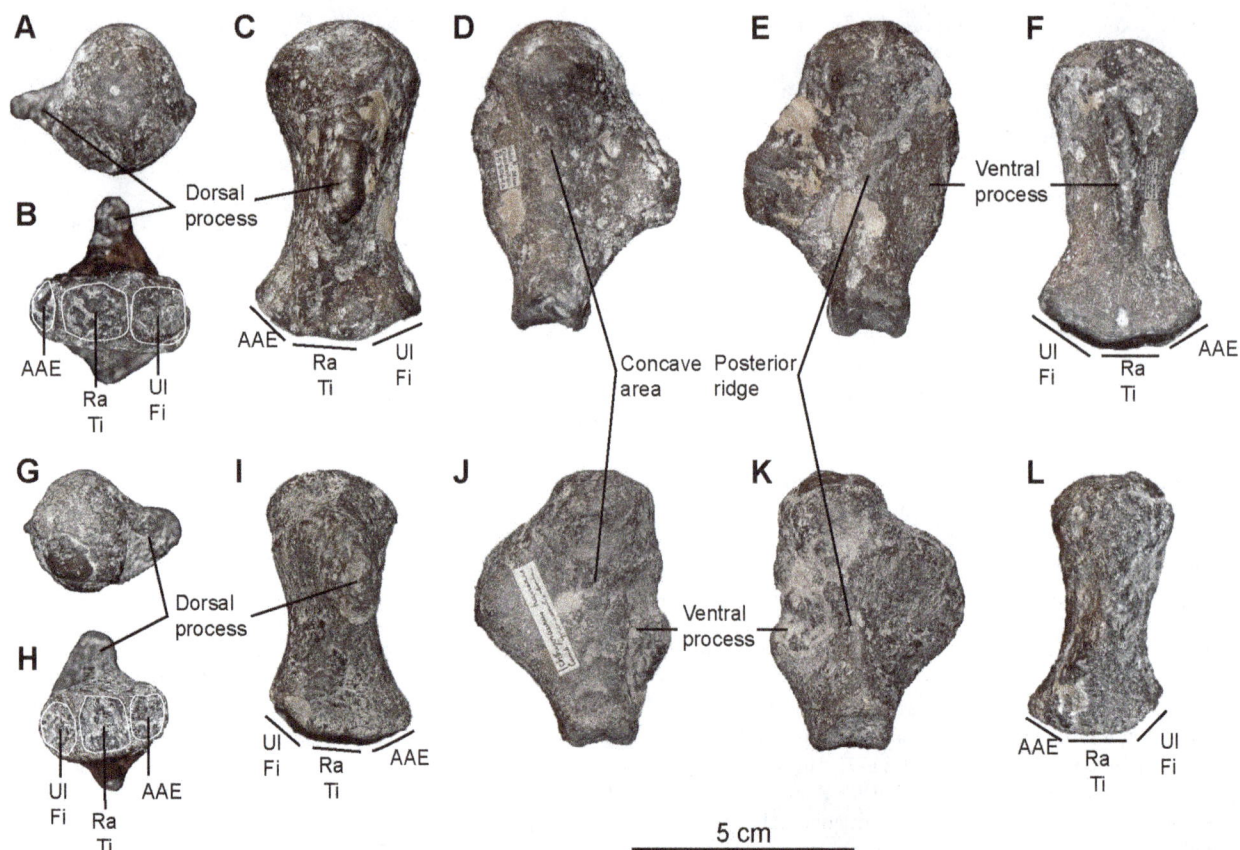

Figure 16. *Cetarthrosaurus walkeri*, propodials. A–F: Holotype (CAMSM B58069), in proximal (A), distal (B), dorsal (C), anterior (D), posterior (E), and ventral (F) views. G–L: referred specimen (CAMSM X50170), in proximal (G), distal (H), dorsal (I), anterior (J), posterior (K), and ventral (L) views. Note the high aspect ratio, the rounded capitulum disconnected from the shaft trochanters, and the high and lamellar dorsal trochanter. Abbreviations: AAE: anterior accessory epipodial element; Fi: fibula; Ra: radius; Ti: tibia; Ul: ulna.

in the upper (early Cenomanian) part of the Cambridge Greensand Member.

Note. The holotype of *C. walkeri* (CAMSM B58069) was described by Seeley [50,106] as a right femur of very unusual shape. Seeley first named *walkeri* as a new species of the genus *Ichthyosaurus* [50]. But his comparison of the propodial with other ichthyosaurs and cetaceans led him to propose a new generic referral for this specimen four years later [106]. Later, this taxon was considered as a mosasaurid (Hulke *in* Lydekker [98]; [3]) and disappeared from the literature. McGowan & Motani's review [69] considered *I. walkeri* as a nomen dubium without discussion and did not mentioned *Cetarthrosaurus*.

During this study, a small right propodial (CAMSM X50170, marked as '*Ichthyosaurus* humerus, Cambridge Greensand, Cambridge') and strikingly similar to CAMSM B589069, was found. It shares all the peculiar features of the holotype of *Cetarthrosaurus walkeri*, but its dorsal surface is less eroded, allowing a better description of that peculiar propodial morphotype. Despite its unusual shape, this propodial is clearly ichthyosaurian (contra Hulke *in* Lydekker [98]). The presence of three distal facets suggests relationship with Ophthalmosauridae, but at least one specimen of the basal baracromian *Stenopterygius* is known to have three distal facets on its femur as well [86]. Moreover, the hemispherical capitulum separated from dorsal and ventral trochanter is unique among post-Triassic ichthyosaurs. Yet, this propodial is diagnostic and, therefore, the taxon *Cetarthrosaurus*

walkeri must be considered as a valid, albeit poorly known, late Albian ichthyosaur.

Description

Cetarthrosaurus is only known from two propodials (Figure 16). Their shape is so unusual that is difficult to decipher the limb they belong to. Accordingly, we describe them as propodials and compare them to humeri and femora of neoichthyosaurians. The shaft of the propodial is constricted and elongated (axial length/anterodistal length = 64.52 mm/33.35 mm = 1.93 in the holotype and 61.19 mm/27.51 mm = 2.22 in CAMSM X50170) and the capitulum is hemispherical. Both the anterior and the posterior surfaces of the shaft are saddle-shaped, but the anterior one is flatter (whereas it is markedly flat or concave in ichthyosaurs [69]). The dorsal trochanter is extremely high: its height is more than 80% the height of the capitulum (even the femur having the largest dorsal trochanter of the CAMSM greensand material [CAMSM B58059] has a ratio of 56.7%, because the capitulum of ophthalmosaurids is usually much larger than that of *C. walkeri*). The dorsal trochanter is oblique, only slightly plate-like (i.e. the dorsal surface is not flat-topped but oblique and bordered by concave areas; this condition is therefore 'intermediate' between basal thunnosaurians and ophthalmosaurids), and extends up to the distal edge of the propodial through a shallow ridge confluent with the dorsal edge of the anterior accessory facet. The ventral trochanter forms a prominent, long, and sheet-like axial ridge bordered by concave areas. Unusually, these trochanters do not

Taxon		Abundance in CBM	
Platypterygius sp.	41	▬▬▬▬▬▬	33%
Ophthalmosaurinae indet.	41	▬▬▬▬▬▬	33%
Sisteronia seeleyi	40	▬▬▬▬▬▬	32%
Cetarthrosaurus walkeri	2	■	2%
	124		

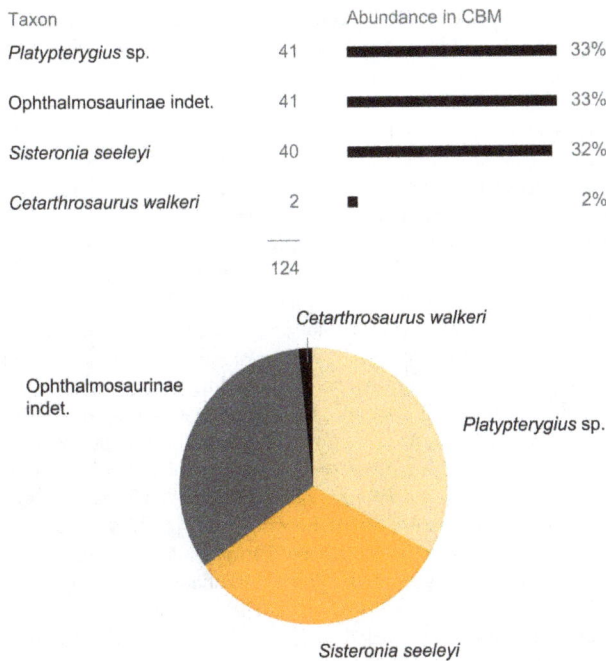

Figure 17. Stage-level taxonomic diversity of Cretaceous ichthyosaurs compared to previous assessments; the number of genera has dramatically increased since year 2002. The position of each stage on the X-axis is proportional to its duration. The grey line represents the generic diversity as of 2002. See Text S7 for the dataset.

merge with the capitulum, as noted by Seeley [50]. Distally, the propodial has three concave facets: a small anterior accessory facet, and two large squared facets for radius/tibia and ulna/ fibula. The radial/tibial facet faces distally and the ulnar/fibular facet faces posterodistally.

Diversity curves

The taxonomic diversity of Cretaceous ichthyosaurs is now significantly higher than hypothesized a few years ago (Figure 17). The Berriasian diversity has been increased because of the recognition of Late Jurassic ichthyosaurs in this stage: *Caypullisaurus bonapartei* [108,109], *Aegirosaurus* sp. (as a Lazarus taxon [1]) and cf. *Ophthalmosaurus* [20]. Despite the description of new fossils from the Valanginian and Hauterivian from western Europe, these stages are still inadequately known, and constitute a 'diversity low point' for the Cretaceous. Indeed, the number of specimens known from this interval is extremely low: RGHP LA 1 is the first diagnostic ichthyosaur reported from the Valanginian [1], and only of handful of ichthyosaur specimens are known from the Hauterivian [19,20,110].

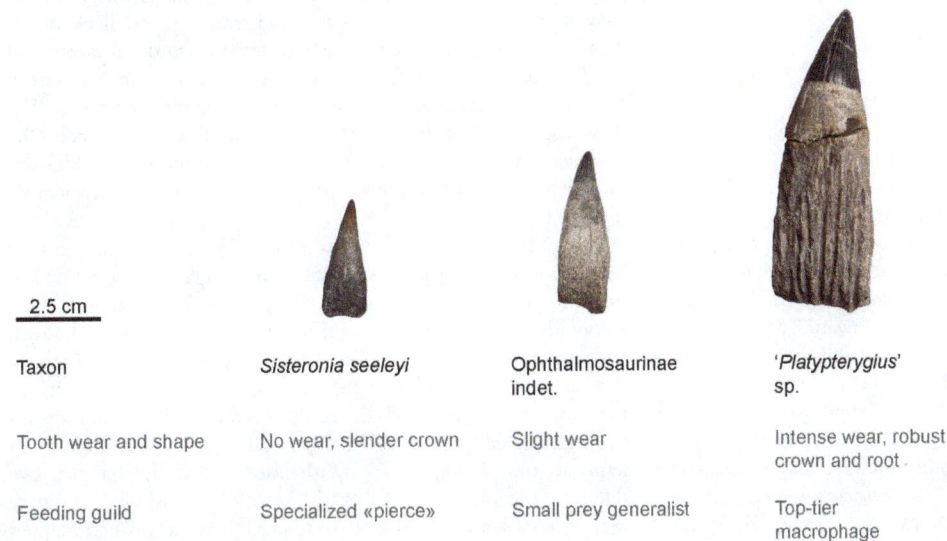

Taxon	*Sisteronia seeleyi*	Ophthalmosaurinae indet.	'*Platypterygius*' sp.
Tooth wear and shape	No wear, slender crown	Slight wear	Intense wear, robust crown and root
Feeding guild	Specialized «pierce»	Small prey generalist	Top-tier macrophage

Figure 18. Evolution of ichthyosaur taxonomic richness during the Jurassic and Cretaceous. The Aptian and Albian are split in two and three substages, respectively. The generic curve (black) considers *Platypterygius* as a single taxon. See Text S7 for the dataset.

Diversity explodes during the Barremian, with the recognition of several platypterygiine ophthalmosaurids such as *Sveltonectes insolitus* and *Simbirskiasaurus birjukovi* from western Russia [21,111], '*P.*' *sachicarum* [112] and '*P.*' *hauthali* [113,114] from South America, in addition to *Malawania anachronus* from Iraq [19]. The diversity diminishes during the Aptian (Figures 17, 18), probably because of sampling and taxonomic biases, because only a handful of diagnostic Aptian ichthyosaurs have been recovered worldwide [95,99]. Then, the diversity becomes very high during the Albian. The generic curve remains rather constant because '*Platypterygius*' was considered as a single entity in this curve; if recent advances regarding the polyphyly and status of *Platypterygius* [18,22,111] are taken into account, it is even possible that the generic taxonomic richness will equal the specific one and reach a value of 10 during the Albian, as suggested by yet unpublished analyses [71]. Splitting the Aptian and the Albian (Figure 18) does not change the picture, but indicates that a high diversity (eight species) is restricted to the late Albian. Comparable parvipelvian ichthyosaur diversity has only been reported in the early Toarcian Lagerstätten of western Europe, where five genera and as many as eleven species have been reported [69,105,115–119], and the Tithonian strata of South America, Germany, England, and Russia, containing seven species and four genera [72,103–105,107,120–128]. The ichthyosaur diversity then severely drops during the Cenomanian and reaches zero by Turonian times.

The taxonomic diversity of Cretaceous ichthyosaurs is now equivalent to or greater than that of their Jurassic ancestors, both at the generic and specific levels, contrary to previous assumptions [4,129,130]. Indeed, the diversity frequently reaches four to five genera and seven to eight species whenever fossil-rich sediments occurring in distinct basins are found, such as the Hettangian–Sinemurian [69,131–135] of western Europe, the Tithonian of England (top of the Kimmeridge Clay Formation; [136,137], Germany [72,100,138,139] and South America [107,108,140], and several periods during the Albian [22]. The extremely abundant material from the Toarcian (possibly several thousands of specimens [141]) certainly biases the record. A 'safer' interpretation of these fluctuating curves is to consider the diversity of ophthalmosaurids was possibly rather constant from their initial Middle Jurassic radiation onwards and only dropped severely at the beginning of the Late Cretaceous.

Discussion

The Albian ichthyosaurs from western Europe

In his Catalogue, Seeley [50] named four new species from the Cambridge Greensand Member: *Ichthyosaurus walkeri*, *Ichthyosaurus doughtyi*, *Ichthyosaurus bonneyi*, and *Ichthyosaurus platymerus*. He did not figure the specimens, nor did he designate holotypes, and only formally described a cast of the holotype specimen of *I. walkeri*. Only *I. walkeri* was subsequently re-described, figured, and made the type species of a new genus, *Cetarthrosaurus* [106], a rare decision at that time. Lydekker [98] and McGowan & Motani [69] considered Seeley's three other species as invalid, being nomina nuda. However, specimens CAMSM B58044, CAMSM B58057, CAMSM B58062 are clearly marked as holotype specimens of *I. doughtyi*, *I. platymerus*, and *I. bonneyi*, respectively; this permits comparison of these specimens with other material and assessment of their validity. Lydekker [98] named *Ophthalmosaurus cantabrigiensis* on the basis of a left humerus (NHMUK 43989) from the Cambridge Greensand Member. McGowan & Motani [69] considered this humerus as indicative of the presence of *Brachypterygius* (*B. cantabrigiensis*) in the Cambridge Greensand Member, because they interpreted the large median facet as a

facet for the intermedium. Yet, they also noted the presence of basioccipitals and humeri referable to *Ophthalmosaurus* in the same member.

Our analysis indicates the presence of three common and distinct taxa represented by numerous diagnostic bones (Table 1): '*Platypterygius*' sp., *Sisteronia seeleyi*, Ophthalmosaurinae indet., and an additional but rare taxon: *Cetarthrosaurus walkeri*. Appendicular bones such as humeri and femora, which appear to be more interspecifically variable within ichthyosaurs, even suggest a higher diversity, and probably reflect the specific diversity. *Ichthyosaurus doughtyi* is an indeterminate ophthalmosaurid, *Ichthyosaurus bonneyi* and *Ichthyosaurus platymerus* are not diagnostic and can be referred to as '*Platypterygius*' sp., and *Brachypterygius cantabrigiensis* is an indeterminate ophthalmosaurine ophthalmosaurid, which supports Lydekker's [98] opinion, given the state of knowledge at his time. Therefore, there is no solid evidence for the presence of *Brachypterygius* in the Cretaceous of Europe. The previous stratigraphic range of *Brachypterygius* (Kimmeridgian–Albian) was one of the main reason why Ensom et al. [142] associated a fragmentary skeleton from the Berriasian of England to this genus; our data suggest this referral is not substantiated.

Platypterygius hercynicus in coeval deposits from northwestern France [18] should be added to the assemblage described above. Many isolated bones from the Cambridge Greensand Member also closely resemble '*P.*' *hercynicus* but cannot be attributed to this species unambiguously. The humerus morphotype 4 morphotype, exhibiting four distal facets has only been reported in '*P.*' *hercynicus* [74,76] and two *Platypterygius* sp. specimens from North America distinct from '*P.*' *americanus* [16,77]. Some basioccipitals (with a raised floor of foramen magnum), teeth, and femora (with three distal facets including one probably for the astragalus) are also identical to that of '*P.*' *hercynicus* (see [74,76]).

Additionally, the two identified humeral morphotypes here referred to '*Platypterygius*' sp. indicate that another large platypterygiine roamed western Europe; this second taxon may correspond to the poorly known species '*P.*' *campylodon*, but the type material of this taxon does not contain postcranial remains (V.F., pers. obs. on CAMSM material), preventing a thorough comparison. However, it should be noted that recent works have highlighted intrageneric or even intraspecific variability in the formation of the distal facets in humeri [13,16] or the fact that the ossification of the humerus may be unrelated to the presence of extrazeugopodial elements. The humerus of *Sveltonectes insolitus* possesses two distal facets, but the forefin also possessed a moon-shaped anterior accessory element that contacted the humerus without imprinting it [21]. This suggests that the number of distal facets (especially the absence/presence of minute anterior and posterior accessory facets) may not be a reliable criterion for assessing taxonomic diversity.

Intuitively, femora would also have a taxonomic signal masked by intraspecific variability and by the degree of perichondral ossification. Within known ophthalmosaurids, femora tend to have a wide diversity of forms, even if relatively few femora are known: each taxon possesses its own morphotype, summarized in Table 3. It is therefore impossible in the current state of our knowledge to have an idea of the variability of these features. Yet, femora from the Cambridge Greensand Member still augment this diversity of femoral morphologies of ophthalmosaurids, by having five different morphotypes (Table 4; Text S6). This disparity may therefore indicate that more than four ichthyosaur taxa co-habited the ecosystem of the Cambridge Greensand Member, as suggested by humerus evidence.

In terms of relative abundances (Figure 19), '*Platypterygius*' sp. represents only 33% (41 specimens) of the assemblage. Ophthal-

Table 3. Overview of the morphological disparity in ophthalmosaurid femora.

Taxon	Facets	Including one for	Capitulum shape	Trochanters
'P. campylodon' (Kiprijanoff material)	2	/	Triangular	High
O. icenicus	2	/	Triangular	Small
Maiaspondylus lindoei	3	Astragalus	?	?
'P.' hercynicus	3	Astragalus	Oblong	Medium
'P.' australis	3	AAE	Rounded	Medium
'P.' americanus	3	PAE	Triangular	High
Sveltonectes insolitus	2	/	Triangular	High

Abbreviations: AAE: anterior accessory epipodial element; PAE: posterior accessory epipodial element. References: 'P. campylodon': Kiprijanoff [91]; O. icenicus: Andrews [80]; M. lindoei: Druckenmiller & Maxwell [22]; 'P.' hercynicus: Kolb & Sander [76]; 'P.' australis: Zammit et al. [13]; 'P.' americanus: Maxwell & Kear [14]; S. insolitus [21].

mosaurinae accounts for 33% of the assemblage (41 specimens as well), *Sisteronia seeleyi* represents 32% (40 specimens) of the assemblage, and *C. walkeri* completes the picture with a relative abundance of 2% (2 specimens). These proportions contradict the popular belief of monogeneric (*Platypterygius*) ichthyosaur assemblages in the Cretaceous (e.g. [4,5]). Moreover, the count of '*Platypterygius*' sp. is probably overestimated relatively to other taxa because this taxon (and possibly *C. walkeri*) is the only one to which one femoral morphotype has been tied, increasing the number of referable specimens, whereas the femora of Cretaceous ophthalmosaurines and *Sisteronia seeleyi* are yet unknown. Similarly, the count for *C. walkeri* may be underestimated, because only two propodials are referable to this poorly known taxon. It is even possible that the femora of *C. walkeri* belong to one of the other ichthyosaur taxa recognized here, but this could only be proven with articulated material. On the subfamilial level, however, Platypterygiinae dominates the assemblage with a relative abundance of 65% versus 33% for Ophthalmosaurinae.

The ichthyosaur assemblage of the Cambridge Greensand Member, containing four co-occurring genera, is the most diversified assemblage ever reported in a single sedimentary body of Cretaceous age. The persistence of '*P.*' *hercynicus* in the latest Albian of northwestern France and possibly Cambridge area and the co-occurrence of large '*Platypterygius*' sp. and *Sisteronia seeleyi* in the Vocontian Basin suggest that western Europe was a diversity hot-spot for Albian–earliest Cenomanian ichthyosaurs, a few million years prior to their final extinction, at the Cenomanian–Turonian boundary [3].

Preliminary assessment of the marine reptile assemblage of a new latest Albian–earliest Cenomanian locality in western Russia (V.F., pers. obs. on SSU material) suggests a similar diversity in this deposit as well, with the presence of at least three ichthyosaur taxa, including platypterygiines and ophthalmosaurines, and with

distinct tooth morphologies. Articulated material is needed for a better understanding of these forms, but this suggests a high diversity in ichthyosaurs of the Albian–Cenomanian boundary in western Russia as well. This situation appears similar to the Lower Albian of Canada, where three to four taxa have recently been recognized [15,16,22,23] and markedly contrasts with the monospecific ichthyosaur assemblages in the Albian of Australia [12,13] and U.S.A. [22], despite the fact that a large number of specimens have been discovered in numerous localities, at least in Australia. Therefore, whereas the taxonomic richness of late Albian ichthyosaurs now reaches eight species (Figure 18); and probably as many genera if *Platypterygius* is split according to recent revisions [18,111]), this diversity shows a strong geographical variability and was not uniformly high worldwide.

Beta diversity is more difficult to assess, as most Cretaceous ichthyosaur localities have yielded a handful of specimens, at best. Even if the Albian record is generally better than that of the rest of the Cretaceous, the Albian ichthyosaur localities are not contemporaneous. Nevertheless, Albian ichthyosaurs appear to have their biogeographical ranges limited to a regional scale; indeed, not a single species is shared between the Australia, North American, Canadian and European provinces, suggesting a high beta diversity. At a smaller geographic scale, ichthyosaur assemblages appear similar, as suggested by the French and eastern England localities described above. It is nevertheless possible that the apparent endemism between the major Albian ichthyosaur provinces is due to poor sampling. Indeed, late Albian ichthyosaurs of western Europe have similarities with the early Albian Canadian assemblage: '*P.*' *hercynicus* is northwestern France and similar '*Platypterygius*' sp. remains in the Cambridge Greensand Member resemble the large but poorly preserved *Platypterygius* sp. described by Maxwell & Caldwell [16], having a similar humerus and forefin. Furthermore, at least one isolated ilium from the

Table 4. Femoral morphotypes recognized in the Cambridge Greensand member.

Morphotype (# of specimens)	Facets	Including one for:	Capitulum shape	Trochanters
FM1 (8)	3	AAE	Triangular	Medium
FM2 (4)	2	/	Triangular	Medium
FM3 (1)	2	/	Rounded	High
FM4 (1)	3	Astragalus?	Rounded	High
FM5 (C. walkeri) (2)	3	AAE	Round, not connected to trochanters	High and lamellar

Abbreviations: AAE: anterior accessory epipodial element; PAE: posterior accessory epipodial element.

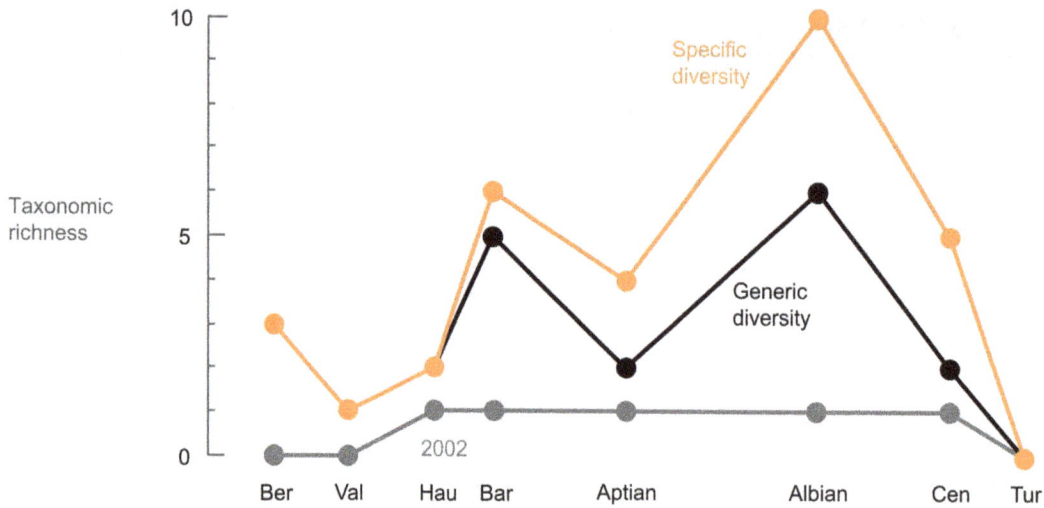

Figure 19. Relative abundance of the taxa recognized in the Cambridge Greensand Member. Platypterygiine taxa are colored in orange, ophthalmosaurine taxa in grey, and insertae sedis in white.

Gault Formation (NHMUK unnumbered) matches that of *Athabascasaurus bitumineus* from the early Albian of Canada [22,23], being markedly recurved posterodorsally. A better knowledge of the teeth and basicranium of the Canadian taxa could help to assess their presence or absence in Eurasian ecosystems.

Ichthyosaur-dominated ecosystems in the late Early Cretaceous of Europe

Tooth size and shape varies greatly among ichthyosaur taxa in the Cambridge Greensand Member. '*Platypterygius*' sp. possesses the largest and most robust teeth: the conical crown is robust, and the numerous apicobasal ridges texturing the crown, the acellular cementum ring and the root likely reinforced the resistance of the teeth under dorsoventral stress, as in corrugated materials. *Sisteronia seeleyi* possesses the smallest and most gracile teeth: the crown is pointed and slender, the tooth lacks conspicuous

apicobasal ridges basally to the crown, and the root is slender and markedly compressed transversely. Ophthalmosaurinae indet. falls in between these extremes.

Wear patterns are similarly contrasted between '*Platypterygius*' sp. and *Sisteronia seeleyi*. '*Platypterygius*' sp. teeth are by far the most worn: the majority of isolated (possibly shed) teeth fall within the most severe category of wear (apex broken and polished), and articulated specimens show a large proportion of functional teeth belonging to this wear category as well (e.g. CAMSM TN283; RGHP PR 1). By contrast, nearly all *Sisteronia seeleyi* teeth are only slightly polished or still have pristine enamel texture on the apex. Articulated rostra of *Sisteronia seeleyi* are currently not available, preventing statistical wear analysis on functional teeth. Similarly, very few ophthalmosaurine teeth occur in the Cambridge Greensand Member, preventing any evaluation of their wear with confidence. Yet, preserved ophthalmosaurine tooth apices belonging to all categories of wear are found within this small assemblage.

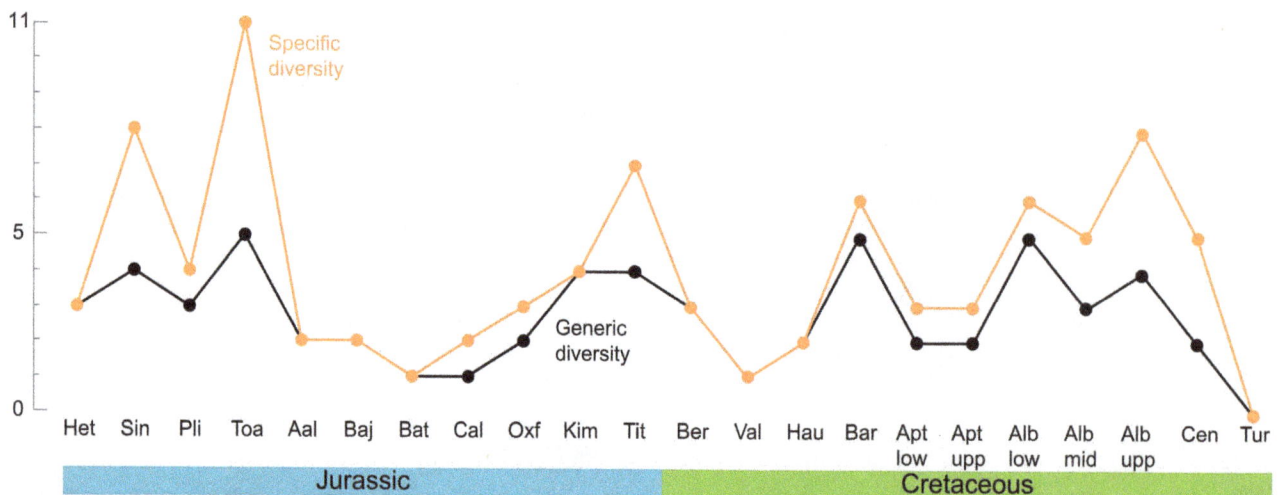

Figure 20. Teeth from the upper part of the Gault Formation and Cambridge Greensand Member (from left to right: CAMSM TN1779 partim; NHMUK R16 partim; NHMUK 47235), illustrating the three feeding guilds colonized by ichthyosaurs in this ecosystem.

This situation, where the most robust teeth are also the most worn and vice-versa, suggests contrasted diet for these taxa. These shape and wear differences also match size differences (Figure 20).

Accordingly, we propose that these ichthyosaurs colonized distinct feeding guilds: '*Platypterygius*' sp. probably belongs to a guild of top-tier predators, possibly feeding on tetrapods (among other prey), given the robust tooth shape and the intense tooth wear with frequent apical tooth breakage and enamel spalling, as already suggested for '*P.*' *australis* [10]. These features are regarded as indicative of such diet in marine crocodyliforms too [143]. Ophthalmosaurines are considered as opportunistic generalists, because their tooth shape and wear closely resembles those in *Aegirosaurus* and adult *O. icenicus*, considered generalists [1,99]. The delicate, slender and unworn teeth of *Sisteronia seeleyi* suggest that it belongs to a 'pierce'-oriented guild, feeding on soft and small prey such as small fishes and neocoleoid cephalopods, according to Massare's criteria [25,28] (Figure 20). These ichthyosaurs therefore occupied up to three feedings guilds within the single ecosystem of the upper Gault Formation/Cambridge Greensand Member, despite the presence of a diversified plesiosaur assemblage including the gigantic pliosaur *Polyptychodon interruptus* [50,68,144]. The presence of ichthyosaurs at several levels of the trophic chain of one ecosystem has not been previously reported from assemblages dating to after the Early Jurassic, when ichthyosaurs dominated the ecosystems of the European archipelago [24,25,119,145] together with several plesiosaur taxa, including large rhomaleosaurids [146–151]. The fact that ichthyosaurs from the late Albian–early Cenomanian deposits of Europe and possibly Russia, like their Early Jurassic ancestors, colonized multiple ecological niches despite the presence of numerous other marine reptile taxa shows that the 'last' ichthyosaurs were still a diversified and important component of the marine ecosystems up to a few millions years prior to their extinction, at least in Europe and Russia.

Conclusions

The thorough analysis of the diversity of the rich ichthyosaur assemblages of middle Albian–earliest Cenomanian of England and southern France yields the following results:

(1) We recognize four taxa as valid: '*Platypterygius*' sp., *Sisteronia seeleyi* gen. et sp. nov., Ophthalmosaurinae indet., and *Cetarthrosaurus walkeri*. We consider *Ichthyosaurus doughtyi*, *Ichthyosaurus bonneyi*, *Ichthyosaurus platymerus*, *Ichthyosaurus angustidens* and *Brachypterygius cantabrigiensis* as invalid; there is no solid evidence for the presence of *Brachypterygius* in the Cretaceous.

(2) Ichthyosaurs occupied several feeding guilds within the mid-Cretaceous ecosystems of western Europe: '*Platypterygius*' sp. likely occupied apex predator along with the large pliosauroid *Polyptychodon interruptus*, *Sisteronia seeleyi* occupied a 'pierce'-oriented guild, and ophthalmosaurine ophthalmosaurids probably occupied a 'generalist/opportunist' guild.

(3) These high taxonomic richnesses and strong ecological presences occur a few million years prior to the final extinction of ichthyosaurs. This indicates that the 'last' ichthyosaurs were diversified and were still a major component of marine ecosystems, contradicting previous views of ichthyosaur evolutionary history. The alpha diversity of ichthyosaur is, however, highly variable between provinces. This new data provides a whole new context to analyze the extinction of ichthyosaurs.

Supporting Information

Text S1 Gault Formation specimens studied here and their assignation, 19 specimens.

Text S2 Upper Greensand Formation specimens studied here and their assignation.

Text S3 Cambridge Greensand Member specimens studied here and their assignation.

Text S4 Marnes Bleues Formation specimens studied here and their assignation.

Text S5 Determination key for isolated elements from the Cambridge Greensand Member.

Text S6 Description of indeterminate femoral morphotypes.

Text S7 Taxa recognized as valid for each stage of the Hettangian–Turonian interval.

Acknowledgments

We warmly thank S. Chapman and P. Barrett at the NHMUK, M. Riley and the staff of the CAMSM, J. Liston and J. Keith Ingham at the GLAHM, M. Evans at the LEICT, Brice Etienne and Thierry Beving at the RGHP, and A. Dreze at the RBINS for access to specimens, logistical help and valuable discussions. This manuscript also benefited from careful and insightful reviews from A. Farke, J. Pardo Pérez and one anonymous reviewer. V.F. was an Aspirant of the Fonds de la Recherche Scientifique-FNRS when this research was conducted.

Author Contributions

Conceived and designed the experiments: VF NB. Performed the experiments: VF. Analyzed the data: VF NB PG. Contributed reagents/materials/analysis tools: VF MG. Wrote the paper: VF.

References

1. Fischer V, Clément A, Guiomar M, Godefroit P (2011) The first definite record of a Valanginian ichthyosaur and its implication for the evolution of post-Liassic Ichthyosauria. Cretaceous Research 32: 155–163.
2. McGowan C (1972) The systematics of Cretaceous ichthyosaurs with particuliar reference to the material from North America. Contributions to Geology 11: 9–29.
3. Bardet N (1992) Stratigraphic evidence for the extinction of the ichthyosaurs. Terra Nova 4: 649–656.
4. Sander PM (2000) Ichthyosauria: their diversity, distribution, and phylogeny. Paläontologische Zeitschrift 74: 1–35.
5. Lingham-Soliar T (2003) Extinction of ichthyosaurs: a catastrophic or evolutionary paradigm? Neues Jahrbuch für Geologie und Paläontologie, Abhandlungen 228: 421–452.
6. Wade M (1984) *Platypterygius australis*, an Australian Cretaceous ichthyosaur. Lethaia 17: 99–113.
7. Wade M (1990) A review of the Australian Cretaceous longipinnate ichthyosaur *Platypterygius* (Ichthyosauria, Ichthyopterygia). Memoirs of the Queensland Museum 28: 115–137.
8. Kear BP (2002) Dental caries in an Early Cretaceous ichthyosaur. Alcheringa 25: 387–390.

9. Kear BP (2003) Cretaceous marine reptiles of Australia: a review of taxonomy and distribution. Cretaceous Research 24: 277–303.

10. Kear BP, Boles WE, Smith ET (2003) Unusual gut contents in a Cretaceous ichthyosaur. Proceedings of the Royal Society of London B Biological Sciences 270: S206–S208.

11. Kear BP (2005) Cranial morphology of *Platypterygius longmani* Wade, 1990 (Reptilia: Ichthyosauria) from the Lower Cretaceous of Australia. Zoological Journal of the Linnean Society 145: 583–622.

12. Zammit M (2010) A review of Australasian ichthyosaurs. Alcheringa 34: 281–292.

13. Zammit M, Norris RM, Kear BP (2010) The Australian Cretaceous ichthyosaur *Platypterygius australis*: a description and review of postcranial remains. Journal of Vertebrate Paleontology 30: 1726–1735.

14. Maxwell EE, Kear BP (2010) Postcranial anatomy of *Platypterygius americanus* (Reptilia: Ichthyosauria) from the Cretaceous of Wyoming. Journal of Vertebrate Paleontology 30: 1059–1068.

15. Maxwell EE, Caldwell MW (2006) A new genus of ichthyosaur from the Lower Cretaceous of Western Canada. Palaeontology 49: 1043–1052.

16. Maxwell EE, Caldwell MW (2006) Evidence for a second species of the ichthyosaur *Platypterygius* in North America: a new record from the Loon River Formation (Lower Cretaceous) of northwestern Canada. Canadian Journal of Earth Sciences 43: 1291–1295.

17. Zammit M (2012) Cretaceous ichthyosaurs: dwindling diversity, or the Empire strikes back? Geosciences 2: 11–24.

18. Fischer V (2012) New data on the ichthyosaur *Platypterygius hercynicus* and its implications for the validity of the genus. Acta Palaeontologica Polonica 57: 123–134.

19. Fischer V, Appleby RM, Naish D, Liston J, Riding JB, et al. (2013) A basal thunnosaurian from Iraq reveals disparate phylogenetic origins for Cretaceous ichthyosaurs. Biology Letters 9: 20130021.

20. Fischer V, Maisch MW, Naish D, Liston J, Kosma R, et al. (2012) New ophthalmosaurids from the Early Cretaceous of Europe demonstrate extensive ichthyosaur survival across the Jurassic–Cretaceous boundary. PLoS ONE 7: e29234.

21. Fischer V, Masure E, Arkhangelsky MS, Godefroit P (2011) A new Barremian (Early Cretaceous) ichthyosaur from western Russia. Journal of Vertebrate Paleontology 31: 1010–1025.

22. Druckenmiller PS, Maxwell EE (2010) A new Lower Cretaceous (lower Albian) ichthyosaur genus from the Clearwater Formation, Alberta, Canada. Canadian Journal of Earth Sciences 47: 1037–1053.

23. Maxwell EE, Druckenmiller PS (2011) A small ichthyosaur from the Clearwater Formation (Alberta, Canada) and a discussion of the taxonomic utility of the pectoral girdle. Paläontologische Zeitschrift 85: 457–463.

24. Godefroit P (1996) Biodiversité des reptiles marins du Jurassique inférieur belgo-luxembourgeois. Bulletin de la Société belge de Géologie 104: 67–76.

25. Massare JA (1987) Tooth morphology and prey preference of Mesozoic marine reptiles. Journal of Vertebrate Paleontology 7: 121–137.

26. Kear BP, Zammit M (2013) *In utero* foetal remains of the Cretaceous ichthyosaurian *Platypterygius*: ontogenetic implications for character state efficacy. Geological Magazine In Press.

27. Zammit M (2011) The Australian Cretaceous ichthyosaur *Platypterygius australis*: understanding its taxonomy, morphology, and palaeobiology: University of Adelaide, South Australia. 106 p.

28. Massare JA (1997) Faunas, behavior, and evolution. In: Callaway JM, Nicholls EL, editors. Ancient Marine Reptiles. San Diego: Academic Press. pp. 401–421.

29. Schubert BW, Ungar PS (2005) Wear facets and enamel spalling in tyrannosaurid dinosaurs. Acta Palaeontologica Polonica 50: 93–99.

30. Thewissen JGM, Sensor JD, Clementz MT, Bajpari S (2011) Evolution of dental wear and diet during the origin of whales. Palaeobiology 37: 655–669.

31. Ross MR (2009) Charting the Late Cretaceous seas: mosasaurs richness and morphological diversification. Journal of Vertebrate Paleontology 292: 409–416.

32. Owen HG (2012) The Gault Group (Early Cretaceous, Albian), in East Kent, S.E. England; its lithology and ammonite biozonation. Proceedings of the Geologists' Association 123: 742–765.

33. Juignet P (1974) La transgression crétacée sur la bordure orientale du Massif armoricain. Aptien, Albien, Cénomanien de Normandie et du Maine. Le stratotype du Cénomanien: Université de Caen. 806 p.

34. Travassac F (2004) Stratigraphie, sédimentologie et géochimie d'une série d'âge barrémien supérieur à albien pro parte du bassin vocontien (SE France) : implications paléoenvironnmentales. Marseille: Ecole doctorale Sciences de l'Environnement d'Aix-Marseille. 54 p.

35. Scott RW (2009) Uppermost Albian biostrigraphy and chronostratigraphy. Notebooks on Geology 2009/03: 1–16.

36. Amédro F (2008) Support for a Vraconnian Stage between the Albien sensu stricto and the Cenomanien (Cretacous System). Notebooks on Geology Memoir 2008/02: 83.

37. Amédro F, Robaszynski F (2008) Zonation by ammonites and foraminifers of the Vraconnian-Turonian interval: A comparison of the Boreal and Tethyan domains (NW Europe/Central Tunisia). Notebooks on Geology, Letter 2008/02: 5.

38. Lehmann J, Heldt M, Bachmann M, Hedi Negra ME (2009) Aptian (Lower Cretaceous) biostratigraphy and cephalopods from north central Tunisia. Cretaceous Research 30: 895–910.

39. Kuhnt W, Moullade M (2007) The Gargasian (Middle Aptian) of La Marcouline section at Cassis-La Bédoule (SE France): Stable isotope record and orbital cyclicity. Notebooks on Geology 2007/02: 1–9.

40. Ogg J, Ogg G, Gradstein FM (2008) A concise geologic timescale. Cambridge.

41. Gradstein FM, Ogg JG, Schmitz M, Ogg G (2012) The Geologic Time Scale 2012. Oxford, Great Britain: Elsevier Science & Technology. 1176 p.

42. Benson RBJ, Butler RJ (2011) Uncovering the diversification history of marine tetrapods: ecology influences the effect of geological sampling biases. In: McGowan AJ, Smith AB, editors. Comparing the geological and fossil records: implications for biodiversity studies. London: Geological Society, Special Publications. pp. 191–208.

43. Butler RJ, Barrett PM, Nowbath S, Upchurch P (2009) Estimating the effects of sampling biases on pterosaur diversity patterns: implications for hypotheses of bird/pterosaur competitive replacement. Palaeobiology 35: 432–446.

44. Butler RJ, Brusatte SL, Andres B, Benson RB (2011) How do geological sampling biases affect studies of morphological evolution in deep time? A case study of pterosaur (Reptilia: Archosauria) disparity. Evolution 66: 147–162.

45. Mannion PD, Upchurch P (2011) A re-evaluation of the 'mid-Cretaceous sauropod hiatus' and the impact of uneven sampling of the fossil record on patterns of regional dinosaur extinction. Palaeogeography Palaeoclimatology Palaeoecology 299: 529–540.

46. Mannion PD, Upchurch P, Carrano MT, Barrett PM (2011) Testing the effect of the rock record on diversity: a multidisciplinary approach to elucidating the generic richness of sauropodomorph dinosaurs through time. Biological Reviews 86: 157–181.

47. Hopson PM, Wilkinson IP, Wood MA (2008) A stratigraphical framework for the Lower Cretaceous of England. British Geological Survey Research Reports RR/08/03: 1–87.

48. Sauvage HE (1882) Recherches sur les reptiles trouvées dans le Gault de l'Est du bassin de Paris. Mémoires de la Société géologique de France, 3e série 2: 21–24.

49. Breton G (2011) Deux nouvelles espèces de crustacés décapodes de l'Albien du Bassin de Paris. Geodiversitas 33: 279–284.

50. Seeley HG (1869) Index of the fossil remains of Aves, Ornithosauria and Reptilia, from the Secondary System of Strata Aranged in the Woodward Museum of the Univeristy of Cambridge; Deighton BC, editor. Cambridge.

51. Martill DM, Unwin DM (2012) The world's largest toothed pterosaur, NHMUK R481, an incomplete rostrum of *Coloborhynchus capito* (Seeley, 1870) from the Cambridge Greensand of England. Cretaceous Research 34: 1–9.

52. Hopson PM (2005) A stratigraphical framework for the Upper Cretaceous Chalk of England and Scotland with statements on the Chalk of Northern Ireland and the UK Offshore Sector. British Geological Survey Research Reports RR/05/01: 1–102.

53. Cookson IC, Hughes NF (1964) Microplankton from the Cambridge Greensand (mid-Cretaceous). Palaeontology 7: 37–59.

54. Barrett PM, Evans SE (2002) A reassessment of the Early Cretaceous reptile 'Patricosaurus merocratus' Seeley from the Cambridge Greensand, Cambridge-shire, UK. Cretaceous Research 23: 231–240.

55. Seeley HG (1876) On an associated series of cervical and dorsal vertebræ of *Polyptychodon*, from the Cambridge Upper Greensand, in the Woodwardian Museum of the University of Cambridge. Quarterly Journal of the Geological Society 32: 433–436.

56. Unwin DM (2001) An overview of the pterosaur assemblage from the Cambridge Greensand (Cretaceous) of Eastern England. Mitteilungen aus dem Museum für Naturkunde in Berlin, Geowissenschaftliche Reihe 4: 189–221.

57. Robaszynski F, Amedro F, Gonzalez-Donoso JM, Linares D (2008) The Albian (Vraconnian)-Cenomanian boundary at the western Tethyan margins (central Tunisia and southeastern France). Bulletin de la Société géologique de France 179: 245–266.

58. Woods MA, Wilkinson GK, Booth KA, Farrant AR, Hopson PM, et al. (2008) A reappraisal of the stratigraphy and depositional development of the Upper Greensand (Late Albian) of the Devizes district, southern England. Proceedings of the Geologists' Association 119: 229–244.

59. Bréhéret J-G (1997) L'Aptien et l'Albien de la Fosse vocontienne (des bordures au bassin) : Evolution de la sédimentation et enseignements sur les événements anoxiques. Société géologique du Nord 25: 614.

60. Wilpshaar M, Leereveld H, Visscher H (1997) Early Cretaceous sedimentary and tectonic development of the Dauphinois Basin (SE France). Cretaceous Research 18: 457–468.

61. Kennedy WJ, Gale AS, Bown PR, Caron M, Davey RJ, et al. (2000) Integrated stratigraphy across the Aptian-Albian boundary in the Marnes Bleues, at the Col de Pré-Guittard, Arnayon (Drôme), and at Tartonne (Alpes-de-Haute-Provence), France: a candidate Global Boundary Stratotype Section and Boundary Point for the base of the Albian Stage. Cretaceous Research 21: 591–720.

62. Accarie H, Beaudoin B, Dejax J, Friès G, Michard J-G, et al. (1995) Découverte d'un dinosaure théropode nouveau (*Genusaurus sisteronis* n. g., n. sp.) dans l'Albien marin de Sisteron (Alpes de Haute-Provence, France) et extension au Crétacé inférieur de la lignée cératosaurienne. Comptes Rendus de l'Académie des Sciences de Paris 320: 327–334.

63. Herrle JO, Mutterlose J (2003) Calcareous nannofossils from the Aptian-Lower Albian of southeast France: palaeoecological and biostratigraphic implications. Cretaceous Research 24: 1–22.

64. Haccard D, Beaudoin B, Gigot P, Jorda M (1989) La Javie. In: BRGM, editor. Carte Géologique de France à 1/50 000.

65. Blainville HMD, de (1835) Description de quelques espèces de reptiles de la Californie, précédée de l'analyse d'un système général d'érpetologie et d'amphibiologie. Nouvelles annales du Muséum d'Histoire naturelle, Paris 4: 233–296.

66. Baur G (1887) On the morphology and origin of the Ichthyopterygia. American Naturalist 21: 837–840.

67. Arkhangelsky MS (2001) The historical sequence of Jurassic and Cretaceous ichthyosaurs. Paleontological Journal 35: 521–524.

68. Lydekker R (1889) Catalogue of the fossil Reptilia and Amphibia in British Museum (Natural History). Part II. containing the orders Ichthyopterygia and Sauropterygia. London: Printed by Orders of the Trustees of the British Museum, London.

69. McGowan C, Motani R (2003) Part 8. Ichthyopterygia; Sues H-D, editor. München: Verlag Dr. Friedrich Pfeil. 175 p.

70. Johnson R (1977) Size independent criteria for estimating relative age and the relationship among growth parameters in a group of fossil reptiles (Reptilia: Ichthyosauria). Canadian Journal of Earth Sciences 14: 1916–1924.

71. Fischer V (2013) Origin, biodiversity and extinction of Cretaceous ichthyosaurs. Liège, Belgium: Université de Liège. 576 p.

72. Bardet N, Fernández M (2000) A new ichthyosaur from the Upper Jurassic lithographic limestones of Bavaria. Journal of Paleontology 74: 503–511.

73. Kirton AM (1983) A review of British Upper Jurassic ichthyosaurs. Newcastle upon Tyne: University of Newcastle upon Tyne. 239 p.

74. Kuhn O (1946) Ein skelett von Ichthyosaurus hercynicus n. sp. aus dem Aptien von Gitter. Berichte der Naturforschenden Gesellschaft Bamberg 29: 69–82.

75. Appleby RM (1961) On the cranial morphology of ichthyosaurs. Proceedings of the Zoological Society of London 137: 333–370.

76. Kolb C, Sander PM (2009) Redescription of the ichthyosaur Platypterygius hercynicus (Kuhn 1946) from the Lower Cretaceous of Salzgitter (Lower Saxony, Germany). Palaeontographica Abteilung A (Paläozoologie, Stratigraphie) 288: 151–192.

77. Adams TL, Fiorillo A (2011) Platypterygius Huene, 1922 (Ichthyosauria, Ophthalmosauridae) from the Late Cretaceous of Texas, USA. Palaeontologia Electronica 14: 19A.

78. Gasparini Z (1988) Ophthalmosaurus monocharactus Appleby (Reptilia, Ichthyopterygia), en las calizas litograpficas tithonianas del area Los Catutos, Nequén, Argentina. Ameghiniana 25: 3–16.

79. McGowan C (1973) The cranial morphology of the Lower Liassic latipinnate ichthyosaurs of England. Bulletin of the British Museum (Natural History) Geology 24: 1–109.

80. Andrews CW (1910) A descriptive catalogue of the Marine Reptiles of the Oxford Clay, part I. London: British Museum of Natural History. 205 p.

81. Maxwell EE (2010) Generic reassignment of an ichthyosaur from the Queen Elizabeth Islands, Northwest Territories, Canada. Journal of Vertebrate Paleontology 30: 403–415.

82. Maxwell EE, Caldwell MW, Lamoureux DO (2011) Tooth histology in the Cretaceous ichthyosaur Platypterygius australis, and its significance for the conservation and divergence of mineralized tooth tissues in amniotes. Journal of Morphology 272: 129–135.

83. Appleby RM (1956) The osteology and taxonomy of the fossil reptile Ophthalmosaurus. Proceedings of the Zoological Society of London 126: 403–447.

84. Nace RL (1939) A new ichthyosaur from the Upper Cretaceous Mowry Formation of Wyoming. American Journal of Science 237: 673–686.

85. Araújo R, Smith AS, Liston J (2008) The Alfred Leeds fossil vertebrate Collection of the National Museum of Ireland–Natural History. Irish Journal of Earth Sciences 26: 17–32.

86. Huene Fv (1922) Die Ichthyosaurier des Lias und ihre Zusammenhänge. Berlin: Verlag von Gebrüder Borntraeger. 114 p.

87. Delair JB (1960) The Mesozoic reptiles of Dorset. Proceedings of the Dorset Natural History and Arhcaeological Society 81: 59–85.

88. Carter J (1846) On the occurence of a new species of Ichthyosaurus in the Chalk. London Geological Journal 1.

89. Carter J (1846) Notice of the jaws of an Ichthyosaurus from the chalk in the neighbourhood of Cambridge. Reports of the British Association for the Advancement of Science 1845: 60.

90. Bardet N (1989) Un crâne d'Ichthyopterygia dans le Cénomanien du Boulonnais. Mémoires de la Société académique du Boulonnais 6: 31 pp.

91. Kiprijanoff W (1881) Studien über die fossilen Reptilien Russlands. Theil 1, Gattung Ichthyosaurus König aus dem severischen Sandstein oder Osteolith der Kreide-Gruppe. Mémoires de l'Académie impériale des Sciences de St-Pétersbourg, VIIe série 28: 1–103.

92. Bardet N (1990) Dental cross-section in Cretaceous Ichthyopterygia: systematic implications. Geobios 23: 169–172.

93. Sirotti A, Papazzoni C (2002) On the Cretaceous ichthyosaur remains from the Northern Apennines (Italy). Bollettino della Societa Paleontologica Italiana 41: 237–248.

94. Blain H-A, Pennetier G, Pennetier E (2003) Présence du genre Platypterygius (Ichthyosauria, Reptilia) dans le Cénomanien inférieur de Villers-sur-Mer (Normandie, France. Echos des falaises 7: 35–50.

95. Broili F (1907) Ein neuer Ichthyosaurus aus der norddeutschen Kreide. Palaeontographica 54: 139–162.

96. Nace RL (1941) A new ichthyosaur from the Late Cretaceous of northeastern Wyoming. American Journal of Science 239: 908–914.

97. Fernández MS, Maxwell EE (2012) The genus Arthropterygius Maxwell (Ichthyosauria: Ophthalmosauridae) in the Late Jurassic of the Neuquén Basin, Argentina. Geobios 45: 535–540.

98. Lydekker R (1888) Note on the classification of the Ichthyopterygia with a notice of two new species. Geological Magazine third series 5: 309–314.

99. Fischer V, Arkhangelsky MS, Uspensky GN, Stenshin IM, Godefroit P (In Press) A new Lower Cretaceous ichthyosaur from Russia reveals skull shape conservatism within Ophthalmosaurinae. Geological Magazine In Press.

100. Scheyer TM, Moser M (2011) Survival of the thinnest: rediscovery of Bauer's (1898) ichthyosaur tooth sections from Upper Jurassic lithographic limestone quarries, south Germany. Swiss Journal of Geoscience 104: S147–S157.

101. McGowan C (1997) The taxonomic status of the Late Jurassic ichthyosaur Grendelius mordax: a preliminary report. Journal of Vertebrate Paleontology 17: 428–430.

102. Pardo-Perez J, Frey E, Stinnesbeck W, Fernández M, Rivas L, et al. (2012) An ichthyosaurian forefin from the Lower Cretaceous Zapata Formation of southern Chile: implications for morphological variability within Platypterygius. Palaeobiodiversity and Palaeoenvironment 92: 287–294.

103. Arkhangelsky MS (1997) On a new genus of ichthyosaurs from the Lower Volgian substage of the Saratov, Volga Region. Paleontological Journal 31: 87–90.

104. Efimov VM (1999) Ichthyosaurs of a new genus Yasykovia from the Upper Jurassic strata of European Russia. Paleontological Journal 33: 92–100.

105. Maisch MW, Matzke AT (2000) The Ichthyosauria. Stuttgarter Beiträge zur Naturkunde Serie B (Geologie und Paläontologie) 298: 1–159.

106. Seeley HG (1873) On Cetarthrosaurus walkeri (Seeley), an ichthyosaurian from the Cambridge Upper Greensand. Quarterly Journal of the Geological Society 29: 505–507.

107. Fernández M (1997) A new ichthyosaur from the Tithonian (Late Jurassic) of the Neuquén Basin (Argentina). Journal of Paleontology 71: 479–484.

108. Fernández M (2007) Redescription and phylogenetic position of Caypullisaurus (Ichthyosauria: Ophthalmosauridae). Journal of Paleontology 81: 368–375.

109. Fernández M (2007) Chapter 11. Ichthyosauria. In: Gasparini Z, Salgado L, Coria RA, editors. Patagonian Mesozoic Reptiles. Bloomington and Indianapolis, Indiana: Indiana University Press. pp. 271–291.

110. Efimov VM (1997) A new genus of ichthyosaurs from the Late Cretaceous of the Ulyanovsk Volga region. Paleontological Journal 31: 422–426.

111. Fischer V, Arkhangelsky MS, Stenshin IM, Uspensky GN, Godefroit P (In Review) The Russian Cretaceous ichthyosaurs Simbirskiasaurus birjukovi and Pervushovisaurus bannovkensis. Journal of Vertebrate Paleontology.

112. Paramo ME (1997) Platypterygius sachicarum (Reptilia, Ichthyosauria) nueva especie del Cretácio de Colombia. Revista Ingeominas 6: 1–12.

113. Fernández M, Aguirre-Urreta MB (2005) Revision of Platypterygius hauthali von Huene, 1927 (Ichthyosauria, Ophthalmosauridae) from the Early Cretaceous of Patagonia, Argentina. Journal of Vertebrate Paleontology 25: 583–587.

114. Huene Fv (1927) Beitrag zur Kenntnis mariner mesozoicher Wirbeltiere in Argentinien. Centralblatt für Mineralogie, Geologie und Paläntologie, B 1927: 22–29.

115. Maisch MW (2008) Revision der Gattung Stenopterygius Jaekel, 1904 emend. von Huene, 1922 (Reptilia: Ichthyosauria) aus dem unteren Jura Westeuropas. Palaeodiversity 1: 227–271.

116. Maisch MW (2001) Neue Exemplare der seltenen Ichthyosauriergattung Suevoleviathan Maisch 1998 aus dem Unteren Jura von Südwestdeutschland. Geologica et Palaeontologica 35: 145–160.

117. Maisch MW (1998) A new ichthyosaur genus from the Posidonia Shale (Lower Toarcian, Jurassic) of Holzmaden, SW-Germany with comments on the phylogeny of post-Triassic ichthyosaurs. Neues Jahrbuch für Geologie und Paläontologie, Abhandlungen 209: 47–78.

118. Maxwell EE (2012) New metrics to differentiate species of Stenopterygius (Reptilia: Ichthyosauria) from the Lower Jurassic of southwestern Germany. Journal of Paleontology 86: 105–115.

119. Martin JE, Fischer V, Vincent P, Suan G (2012) A longirostrine Temnodontosaurus (Ichthyosauria) with comments on Early Jurassic ichthyosaur niche partitioning and disparity. Palaeontology 55: 995–1005.

120. Gasparini Z, De La Fuente M, Fernández M (1995) Sea reptiles from the lithographic limestones of the Neuquén Basin, Argentina. II international symposium on lithographic limestones. Lleida - Cuenca (Spain): Ediciones de la Universidad Autónoma de Madrid. pp. 81–84.

121. Spalletti L, Gasparini Z, Veiga G, Schwarz E, Fernández M, et al. (1999) Facies anóxicas, procesos deposicionales y herpetofauna de la rampa marina titoniano-berriasiana en la Cuenca Neuquina (Yesera del Tromen), Neuquén, Argentina. Revista Geológica de Chile 1: 109–123.

122. Buchy MC, Lopez Oliva JG (2009) Occurrence of a second ichthyosaur genus (Reptilia; Ichthyosauria) in the Late Jurassic Gulf of Mexico. Boletin de la Sociedad Geologica Mexicana 61: 233–238.

123. Buchy M-C (2010) First record of *Ophthalmosaurus* (Reptilia: Ichthyosauria) from the Tithonian (Upper Jurassic) of Mexico. Journal of Paleontology 84: 149–155.

124. Arkhangelsky MS (1998) On the ichthyosaurian fossil from the Volgian stage of the Saratov Region. Paleontological Journal 32: 192–196.

125. Arkhangelsky MS (2000) On the ichthyosaur *Otschevia* from the Volgian Stage of the Volga Region. Paleontological Journal 34: 549–552.

126. Arkhangelsky MS (2001) On a new ichthyosaur of the genus *Otschevia* from the Volgian Stage of the Volga Region near Ulyanovsk. Paleontological Journal 35: 629–634.

127. Efimov VM (1998) An ichthyosaur, *Otschevia pseudoscythica* gen. et sp. nov. from the Upper Jurassic strata of the Ulyanovsk Region (Volga Region). Paleontological Journal 32: 82–86.

128. Efimov VM (1999) A new family of ichthyosaurs, the Undorosauridae fam. nov. from the Volgian stage of the European part of Russia. Paleontological Journal 33: 174–181.

129. Bardet N (1994) Extinction events among Mesozoic marine reptiles. Historical Biology 7: 313–324.

130. Bardet N (1995) Evolution et extinction des reptiles marins au cours du Mésozoïque. Palaeovertebrata 24: 177–283.

131. McGowan C (1974) A revision of the longipinnate ichthyosaurs of the Lower Jurassic of England, with description of the new species (Reptilia, Ichthyosauria). Life Science Contributions, Royal Ontario Museum 97: 1–37.

132. McGowan C (1974) A revision of the latipinnate ichthyosaurs of the Lower Jurassic of England (Reptilia, Ichthyosauria). Life Science Contributions, Royal Ontario Museum 100: 1–30.

133. McGowan C (1989) *Leptopterygius tenuirostris* and other long-snouted ichthyosaurs from the English Lower Lias. Palaeontology 32: 409–427.

134. McGowan C (1996) Giant ichthyosaurs of the Early Jurassic. Canadian Journal of Earth Sciences 33: 1011–1021.

135. Godefroit P (1993) Les grands ichthyosaures sinémuriens d'Arlon. Bulletin de l'Institut Royal des Sciences Naturelles de Belgique Sciences de la Terre 63: 25–71.

136. Delair JB (1972) Some recent discoveries of Kimmeridgian reptiles at Swindon. Wiltshire Archaeological and Natural History Magazine 67: 12–15.

137. Taylor MA, Benton MJ (1986) Reptiles from the Upper Kimmeridge Clay (Kimmeridgian, Upper Jurassic) of the vicinity of Egmont Bight, Dorset. Proceedings of the Dorset Natural History and Archaeological Society 107: 121–125.

138. Billon-Bruyat J-P, Lécuyer C, Martineau F, Mazin J-M (2005) Oxygen isotope compositions of Late Jurassic vertebrate remains from lithographic limestones of western Europe: implications for the ecology of fish, turtles, and crocodilians. Palaeogeography Palaeoclimatology Palaeoecology 216: 359–375.

139. Buffetaut E (1994) Tetrapods from the Late Jurassic and Early Cretceous lithographic limestones of Europe: a comparative review. Geobios, mémoire spécial 16: 259–265.

140. Gasparini Z, Fernández M (2005) Jurassic marine reptiles of the Neuquén Basin: records, faunas and their palaeobiogeographic significance. In: Veiga GD, Spalletti LA, Howell JA, Schwarz E, editors. The Neuquén Basin, Argentina: A case study in sequence stratigraphy and basin dynamics. London: Geological Society, special Publications. pp. 279–294.

141. McGowan C (1991) Dinosaurs, Spitfires, and Sea Dragons: Harvard University Press. 365 p.

142. Ensom PC, Clements RG, Feist-Burkhardt S, Milner AR, Chitolie J, et al. (2009) The age and identity of an ichthyosaur reputedly from the Purbeck Limestone Group, Lower Cretaceous, Dorset, southern England. Cretaceous Research 30: 699–709.

143. Young MT, Brusatte SL, de Andrade MB, Desojo JB, Beatty BL, et al. (2012) The cranial osteology and feeding ecology of the metriorhynchid crocodylomorph genera *Dakosaurus* and *Plesiosuchus* from the Late Jurassic of Europe. PLoS ONE 7: e44985.

144. Owen R (1851–64) A monograph on the fossil Reptilia of the Cretaceous formations. London: The Palæntological Society.

145. Fischer V, Guiomar M, Godefroit P (2011) New data on the palaeobiogeography of Early Jurassic marine reptiles: the Toarcian ichthyosaur fauna of the Vocontian Basin (SE France). Neues Jahrbuch für Geologie und Paläontologie 261: 111–127.

146. Benson RB, Ketchum HF, Noè LF, Gómez-Pérez M (2011) New information on Hauffiosaurus (Reptilia, Plesiosauria) based on a new species from the Alumn Shale member (Lower Toarcian: Lower Jurassic) of Yorkshire, UK. Palaeontology 54: 547–571.

147. Benson RB, Evans M, Druckenmiller PS (2012) High diversity, low disparity and small body size in plesiosaurs (Reptilia, Sauropterygia) from the Triassic–Jurassic boundary. PLoS ONE 7: e31838.

148. Vincent P, Benson RBJ (2012) *Anningasaura*, a basal plesiosaurian (Reptilia, Plesiosauria) from the Lower Jurassic of Lyme Regis, United Kingdom. Journal of Vertebrate Paleontology 32: 1049–1063.

149. Vincent P, Bardet N, Mattioli E (2013) A new pliosaurid from the Pliensbachian (Early Jurassic) of Normandy (Northern France). Acta Palaeontologica Polonica 58: 471–485.

150. Vincent P (2008) Les Plesiosauria (Reptilia, Sauropterygia) du Jurassique inférieur : systématique, anatomie, phylogénie et paléoécologie. Paris: Museum national d'Histoire naturelle de Paris. 577 p.

151. Smith AS (2007) Anatomy and Systematics of the Rhomaleosauridae (Sauropterygia: Plesiosauria): University College Dublin. 278 p.

9

Land Snails as a Diet Diversification Proxy during the Early Upper Palaeolithic in Europe

Javier Fernández-López de Pablo[1,2]*, Ernestina Badal[3], Carlos Ferrer García[4], Alberto Martínez-Ortí[5], Alfred Sanchis Serra[4]

1 Institut Català de Paleoecologia Humana i Evolució Social, Zona Educacional 4 Campus Sescelades (Edifici W3), Tarragona, Spain, 2 Àrea de Prehistòria, Universitat Rovira i Virgili (URV), Tarragona, Spain, 3 Departament de Prehistòria i Arqueologia, Facultat de Geografia i Història, Universitat de València, València, Spain, 4 Museu de Prehistòria de València, SIP (Servei d'Investigació Prehistòrica), Diputació de València, València, Spain, 5 Museu Valencià d'Història Natural & i\ Biotaxa, Valencia, Spain

Abstract

Despite the ubiquity of terrestrial gastropods in the Late Pleistocene and Holocene archaeological record, it is still unknown when and how this type of invertebrate resource was incorporated into human diets. In this paper, we report the oldest evidence of land snail exploitation as a food resource in Europe dated to 31.3-26.9 ka yr cal BP from the recently discovered site of Cova de la Barriada (eastern Iberian Peninsula). Mono-specific accumulations of large *Iberus alonensis* land snails (Ferussac 1821) were found in three different archaeological levels in association with combustion structures, along with lithic and faunal assemblages. Using a new analytical protocol based on taphonomic, microX-Ray Diffractometer (DXR) and biometric analyses, we investigated the patterns of selection, consumption and accumulation of land snails at the site. The results display a strong mono-specific gathering of adult individuals, most of them older than 55 weeks, which were roasted in ambers of pine and juniper under 375°C. This case study uncovers new patterns of invertebrate exploitation during the Gravettian in southwestern Europe without known precedents in the Middle Palaeolithic nor the Aurignacian. In the Mediterranean context, such an early occurrence contrasts with the neighbouring areas of Morocco, France, Italy and the Balkans, where the systematic nutritional use of land snails appears approximately 10,000 years later during the Iberomaurisian and the Late Epigravettian. The appearance of this new subsistence activity in the eastern and southern regions of Spain was coeval to other demographically driven transformations in the archaeological record, suggesting different chronological patterns of resource intensification and diet broadening along the Upper Palaeolithic in the Mediterranean basin.

Editor: Nuno Bicho, Universidade do Algarve, Portugal

Funding: The fieldwork research and the radiocarbon analyses were supported by private funds provided by the Fundación Adendia in the framework of the research project named "El Poblamiento inicial de Benidorm y la Marina Baixa (Alicante)". JFL is supported by a Ramón y Cajal program postdoctoral research grant (Ref. RYC-2011-09363) of the MINECO Spanish Ministry and the Consolidated Research Groug (Ref. SGR-2014-900) "Group d'anàlisis de processos socioecológics, canvis culturals i dinàmiques de Població a la Prehistòria". The MINECO Spanish Ministry also funded the research projects "Paleolítico Medio final y Paleolítico Superior inicial en la región central mediterránea ibérica (Valencia y Murcia)" (Ref. HAR2012-32703) and "Paleoflora ibérica en un contexto de complejidad: interacciones fisiográficas, ecológicas y evolutivas" (Ref.2012CGL-34717) that supported the post-excavation palaeobotanical analyses and the cost of a radiocarbon determination. The funders had no role in the study design, data collection and analysis, decision to publish or the preparation of the manuscript.

Competing Interests: The authors have declared that no competing interests exist.

* Email: jfernandez@iphes.cat

Introduction

Diet change is a widely debated research topic of the Middle to Upper Palaeolithic transition. Studies on vertebrate prey mobility, size and body biomass suggest that, in many areas of Europe, the first anatomically modern humans (AMH) had a broader diet than Neanderthals, who mainly focused on large- and medium-size herbivores [1–3]. However, this view has been called into a question by the increasing body of archaeological evidence indicating that Neanderthals' subsistence also relied on a varied range of resources including plants, fish, birds, shellfish, tortoises, marine mammals and rabbits [4–15]. In this context, terrestrial molluscs were not believed to have been of any importance in the study of the dietary change and nutritional ecology during the Middle to Upper Palaeolithic transition. Unlike the increasing evidence for the consumption of marine molluscs amongst the

Neanderthals, there is a no clear signal of land snail exploitation during the Middle Palaeolithic, where terrestrial molluscs are considered intrusive in archaeological contexts, accumulated by thanatocoenoses or transported by non-human predators. Furthermore, the use of land snails as a food resource has been openly questioned in several Early Upper Palaeolithic contexts on the grounds of taphonomic and spatial analyses [16]. However, as posed by Lubell [17–18], the environmental interest of Late Pleistocene land snails and the paucity of specific studies focused on taxonomic, taphonomic, quantitative and biometric studies have prevented an understanding of the beginning, context and specific modalities of this type of subsistence activity.

In this paper, we report new evidence of land snail consumption from the Gravettian archaeological site of Cova de la Barriada (Benidorm, Spain) in the southeastern Iberian Peninsula, dated to 31.3–26.9 yr cal BP. This site has yielded mono-specific concentrations of large *Iberus alonensis* (Ferussac 1821) land snails associated with occupational features, lithic artefacts and mammalian faunal assemblages accumulated by humans. Through taphonomic and biometric analyses, we will investigate the patterns of selection and cultural accumulation of this species of edible land snail at the site. This case study illustrates new patterns of economic diversification during the Early Upper Palaeolithic in the eastern and southern Iberian Peninsula, which is not documented in other European and circum-Mediterranean regions until the Late Upper Palaeolithic, more than 10,000 years later.

Materials and Methods

1. Site description

Cova de la Barriada is formed by two connected rockshelters, so-called lower and upper rockshelters, respectively, located at the base of a tectonic escarpment of Mesozoic limestone on the western slope of the Serra Gelada mountain (Figure 1). The site is oriented towards the NW at 180 m.a.s.l.

Excavation was undertaken in January 2011 and consisted of a series of three test pits to evaluate the preservation of the sedimentary fill. Archaeological Fieldwork permit (Ref.2010/1023-A) was issued by the Dirección General de Patrimonio (Generalitat Valenciana, Valencia, Spain) to Javier Fernández-López de Pablo (Permit number: 2010/1023-A). In the lower rockshelter, Test Pit 2, which is 2 m², revealed a thin remnant Holocene deposit overlying the limestone bedrock. Despite the lack of lithic and faunal assemblages, two different anthropic accumulations of land snails (*Sphincterochila candidissima*) and marine molluscs (*Cerastoderma glaucum* and *Patella* sp) were documented in this deposit, suggesting an Early to Middle Holocene age (likely Mesolithic).

In the upper rockshelter, the archaeological fill was formed by colluvial sedimentation that, for the most part, has been eroded away by water (karstic) erosion and subsequent transport as well as recent husbandry and looting activities. Test Pit 3, which is 1 m², yielded archaeological materials in clearly secondary position from the dismantled Late Pleistocene deposits.

Only Test Pit 1, located at the south of the upper rockshelter, right outside of the roofed area, preserved *in situ* archaeological levels with combustion structures and associated lithic, faunal and land snail assemblages. The archaeo-sedimentary deposit is formed by a succession of three main stratigraphic units, subdivided into several subunits with different contents of boulders, angular blocks and gravels in a matrix of sands and silts (Figure 2). From top to bottom, the stratigraphic sequence is described as follows:

Unit I is composed of five subunits. Subunit I.1 is a calcareous flowstone of horizontal geometry. Subunits I.2 to I.5 are mainly composed by very pale brown (10 YR 7/3 and 7/4) thin fraction (70%) and angular heterometric gravels. Subunit I.5, which overlies Unit II with an angular unconformity, yielded archaeological evidence (artefacts and features) that correspond to Archaeological Unit A.

Unit II is a 10-cm thick deposit of massive structure containing a heterometric thick fraction of gravitational boulders and angular blocks and gravels in a matrix of very pale brown (10 YR 8/3) thin fraction. It dips E-W with an inclination of 12–10°. This stratigraphic unit corresponds to Archaeological Unit B, containing both artefacts and features.

Unit III is composed of five different subunits. Subunits III.1 and III.3 contain 90% of greyish orange (10 YR 7/4) sands and silts partially separated by a laterally discontinuous subunit of

gravels (III.2). The underlying subunit III.4 (10 YR 8/3 and 8/4) is predominantly composed of a thick fraction (boulders and angular blocks), whereas subunit III.5 is mainly formed by massive structure sands and silts. Occupational evidence (Archaeological Unit C) is restricted to subunits III.1 and III.3.

Archaeological units A, B and C have yielded Early Upper Palaeolithic artefacts, faunal assemblages and combustion structures whose basic morphological and dimensional attributes are presented in Table 1. Despite the partial conservation of the combustion structures, most of them (EC-1, EC-2, EC-3, EC-4 and EC-5) have a flat section associated with heterogeneous carbonaceous lenses and fire-cracked limestone blocks. In contrast, combustion structure BM, which was partially documented because it extended outside the limits of the test pit, has a shallow pit morphology and a concave section containing homogeneous carbonaceous sediments with abundant charcoal. On the other hand, combustion structure EC-6 has an irregular concave section associated with burnt and fire-cracked limestone blocks.

A series of AMS radiocarbon dates from individual and taxonomically determined charcoal samples recovered from the combustion structures were produced to assess the chronology of the stratigraphic sequence (Table 2). The samples are charcoal of *Pinus nigra* from EC-1 level A, Fabaceae charcoal from EC-5 level B and *Juniperus* sp. charcoal from EC-6 level C. All samples were plotted at the time of excavation and analysed at the Beta Analytic Laboratories in London. Calibration was performed using Oxcal v.4.1.3 [19] and the Intcal13 calibration curve [20].

Radiocarbon dates from levels A, B and C yielded significantly different chronologies in accordance with their stratigraphic position, suggesting two hiatuses between archaeological units C and B and archaeological units B and A. The chronological gaps between the above-mentioned levels are consistent with the erosive contact documented between subunits I.A and II and subunits III.1 and III.3.

Figure 3 represents a correlation between the radiocarbon chronology of Levels A and B of Cova de la Barriada and the AMS radiocarbon dates from the well-known Gravettian units of the Nerja and Cendres caves, the regional reference sequences in the Iberian Mediterranean region for this period [21–22]. In addition, we compared the summed probability chronological distributions with the global climatic ^{18}O GISP2 curve [23] and the regional variations of sea surface temperatures obtained in the Alborán Sea [24]. According to this tentative correlation, Level B falls within the accepted chronology of the H3 Heinrich event in the Mediterranean Sea, whereas Level A appears to fall in Greenland Interstadial 3.

At the time of the Early Upper Palaeolithic occupations in Cova de la Barriada (30–25 ka), the Mediterranean sea level was 90–100 m lower than today, implying a distance of approximately 20 km between the site and the shore line on the basis of local data on marine floor topography [25].

The archaeological and paleontological information presented in this paper involved the direct analysis of 321 paleobotanical specimens, 489 vertebrate specimens and 1484 invertebrate (land snails) specimens.

2. Contextual information

A set of Upper Palaeolithic artefacts were recovered in association with the land snails and faunal assemblages in Test Pit 1. Despite the paucity of the lithic assemblages (n = 39), two dihedral-deviated burins were documented at Level A, and two splintered pieces were documented in Levels A and B. In addition, three small umbo-pierced marine shells of *Glycimeris* sp. were recovered in Levels A (n = 2) and C (n = 1).

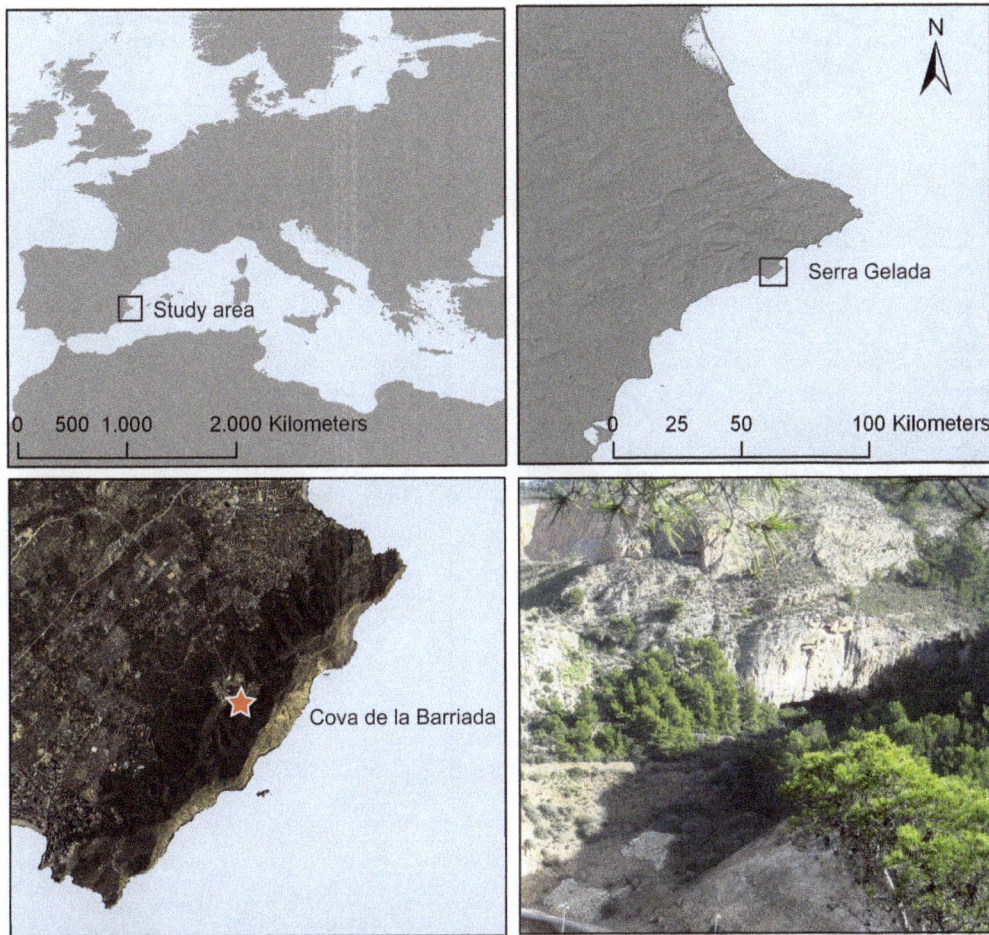

Figure 1. Study area and site location. Bottom right: general view of Cova de la Barriada site.

The palaeobotanical evidence recovered comprises 328 charcoal remains of 9 different taxa (Table 3). They represent a genuine western Mediterranean Late Pleistocene landscape, dominated by cryophilous pines (*Pinus nigra and/or Pinus sylvestris*) and junipers (*Juniperus* sp.), which, in turn, are the most common taxa documented as fuel in the Upper Palaeolithic and other woody plants such as rosemary (*Rosmarinus officinalis*), *Cistus* sp. Fabaceae and Lamiaceae [26–28]. In summary, the taxonomic association identified throughout the archaeological sequence indicates a cold ecology with the punctual presence of genuine Mediterranean flora such as the rosemary in level A. The present-day distribution of the *Pinus nigra* and *Pinus sylvestris* forests are found at 1500 m.a.s.l., from medium up to high altitudinal mountain ranges, associated with mean annual precipitation regimes of 500–1000 mm and temperatures between 8 and 13 °C. According to these parameters, we could suggest similar palaeoenvironmental conditions during the Gravettian occupation of the site, as it is also supported by the more complete antracological sequence of Cova de les Cendres, located just 25 km away northward [26].

The spectrum of taxa represented in the hearth structure (BM and EC-6) is very narrow: four amongst the 48 charcoal fragments recovered in BM and just two from the 66 charcoal fragments in fireplace EC-6. This might be explained by both the low number of charcoals from each fireplace and their short-term accumulation as a fuel of the last burning episode, thus, reflecting a punctual

harvest of firewood. In any case, the flora identified in the combustion structures is in ecological agreement with that recovered from the archaeological units as a result of a longer accumulation process. Both charcoals from fireplaces and archaeological units are the result of human agency.

The faunal assemblage of Test Pit 1 is composed of 489 elements. The ratio of identified specimens at the taxonomic level (NISP) is low, varying between 14.6% in Level A and 2.7% in Level C. Four species of ungulates have been identified: aurochs (*Bos primigenius*), horse (*Equus ferus*), red deer (*Cervus elaphus*) and the Spanish ibex (*Capra pyrenaica*). In addition, lagomorphs are represented throughout the archaeological sequence.

The taphonomic analysis points to both anthropogenic accumulation and post-depositional breakage, with a very marginal contribution by non-human predators. Despite the lack of cut marks, partially explained by the high occurrence of post-depositional calcareous concretions, percussion marks have been clearly identified on diaphyseal fragments of ibex (1), medium-size (2) and large-size (1) ungulates in Levels A and B, as well as a human bite mark on a leporid bone (Table 4).

Digestive corrosion and alterations indicating the intervention of non-human predators are marginal (just one leporid fragment), suggesting an anthropogenic origin for almost all the osteological remains. Despite the small size of the zooarchaeological dataset, the spectrum of identified species, including red deer, Spanish ibex, horse and aurochs, is consistent with the representation

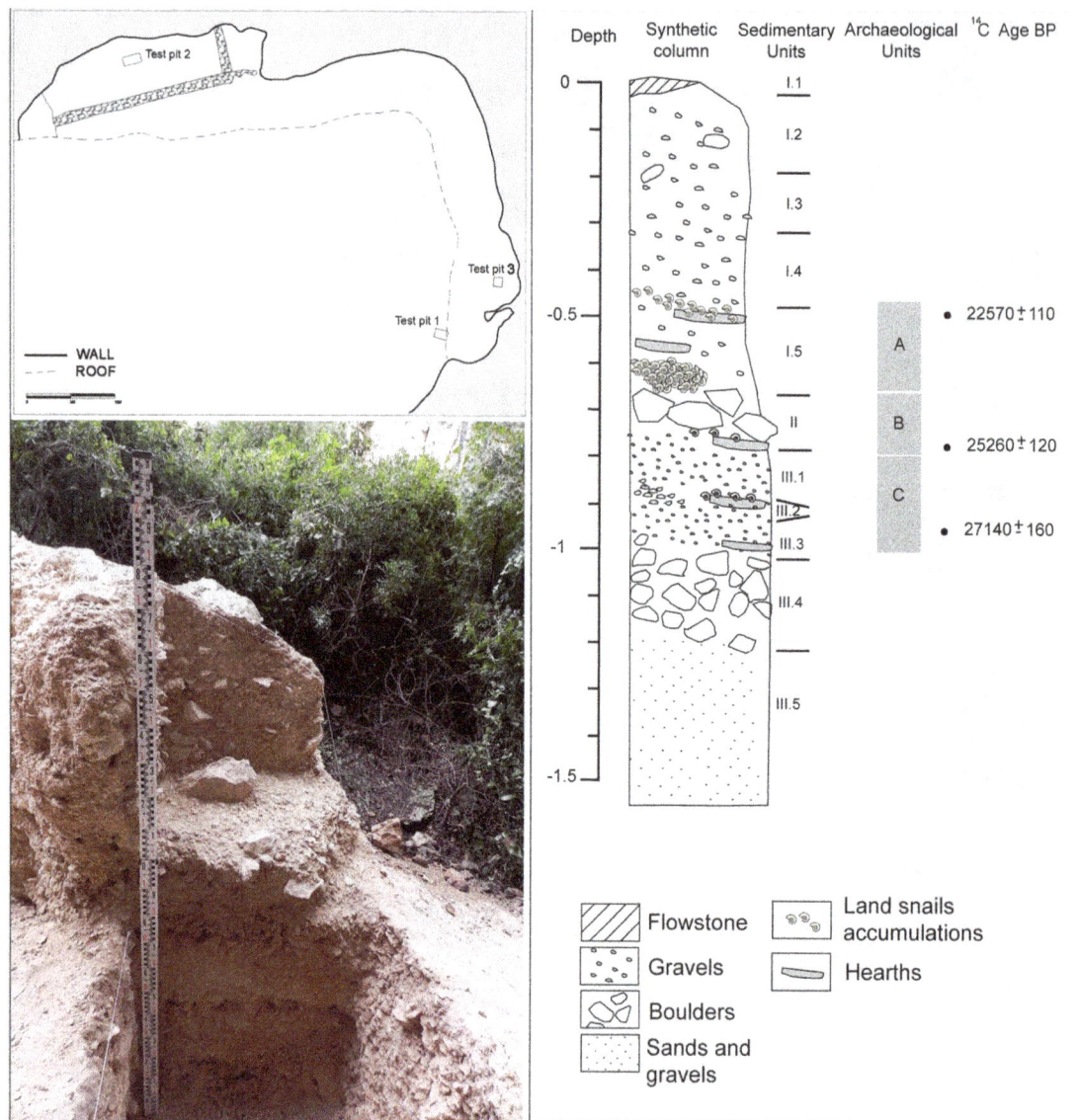

Figure 2. Up left: plan of Cova de la Barriada site with the location of test pits. Bottom left: Test pit 1, once excavated. Right: Synthetic column of the test pit 1 stratigraphy.

Table 1. Descriptive attributes of the combustion structures documented in test pit 1.

Combustion structures	Sedimentary	Level	Section	Area	Slope	¹⁴C Age BP
	Sub-unit			conserved*(m2)	(degrees)	
EC-1	I.5	A	Flat	0.1644	11	22750±110
EC-2	I.5	A	Flat	0.0843	7	
BM	I.5	A	Concave	0.0805	7	
EC-3	II	B	Flat	0.0997	12	
EC-5	II	B	Flat	0.0320	5	25260±120
EC-4	III.1	C	Flat	0.8669	5	
EC-6	III.3	C	Concave	0.1186	4	27140±160

*Area conserved denotes the surface of the carbonaceous area.

Table 2. Radiocarbon dates of the test pit 1 of Cova de la Barriada calibrated with Oxcal 4.2 [19] and the Intcal 13 calibration curve [20].

Level	Context	Sample	Ref.Lab	C$^{12/13}$	^{14}C Age BP	2 Sigma Cal BP
A	EC-1	*Pinus nigra*	Beta-296222	−21.9‰	22750±110	27398–26712
B	EC-2	Fabaceae	Beta-296223	−22.5‰	25260±120	29642–28958
C	EC-3	*Juniperus* sp.	Beta-362534	−20.5‰	27140±160	31342–30897

pattern found in other Mediterranean Gravettian contexts such as Levels XIV and XVI of Cova de les Cendres, located just 25 km to the north on the modern coast line [27]. These ungulates suggest the human exploitation of different ecotones such as littoral plains, open forests and abrupt-rocky slopes. On the other hand, the representation of lagomorphs at the site fits with the robust body of datasets indicating that the rabbit is the most abundant taxa found in the Iberian Mediterranean Region throughout the Upper Palaeolithic, especially from the Gravettian onwards [29–30].

3. Methods

3.1. Taxonomic identification. The land snails analysed in this study −1484 number of identified specimens (NISP) and 832 number of minimal individuals(MNI)- were recovered during the excavations of Test Pit 1, using dry sieving through 5 and 1 mm

mesh. Archaeological land snail specimens were examined and compared with published materials in the regional taxonomic literature [31,32] and reference gastropod collections at the Natural History Museum of Valencia and the Department of Zoology at the University of Valencia. The quantification procedures were based on the NISP (excluding the whorl fragments smaller than 3 mm) with the MNI based on apical counts [33]. The land snails and archaeological materials presented in this study are deposited in the MARQ Museum (Alicante, Spain). The fieldwork excavation memory (with plans, stratigraphic sections, photographs and drawings) and all the original spreadsheets with the charcoal raw counts, faunal descriptions, land snail morphometrics are deposited at the Institut Català de Paleoecologia Humana i Evolució Social (Tarragona, Spain).

3.2. Taphonomic analysis. To assess the land snail taphonomy and breakage patterns, we established four main

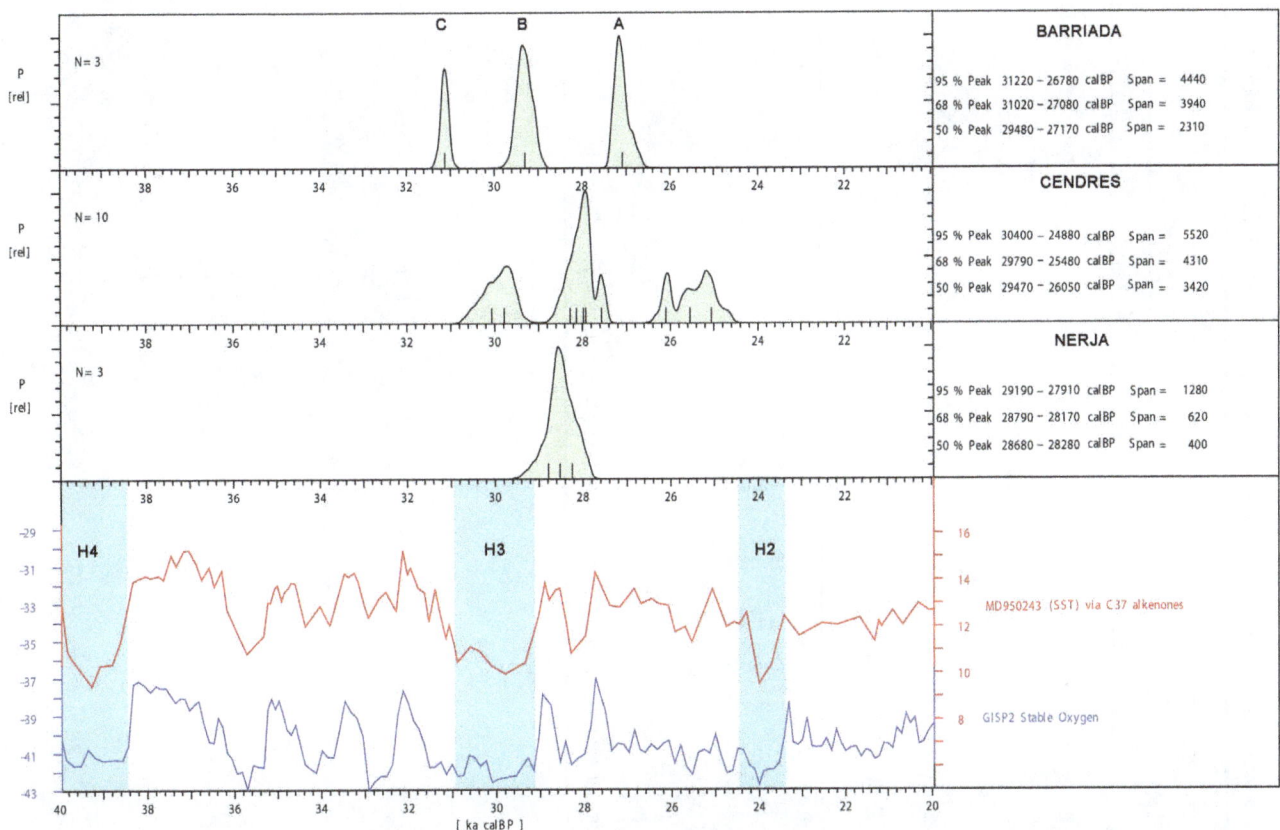

Figure 3. Top: Cumulative calibrated dating probability of the radiocarbon dates from the levels A and B of Cova de la Barriada and the Gravettian units of Nerja [21] and Cendres sites [22] plotted with CalPal (vers. October 2013) [89] using the IntCal13 calibration curve [20]. Bottom: δ^{18}O variation from the GISP2 curve [23] and Sea Surface Temperatures obtained from alkenonne data from the Alborán Sea [24].

Table 3. Charcoal assemblages recovered in the Gravettian levels of Cova de la Barriada.

	Level A		BM		Level B		Level C	
	n	%	n	%	n	%	n	%
Pinus nigra	83	61.02	27	56.25	64	90.14	2	3.03
Juniperus sp.	28	20.59	19	39.58	4	5.63	64	96.97
Rosmarinus officinalis	2	1.47						
Fabaceae	13	9.56			1	1.41		
Lamiaceae	1	0.74			2	2.82		
Cistus sp.			1	2.08				
Conifera	5	3.68	1	2.08				
Bark	4	2.94						
Total	136	100	48	100	71	100	66	100

Table 4. Faunal assemblages (NISP) from the test pit 1 of Cova de la Barriada.

	Level A		BM		Level B		Level C	
	n	%	n	%	n	%	n	%
Bos primigenius		0					1	0.90
Capra pyrenaica	1	0.54			3	1.05		
Cervus elaphus	2	1.08			3	1.05		
Equus ferus					1	0.39		
Leporidae	24	12.97	3	33.33	17	5.98	2	1.80
Total det	27	14.59	3	33.33	24	8.54	3	2.70
Large size	2	1.08			3	1.05	3	2.70
Medium size	83	44.86			225	79.22	104	93.69
Small size	57	30.81	6	66.66	21	7.39	1	0.90
Undetermined	16	8.64			11	3.87		
Total undet	158	85.41	6	66.66	257	90.49	108	97.29
Total	185	100	9	100	284	100	111	100

Figure 4. SEM microscope image showing the characteristic reticular sculpture of the genus *Iberus* shells.

fragmentation categories: 1. complete (uncrushed) snails, 2. partially crushed (up to the 50% of the whole shell) snails, 3. snail fragments (between 10–50% of the whole shell) and 4. snail debris or small snail fragments <10% of the whole shell. Small whorl fragments, of less than 3 mm, are very frequent and were not quantified since most of them were produced during the excavation and by post-depositional diagenetic processes.

Micro-DXR analyses were performed on a sample of two live specimens and four fossil land snails to compare the aragonitic-calcitic composition between present-day snails without burning traces and fossil shells recovered in the combustion areas. Analyses were made using a Bruker-AXS D8-Discover diffractometer equipped with a parallel incident beam (Göbel mirror), vertical θ-θ goniometer, XYZ motorised stage and a GADDS (General Area Diffraction System) in the Servei de Recursos Científics i Tècnics (SRCT) at Universitat Rovira i Virgili (URV) in Tarragona.

3.3. Biometric analysis. Biometric analysis based on maximum height and diameter measures of all the complete specimens was undertaken to investigate the age selection patterns of land snail accumulations. All measurements were obtained using digital callipers and statistically analysed with the R software package [34]. In addition, to estimate the age of the archaeological assemblages, we compared the aforementioned measurements with those obtained from five different populations of modern specimens born and raised in the laboratory [35].

Results

1. Taxonomic analysis and quantification

Taxonomic identification of land snails, based on macroscopic and microscopic analyses, revealed that *Iberus alonensis* is the only species of land snail in the Upper Palaeolithic levels of the site. As SEM images clearly show (Figure 4), the shell surface displays a

Table 5. *Iberus alonensis* accumulations in the test pit 1 of Cova de la Barriada.

	Volume	NISP		MNI		NISP/m³	MNI/m³
	m	n	%	n	%	n	n
Level A	0.1415	733	49.39	418	50.24	5180	2954
BM	0.0085	356	23.99	180	21.63	41882	21176
Level B	0.0907	229	15.43	133	15.99	2240	1466
Level C	0.0946	166	11.19	101	12.14	1754	1067
Total	0.3353	1484	100	832	100	51056	2663

clear diagnostic reticular sculpture that is characteristic [36] of *I. alonensis*. That is the only species of the genus *Iberus* living in this area. It inhabits limestone Mediterranean environments with different vegetation formations, from open forests of pine and oak to xerophytic bushes [31–32]. Its life span ranges from ten to twelve years, and it displays two reproductive cycles during spring and fall.

Depending on the substrate, it estivates and hibernates buried or hidden in crevices. The occurrence of *I. alonensis* land snails has been reported, with very different frequencies, at several Late Pleistocene and Early Holocene archaeological sites in the southern and eastern Iberian Peninsula, where it has been interpreted as food remains [37–41].

Table 5 details the quantification of *I. alonensis* shells found in the Palaeolithic levels of Cova de la Barriada along with the excavated volume of the archaeological deposits. Almost 73% come from Level A and combustion structure BM, even though *I. alonensis* concentrations are also found in Levels B and C. If the shell quantitative values are normalised by volume of excavated sediment (NISP and MNI/m³), the picture that emerges is completely different. The combustion structure BM yields significantly much higher NISP and MNI values than Levels A, B, and C, whereas the previous quantitative differences between Levels A, B and C are less significant now.

According to the higher density values of *I. alonensis* shells, as well as the feature morphology and the anthracological record, the BM hearth structure can be interpreted as a specific-purpose cooking pit for roasting land snails. In contrast, the density values found at Levels A, B and C suggest a different accumulation pattern as waste around combustion and habitation areas (Figure 5).

2. Taphonomic results

Table 6 presents the breakage-size categories of *I. alonensis* land snails grouped by levels and reported as absolute and relative values regarding the NISP. Whole specimens and fragments preserving between 10–50% of the shell represent the most common classes, followed by fragments smaller than the 10%. Fragments larger than the 50% of the shell are the less common category in all the levels except for level C.

The high percentage of whole specimens and its mono-specific representation pattern is common in cultural accumulations and very rare in those land snail concentrations produced by small carnivores and birds [42]. Several studies have reported different patterns of shell damage produced by birds and mammals. Allen [43] reports a worldwide list of 67 species of birds eating land snails. However, just few works have provided detailed and comprehensive descriptions of the resulting shell breakage patterns. Turdidae birds -mainly the blackbird and the song thrush- produce marks of strikes on the left part of the last whorl [44]. In the case of the thrush song, as a result of using a stone anvil on which to break open the shells held by the peristome [42]. Other bird families with descriptions about land snails shell breakage patterns are the wekas (*Galliralus australis*) and the parrots[45]. Even though both species were not present in Europe during the Late Pleistocene, we can assume similar or close-related patterns of damage produced by other potential bird species as a result of the mechanical breakage of shells using the beak. The wekas produces pecking marks in the spire and early whorls as well as the base of a shell, making a hole through the centre [45]. On the other hand, parrots leave pairs of vertical scratches around the side of a shell and, also, can produce the removal of the early spirals [45].

Figure 5. Cultural accumulations of *l. alonensis* land snails. (A) Spatial association of land snails to carbonaceous sediments and fire-cracked stones of combustion structure EC-1, (B) spatial association of land snails, bone fragments and lithic artifacts of level C, (C) combustion structure BM containing land snails and homogenous carbonaceous sediments; (D) land snails found into the combustion structure BM, (E) morphology of the combustion structure BM once excavated (maximum length 41 cm), (F) burning traces of *l. alonensis* found into the filling sediment of combustion structure BM.

In contrast, the number mammal species displaying predatory behaviour on land snails is considerably lower, 29 species[43]. Most of the literature about shell damage patterns mainly focuses on the feral pig, the badger (*Meles meles*), the europeam hedgehog (*Erinaceus europaeus*), the mice (*Apodemus sylvaticus*) and the ship rat (*Rattus rattus*). These species produce different patterns of damage depending on whether they eat the whole shell or they bite and/or gnaw part of the shell for accessing to flesh. Feral pigs exhibit a characteristic damage around the shell periphery and the flattened halves of the shells, often associated to teeth marks impressed as shatter points [45]. It is also reported the consumption of whole snails amongst feral pigs, a practice which leaves small fragments of shells in the faeces. On the other hand, badgers and hedgehogs produce a significant destruction of land snail shells in small fragments, with a differential preservation of the columnella [42].

Table 6. Fragment conservation cathegories of *I. alonensis* land snails.

	Level A		BM		Level B		Level C	
	n	%	n	%	n	%	n	%
Whole	302	41.2	112	31.46	67	29.26	42	25.3
>50%	82	11.19	18	5.06	27	11.79	45	27.11
10–50%	164	22.37	157	44.1	80	34.93	50	30.12
<10%	181	24.69	69	19.38	55	24.02	29	17.47
Total	733	100	356	100	229	100	166	100

Rodents such as the mice or rat break land snail shells in different ways. Because his small jaw size and mouth gape, the mice uses to concentrate his predatory activity on small or medium size land snails, mainly producing tooth bites on the aperture [44]. In contrast, the ship rat gnaws through the side of the shell and the inner whorls of the helicid snails [45], and can produce very distinctive damage patterns as the cracking of the lower part of the shell without affecting the lip or the proto-conche [46].

The shell damage produced by the above mentioned land snail predators is different from that observed in the archaeological specimens found at the Gravettian levels of Cova de la Barriada. First, it should be noted the high percentage of complete specimens, a fragmentation category seldom found in land snail assemblages produced by birds, insectivore mammals and rodents. Second, some big fragments (>50%) have fractures on the aperture, a pattern that can be created by small land snail predators such as the mice. However, small fractures -including those found on the aperture- can also be attributed to human manipulation during the land snail consumption, post-depositional damage caused by trampling and, even, the excavation process. Our next land snail fragmentation categories – medium size (10–50%) and small size (<10%)- overlaps with the size distributions produced by different avian and mammal predators which intensively break the land snail shells such as the Turdidae birds and, especially, the European hedgehog. However, the sum of both fragmentation categories found in the combustion structure BM, whose anthropic origin casts no doubts, reaches the 64%. Thus, medium and small size fragments can also be found in archaeological contexts completely produced by humans and affected by non-predatory post-depositional breakage such as trampling and/or sediment compaction. It is interesting to note that those highly diagnostic fractures produced by rats described by Moreno-Rueda [46] are absent in the archaeological assemblage presented here.

The taxonomic composition of land snail accumulations is an important variable to differenciate its anthropic or natural origin. Badgers tend to produce low-density accumulations of land snails with different taxonomic representation and a very diverse fragmentation pattern [42]. In contrast, land snail accumulations produced by Turdidae birds are quantitatively more abundant, displaying a narrow taxonomic selection (one or two species) and a very distinctive breakage pattern in the apex and the whorls, with no conservation of complete individuals [42]. The mono-specific composition of the land snail assemblage is a consistent argument against the natural accumulations produced by non-human predators. Different studies have shown that predation produced by birds and mammals focus on a restricted number of species, with a marked selection of birds toward large land snails [43,44]. However, the taxonomic composition found in birds and mammal accumulations is not mono-specific, as supported by the quantitative evidence provided by Rosin on birds and mouses [44] and Faus on rat nests [47].

Finally, in the Iberian Peninsula, land snails are found in low-density accumulations in red fox lairs [48]. However, the origin of such accumulations is due to thanatocoenoses processes according to the documented high taxonomic and biometric diversity. As we will see in the next section, the biometric analysis of *I. alonensis* shells from Cova de la Barriada does not support a similar accumulation process to explain the presence of land snail at the site.

The relative frequencies of whole shells found throughout the archaeological sequence display a clear decreasing pattern from top to bottom, suggesting the occurrence of post-depositional mechanical breakage due to sediment compaction.

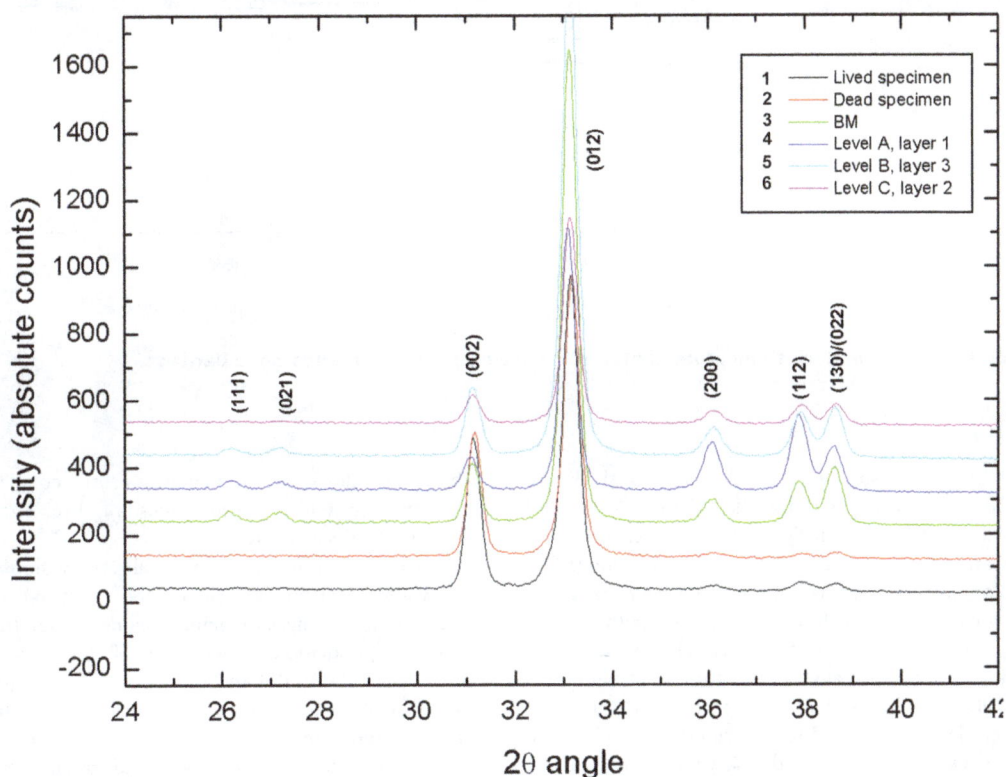

Figure 6. Up: X-ray diffractograms of modern (1–2) and fossil (4–6) *I. alonensis* shells. Difractograms have been moved vertically to avoid overlapped data. Vertical labels indicate the Miller indexes for aragonite diffraction peaks.

The results of the X-Ray diffraction analysis for each sample are shown in Figure 6. All samples presented aragonite ($CaCO_3$) as the only mineral phase. The experimental relative intensities for aragonite do not match with the expected ones (when compared with the pattern 41–1475 of the ICDD database) because of the preferred orientation of aragonite crystals in the shell, which is very common in biogenic structures. According to J.E. Parker et al. [49], the samples analysed did not reach a temperature over $\approx 375°C$, where the irreversible transformation of aragonite-calcite is expected to start in biogenic structures. However, the samples analysed presented some changes on the texture of aragonite that can be seen from the relative intensities of the peaks in the

diffractograms. For samples BM, Level A and Level B, the peaks (111) and (021) are well detected, whereas for the other samples, such peaks have almost no intensity. It is well known that the annealing process introduces changes in the texture of bulk materials together with crystal phase transformation [50–51]. Unfortunately, no information has been found on how the aragonite texture in the snail shell changes as a function of the annealing treatment.

3. Biometric analysis

Land snail shell maximum height and width measurements were recorded using a digital calliper. The descriptive statistics of

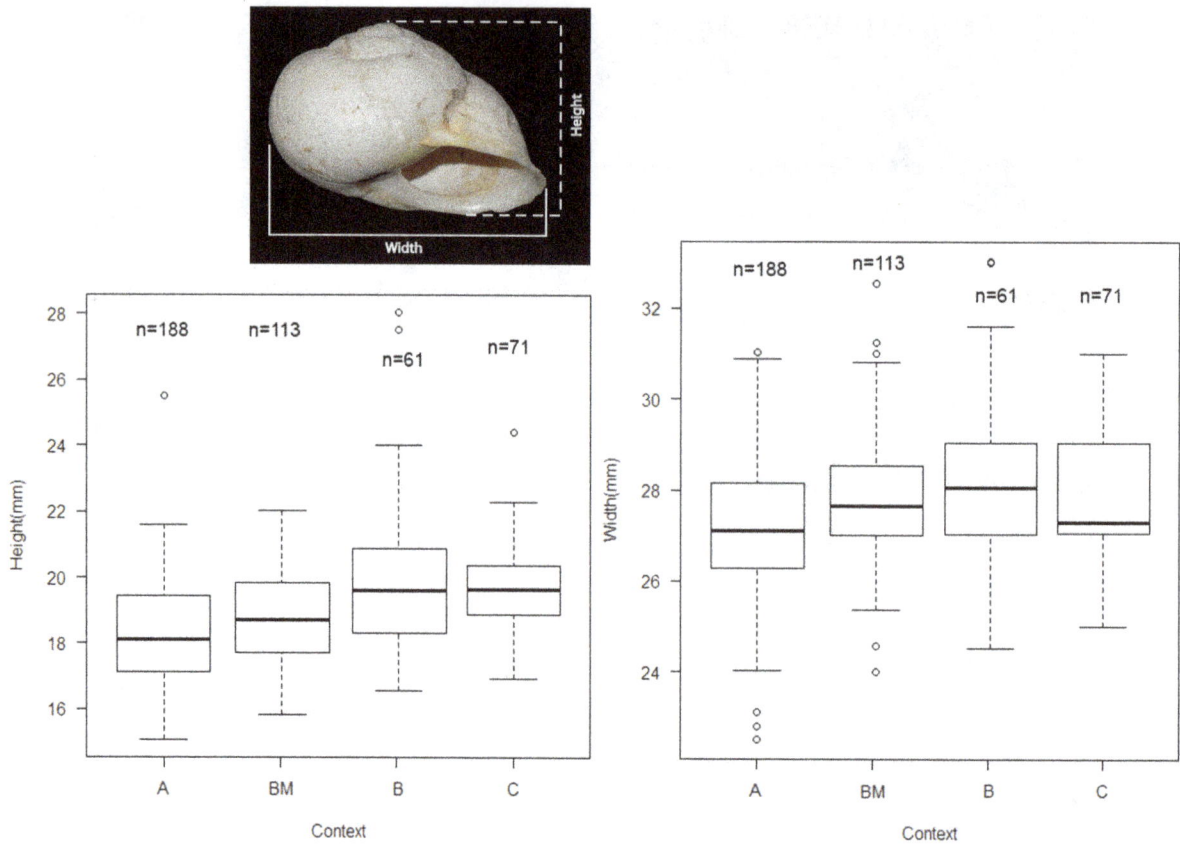

Figure 7. Boxplot of height and width measurementes of *I. alonensis* shells from Cova de la Barriada.

the width and height measurements for *I. alonensis* shells are reported in Table 7 and Figure 7 for the different levels and combustion structure BM: 432 shells were suitable for analysis, providing a mean value of 27.62 (±1.48) mm width and 18.79 (±1.67) height for all the specimens. The skewness values of all the snails indicate a closely normal distribution shape for the widths and a slightly skewed distribution of heights. The variance and standard deviation values indicate a narrow range of land snail widths (Std = 1.48), which is consistent with with shell size distribution recorded in archaeological records [52–55]. The higher degree of variation found in the skewness and standard deviation values of heights (Std = 1.67) is partially explained by the outliers found in Level B. However, if these outliers are removed, the variance and standard deviations values are much closer to those found in the other levels.

The minimal variation in width and height values makes possible a tentative essay of age selection patterns on the basis of modern data of shell growth as it has been demonstrated that shell width and land snail age are positively correlated [35]. Figure 8 represents the width measures of five different populations of land snails whose progenitors were harvested at different localities in southeastern Spain, although reproduced and raised in the laboratory under controlled feeding and temperature conditions. Though the ideal comparative scenario would be that of using growth rates from wild specimens, such a procedure is impossible to achieve given the current-day overharvesting of *I. alonensis* and/or their replacement from many habitats by other gastropods. The use of different populations of individuals raised in the

laboratory allows us to establish a first comparative analysis between controlled growth rates of live size variation of archaeological specimens.

The x-axis is a five-week temporal series within the first year of life, whereas the boxplots record the mean and the distribution values of width measurements obtained from the five modern reference populations, each one of them composed of 10 individuals. The pink band represents the interquartile range of the archaeological specimens and the dashed line the mean value. The distribution of width values of the archaeological specimens is significantly different from those individuals younger than 45 weeks found in the modern reference collection. On the other hand, individuals of 45 to 55 weeks age display a partial overlap with the width values exhibited by the fossil specimens, even though the mean values are consistently different. Despite the fact that the relationship between age and shell size of modern and Late Pleistocene *I. alonensis* might not be exactly the same, given the distinctive environmental conditions at the end of MIS 3 [56], both observations, the narrow range of the fossil widths and the significant differences regarding the shell size of land snails younger than 45 weeks, suggest a strong selection pattern where immature individuals were ruled out. In addition, the fragmentation pattern observed in the archaeological specimens supports this interpretation. The lack of juvenile (and more fragile) shells seems not to be explained by sediment compaction because, in each level, the "whole shell" category is more frequent than the small fragments <10%.

Table 7. *I. alonensis* sample and descriptive statistics of width and height measuruments through time.

	Width (mm)					Height (mm)				
	A	BM	B	C	Total	A	BM	B	C	Total
n	188	113	61	70	432	188	113	61	70	432
Mean	27.29	27.77	28.21	27.92	27.62	18.37	18.7	19.83	19.58	18.79
Median	27.11	27.66	28.06	27.29	27.43	18.12	18.71	19.61	19.63	18.72
Maximum	31.03	32.55	33.03	31.02	33.03	25.5	22.02	28.05	24.40	28.05
Minimum	22.52	24.01	24.53	25.02	22.52	15.07	15.84	16.56	16.93	15.07
Variance	2.01	2.81	1.89	1.84	2.19	2.23	1.77	4.6	2.47	2.79
Std	1.41	1.37	1.67	1.35	1.48	1.49	1.33	2.14	1.57	1.67
Skewness	-0.23	0.53	0.58	0.3	0.21	0.68	0.08	1.57	0.47	1.07
Kurtosis	3.58	3.96	3.75	2.35	4.01	4.31	2.44	6.91	3.7	6.58
1st Quartile	26.28	27.01	27.02	27.06	26.82	17.12	17.73	18.31	18.87	17.55
3rd Quartile	28.16	28.54	29.04	29.04	28.52	19.44	19.83	20.82	20.36	19.92

Discussion

1. Land snail accumulations and cultural selection patterns

Anthropogenic accumulations of large snails dated during the Gravettian period (31.3–26.9 kyr cal BP) have been reported on the basis of taxonomic, taphonomic and biometric data and their association with occupational features, lithic and faunal assemblages. Land snails found in the Gravettian levels of Cova de la Barriada reveal the existence of mono-specific gathering strategies focused on large pulmonate gastropods during the Early Upper Palaeolithic. Taphonomic analyses indicate significant differences between the breakage patterns documented for the Gravettian levels of Cova de la Barriada and those produced by other land snail accumulators such as birds, hedgehogs, badgers, mices and rats. The most significant difference is the high representation of complete shells. Such a pattern is very uncommon in the shell remains left by other land snail predators, which systematically destroy part of the shell spiral, the sides or the aperture for gaining access to flesh. As it has been observed before, mechanical breakage due to sedimentary compaction was responsible for some land-snail fractures, as reported with other zooarchaeological remains. However, the high frequencies of complete specimens and their clear spatial association with combustion structures advocate for the cultural origin of the land snail accumulations as food remains. Furthermore, the micro-DXR analysis indicates differences in the aragonitic composition between live and fossil specimens, suggesting that the archaeological samples were heated under controlled conditions (below 375°C) given the null conversion of aragonite into calcite. Recent experiments on the prehistoric cooking process of the large edible land snails *Helix* sp. indicate that roasting land snails on embers between 5-8 minutes is the most plausible technique documented at Pupicina cave, on the eastern Adriatic area, leaving barely visible traces of burning [57]. In addition, our analytical protocol using DXR indicates that because of post-depositional bleaching, fossil specimens without clear macroscopic evidence of burning were heated as well, given the presence of different aragonitic signals not found in modern specimens. Furthermore, charcoal data from the combustion structure BM, which delivered 112 complete land snails, provides fresh evidence about the fuel employed in the cooking process of the land snails: *Pinus nigra/sylvestris* and juniper charcoals, most of them between 2 and 15 timber rings. The structure BM is different from other examples of combustion structures used to cook land snails dated to the Capsian in Algeria. The Capsian examples consist on hearth pits delimited by vertical stone slabs and filled by two layers of heated stones between which the land snails are boiled by heating radiation [58].

On the other hand, the biometric analysis of *I. alonensis* shell widths and heights indicates a narrow selection pattern of land snail sizes. The comparative analysis between the archaeological specimens and modern land snails raised in the laboratory suggest that most of the gathered individuals were older than one year, and those younger than 45 weeks were not gathered at all. Such a strong age selection pattern suggest sustainable exploitation of *I. alonensis* based on knowledge of its reproductive cycle.

2. Cultural and economic implications of land snail consumption during the Early Upper Palaeolithic

The body of archaeological evidence about the consumption of edible land snails by humans prior to the Late Glacial Maximum is scarce, and its interpretation as food remains is rarely accepted for such an old chronology [17–18]. In Table 8 and Figure 9, we present a list of Upper Palaeolithic contexts recording the first cite

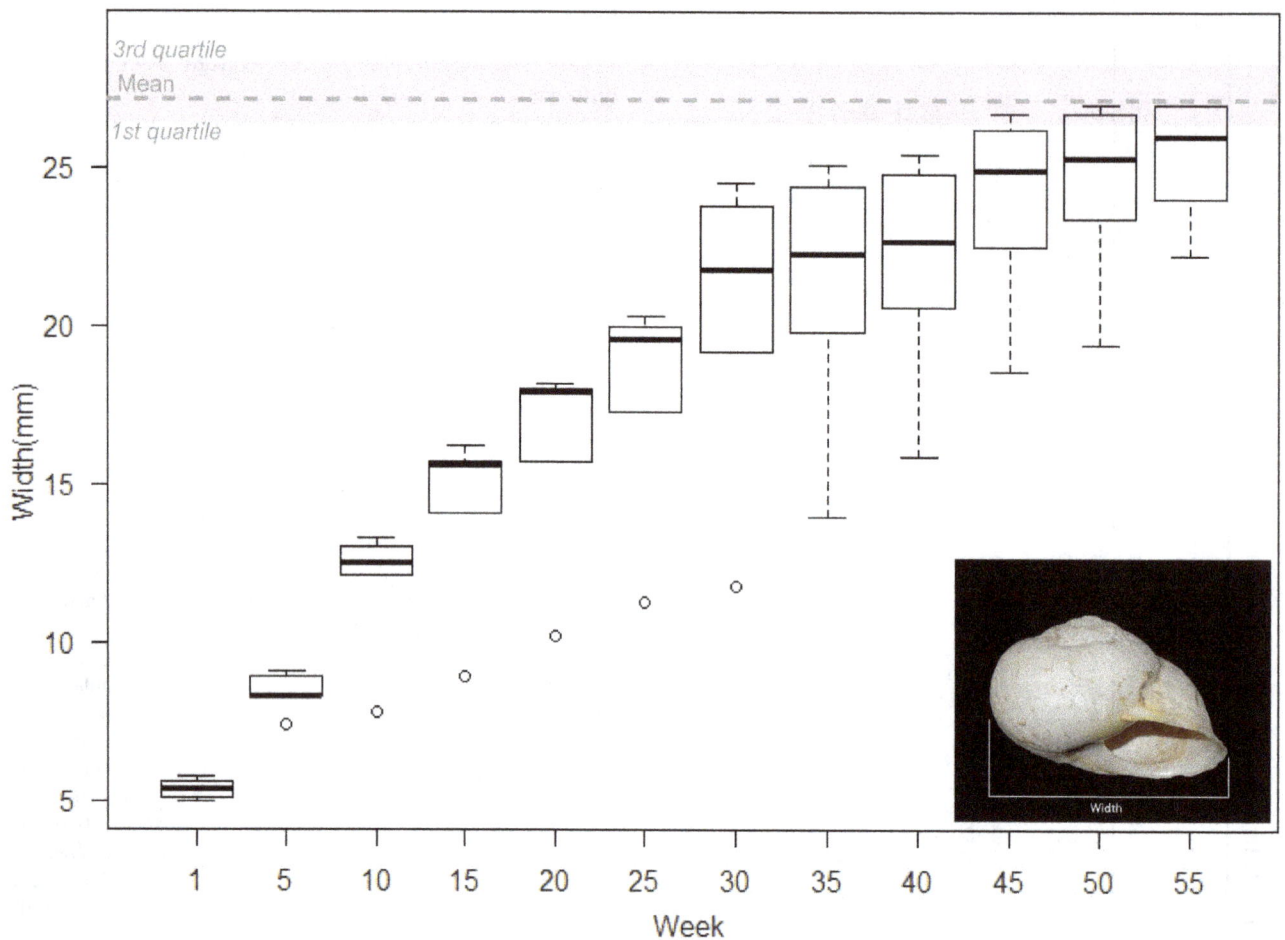

Figure 8. Boxplot of width measurements from 5 modern reference populations of *I. alonensis* during their first year of live (raw data obtained from [36]**.** The velvet band represents the interquartile range of the archaeological specimens found at the Gravettian levels of Cova de la Barriada.

about the use of terrestrial gastropods as a food resource, with the corrected chronology and taxonomic identification.

The first acknowledged evidence of land snail consumption by humans [17] comes from the Mumba-Höle site (Tanzania), a reference archaeological sequence of East Africa. Melhman [59] reported the existence of middens of the large land snail *Achatina* sp. in both beds III-lower and V dated to 31070±500. It is worth noting that such accumulations occurred in association with rich artefact assemblages, oyster shell beds and crescent microliths. Melhman interpreted the lithic industry of level V as a Middle Stone Age (MSA) and Later Stone Age (LSA) transitional industry. However, new excavations at this site and the production of new dates allow revision of the previous cultural interpretation of level V as a transitional industry to a full LSA lithic assemblage [60]. Recent contributions on the chronology of the archaeological sequence[61], through optically stimulated luminescence dating of quartz and feldspar grains, yielded however, older ages for levels III inf (36.8±3.4 kyrs) and V upper (49.1±4.3 kyrs). The specific study of the land snail assemblages remains unpublished.

In the Levant, the Ksar 'Aqil (Beirut) site yielded land-snail accumulations in Level 3b of Tixier's excavations [62], dated to between 22,000 and 23,000 BP uncalibrated [63]. It is difficult to maintain, with the current data, the older use of land snails as food resource in this region. For instance, further north, at Üçagizli Cave (Hatay, Turkey), a well-documented archaeological se-

quence of the Initial Upper Palaeolithic and the Ahmarian dated to 44–33 kyrs cal BP [64], there were found several land snail species, including *Pomatias elegans*, *Xeropicta* sp. and *Helix* cf. *engaddensis*, which accumulated through natural processes. Despite the significant representation of the small game at the site, with several species of fish, leporids, birds, tortoises and shellfish, there is no evidence for the use of pulmonate gastropods as food.

In the Argolid, Franchthi Cave and Klissoura Cave 1 have been thoroughly investigated, yielding a different chronological pattern for the consumption of *Helix figulina*, the main edible land snail species documented in the Late Pleistocene and Early Holocene archaeological deposits of this region. At Klissoura Cave 1, *Helix figulina* land snails occur from the Upper Palaeolithic to the Mesolithic. The shells found in the Uluzzian and the Early Aurignacian units (Layers V and IV) lack evidence of human modification [65]. In contrast, land snails become moderately abundant from the Upper Aurignacian (Level IIIc), with evidence of high percentages of broken lips, interpreted as evidence of human consumption, growing exponentially up to the Mesolithic (Layers 3–5). No radiocarbon information is available for Layer IIIc, whose chronology spans between the dates of the underlying Middle Aurignacian units (IIIe-IIIg), dated to 35–36 kyrs cal BP, and the overlying Upper Palaeolithic levels (III″), dated to c. 28 kyrs cal BP [66]. The antiquity of land snail exploitation

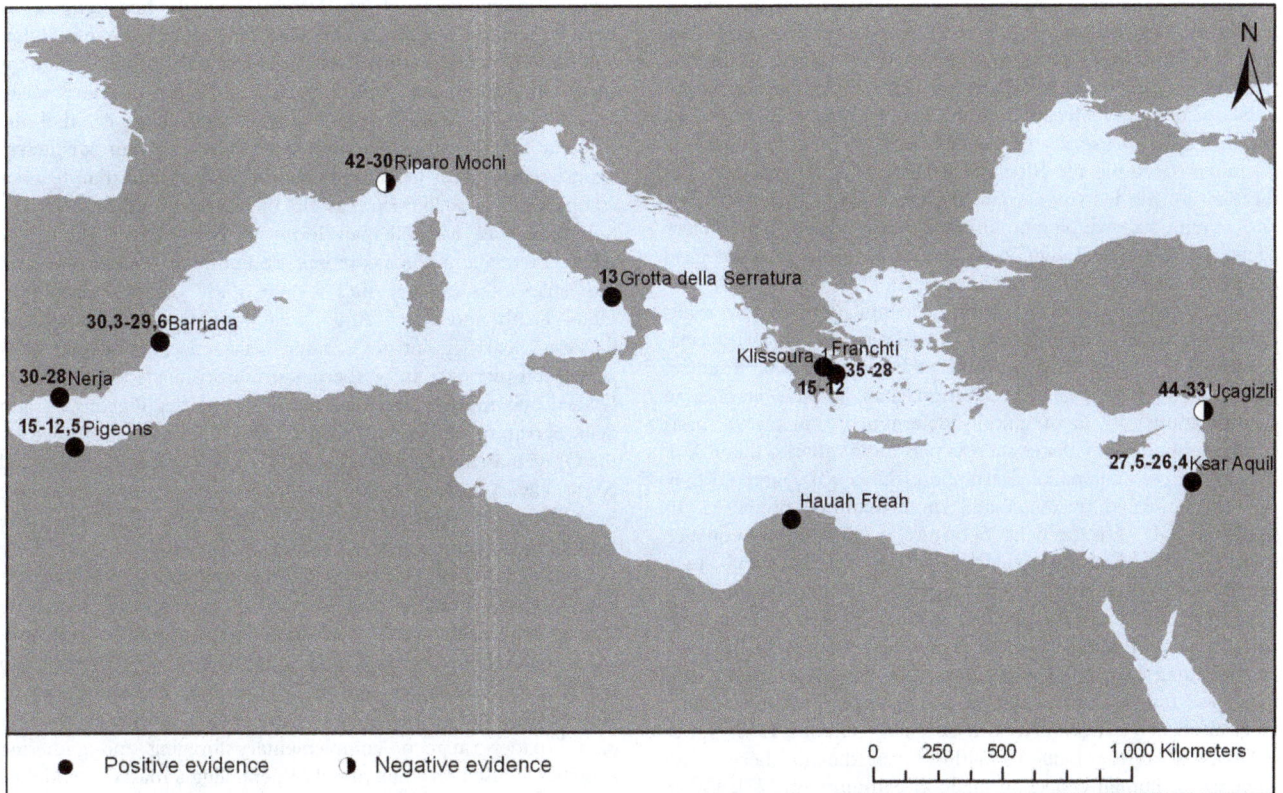

Figure 9. Map of Upper Palaeolithic sites of the Mediterranean basin displaying the earliest occurrence of land snails as food remains. Bold numbers indicate the calibrated chronology in ka yrs (see also table 8). The black & white symbols denote Early Upper Palaeolithic sites with a consistent pattern of small prey exploitation including shellfish and lacking evidence of land snail consumption.

documented at this site is not mirrored at the neighbouring Franchthi cave, where food debris of *Helix figulina* occurs much later, in the Epigravettian units (S2-T3) dated to 15–12 kyr cal BP [67].

At Haua Fteah cave in Cyrenaica, land snail accumulations interpreted as food debris were reported in the pre-Aurignacian, Iberomaurisian, Capsian and Neolithic layers of McBurney

excavations [68–69]. New excavations [70–71] have documented land snails throughout the whole archaeological sequence. Detailed quantitative information about land snail occurrence is available for the uppermost part of the sequence, Layers III-X, from the Historic period up to the Libyco-Capsian [72]. At these layers, as well as in the neighbour site of Hagfet al-Gama, the shells of *Helix melanostoma*, *Rumina decollata*, and *Trochoidea*

Table 8. Earliest evidence of systematic conssumption of land snails.

	Site/Level	Culture	ka yrs cal BP	Species	Reference
East Africa	Mumba/Vupper	Later Stone Age	49.1±4.3*	*Achatina* sp.***	[59–61]
	Mumba/III lower	Later Stone Age	36.8±3.4*	*Achatina* sp.***	[59–61]
Levant	Ksar 'Aqil/3b	Late Upper Paleolithic	27.5–26.4	undetermined	[17,62–63]
Greece	Klissoura 1/IIIc	Upper Aurignacian	35–28**	*Helix figulina*	[65–66]
	Franchthi/S2-T3	Epigravettian	15–12	*Helix figulina*	[67]
Italy	G. della Serratura/8G	Late Epigravettian	14.1–13.7	*Helix* cf. *ligata*	[52]
North Africa	Taforalt/grey series	Iberomaurisian	15–12.5	*Theba* sp., *Cornu aspersum*, *Otala punctata*	[80]
	Haua Fteah/XVI-XXV	Dabban	43–17	*Helix melanostoma*	[70–71]
Iberia	Nerja/IV	Gravettian	27.9–26.9	*Iberus alonensis*	[80–81]
	Barriada/C, B, A	Gravettian	31.3–26.7	*Iberus alonensis*	This study

*Optically stimulated luminiscence dating of quartz and feldespar grains.
**dating estimates inferred by the chronology from the layers above and below.
***Lack of detailed quantitative and evidence.

cretica are thought to represent food debris. For the earlier prehistoric occupations at Hauah Fteah, preliminary data indicates land snails were used as food resources during the Oranian (c. 14–10 ka cal BP) to a lesser extent than during the Lybico-Capsian, whereas during the Dabban period (c. 40–15 ka cal BP) their use seems occasional, and no food molluscs are documented during the Mousterian layers [70]. Future works at this site are needed to elucidate the beginning of the systematic and continuous exploitation of land snails along the Dabban occupations, whose radiocarbon record covers a chronological span of more than 25,000 calendric years [71].

In the Balkans and Italy, cultural accumulations of land snails have been reported from the Late glacial and the Early Holocene onwards [52,73–76] but not during earlier periods. For instance, Level G of Riparo Mochi (Aurignacian) yielded land snail accumulations (not taxonomically determined), but on the basis of the taphonomic evidence such as perforation marks, the lack of burning traces and spatial distribution, Stiner [3] suggested they were accumulated by avian and small mammal predators. In southern Italy, Grotta della Serratura, a long cave sequence spanning the Gravettian, Evolved Epigravettian and Late Epigravettian occupations is paradigmatic. Despite the presence of marine and terrestrial molluscs throughout the sequence, the consumption of *Helix* cf. *ligata* is just documented in the Late Epigravettian layers [75]. In contrast, in Sicily, the pulmonate gastropod *Eobania vermiculata* is found throughout the archaeological sequence from Grotta d'Oriente, from the Late Upper Palaeolithic to the Late Mesolithic, even though there is no evidence of human collection for food consumption [77]. Other Late Glacial and Early Holocene Italian sites such as Grotta della Madonna a Praia a Mare and Grotta del Mezzogiorno have provided abundant evidence of *Helix* cf. *ligata* land snail shells [78–79]. However, its interpretation as food remains is still inconclusive because the malacological assemblages come from old excavations.

In the Magreb, at Grotte des Pigeons, recent malacological studies [80] demonstrate an intensive exploitation of land snails associated to the grey series of Iberomaurisian affiliation dated to 15–12.5 kyrs cal BP. In contrast, the land snails found in the underlying yellow series were not the result of human agency. At the same site, a few large specimens of *Otala punctata* were reported in the Aterian levels, associated with thin ash lenses dating to around 80,000 BP, suggesting "a very small scale exploitation"[80:13] even though no detailed description and quantification are available for such an older occurrence.

Finally, at Cueva de Nerja (south Iberian Peninsula), accumulations of *Iberus alonensis* are found throughout the archaeological sequence, from the Gravettian to the Chalcolithic. The oldest occurrence is documented in Levels 13 and 11 of the Vestíbulo chamber, where continental gastropods are dominant during the Gravettian and during the subsequent Solutrean occupations [81–82].

According to this review, the exploitation of edible land snails is a subsistence activity restricted to anatomically modern humans post-dating the Middle to Upper Palaeolithic transition, geographically restricted to some of the warmest European regions

during the final stages of the MIS 3. During the beginning of the Early Upper Palaeolithic, edible land snails appear have accumulated in archaeological deposits through natural processes and/or non-human predators in the Eastern and Central Mediterranean as well as the Magreb. Our study reveals, however, that the exploitation of large edible land snails in the Iberian Mediterranean region is clear from the beginning of the Gravettian techno-complex c. 31.3–26.6 kyr cal BP onwards as indicated by the Solutrean and Magdalenian levels of Nerja cave. While the information about the taxonomic composition, taphonomy and biometrics of some key sites for the Early Upper Palaeolithic, Haua Fteah and Ksar 'Aqil, is rather limited, the remaining datasets clearly point to a later occurrence, associated with posterior Epigravettian or Iberomaurisian contexts after the Late Glacial maximum as documented in Grotte des Pigeons, Grotta della Serratura or Franchthi Cave. The main question posed by the Gravettian accumulations of land snails found at Barriada and Nerja caves addresses the origin of such a new economic behaviour, not reported in the Iberian Peninsula during the Middle Palaeolithic nor the Aurignacian periods.

Cova de la Barriada suggests that the exploitation of *I. alonensis* land snails by humans entailed a rigid mono-specific gathering strategy and a clear age-size selection pattern, culturally transmitted and maintained through generations. Their incorporation into the human subsistence systems during the beginning of the Gravettian could be interpreted as a regionally specific component of a broader range of complementary foraging and gathering activities focused on small prey [29], entailing a marked gender or age-related division of labour operating in a context of population growth [83].

Considering both the exploitation of marine molluscs during the Middle Palaeolithic and the lack of evidence on land snail consumption in Early Upper Palaeolithic contexts prior to the 31 ka yrs cal BP, there are no arguments to interpret the incorporation of land snails into the human diet in terms of different cognitive capabilities between Neanderthals and Anatomic Modern Humans. Rather, as the eastern and southern Iberian record indicates, the appearance of this new subsistence activity was coeval with other demographically driven transformations such as the significant increase of the number of sites [84–86], and beginning of the production of portable art [87–88].

Acknowledgments

We gratefully acknowledge to Arpa Patrimonio S.L. their assistance during the fieldwork, to the manuscript reviewer's for their helpful comments and suggestions, to Marina Lozano (IPHES) for her assistance on the SEM images and to Francesc Gispert (SRCT-URV) for his assistance on the Micro-DXR analyses.

Author Contributions

Conceived and designed the experiments: JFL. Performed the experiments: JFL EB CFG ASS. Analyzed the data: JFL EB CFG ASS AMO. Contributed reagents/materials/analysis tools: JFL EB CFG ASS AMO. Contributed to the writing of the manuscript: JFL EB CFG ASS.

References

1. Hockett B, Haws J (2005) Nutritional ecology and the human demography of Neandertal extinction. Quaternary International 137: 21–34

2. Richards M, Pettit P, Stiner MC, Trinkaus E (2001) Stable isotope evidence for increasing dietary breadth in the European mid-Upper Paleolithic. PNAS 98(11): 6528–6532.

3. Stiner MC, Munro ND, Surovell TA, Tchernov E, Bar-Yosef O (1999) Paleolithic population growth pulses evidenced by small animal exploitation. Science 283: 190–194.

4. Blasco R (2008) Human consumption of tortoises at level IV of Bolomor Cave (Valencia, Spain). Journal of Archaeological Science 35: 2939–2848.

5. Blasco R, Fernández Peris J (2009) Middle Pleistocene bird consumption at level XI of Bolomor Cave (Valencia, Spain). Journal of Archaeological Science 36: 2213–2223.

6. Blasco R, Fernández Peris J (2012) A uniquely broad spectrum diet during the Middle Pleistocene at Bolomor Cave (Valencia, Spain). Quaternary International 252: 16–31.

7. Blasco R, Fernández-Peris J (2012) Small and large game: human use of diverse faunal resources at level IV of Bolomor Cave (Valencia, Spain). Comptes Rendus Palevol 11: 265–282.

8. Brown K, Fa DA, Finlayson G, Finlayson C (2011) Small game and marine resource exploitation by Neanderthals: the evidence from Gibraltar. In: Bicho, N., Haws, J. (Eds.), Trekking the Shore: Changing coastlines and the antiquity of coastal settlement. Interdisciplinary contribution to archaeology. Springer, New York, pp. 247–271.

9. Colonese AC, Mannino MA, Bar-Yosef D, Mayer DE, Fa D, et al. (2011) Marine mollusc exploitation in Mediterranean prehistory: An overview. Quaternary International 239: 86–103.

10. Cortés-Sánchez M, Morales-Muñiz A, Simón-Vallejo MD, Lozano-Francisco MC, Vera-Peláez JL, et al. (2011) Earliest Known Use of Marine Resources by Neanderthals. PLOS ONE 6(9): e24026. doi:10.1371/journal.pone.0024026

11. Henry AG, Brooks AS, Piperno DR (2011) Microfossils in calculus demonstrate consumption of plants and cooked foods in Neanderthal diets (Shanidar II, Iraq; Spy I and II, Belgium). PNAS 108(2):486–491

12. Sanchis Serra A (2012) Los lagomorfos del Paleolítico medio en la vertiente mediterránea ibèrica. Humanos y otros predadores como agentes de aporte y alteración de los restos óseos en yacimientos arqueológicos. Trabajos Varios del SIP 115.

13. Sanchis Serra A, Fernández Peris J (2008) Procesado y consumo antrópico de conejo en la Cova del Bolomor (Tavernes de la Valldigna, Valencia). El nivel XVIIc (ca. 350 ka). Complutum 19 (1): 25–46.

14. Stringer CB, Finlayson JC, Barton RNE, Fernández-Jalvo Y, Cáceres I, et al. (2008) Neanderthal exploitation of marine mammals in Gibraltar. PNAS 105 (38):14319–14324.

15. Zilhão J, Angelucci D, Badal-García E, d'Errico F, Daniel F, et al. (2010) Symbolic Use of Marine Shells and Mineral Pigments by Iberian Neandertals. PNAS 107 (3): 1023–1028.

16. Stiner MC (1999) Trends in Paleolithic mollusk exploitation at Riparo Mochi (Balzi Rossi, Italy): food and ornaments from the Aurignacian through Epigravettian. Antiquity 73(282): 735–754.

17. Lubell D (2004) Prehistoric edible land snails in the circum-Mediterranean: the archaeological evidence. In: Brugal JP, Desse, J, editors, Petits animaux et sociétés humaines. Du complément alimentaire aux ressources utilitaires (XXIV rencontres internationales d'archéologie et d'historie d'Antibes). Antibes: Editions APDCA: 77–98.

18. Lubell D, Barton N (2011) Gastropods and humans in the Late Paleolithic and Mesolithic of the western Mediterranean basin. Quaternary International 244: 1–4.

19. Bronk Ramsey C (2009) Bayesian analysis of radiocarbon dates. Radiocarbon 51/1, 337–360.

20. Reimer PJ, Bard E, Bayliss A, Beck JW, Blacwell PG, et al. (2013) INTCAL 13 and MARINE 13 Radiocarbon age calibration curves 0-50,000 years Cal BP. Radiocarbon 55/4, 1869–1887.

21. Jordá J, Maestro A, Aura JE, Álvarez E, Avezuela B, et al. (2011) Evolución paleogeográfica, paleoclimática y paleoambiental de la costa meridional de la Península Ibérica durante el Pleistoceno superior. El caso de la Cueva de Nerja (Málaga, Andalucía, España). Bol. R. Soc. Esp. Hist. Nat. Sec. Geol., 105 (1–4): 137–147.

22. Villaverde V, Román D (2013) El Graventiense de la vertiente mediterránea Ibérica: panorama y perspectivas. In: de las Heras C, Laheras JA, Arrizalabaga A, de la Rasilla M, editors. Pensando el Graventiense: nuevos datos para la región cantábrica en su contexto peninsular y europeo. Santander: Museo Nacional y Centro de Investigación de Altamira. pp. 34–54.

23. Stuiver M, Grootes PM (2000) GISP2 oxygen isotope ratios. Quaternary Research 53: 277–284.

24. Cacho I, Grimalt JO, Canals M, Sbaffi L, Shackleton NJ, et al. (2001) Variability of the western Mediterranean Sea Surface temperatures during the last 25,000 years and its connection with the northern hemisphere climatic changes. Paleoceanography 16: 40–52.

25. Rey J, Fumanal MP (1996) Cuaternario submarino frente a la Serra Gelada (Alicante) y sus implicaciones eustático-paleogeográficas. Cuadernos de Geografía, 60: 243–258.

26. Badal E, Carrión Y (2001) Del Glaciar al Interglaciar: los paisajes vegetales a partir de los restos carbonizados hallados en las cuevas de Alicante. In: Villaverde V, editor. De Neandertales a cromañones. El inicio del poblamiento humano en las tierras valencianas. Valencia: Universitat de Valencia pp. 21–41.

27. Villaverde V, Román D, Martínez R, Badal García E, Pérez Ripoll M, et al. (2010) El Paleolítico superior en el País Valenciano. Novedades y perspectivas. El Paleolítico superior peninsular: novedades del siglo XXI:[homenaje al profesor Javier Fortea]. Barcelona, pp. 85–114.

28. Barton CM, Villaverde V, Zilhão J, Aura JE, García O, et al. (2013) In glacial environments beyond glacial terrains: Human eco-dynamics in Late Pleistocene Mediterranean Iberia. Quaternary International (2013), http://dx.doi.org/10.1016/j.quaint.2013.05.007

29. Fa J, Stewart JR, Lloveras Ll, Vargas JM (2013) Rabbits and hominin survival in Iberia, Journal of Human Evolution 64 (4): 233–241.

30. Rodríguez-Hidalgo J, Saladié P, Canals A (2011) Following the White Rabbit: A Case of a Small Game Procurement Site in the Upper Palaeolithic (Sala de las Chimeneas, Maltravieso Cave, Spain). International Journal of Osteoarchaeology. (wileyonlinelibrary.com) DOI: 10.1002/oa.1238

31. Martínez-Ortí A (1999) Moluscos terrestres testáceos de la Comunidad Valenciana. Valencia: Universitat de València.

32. Martínez-Ortí A, Robles F (2003) Los moluscos continentales de la Comunidad Valenciana, 11. Conselleria de Territori i Habitatge, Generalitat Valenciana, Colección Biodiversidad.

33. Claassen C (1998) Shells. Cambridge Manuals in Archaeology. Cambridge University Press. 266 p.

34. R Core Team (2013) R: A language and environment for statistical computing. R Foundation for Statistical Computing. Vienna, Austria. URL http://www.R-project.org/

35. Rodriguez-Perochena I (2006) Estudio experimental de la reproducción y crecimiento de Iberus gualterianus gualterianus (Linnaeus) e Iberus gualterianus alonensis (Férussac) (Gastropoda, Helicidae) en condiciones de laboratorio. Unpublsihed Dissertation. Universidad Complutense de Madrid. 340 p.

36. Martínez-Ortí A (1999) Sobre la ornamentación de la concha de Pseudotachea splendida (Draparnaud, 1801) (Mollusca Gastropoda: Helicidae). Malacologic@: 1–4.

37. Fernández-López de Pablo J, Gómez-Puche M, Martínez-Ortí A (2011) Systematic consumption of non-marine gastropods at open-air Mesolithic sites in the Iberian Mediterranean region. Quaternary International 244(1): 45–53.

38. Jordá JF, Avezuela B, Aura JE, Martín-Escorza C (2011) The gastropod fauna of the Epipaleolithic shell midden in the Vestibulo chamber of Nerja Cave (Málaga, southern Spain). Quaternary International 244(1): 27–36.

39. Lloveras Ll, Nadal J, Argüelles PG, Fullola JM, Estrada A (2011) The land snail midden from Balma de Gai (Barcelona, Spain) and the evolution of terrestrial gastropod consumption during the la the Paleolithic and Mesolithic in eastern Iberia. Quaternary International 244(1): 37–44.

40. Martínez M (1997) El final del Paleolítico en las tierras bajas del Sudeste español. In: Fullola JM, Soler N, editors. El món mediterràni després del Pleniglacial 18000–12000 BP. Girona: Museu d'Arqueologia de Catalunya. pp. 345.

41. Vilaseca S (1991) El conchero del Camping Salou (Cabo Salou, provincia de Tarragona). Trabajos de Prehistoria 28: 63–92.

42. Estrada A, Nadal J, Lloveras L, Valenzuela S, García-Argüelles P (2009) Acumulaciones de gasterópodos terrestres en yacimientos epipaleolíticos. Aproximación tafonómica al registro fósil de la Balma de Gai (Moià, Barcelona). Barcelona: Monografies del SERP-Universitat de Barcelona. 7. pp. 83–91.

43. Allen JA (2004) Avian and mammalian predators of terrestrial gastropods. In: Barker GM. editor. Natural enemies of terrestrial molluscs. New Zeeland: CABI Publishing. pp.1–36.

44. Rosin ZM, Olborska P, Surmacki A, Tryjanowski P (2011) Differences in predatory pressure on terrestrial land snails by birds and mammals. Journal of Biosciences 36(4): 691–699.

45. Meads MJ, Walker KJ, Elliot GP (1984) Status, conservation and management of the land snails of the genus Powelliphanta (Mollusca: Pulmonata). New Zealand Journal of Zoology 11(3): 277–306.

46. Moreno-Rueda G (2009) Disruptive selection by predation offsets stabilizing selection on shell morphology in the land snail Iberus g. gualterianus. Evolutionary Ecology 23: 463–471.

47. Faus FV (1988) Contribución al conocimiento de la malacofagia de Rattus rattus (Linnaeus, 1758). Mediterránea. Series de Estudios Biológicos 10: 19–27.

48. Sanchis Serra A, Pascual Benito J (2011) Análisis de las acumulaciones óseas de una guarida de pequeños mamíferos carnívoros (Sitjar Baix, Onda, Castelló): implicaciones arqueológicas. Archaeofauna: International Journal of Archaeozoology 20: 47–71.

49. Parker JE, Thompson SP, Lennie AR, Potter J, Tang CC (2010) A study of the aragonite-calcite transformation using Raman spectroscopy, synchrotron powder diffraction and scanning electron microscopy. CrystEngComm 12: 1590–1599.

50. Cullity BD (1978) Elements of X-ray diffraction. Addison-Wesley Publishing Company, Inc.

51. Wenk HR (1985) Preferred Orientation in Deformed Metals and Rocks: An introduction to modern Texture Analysis. Academic Press, Inc.

52. Colonese AC (2005) Land snails exploitation: shell morphometry of Helix cfr. ligata (Müller, 1774) (Gastropoda, Pulmonata) from Late Epigravettian layers of Grotta della Serratura (Salerno). Rivista di Scienze Preistoriche-Supplemento 1: 271–284.

53. Fernández-López de Pablo J, Gómez-Puche M, Martínez-Ortí A (2011) Systematic consumption of non-marine gastropods at open-air Mesolithic sites in the Iberian Mediterranean region. Quaternary International 244(1): 45–53.

54. Gutiérrez-Zugasti I (2011) Early Holocene land snail exploitation in northern Spain: the case of La Fragua cave. Environmental Archaeology 16: 36–48.

55. Lloveras Ll, Nadal J, Argüelles PG, Fullola JM, Estrada A (2011) The land snail midden from Balma de Gai (Barcelona, Spain) and the evolution of terrestrial gastropod consumption during the la the Paleolithic and Mesolithic in eastern Iberia. Quaternary International 244(1): 37–44.

56. Jiménez-Espejo FJ, Martínez-Ruiz F, Finlayson C, Paytan A, Sakamoto T, et al. (2007) Climate forcing and Neanderthal extinction in Southern Iberia: insights from a multiproxy marine record. Quaternary Science Reviews 26: 836–852.

57. Rizner M, Vukosavljevic N, Miracle P (2009) The paleoecological and paleodietary significance of edible land snails (*Helix sp.*) across the Pleistocene-Holocene transition on the eastern Adriatic coast. In: McCarthan SB, Schulting R, Warren G, Woodman P editors. Mesolithic Horizons. Papers Presented at the Seventh International Conference on the Mesolithic in Europe, Belfast 2005, vol.II. Oxbow, Oxford, pp. 527–532.

58. Amara A (2011) Des structures de cuisson d'hélicidés dans les dépôts archéologiques de Koudiet Djerad, Ksar Chellala (Région de Tiaret – Algérie). In: Actes du Colloque International Préhistoire Maghrébine, Tome I. Alger: Travaux du CNRPAH, Nouvelle série N° 11, pp. 293–297.

59. Mehlman MJ (1979) Mumba-Höhle revisited: the relevance of a forgotten excavation to some current issues in East African prehistory. World Archaeology 11: 80–94.

60. Diez-Martín F, Domínguez-Rodrigo M, Sánchez P, Mabulla AZP, Tarriño A, et al. (2009) The Middle to Later Stone Age Technological Transition in East Africa. New data from Mumba Rockshelter Bed V (Tanzania) and their implications for the origin of modern human Behavior. Journal of African Archaeology 7(2): 147–173.

61. Gliganic LA, Jacobs Z, Roberts RG, Domínguez-Rodrigo M, Mabulla AZP (2012) New ages for Middle and Later Stone Age deposits at Mumba rockshelter, Tanzania: Optically stimulated luminescence dating of quartz and feldspar grains. Journal of Human Evolution, 62(4): 533–547

62. Tixier J (1970) L'abri sous-roche de Ksar 'Aqil: la campagne de fouilles 1969. Bulletin du Musée de Beyrouth, 23: 173–191.

63. Mellars P, Tixier J (1989) Radiocarbon-accelerator dating of Ksar 'Aqil (Lebanon) and the chronology of the Upper Palaeolithic sequence in the Middle East. Antiquity 63:761–768.

64. Kuhn SL, Stiner MC, Güleç E, Özer I, Yilmaz H, et al. (2009) The early Upper Paleolithic occupations at Üçağızlı Cave (Hatay, Turkey). Journal of Human Evolution 56(2): 87–113.

65. Starkovich B, Stiner MC (2010) Upper Paleolithic animal exploitation at Klissoura Cave 1 in southern Greece: Dietary trends and animal taphonomy. *Eurasian Prehistory* 7 (2): 107–132.

66. Khun SL, Pigati J, Karkanas P, Koumouzelis M, Kozlowski JK, et al. (2010) Radiocarbon dating results for the Early Upper Paleolithic of Klissoura Cave 1. Eurasian Prehistory, 7 (2): 37–46.

67. Stiner MC, Munro ND (2011) On the Evolution of Paleolithic Diet and Landscape at Franchthi Cave (Peloponnese, Greece). Journal of Human Evolution 60: 618–636.

68. Hey RW (1967) Land-snails. In: McBurney CBM, editor, The Haua Fteah (Cyrenaica) and the Stone Age of the South-East Mediterranean. London: Cambridge University Press. pp. 358.

69. Klein RG, Scott K (1986) Re-analysis of faunal assemblages from the Haua Fteah and other Late Quaternary archaeological sites in Cyrenaïcan Libyca. Journal of Archaeological Science 13: 515–542.

70. Barker G, Bennet P, Farr L, Hill E, Hunt C, et al. (2012) The Cyrenaican Prehistory Project 2012: the fifth season of investigations of the Haua Fteah cave. Libyan Studies 43: 115–136.

71. Douka K, Jacobs Z, Lane C, Grün R, Far L, et al. (2014) The chronostratigraphy of the Haua Fteah cave (Cyrenaica, northeast Libya). Journal of Human Evolution 66: 39–63.

72. Hunt CO, Reynolds TG, El-Rishi HA, Buzanian A, Hill E, et al. (2011) Resource pressure and environmental change on the North African littoral: Epipaleolithic to Roman gastropods from Cyrenaica, Libya. Quaternary International 244: 15–26.

73. Girod A (2011) Land snails from Late Glacial and Early Holocene Italian sites. Quaternary International 244: 105–116.

74. Miracle P (2001) Feast or famine? Epipaleolithic subsistence in the northern Adriatic basin. Documenta Praehistorica XXVIII: 177–197.

75. Martini F, Colonese AC, Di Giuseppe Z, Ghinassi M, Lo Vetro D, et al. (2009) Human-environment relationships in SW Italy during Late glacial-early Holocene transition: some examples from Campania, Calabria and Sicily. Méditerranée: Revue geographique des pays mediterraneens 112: 89–9

76. Mussi M, Lubell D, Arnoldus-Huyzendveld A, Agotini S, Coubray S (1995) Holocene land snail exploitation in the highlands of Central Italy and Eastern Algeria: a comparison. Préhistoire européenne 7: 169–189.

77. Colonese AC, Zanchetta G, Drysdale RN, Fallick AE, Manganelli G, et al. (2011) Stable isotope composition of Late Pleistocene-Holocene *Eobania vermiculata* (Müller, 1774) (Pulmonata, Stylommatophora) shells from the Central Mediterranean basin: Data from Grotta di Oriente (Favignana, Sicily). Quaternary International 244: 76–87.

78. Durante S, Setepassi F (1972) I molluschi del giacimento quaternario della Grotta della Madonna Praia a Mare (Calabria). Quaternaria, XVI: 255–269.

79. Colonese AC, Tozzi C (2010) I reperti malacologici di Grotta del Mezzogiorno (Salerno): implicazioni culturali e paleoecologiche. Proceedings of the V Convegno Nazionale di Archeozoologia, Rovereto, Italy, pp. 93–96.

80. Taylor VK, Barton RNE, Bell M, Bouzougar S, Collcutt S, et al. (2011) The Epipaleolithic (Iberomaurusian) at Grotte des Pigeons (Taforalt), Morocco: A preliminary study of the land Mollusca. Quaternary International 244(1): 5–14.

81. Aura JE, Jordá JF, Pérez M, Badal M, Avezuela B, et al. (2013) El corredor costero meridional: los cazadores gravetienses de la cueva de Nerja (Málaga, España). In: de las Heras C, Laheras JA, Arrizalabaga A, de la Rasilla M, editors. Pensando el Gravetiense: nuevos datos para la región cantábrica en su contexto peninsular y europeo. Santander: Museo Nacional y Centro de Investigación de Altamira. pp.104–113.

82. Jordá JF, Avezuela B, Aura JE, Martín-Escorza C (2011) The gastropod fauna of the Epipaleolithic shell midden in the Vestibulo chamber of Nerja Cave (Málaga, southern Spain). Quaternary International 244(1): 27–36.

83. Stiner MC, Kuhn SL (2006) What's a mother to do? A hypothesis about the division of labour and modern human origins. Current Anthropology 47(6): 953–980.

84. Gamble C, Davies W, Pettitt P, Hazelwood L, Richards M (2005) The archaeological and genetic foundations of the European population during the Late Glacial: implications for 'agricultural thinking'. Cambridge Archaeological Journal 15: 193–223

85. Bocquet-Appel JP, Demars PY, Noiret L, Dobrowsky D (2005) Estimates of Upper Paleolithic metapopulation size in Europe from archaeological data. *Journal of Archaeological Science* 32, 1656–1668.

86. Schmidt I, Bradtmöler M, Kehl M, Pastoors A, Tafelmaier Y, et al. (2012) Rapid climate change and variability of settlement patterns in Iberia during the Late Pleistocene. Quaternary International 274: 179–204.

87. Villaverde V (1994) Arte Paleolítico de la cova del Parpalló estudio de la coleción de plaquetas y cantos grabados y pintados. 2.vol. Valencia: Diputación Provincial.

88. Bicho N, Carvalho AF, González-Sainz C, Sanchidrián JL, Villaverde V, et al. (2007) The Upper Paleolithic Rock Art of Iberia. Journal of Archaeological Method and Theory 14 (1): 81–151.

89. Weninger B, Jöris O (2008) A 14C age calibration curve for the last 60 ka: the Greenland-Hulu U/Th timescale and its impact on understanding the Middle to Upper Paleolithic transition in Western Eurasia. Journal of Human Evolution 55:772–781.

10

Tree Migration-Rates: Narrowing the Gap between Inferred Post-Glacial Rates and Projected Rates

Angelica Feurdean[1,2*], **Shonil A. Bhagwat**[3,4,5,6], **Katherine J. Willis**[5,6,7], **H. John B Birks**[3,7,8], **Heike Lischke**[9], **Thomas Hickler**[1,10]

1 Senckenberg Research Institute and Natural History Museum and Biodiversity and Climate Research Center, Frankfurt am Main, Germany, **2** Romanian Academy "Emil Racoviţă" Institute of Speleology, Cluj Napoca, Romania, **3** Department of Geography, The Open University, Milton Keynes, United Kingdom, **4** School of Geography and the Environment, University of Oxford, Oxford, United Kingdom, **5** Long-Term Ecology Laboratory, Biodiversity Institute, Department of Zoology, University of Oxford, Oxford, United Kingdom, **6** Biodiversity Institute, Oxford Martin Institute, Department of Zoology, University of Oxford, Oxford, United Kingdom, **7** Department of Biology, University of Bergen, Bergen, Norway, **8** Environmental Change Research Centre, University College London, London, United Kingdom, **9** Dynamic Macroecology, Landscape Dynamics, Swiss Federal Institute for Forest, Snow and Landscape Research WSL, Birmensdorf, Switzerland, **10** Biodiversity and Climate Research Centre and Senckenberg Gesellschaft für Naturforschung and Goethe University, Frankfurt am Main, Germany

Abstract

Faster-than-expected post-glacial migration rates of trees have puzzled ecologists for a long time. In Europe, post-glacial migration is assumed to have started from the three southern European peninsulas (southern refugia), where large areas remained free of permafrost and ice at the peak of the last glaciation. However, increasing palaeobotanical evidence for the presence of isolated tree populations in more northerly microrefugia has started to change this perception. Here we use the Northern Eurasian Plant Macrofossil Database and palaeoecological literature to show that post-glacial migration rates for trees may have been substantially lower (60–260 m yr^{-1}) than those estimated by assuming migration from southern refugia only (115–550 m yr^{-1}), and that early-successional trees migrated faster than mid- and late-successional trees. Post-glacial migration rates are in good agreement with those recently projected for the future with a population dynamical forest succession and dispersal model, mainly for early-successional trees and under optimal conditions. Although migration estimates presented here may be conservative because of our assumption of uniform dispersal, tree migration-rates clearly need reconsideration. We suggest that small outlier populations may be a key factor in understanding past migration rates and in predicting potential future range-shifts. The importance of outlier populations in the past may have an analogy in the future, as many tree species have been planted beyond their natural ranges, with a more beneficial microclimate than their regional surroundings. Therefore, climate-change-induced range-shifts in the future might well be influenced by such microrefugia.

Editor: David Nogues-Bravo, University of Copenhagen, Denmark

Funding: The macrofossil database was funded through National Environment Research Council grant number NE/D001578/1, awarded to MEE and KJW. AF and TH acknowledge funding from the German Research Foundation (grant FE-1096/2-1), TH from the Biodiversity and Climate Research Center, and HL from the Swiss National Science foundation (grant 315230_122434). The funders had no role in study design, data collection and analysis, decision to publish, or preparation of the manuscript.

Competing Interests: The authors have declared that no competing interests exist.

* E-mail: angelica.feurdean@senckenberg.de, angelica.feurdean@gmail.com

Introduction

Estimating rates of tree migration is critical for understanding how species range distributions are shaped by past expansion and contraction, and how species might respond in the future to climate and land-use changes. Plant range-shifts are primarily determined by climate, but life-history traits (rate of establishment, growth, survival, dispersal ability, etc.) are also important [1,2]. Migration rates of European tree species in response to past climate changes have generally been estimated by assuming that these species persisted during the last glacial maximum (LGM) in southern Europe (southern refugia) with their northernmost distributions at approximately 40–45°N latitude [3,4,5,6]. As a consequence, it has often been assumed that trees dispersed rapidly (100–1000 m yr^{-1}) via long-distance dispersal in response to climate warming during the early post-glacial [3,7]. The apparent mismatch between observed seed dispersal distances and estimates based on ecological and seed dispersal processes during the Holocene (post-glacial) has often been referred to as Reid's paradox of rapid plant migration [8]. Although palaeoecological, theoretical, and modelling studies have shown that long-distance dispersal could explain rapid migration rates [7,9,10], the possibility of such a rapid migration-capacity has been challenged [11].

A growing body of paleoecological, genetic, and climate-modelling literature, particularly from previously less-studied regions, such as eastern Europe and northern Asia, suggests a more northerly glacial survival for both early- and mid-successional tree species [12–28]. Genetic studies confirm the importance of advancing leading-edge populations for colonization [16,29] but it is still unclear how widespread this phenomenon was. However, it is now increasingly acknowledged that these northerly populations might have acted as source populations for post-glacial expansion, in addition to the populations in more southerly locations. This has led to a paradigm shift from colonization via long-distance dispersal to rapid colonization via dispersal from local scattered populations [21,26,30]. Considering migration from northern refugia at the end of LGM implies that populations were closer to their present range-limits than estimated under the assumption of dispersal solely from the south, and therefore species migration rates are likely to have been lower than previously assumed [21].

Here we use fossil palaeoecological data to estimate post-glacial migration rates for eight tree taxa (including both shade-intolerant and shade-tolerant trees) taking into account re-population from northern refugia and thus assuming re-colonization via local scattered populations. These eight taxa have wide geographical ranges today in Europe and thus can be presumed to be relatively hardy and to have wide ecological tolerances. We compare these rates to estimates that assume re-colonization only from southern refugia (south of 40–45°N), and show how migration speeds are over-estimated by assuming southern refugia as the only population source. Finally, our post-glacial estimates are compared to maximum migration rates and those projected for the future with a process-based forest succession and dispersal model [31].

We find that post-glacial migration rates for trees may have been substantially lower than those estimated by assuming migration from southern refugia only. We suggest that small outlier populations may be a key factor in understanding past migration rates and in predicting potential future range-shifts.

Materials and Methods

We determined from the Northern Eurasian Plant Macrofossil Database [24] and relevant palaeoecological literature, the northernmost geographical distributions of eight tree taxa at the end of the LGM (approximately 18,000 years ago) and the point in time when these trees reached their modern northern limits (Fig. 1; Table 1). Five out of eight tree taxa, however, are only identified to the generic level and could thus involve species with different geographical distributions, climate requirements, and dispersal modes. Compared to previous attempts at determining range shifts, we increased the taxonomic and spatial resolution by mainly using plant macrofossil remains for our migration-rate estimates. This is because pollen alone does not provide unambiguous evidence for the local presence of a species due to the problems of long-distance pollen dispersal.

The present-day northern limits for each species were based on *Atlas Florae Europaeae*. Two sets of values were computed for each tree taxon: (1) northern migration rate, taking into account that the tree was present in northern refugia (>45°N) and spread to its post-glacial northern limit from these locations, and (2) southern migration rate, assuming that the tree only spread from southern refugia located between 40 and 45°N latitude to its post-glacial northern limit (Figs. 1, 2, Table 1). In the case of southern refugia, the maximum northern limit for cold-tolerant deciduous and coniferous tree species was estimated at 45°N, whereas for temperate deciduous trees the limit is 40°N [4,5,6]. The distances (in km) between the start of the migration and the northern range-limit locations were estimated linearly assuming a uniform spread from 18,000 yr BP until the time when a taxon reached its northern distribution (Figs. 1, 2, Table 1). Over-estimation (as percentages) of the migration rates was calculated as the ratio between the southern (2) and northern migration rates (1) multiplied by 100 (Table 2). In addition, we have estimated rates of migration assuming that superimposed on this long-term expansion of tree populations, tree movements could have been halted during the two major cold periods: Heinrich Event1, HE1

(lasting ~4000 years) and the Younger Dryas, YD (~1000 years) (Table 2)).

Results

Our analysis shows that: 1) the spreading-rate estimates from northern refugia are substantially lower than those that assume colonization only from the south, and 2) early-successional trees (*Betula, Pinus, Alnus*) migrated faster than mid- and late-successional ones (*Picea, Abies alba, Quercus, Carpinus betulus, Fagus sylvatica*) (Fig. 2, Table 2). The early-successional tree migration rates that assume spreading from the north vary from 100 to 260 m yr^{-1}, whereas migration-rates that assume spreading from the south vary from 225 to 540 m yr^{-1} (Fig. 2A, Table 2). The mid- and late-successional tree migration rates that assume spreading from the north vary between 60 and 170 m yr^{-1} whereas migration-rates that assume spreading from the south range between 115 and 385 m yr^{-1} (Fig. 2B, Table 2). The northerly estimates are lower than those obtained by assuming survival in southern Europe only, with over-estimates ranging from 120 to 540 m yr^{-1} (Table 2). The northerly migration estimates assuming no movement of taxa during the Heinrich Event1 and the Younger Dryas range between 0 and 515 m yr^{-1} for early-successional trees and between 90 and 735 m yr^{-1} for mid- and late-successional trees (Table 2).

Discussion

The post-glacial migration-rate estimates assuming colonization from northern populations suggest that the rates for both early- and mid- to late-successional trees are much lower than previously estimated, and that early-successional trees generally migrated faster than mid- and late-successional ones (Fig. 2). Our results also show that many taxa, in particular early- and mid-successional trees, reached their modern northern distribution in the early Holocene, a time of rapid climate changes (Table 1). The ability of early-successional pioneer taxa to persist in the harsh cold and dry LGM climate and to migrate faster than mid- and late-successional taxa would be expected as a result of differences in life history (fast growth, large seed production, far-distance dispersal) and greater stress tolerance to large amplitude temperature change, drought, etc [3,9,22,32]. Generation time might also explain differences in migration rates among taxa with similar dispersal properties. For example, the early-successional trees *Pinus, Betula*, and *Alnus* first set seed at an age of 10–20 years, whereas *Picea abies* first sets seed at 30–40 years or even after 50 years [33]. Thus, although *Pinus* and *Picea* have similar seed dispersal properties, *Pinus* is able to spread much faster than *Picea* does (Fig. 2, Table 1).

Post-glacial migration rates similar to our fossil estimates have been derived from climate-driven modelled refugia at a spatial resolution of *ca* 16 km, and range between 35 and 380 m yr^{-1}, with generally higher values for shade-intolerant trees (*Betula pendula, B. pubescens, Pinus sylvestris*) and for *Picea abies* than for other shade-tolerant species [2,19]. These authors also suggest that accessibility from refugia explains to a large extent the post-glacial range shifts for many species, in particular those with a limited dispersal ability. However, they do not use fossil data as evidence of actual refugia [2,19]. Fossil migration estimates of 250 m yr^{-1} for *Picea abies* and 100 m yr^{-1} for *Fagus sylvatica* have been recently obtained for southern Scandinavia based on pollen records and model simulation output [34] and are lower than those previously obtained of *ca.* 500 m yr^{-1} for the same region [3]. Using a similar approach as above, range displacements (contraction and expansion) between −170 and 270 m yr^{-1} have been reported from North America [35], which are higher than those previously

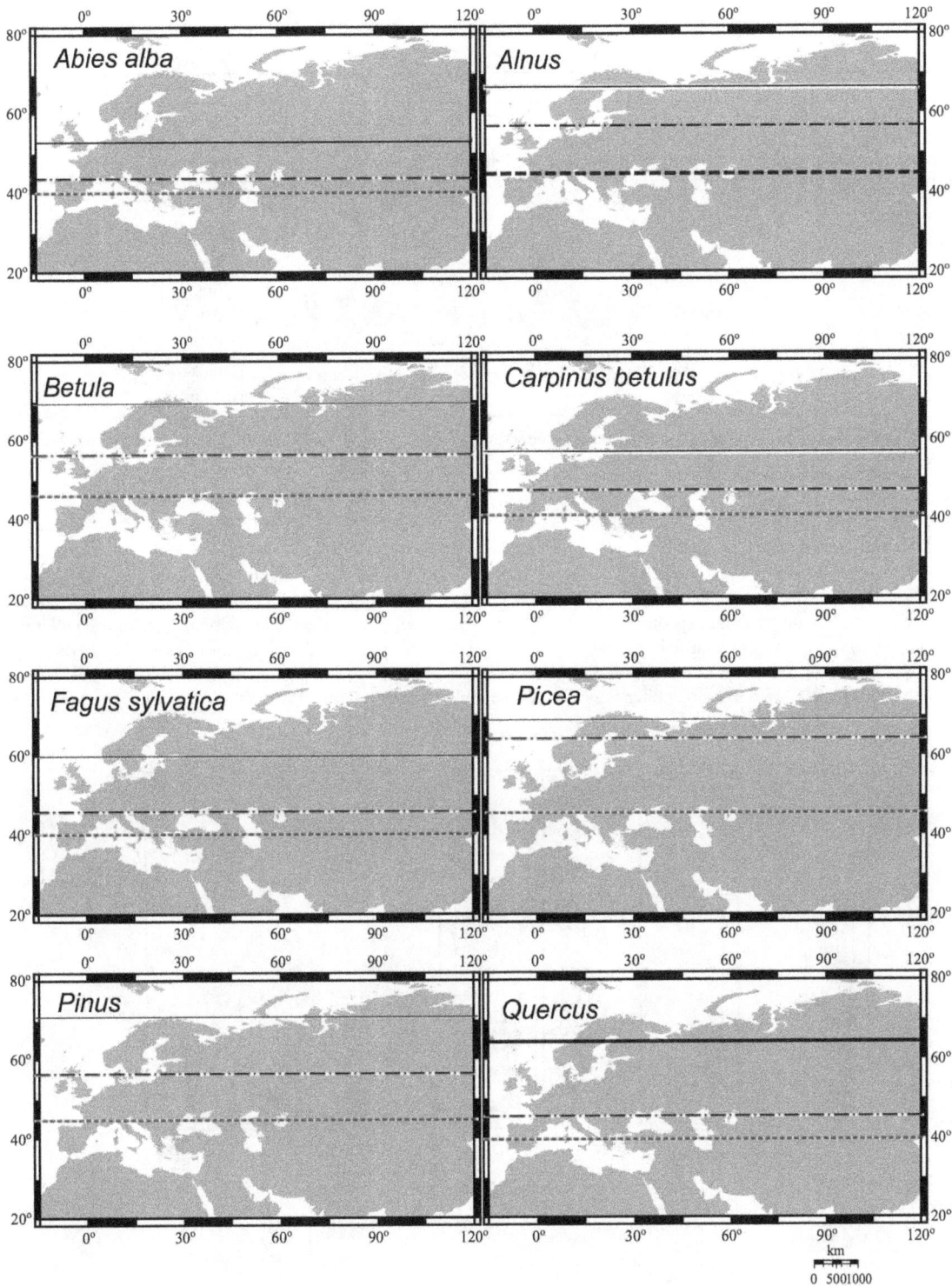

Figure 1. The northern range limit that a tree taxon has reached (line) either from southern refugia (dotted line) located 40–45°N or from northern refugia (dashed line) (see details in Table 1).

obtained (<100 m yr^{-1}) in this region based on phylogeograph-raphic data [29]. It is therefore evident that most species were generally only capable of migration rates less than 260 m yr^{-1}. Some late-successional taxa such as *Picea*, *Abies alba*, and *Quercus* that spread quickly in the early Holocene (Fig. 2B) under conditions of low competition and low human impact should more accurately reflect their intrinsic rates of spread than other late-succession species such as *Fagus sylvatica* and *Carpinus betulus* that spread during the late Holocene and were probably dependent, to some degree, on anthropogenic disturbance of the already established forests. Our estimates indeed show fast migration rates for these species, namely *Abies alba* (170 m yr^{-1};

Table 1. List of tree taxa for which fossil evidence (pollen, plant macrofossils, charcoal) exists for their survival at 18,000 cal yr BP north of 40°N.

Species	Distance from southern refugia (km) to the present day limit	Distance northern refugia (km) to the present day limit	Time of arrival at northern limit (cal yr BP)	References
Abies alba	1340	1100	11,500	[14]
Alnus (tree)	2450	1100	7000	[24]
Betula (tree)	2700	1300	13,000	[24]
Carpinus betulus	1850	1000	2000	[15]
Fagus sylvatica	2250	1500	1000	[16,47]
Picea	2700	500	11,000	[24]
Pinus (tree)	2900	1550	10,000	[24]
Quercus (temperate)	2650	2000	6000	[5]

The distance (in km) from the perceived southern location and northern locations, respectively, and the time (calibrated years BP) when each species reached the present-day northern range limit is also given.
Cal yr BP = calibrated years before present (AD 1950).

max. 735 m yr^{-1}) and *Quercus* (165 m yr^{-1}; max. 285 m yr^{-1}), except for *Picea* (70 m yr^{-1}; max. 250 m yr^{-1}), suggesting that their northern populations responded rapidly to climate change and were therefore not greatly affected by migrational lags during the post-glacial. Models of post-glacial population expansion indicate that initially tree species spread rapidly as low-density populations or isolated individuals in an advancing wave front reaching their northern limit ahead of mass colonization [34–37]. Later, when climate conditions became suitable and more stable, and there were higher population densities, species migrated more slowly due to increased competition from already present populations [19,33,35,36]. A reduction in migration rates is also predicted to have occurred towards the species distributional limits because of less suitable climate [19,35–38]. It should, however, be noted that

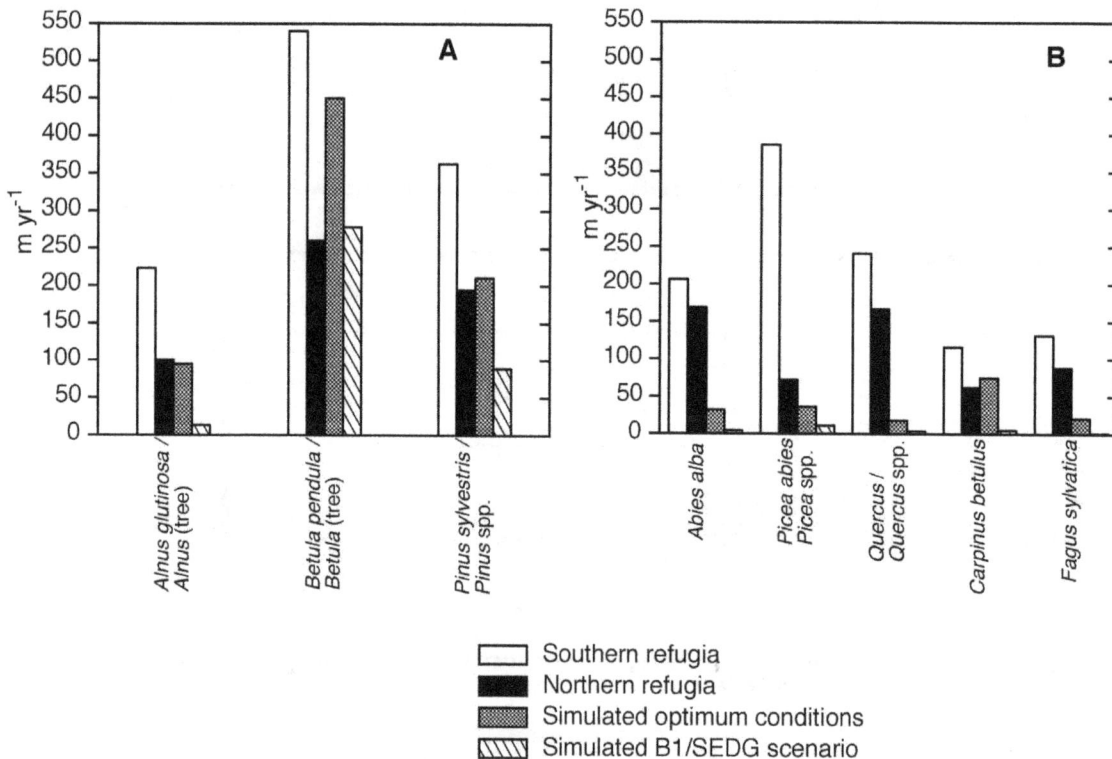

Figure 2. Post-glacial migration-rate estimates for (A) early-successional trees and (B) mid- to late-successional trees assuming colonization from southern and northern refugia, and comparisons with projected mean migration rates from a process-based model [from 31]. When no exact species name is available in the fossil record the genus name is used. *Picea* spp. includes *P. abies* and *P. obovata*, *Pinus* spp. includes *P. sylvestris* and *P. sibirica*, and *Quercus* spp. could include several species although we expect mainly temperate species such as *Q. petraea*, *Q. robur*, and *Q. pubescus*.

Table 2. Southern post-glacial migration-rate estimates (m yr^{-1}) assuming that species spread to their present-day northern limit from the south (40–45°N latitude), and northern migration rates assuming that species spread to their present-day northern limit from their northernmost refugia.

Species	Southern fossil estimates	Northern fossil estimates	Over-estimates (fossil)	Projected mean rates (CC-Scenario B1/SEDG)	Projected max. rates (under optimal conditions)
Early succesional					
Alnus glutinosa/Alnus (tree)	225 *(410)*	100 *(185)*	225	13.6±15.3	95
Betula pendula/Betula (tree)	540 *(0)*	260 *(0)*	210	278.2±103.8	450
Pinus sylvestris/Pinus (tree)	360 *(970)*	195 *(515)*	185	89±35.4	210
Mid- to late succesional					
Abies alba	205 *(895)*	170 *(755)*	120	4.5±6.1	32
Carpinus betulus	115 *(170)*	60 *(90)*	185	5.0±6.5	75
Fagus sylvatica	130 *(190)*	90 *(125)*	150	1.1±1.7	21
Picea abies	385 *(1350)*	70 *(250)*	540	11.2±6.1	36
Quercus petraea/ Quercus spp.	220 *(380)*	165 *(285)*	130	3.6±4.5	18

Estimated rates of migration assuming no movement of taxa during the two major cold periods: Heinrich Event1 (lasting ~4000 years) and the Younger Dryas (lasting ~1000 years) are given in italics. Projected mean migration rates (m yr^{-1}) for several tree species for 2100 using B1/SEDG greenhouse gas emission/land-use change scenarios, as well as the maximum rates derived from weak competition and optimum temperature conditions [31] are also given. When no exact species name is available in the fossil record the genus name is used.

in the case of *Picea*, for which our calculations are based on the Eurasian macrofossil data-base and thus on northern locations, migration rates are lower than the estimates for *Picea abies* [34–36], which are based on European pollen and macrofossil data. The migration-rate estimates for *Picea* involve two potential species, namely *Picea abies* and *P. obovata*; the latter tolerates colder temperatures and has a more northern and eastern geographical distribution than *P. abies*, which could have lowered the overall migration rates for *Picea*.

Our migration estimates are calculated assuming a uniform spread from 18,000 yr BP until the time when the taxon reached its northern distribution, despite the possibility that superimposed on this long-term expansion of tree populations, tree movements could have been halted by two major cold periods: Heinrich Event 1 and the Younger Dryas [39–40]. Nevertheless, our migration estimates that assume no movement during HE1 and YD (5000 years) (Table 2) change little for taxa that reached their modern northern distribution during the late Holocene (*Fagus sylvatica, Carpinus betulus*). However, our estimates under this scenario (Table 2) show high migration rates for most taxa that had already reached their modern northern distribution during the early Holocene (*Pinus, Abies, Betula, Picea*), at times of rapid climate changes. This probably reflects tree spread at low-densities in an advancing wave front ahead of mass colonization at the late-glacial/Holocene transition. Overall, we think, that assuming uniform migration rates over the entire time interval, as opposed to no movement for a period of 5000 years, appears to provide more realistic migration-rate estimates as this procedure balances fast northward movements during periods of rapid climate change and slower northward movements during periods of cold or stable climate conditions and at higher population density. Estimates of latitudinal taxa displacement from North America show dynamic changes between 16 and 12 k yr BP (expansion, contraction, stagnation), predominant fast northward expansion between 12 and 7 k yr BP and overall lower migration rates during warm and/ or stable conditions occurring between 7 and 1 k yr BP [35]. Like with any fossil estimates, the spatial and temporal scales of fossil data considered for our analysis cannot provide exact locations of

all northerly refugia or of the time when a taxon reached its modern northern limit. Our post-glacial migration rates should therefore be regarded as approximate estimates. Migration rates are also integrated over the varying biotic and abiotic conditions along the different migrational paths. The presence of large European mountains chains (Pyrenees, Alps, Carpathians) were previously found to not have acted as geographical barriers in the case of the spread of *Fagus sylvatica* in Europe during the last glacial–postglacial, but rather to have facilitated its survival and spread [16]. Topography might therefore be less important for many other trees considered here.

We compared our post-glacial migration-rate estimates with simulated migration rates under i) optimal conditions of competition and climate, and ii) conditions of future climate change and land-use scenarios that have been simulated using a tree migration meta-model [31]. This meta-model had been regressed against simulations of the spatio-temporal forest landscape model TreeMig [41] under various climate, competition, and fragmentation conditions and was used to constrain migration distances in an empirical species distribution model at the European scale under future land-use and resulting fragmentation and climate change. This approach thus accounts for climate influence, landscape pattern, habitat suitability, population dynamics, competition, and seed dispersal (for details see 31), unlike other models projecting the impact of 21st century climate and landscape fragmentation on species range-shifts, which commonly use two extreme scenarios, i.e. unlimited or no dispersal [42,43,44]. The simulated rates fit better with our estimated "northern" than with the "southern" post-glacial migration rates (Fig. 2, Table 2) or with previous pollen-derived estimates [3]. The fact that the simulated rates for shade-intolerant species (mean value for future rates of 155 m yr^{-1}) are of the same magnitude as those derived for the post-glacial rates implies that this is an important step towards understanding how some trees migrate in response to climate change and habitat fragmentation. Nevertheless, the simulated migration rates for shade-tolerant tree species (mean rate for the future is 15 m yr^{-1}) are still an order of magnitude slower than the post-glacial migration rates, although the optimum simulated values (the maximum rate a species could move without competition or with weak competition only and without habitat fragmentation) are closer to the post-glacial estimates (Fig. 2, Table 2). This might indicate a weak influence of competition and fragmentation during the Holocene compared to the future. In addition, the reason that the simulated migration rates for shade-tolerant tree species are generally lower than the estimated post-glacial migration rates further suggests that glacial refugia for shade-tolerant tree species were farther north than currently known. Interestingly, *Picea*, whose post-glacial migration rate based on fossil data [this study, 34, 36] and hindcast model results [2,19] suggests a rather fast migration (70–250 m yr^{-1}), yields much slower rates of 10 m yr^{-1} in the future and 36 m yr^{-1} with the optimal dispersal simulations (Fig. 2). One potential reason is that, in the simulation model, *Picea abies* is assigned to the class of "medium dispersers" instead of "long-distance dispersers", despite its (particularly at high latitudes) small seeds [45].

In summary, deriving insights on how species range distributions were shaped by expansions during the post-glacial represents a contribution to our understanding of tree-species migration rates and their likely response to future climate and land-use changes. We demonstrate that post-glacial migration estimates assuming colonization from local northern populations are much lower than previously estimated assuming colonization from the south only. Thought migration estimates presented here may be conservative because of our assumption of uniform dispersal, tree migration-

rates clearly need reconsideration (Figs. 1, 2, Table 2). Although these post-glacial migration-rate estimates might change in the future when more fossil evidence becomes available, palaeobotanical data suggest that populations persisting in northern Europe during the LGM are less likely to be limited in their distribution by migrational lag. We suggest that allowing for small outlier populations is crucial for understanding past tree migration-rates and for simulating potential future changes, but this has rarely been done. The importance of outlier populations in the past may have an analogy in the future, as many tree species have been planted in small populations beyond their natural ranges, for example in parks [46], usually with a slightly more beneficial micro-climate than their surroundings. Therefore, climate-change induced range-shifts in the future might well be influenced by such microrefugia.

Acknowledgments

H.J.B.B. acknowledges help from Cathy Jenks. T. Giesecke and D. Magri are gratefully acknowledged for providing constructive comments on early draft of the paper.

Author Contributions

Conceived and designed the experiments: AF SAB KJW TH. Analyzed the data: AF SAB. Wrote the paper: AF TH. Contributed to brainstorming and writing: AF SAB KJW HJBB HL TH.

References

1. Higgins SI, Clark JS, Nathan R, Hovestadt T, Schurr F, et al. (2003) Forecasting plant migration rates: managing uncertainty for risk assessment. J Ecol 91: 341–347.

2. Normand S, Ricklefs RE, Skov F, Bladt J, Tackenberg O, et al. (2011) Postglacial migration supplements climate in determining plant species ranges in Europe. Proc R Soc Lond B 278: 3644–3653.

3. Huntley B, Birks HJB (1983) *An Atlas of Past and Present Pollen Maps for Europe: 0–13000years ago*. Cambridge University Press, Cambridge.

4. Bennett KD, Tzedakis PC, Willis KJ (1991) Quaternary refugia of North European trees. J. Biogeogr 18: 103–115.

5. Petit RJ, Brewer S, Bordács S, Burg K, Cheddadi R, et al. (2002) Identification of refugia and post-glacial colonisation routes of European white oaks based on chloroplast DNA and fossil pollen evidence. For Ecol Manag 156: 49–74.

6. Tzedakis PC, Lawson IT, Frogley MR, Hewitt GM, Preece RC (2002) Buffered tree population changes in a Quaternary refugium: evolutionary implications. Science 297: 2044–2047

7. Ritchie JC, MacDonald GM (1996) The patterns of post-glacial spread of white spruce. J Biogeogr 13: 527–540.

8. Clark JS, Fastie C, Hurtt G, Jackson ST, Johnson C, et al. (1998) Reid's paradox of rapid plant migration. BioScience 48: 13–24.

9. Clark JS (1998) Why trees migrate so fast: Confronting theory with dispersal biology and the paleo record. Am Nat 152: 204–224.

10. Nathan R, Katul G, Horn HS, Thomas SM, Oren R, et al. (2002) Mechanisms of long-distance dispersal of seeds by wind. Nature 418: 409–413.

11. Clark JS, Lewis M, Horvath L (2001) Invasion by extremes: population spread with variation in dispersal and reproduction. Am Nat 157: 537–554.

12. Willis KJ, Rudner E, Sumegi P (2000) The full-glacial forests of central and southeastern Europe. Quat Res 53: 203–213.

13. Kullman L (2002) Rapid recent range-margin rise of tree and shrub species in the Swedish Scandes. J Ecol 90: 68–77.

14. Terhurne-Berson R, Litt T, Cheddadi R (2004) The spread of *Abies* throughout Europe since the last glacial period: combined macrofossil and pollen data. Veg Hist & Archaeobot 13: 257–268.

15. Willis KJ, van Andel TH (2004) Trees or no trees? The environments of central and eastern Europe during the Last Glaciation. Quat Sci Rev 23: 2369–2387.

16. Magri D, Vendramin GG, Comps B, Dupanloup I, Geburek T, et al. (2006) A new scenario for the Quaternary history of European beech populations: palaeobotanical evidence and genetic consequences. New Phytol 117: 199–221.

17. Feurdean A, Wohlfarth B, Björkman L, Tantau I, Bennike O, et al. (2007) The influence of refugial populations on lateglacial and early vegetational changes in Romania. Rev Palaeobot & Palynol 145: 305–320.

18. Svenning JC, Skov F (2007a) Ice age legacies in the geographical distribution of tree species richness in Europe. Global Ecol & Biogeogr 16: 234–245.

19. Svenning JC, Skov F (2007b) Could the tree diversity pattern in Europe be generated by postglacial dispersal limitation? Ecol Lett 10: 453–460.

20. Svenning JC, Normand S, Kageyama M (2008) Glacial refugia of temperate trees in Europe: insights from species distribution modelling. J Ecol 96: 1117–1127.

21. Birks HJB, Willis KJ (2008) Alpines, trees, and refugia in Europe. Plant Ecol & Divers 1: 147–160.

22. Bhagwat SA, Willis KJ (2008) Species persistence in northerly glacial refugia of Europe: a matter of chance or biogeographical traits? J Biogeogr 35: 464–482.

23. Kuneš P, Pelánková B, Chytrý M, Jankovská V, Pokorný P, et al. (2008) Interpretation of the last-glacial vegetation of eastern-central Europe using modern analogues from southern Siberia. J Biogeogr 35: 2223–2236.

24. Binney HA, Willis KJ, Edwards ME, Bhagwat SA, Anderson PA, et al. (2009) The distribution of late-Quaternary woody taxa in northern Eurasia: evidence from a new macrofossil database. Quat Sci Rev 28. 2445–2464.

25. Allen JRM, Hickler T, Singarayer JS, Sykes MT, Valdes PJ, et al. (2010) Last glacial vegetation of northern Eurasia. Quat Sci Rev 29: 2604–2618.

26. Stewart JR, Lister AM, Barnes J, Dalen L (2010) Refugia revisited: individualistic responses of species in space and time. Proc R Soc Lond B 277: 661–671.

27. Väliranta M, Kaakinen A, Kuhry P, Kultti S, Salonen JS, et al. (2011) Scattered late-glacial and early-Holocene tree populations as dispersal nuclei for forest development in NE European Russia. J Biogeogr 38: 922–932.

28. Ohlemuller R, Huntley B, Normand S, Svenning JC (2012) Potential source and sink locations for climate-driven species range shifts in Europe since the Last Glacial Maximum. Global Ecol & Biogeogr 21: 152–163.

29. McLachlan JS, Clark JS, Manos PS (2005) Molecular indicators of tree migration capacity under rapid climate change. Ecology 86: 2088–2098.

30. Stewart JR, Lister AM (2001) Cryptic northern refugia and the origins of the modern biota. Trends Ecol Evol 16: 608–613.

31. Meier ES, Lischke H, Schmatz DR, Zimmermann NE (2012) Climate, competition and connectivity affect future migration and ranges of European trees. Global Ecol Biogeogr 21: 164–178.

32. Feurdean A, Tămaş T, Tanţău I, Fărcaş S (2012) Elevational variation in regional vegetation responses to late-glacial climate changes in the Carpathians. J Biogeogr 29: 258–271.

33. Lischke H, Löffler T (2006) Intra-specific density dependence is required to maintain diversity in spatio-temporal forest simulations with reproduction. Ecol Model 198: 341–361.

34. Bialozyt R, Bradley LR, Bradshaw RHW (2012) Modelling the spread of *Fagus sylvatica* and *Picea abies* in southern Scandinavia during the late Holocene. J Biogeogr 39: 665–675.

35. Ordonez A, Williams JW (2013) Climatic and biotic velocities for woody taxa distribution over the last 16 000 years in eastern North America. Ecol Lett 16:773–781.

36. Giesecke T (2005) Moving front or population expansion: How did *Picea abies* (L.) Karst. become frequent in central Sweden? Quat Sci Rev 24: 2495–2509.

37. Giesecke T, Bennett KD (2004) The Holocene spread of *Picea abies* (L.) Karst. In Fennoscandia and adjacent areas. J Biogeogr 31: 1523–1548.

38. Birks HJB (1989) Holocene isochrone maps and patterns of tree-spreading in the British Isles. J. Biogeogr 16: 503–540.

39. Naughton F, Sánchez Gofii MF, Kageyama M, Bard E, Cortijo E, et al. (2009) Wet to dry climatic trend in north western Iberia within Heinrich events. Earth Planet Sci Lett 284: 329–342.

40. Blockley SPE, Lane CS, Hardiman M, Rasmussen SO, Seierstad IK, et al. (2012) Synchronisation of palaeoenvironmental records over the last 60,000 years, and an extended INTIMATE1 event stratigraphy to 48,000 b2k. Quat Sci Rev 36: 2–10.

41. Lischke H, Zimmermann NE, Bolliger J, Rickebusch S, Loffler TJ (2006) TreeMig: A forest-landscape model for simulating spatio-temporal patterns from stand to landscape scale. Ecol Model 199: 409–420.

42. Thomas CD, Cameron A, Green RE, Bakkenes M, Beaumont LJ, et al. (2004) Extinction risk from climate change. Nature 427: 145–148.

43. Thuiller W, Lavorel S, Araujo MB, Sykes MT, Prentice IC (2005) Climate change threats to plant diversity in Europe. Proc Nat Acad Sci USA 102: 8245–8250.

44. Hickler T, Vohland K, Feehan J, Miller P, Fronzek S, et al. (2012) Projecting tree species – based climate- driven changes in European potential natural vegetation with a generalized dynamic vegetation model. Global Ecol & Biogeogr 21: 50–63.

45. Helmisaari H, Nedialko N (1989) Survey of ecololgical characteristics of boreal tree species in Fennoscandia and the USSR. IIASA Working Papers, vol 89–65. IIASA, Laxenburg.

46. Van der Veken S, Hermy M, Vellend M, Knapen A, Verheyen K (2008) Garden plants get a head start on climate change. Front Ecol & Environ 6: 212–216.

47. Giesecke T, Hickler T, Kunkel T, Sykes MT, Bradshaw RHW (2007) Towards an understanding of the Holocene distribution of Fagus sylvatica L. J Biogeogr 34: 118–131.

11

The Earliest Giant *Osprioneides* Borings from the Sandbian (Late Ordovician) of Estonia

Olev Vinn[1]*, Mark A. Wilson[2], Mari-Ann Mõtus[3]

1 Department of Geology, University of Tartu, Tartu, Estonia, 2 Department of Geology, The College of Wooster, Wooster, Ohio, United States of America, 3 Institute of Geology, Tallinn University of Technology, Tallinn, Estonia

Abstract

The earliest *Osprioneides kampto* borings were found in bryozoan colonies of Sandbian age from northern Estonia (Baltica). The Ordovician was a time of great increase in the quantities of hard substrate removed by single trace makers. Increased predation pressure was most likely the driving force behind the infaunalization of larger invertebrates such as the *Osprioneides* trace makers in the Ordovician. It is possible that the *Osprioneides* borer originated in Baltica or in other paleocontinents outside of North America.

Editor: Andrew A. Farke, Raymond M. Alf Museum of Paleontology, United States of America

Funding: O.V. is indebted to the Sepkoski Grant program (Paleontological Society), Estonian Science Foundation grant ETF9064, Estonian Research Council grant IUT20-34 and a target-financed project from the Estonian Ministry of Education and Science (SF0180051s08; Ordovician and Silurian climate changes, as documented from the biotic changes and depositional environments in the Baltoscandian Palaeobasin) for financial support. M-A.M was also supported by a target-financed project from the Estonian Ministry of Education and Science (SF0140020s08). The funders had no role in study design, data collection and analysis, decision to publish, or preparation of the manuscript.

Competing Interests: The authors have declared that no competing interests exist.

* E-mail: olev.vinn@ut.ee

Introduction

The oldest macroborings in the world are the small simple holes of *Trypanites* reported in Early Cambrian archaeocyathid reefs in Labrador [1,2]. The next oldest macroborings are found in carbonate hardgrounds of Early Ordovician age [3,4,5,6]. There was a great increase in bioerosion intensity and diversity in the Ordovician, now termed the Ordovician Bioerosion Revolution [7]. In the Middle and Late Ordovician, shells and hardgrounds are often thoroughly riddled with holes, most of them attributable to *Trypanites* and *Palaeosabella* [8]. In addition, Ordovician bioerosion trace fossils include bivalve borings (*Petroxestes*), bryozoan etchings (*Ropalonaria*), sponge borings (*Cicatricula*), *Sanctum* (a cavernous domichnium excavated in bryozoan zoaria by an unknown borer) and *Gastrochaenolites* [8,9]. Bioerosion was very common in the Middle Paleozoic, especially in the Devonian [10]. Later in the Mesozoic bioerosion intensity and diversity further increased [6,9,11,12], and deep, large borings became especially common [13].

The bioerosion trace fossils of Ordovician of North America are relatively well studied [14,15,16]. In contrast, there is a limited number of works devoted to the study of bioerosional trace fossils in the Ordovician of Baltica. The earliest large boring occurs in the Early to Middle Ordovician hardgrounds and could belong to *Gastrochaenolites* [4,17]. Abundant *Trypanites* borings are known from brachiopods of the Arenigian [18] and Sandbian [19]. Wyse Jackson and Key [20] published a study on borings in trepostome bryozoans from the Ordovician of Estonia. They identified two ichnogenera, *Trypanites* and *Sanctum*, in bryozoans of Middle and Upper Ordovician strata of northern Estonia.

The aims of this paper are to: 1) determine whether the shafts in large Sandbian bryozoans belong to previously known or a new bioerosional ichnotaxon for the Ordovician; 2) determine the systematic affinity of the trace fossil; 3) discuss the ecology of the trace makers; 4) discuss the paleobiogeographic distribution of the trace fossil; and 4) discuss the occurrence of large borings during the Ordovician Bioerosion Revolution.

Geological Background and Locality

During the Ordovician, the Baltica paleocontinent migrated from the temperate to the subtropical realm [21,22]. The climatic change resulted in an increase of carbonate production and sedimentation rate on the shelf during the Middle and Late Ordovician. In the Upper Ordovician the first carbonate buildups are recorded, emphasizing a striking change in the overall character of the paleobasin [23].

The total thickness of the Ordovician in Estonia varies from 70 to 180 m [23]. The Ordovician limestones of Estonia form a wide belt from the Narva River in the northeast to Hiiumaa Island in the northwest [23]. In the Middle Ordovician and early Late Ordovician, the slowly subsiding western part of the East-European Platform was covered by a shallow, epicontinental sea with little bathymetric differentiation and an extremely low sedimentation rate. Along the extent of the ramp a series of grey calcareous - argillaceous sediments accumulated (argillaceous limestones and marls), with a trend of increasing clay and decreasing bioclasts in the offshore direction [22].

The material studied here was collected from the Hirmuse Creek (Fig. 1) and Alliku Ditches (Fig. 1) of Sandbian age (Haljala

Stage) (Fig. 2). Hirmuse Creek is located in Maidla parish of Ida-Viru County. Clayey and skeletal limestones with interlayers of marls are exposed on the creek bed and in its banks. The fossil assemblage includes algae (*Mastopora*), brachiopods (*Clinambon, Leptaena, Platystrophia, Apartorthis, Porambonites, Pseudolingula*), conulariids, gastropods (*Lesueurilla, Holopea, Bucanella, Pachydictya, Megalomphala, Cymbularia*), ichnofossils (*Amphorichnus, Arachnostega*), sponges, receptaculitids (*Tettragonis*), rugosans (*Lambelasma*), bryozoans, and asaphid trilobites. The Alliku Ditches are located in Harju County near the village of Alliku. Clayey limestones with interlayers of marls are exposed here. The fauna includes algae, brachiopods, bryozoans (*Aluverina, Annunziopora, Batostoma, Ceramoporella, Coeloclema, Constellaria, Corynotrypa, Crepipora, Dekayella, Diazipora, Diplotrypa, Enallopora, Esthoniophora, Graptodictya, Hallopora, Hemiphragma, Homotrypa, Homotrypella, Kukersella, Lioclemella, Mesotrypa, Monotrypa, Nematopora, Nematotrypa, Oanduella, Moyerella, Pachydictya, Phylloporina, Prasopora, Proavella, Pseudohornera, Rhinidictya, Stictoporella*), echinoderms, gastropods, ostracods, rugosans, receptaculitids and trilobites according to Rõõmusoks [24].

No permits were required for the described study, which complied with all relevant Estonian regulations, as our study did not involve collecting protected fossil species. Three described bryozoan specimens with the *Osprioneides* borings are deposited at the Institute of Geology, Tallinn University of Technology (GIT), Ehitajate tee 5, Tallinn, Estonia, with specimen numbers GIT-398-729, GIT 665-18 and GIT 665-19.

Results

Numerous unbranched, single-entrance, large deep borings with oval cross sections were found in three large trepostome bryozoan colonies (Figs. 3, 4, 5, 6). The borings are vertical to subparallel to the bryozoan surfaces and have a tapered to rounded terminus. Several borings have lost their roofs due to erosion. The boring apertures' minor axis is 2.7 to 7.0 mm (M = 5.05, sd = 1.34, N = 12) and major axis is 7.0 to 15.0 mm (M = 10.37, sd = 2.60, N = 12) long. The axial ratio (major axis/minor axis) of the borings ranges from 1.60 to 2.59 (M = 2.08, sd = 0.29, N = 12). Three

completely preserved borings are 25 mm (aperture 12×6 mm), 28 mm (aperture 9×4.5 mm), and 32 mm (aperture 13×6 mm) deep. Two unroofed borings have depths of 35 mm and 50 mm. The borings are abundant in the studied samples (Figs. 3, 4, 5, 6). They occasionally truncate each other, which somewhat resembles a branching pattern. There are no linings or septa inside the borings. The growth lamellae of the bryozoans show no reactions around the borings. Small *Trypanites* borings occur inside the large boring with oval cross section. The apertures of the large borings occur on both the upper and lower surfaces of the bryozoans (the upper and lower surface of bryozoans was determined by looking at skeletal growth).

Discussion

Taxonomic Affinity of the Borings and the Possible Trace Maker

The borings in these bryozoans resemble somewhat *Petroxestes* known from Late Ordovician bryozoans and hardgrounds of North America [14]. Both are of unusually large size for Ordovician borings, and both have oval-shaped apertures. However, in *Petroxestes* the aperture width is much greater than the boring's depth. In contrast, the depth of the borings in bryozoans is much greater than their apertural width. Unlike *Petroxestes*, the Sandbian borings examined here have a tapering terminus and somewhat sinuous course. The axial ratio of *Petroxestes* borings aperture (major axis/minor axis) is also much greater than observed in these borings.

The other similar large Palaeozoic boring is *Osprioneides*, which is known from the Silurian of Baltica, Britain and North America [13]. We assign borings in the bryozoans studied here to *Osprioneides kampto* because of their similar general morphology. They have a single entrance, an oval cross section, and significant depth similar to *Osprioneides kampto*. Their straight, curved to somewhat sinuous shape also resembles that of *Osprioneides*. Both *Osprioneides* and these borings in bryozoans have a tapered to rounded terminus.

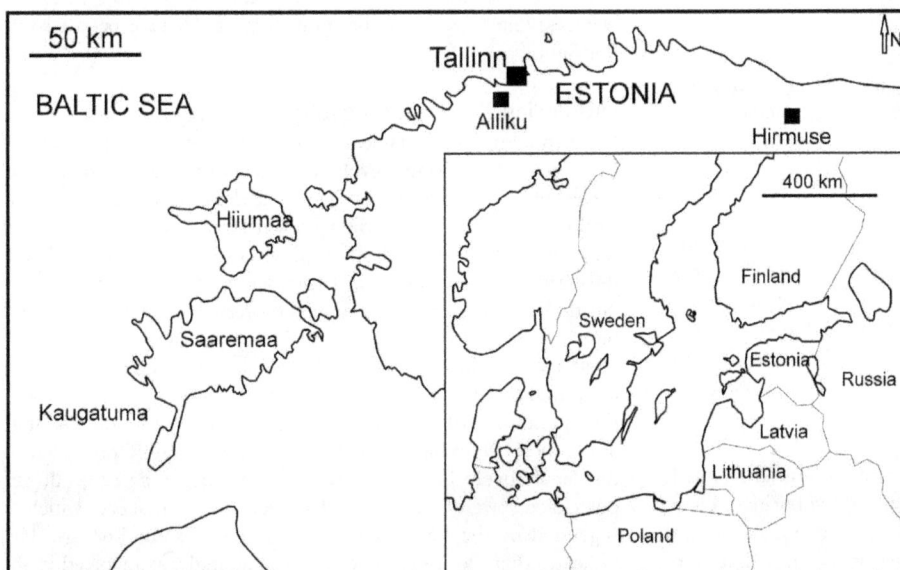

Figure 1. Locality map. Location of Hirmuse Creek and Alliku Ditches in North Estonia.

Series	Stage	Regional Stage
Upper Ordovician	Hirnant.	Porkuni
Upper Ordovician	Katian	Pirgu
Upper Ordovician	Katian	Vormsi
Upper Ordovician	Katian	Nabala
Upper Ordovician	Katian	Rakvere
Upper Ordovician	Katian	Oandu
Upper Ordovician	Sandbian	Keila
Upper Ordovician	Sandbian	Haljala
Upper Ordovician	Sandbian	Kukruse
Middle Ordovician	Darriwilian	Uhaku
Middle Ordovician	Darriwilian	Lasnamäe
Middle Ordovician	Darriwilian	Aseri
Middle Ordovician	Darriwilian	Kunda

Osprioneides kampto (marked at Haljala)

Figure 2. Stratigraphic scheme. The Middle and Upper Ordovician in Estonia. Location of *Osprioneides kampto* borings. Modified after Hints et al. (2008).

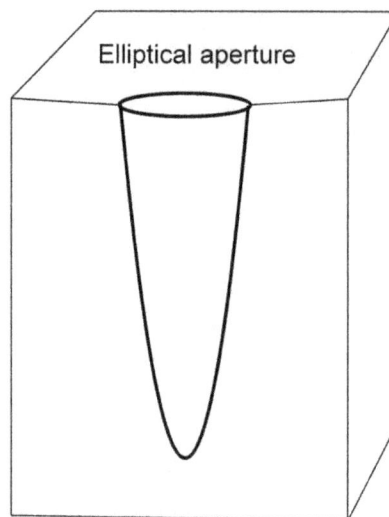

Figure 3. *Osprioneides kampto.* Schematic line drawing showing a straight boring.

Elliptical aperture

Most likely the *Osprioneides* trace maker was a soft-bodied animal similar to polychaete worms that used chemical means of boring as suggested by Beuck et al. [13]. This is supported by the slightly curved to sinuous course of several borings and their variable length. The presence of a tapered terminus in *Osprioneides* means bivalves were very unlikely to have been the trace makers.

Paleoecology and Taphonomy

Osprioneides borings were made post mortem because the growth lamellae of the bryozoan do not deflect around the borings. There are also no signs of skeletal repair by the bryozoans. Several *Osprioneides* borings truncate other *Osprioneides* borings that were likely abandoned by the trace maker by that time. Similarly, empty *Osprioneides* borings were colonized by *Trypanites* trace makers. This indicates that the *Osprioneides* borings may have appeared relatively

early in the ecological succession. Overturning of the bryozoan zoaria can explain the occurrence of *Osprioneides* borings apertures on both upper and lower surfaces. There is no sign of encrustation on the walls of the studied *Osprioneides* borings, suggesting relatively rapid burial of the host bryozoans shortly after the *Osprioneides* colonization.

It is likely that *Osprioneides* trace makers were suspension feeders similar to the *Trypanites* animals due to their stationary life mode [25]. Bryozoan skeletons may have offered them protection against predators and a higher tier for suspension feeding. Previously known host substrates of *Osprioneides* comprise stromatoporoids and tabulate corals. This new occurrence of *Osprioneides* borings in large bryozoans shows that the trace maker possibly selected its substrate only by size of skeleton because the traces are not found in smaller fossils. However, they are not found in any Ordovician hard-grounds that provide more area than do the bryozoan colonies. Wyse Jackson and Key [20] suggest that large bryozoan colonies were exploited by borers because they would have been easy to bore into.

Ordovician Bioerosion Revolution

Morphological diversification was not the only result of the Ordovician Bioerosion Revolution. Most of the large bioerosional traces of the Paleozoic had their earliest appearances in the Ordovician [7,8]. The earliest known large borings are those of *Gastrochaenolites* from the Early Ordovician of Baltica [4,17]. Later, during the Late Ordovician, large *Petroxestes* borings appeared in North America. At the same time the *Osprioneides* borings described here appeared in Baltica. Thus the Ordovician was also the time of great increase in quantities of hard substrate removed by single trace makers. The biological affinities of Ordovician *Gastrochaeno-lites* are not known [8], but it may have been a soft-bodied animal. The Late Ordovician *Petroxestes* was almost certainly produced by the facultatively boring bivalve *Corallidomus scobina* [26]. Boring polychaetes were the likely *Osprioneides* trace makers, which is suggested by the somewhat sinuous shape of some borings. This indicates that more than one group of animal was involved in the appearance of large bioerosional traces during the Ordovician Bioerosion Revolution. Increased predation pressure [27] was most likely the driving force behind the infaunalization of larger

Figure 4. *Osprioneides kampto* **borings (Os).** A bryozoan from Hirmuse Creek, Sandbian, Upper Ordovician, Estonia. Tr – *Trypanites* borings. GIT 398–729.

invertebrates such as the *Osprioneides* trace makers in the Ordovician. On the other hand in echinoids, for example, infaunalization was presumably the result of colonization of unoccupied niche space [28].

Paleobiogeography

Osprioneides is a relatively rare fossil compared to the abundance of *Trypanites* in the Silurian of Baltica [29]. In the Silurian, *Osprioneides* borings also occur outside of Baltica. They are known from the Llandovery of North America and Ludlow of the Welsh Borderlands [30]. *Osprioneides* is presumably absent in the Ordovician of North America because Ordovician bioerosional trace fossils of North America are relatively well studied [15,16]. Thus, it is possible that the *Osprioneides* trace maker originated in Baltica or elsewhere and migrated to North America in the Silurian. This may well be connected to the decreased distance between Baltica and Laurentia (the closing of the Iapetus Ocean) and the loss of provinciality of faunas in the Silurian.

Figure 5. *Osprioneides kampto* **borings (Os).** A bryozoan from Hirmuse Creek, Sandbian, Upper Ordovician, Estonia. Tr – *Trypanites* borings. GIT 665-18.

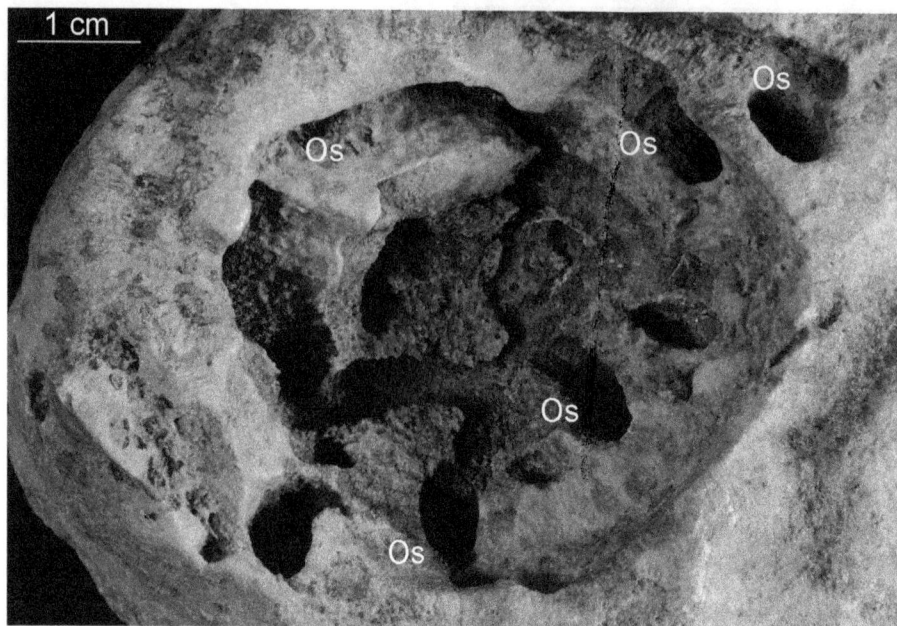

Figure 6. *Osprioneides kampto* **borings (Os).** A bryozoan from Hirmuse Creek, Sandbian, Upper Ordovician, Estonia. Tr – *Trypanites* borings. GIT 665-19.

Acknowledgments

We are grateful to Ursula Toom for finding the specimens among the old collections of the Institute of Geology, Tallinn University of Technology. Ursula Toom, Prof. Dimitri Kaljo, Dr. Linda Hints, Dr. Helje Pärnaste from the Institute of Geology, Tallinn University of Technology and Dr. Mare Isakar from the Geological Museum of the University of Tartu Natural History Museum are thanked for identifications of the associated fossils. We are grateful to G. Baranov from Institute of Geology at TUT for technical help with images. We are grateful to Dr. Harry Mutvei and an anonymous reviewer for the constructive reviews.

Author Contributions

Conceived and designed the experiments: OV MAW MAM. Performed the experiments: OV MAW MAM. Analyzed the data: OV MAW MAM. Contributed reagents/materials/analysis tools: OV MAW MAM. Contributed to the writing of the manuscript: OV MAW MAM.

References

1. James NP, Kobluk DR, Pemberton SG (1977) The oldest macroborers: Lower Cambrian of Labrador. Science 197: 980–983.

2. Kobluk DR, James NP, Pemberton SG (1978) Initial diversification of macroboring ichnofossils and exploitation of the macroboring niche in the lower Paleozoic. Paleobiology 4: 163–170.

3. Palmer TJ, Plewes CR (1993) Borings and bioerosion in the fossil record: Geology Today 9: 138–142.

4. Ekdale AA, Bromley RG (2001) Bioerosional innovation for living in carbonate hardgrounds in the Early Ordovician of Sweden. Lethaia 34: 1–12.

5. Dronov AV, Mikuláš R, Logvinova M (2002) Trace fossils and ichnofabrics across the Volkhov depositional sequence (Ordovician, Arenigian of St. Petersburg Region, Russia). J Czech Geol Soc 47: 133–146.

6. Taylor PD, Wilson MA (2003) Palaeoecology and evolution of marine hard substrate communities. Earth Sci Rev 62: 1–103.

7. Wilson MA, Palmer TJ (2006) Patterns and processes in the Ordovician Bioerosion Revolution. Ichnos 13: 109–112.

8. Wilson MA (2007) Macroborings and the evolution of bioerosion. In Miller III, W (ed.). Trace fossils: concepts, problems, prospects. Amsterdam: Elsevier. 356–367.

9. Bromley RG (2004) A stratigraphy of marine bioerosion, p.455–481. In McIlroy, D. (ed.) The application of ichnology to palaeoenvironmental and stratigraphical analysis. Geological Society of London Special Publications, 228p.

10. Zatoń M, Zhuravlev AV, Rakociński M, Filipiak P, Borszcz T, et al. (2014) Microconchid-dominated cobbles from the Upper Devonian of Russia: Opportunism and dominance in a restricted environment following the Frasnian-Famennian biotic crisis. Palaeogeogr Palaeoclimat Palaeoecol 401: 142–153.

11. Zatoń M, Machocka S, Wilson MA, Marynowski L, Taylor PD (2011) Origin and paleoecology of Middle Jurassic hiatus concretions from Poland. Facies 57: 275–300.

12. Zatoń M, Wilson MA, Zavar E (2011) Diverse sclerozoan assemblages encrusting large bivalve shells from the Callovian (Middle Jurassic) of southern Poland. Palaeogeogr Palaeoclimat Palaeoecol 307: 232–244.

13. Beuck L, Wisshak M, Munnecke A, Freiwald A (2008) A giant boring in a Silurian stromatoporoid analysed by computer tomography. Acta Palaeont Pol 53: 149–160.

14. Wilson MA, Palmer TJ (1988) Nomenclature of a bivalve boring from the Upper Ordovician of the Midwestern United States. J Paleont 62: 306–308.

15. Erickson JM, Bouchard TD (2003) Description and interpretation of Sanctum laurentiensis, new ichnogenus and ichnospecies, a domichnium mined into Late Ordovician (Cincinnatian) ramose bryozoan colonies. J Paleont 77: 1002–1010.

16. Tapanila L, Copper P (2002) Endolithic trace fossils in the Ordovician-Silurian corals and stromatoporoids, Anticosti Island, eastern Canada. Acta Geol Hisp 37: 15–20.

17. Vinn O, Wilson MA (2010a) Early large borings from a hardground of Floian-Dapingian age (Early and Middle Ordovician) in northeastern Estonia (Baltica). Carnets Géol CG2010_L04.

18. Vinn O (2004) The earliest known Trypanites borings in the shells of articulate brachiopods from the Arenig (Ordovician) of Baltica. Proc Est Acad Sci Geol 53: 257–266.

19. Vinn O (2005) The distribution of worm borings in brachiopod shells from the Caradoc Oil Shale of Estonia. Carnets Géol CG2005_A03.

20. Wyse Jackson PN, Key MM Jr (2007) Borings in trepostome bryozoans from the Ordovician of Estonia: two ichnogenera produced by a single maker, a case of host morphology control. Lethaia 40: 237–252.

21. Torsvik TH, Smethurst MA, van der Voo R, Trench A, Abrahamsen N, et al. (1992) Baltica. A synopsis of Vendian–Permian palaeomagnetic data and their palaeotectonic implications. Earth Sci Rev 33: 133–152.

22. Nestor H, Einasto R (1997) Ordovician and Silurian carbonate sedimentation basin. 192–204. In A. Raukas and A. Teedumäe (eds.), Geology and mineral resources of Estonia. Estonian Academy Publishers, Tallinn, 436 pp.

23. Mõtus MA, Hints O (2007) Excursion Guidebook. In 10th International Symposium on Fossil Cnidaria and Porifera. Excursion B2: Lower Paleozoic geology and corals of Estonia. August 18–22, 2007. Institute of Geology at Tallinn University of Technology, 66 p.

24. Rõõmusoks A (1970) Stratigraphy of the Viruan series (Middle Ordovician) in northern Estonia. University of Tartu, Tallinn, 346 pp.

25. Nield EW (1984) The boring of Silurian stromatoporoids – towards an understanding of larval behaviour in the Trypanites organism. Palaeogeogr Palaeoclimat Palaeoecol 48: 229–243.

26. Pojeta J, Jr, Palmer J (1976) The origin of rock boring in mytilacean pelecypods. Alcheringa 1: 167–179.

27. Huntley JW, Kowalewski M (2007) Strong coupling of predation intensity and diversity in the Phanerozoic fossil record. PNAS 38: 15006–15010.

28. Borszcz T, Zatoń M (2013) The oldest record of predation on echinoids: evidence from the Middle Jurassic of Poland. Lethaia 46: 141–145.

29. Vinn O, Wilson MA (2010) Occurrence of giant borings of Osprioneides kampto in the lower Silurian (Sheinwoodian) stromatoporoids of Saaremaa, Estonia. Ichnos 17: 166–171.

30. Newall G (1970) A symbiotic relationship between Lingula and the coral Heliolites in the Silurian. Geol J Spec Issue 3: 335–344.

12

Mean Annual Precipitation Explains Spatiotemporal Patterns of Cenozoic Mammal Beta Diversity and Latitudinal Diversity Gradients in North America

Danielle Fraser[1,2]*, **Christopher Hassall**[1,3], **Root Gorelick**[1,4,5], **Natalia Rybczynski**[1,2]

1 Department of Biology, Carleton University, Ottawa, Ontario, Canada, **2** Palaeobiology, Canadian Museum of Nature, Ottawa, Ontario, Canada, **3** School of Biology, University of Leeds, Leeds, United Kingdom, **4** Department of Mathematics and Statistics, Carleton University, Ottawa, Ontario, Canada, **5** Institute of Interdisciplinary Studies, Carleton University, Ottawa, Ontario Canada

Abstract

Spatial diversity patterns are thought to be driven by climate-mediated processes. However, temporal patterns of community composition remain poorly studied. We provide two complementary analyses of North American mammal diversity, using (i) a paleontological dataset (2077 localities with 2493 taxon occurrences) spanning 21 discrete subdivisions of the Cenozoic based on North American Land Mammal Ages (36 Ma – present), and (ii) climate space model predictions for 744 extant mammals under eight scenarios of future climate change. Spatial variation in fossil mammal community structure (β diversity) is highest at intermediate values of continental mean annual precipitation (MAP) estimated from paleosols (~450 mm/year) and declines under both wetter and drier conditions, reflecting diversity patterns of modern mammals. Latitudinal gradients in community change (latitudinal turnover gradients, aka LTGs) increase in strength through the Cenozoic, but also show a cyclical pattern that is significantly explained by MAP. In general, LTGs are weakest when continental MAP is highest, similar to modern tropical ecosystems in which latitudinal diversity gradients are weak or undetectable. Projections under modeled climate change show no substantial change in β diversity or LTG strength for North American mammals. Our results suggest that similar climate-mediated mechanisms might drive spatial and temporal patterns of community composition in both fossil and extant mammals. We also provide empirical evidence that the ecological processes on which climate space models are based are insufficient for accurately forecasting long-term mammalian response to anthropogenic climate change and inclusion of historical parameters may be essential.

Editor: Alistair Robert Evans, Monash University, Australia

Funding: D. Fraser was supported by a Natural Science and Engineering Research Council of Canada (NSERC) postgraduate scholarship, a Fulbright Traditional Student Award, a Mary Dawson Pre-Doctoral Fellowship grant, an Ontario Graduate Scholarship (OGS), and a Koningstein Scholarship for Excellence in Science and Engineering. C. Hassall was supported by an Ontario Ministry of Research and Innovation Postdoctoral Fellowship. R. Gorelick was supported by an NSERC Discovery Grant (#341399). N. Rybczynski was supported by an NSERC Discovery Grant (#312193). The funders had no role in study design, data collection and analysis, decision to publish, or preparation of the manuscript.

Competing Interests: The authors have declared that no competing interests exist.

* Email: danielle_fraser@carleton.ca

Introduction

Terrestrial species from all major taxonomic groups show dramatic changes in richness and diversity across the landscape [1]. One of the fundamental goals in ecology is therefore to ascertain why there are more species in some places than in others. A satisfactory answer would identify and disentangle the drivers of biodiversity at all spatial scales, from the microhabitat to the globe, as well as explain changes through time. Attempts to provide such an answer have produced many studies of species richness patterns and community composition in extant organisms [1–8]. Prime examples are the numerous studies of latitudinal richness gradients (LRGs), which have been observed in many terrestrial groups including angiosperms, birds, mammals, insects and other invertebrates. The best supported hypotheses show that richness declines toward the poles in correlation with reductions in precipitation, temperature, and net primary productivity [9]. Correlation of global climate with animal richness over the past

65 Ma, specifically a decline in richness as climates cooled, similarly supports a link between diversity and climate [10–12]. However, of the spatial and temporal dimensions of diversity, spatial patterns of community differences ("β diversity") are infrequently studied despite considerable variation on both local and regional scales [2,13,14] and their influential role in the structuring of continental-scale richness patterns including LRGs [3,4].

β diversity has been defined most broadly as the differentiation in community composition (i.e. the species that make up the community) among regions or along environmental gradients [15]. Similar to LRGs, β diversity generally declines from the tropics to the poles in correlation with climate [2]. However, temporal changes in β diversity remain poorly studied despite their potential power for illuminating the drivers of past and present richness patterns and importance in modern conservation [16–18]. This study therefore tests the hypothesis that climatic influences on mammalian β diversity apply equally to temporal patterns, i.e. that the underlying ecological processes are "ergodic" (dynamic processes that are the same in both time and space).

The mid to late Cenozoic (36 Ma to present) has been a time of dramatic mammalian diversity change, shaped in part by the transition from the productive ice-free ecosystems of the early to mid Cenozoic to the more temperate glaciated ecosystems of the late Cenozoic. Under these changing climatic conditions, mammalian communities show dramatic reductions in richness, changes in community composition, and morphology [10,19–24]. The most dramatic changes occurred at high latitudes, where ecosystems transitioned from *Metasequoia* forests during the early to mid Cenozoic [25,26] to boreal-type forests during the later Cenozoic and to modern tundra [27]. Associated with Cenozoic climate change, were changes in latitudinal climate gradients; overall, the intensity of latitudinal climate gradients increased toward the present, reflecting disproportionate polar cooling due to the formation of permanent Arctic glaciation [28,29]. We therefore predict that latitudinal diversity gradients increased in strength under cooler, less productive environmental conditions just as modern LRGs are steeper in temperate than in tropical regions. Further, we predict that β diversity declined under cooler, less productive environmental conditions just as modern β diversity declines toward the poles [2,7].

Quaternary (2.6 Ma to present) climates have been cool relative to the majority of the late Cenozoic. Recently, however, high latitudes have experienced disproportionate increases in annual temperature (up to 2°C to date), increases in plant primary productivity, and loss of large areas of perennial ice under anthropogenic global warming [30]. Flora and fauna have responded through shifts in phenology [31], *in situ* evolution [32], and, in some cases, extinction [33]. However, perhaps the most often recorded response is the climatically-correlated pattern of extirpations and colonization that manifest as shifts in the location of a species' geographic range. Distributional studies over ecological timescales (<100 yrs) have recorded dramatic poleward range shifts and expansions for a wide range of terrestrial taxa in response to northern warming [34,35]. Projections (i.e. Special Report on Emissions Scenarios) for the next 100 years predict levels of global warming similar to the middle Miocene (+6°C) − a time of reduced or absent perennial Arctic glaciation [36,37] − or warmer (+11°C for the most extreme case; Table S1). We therefore expect continued range expansion, extinction, evolution, and community level changes among North American animals and plants.

A common approach to predicting the long-term outcomes of climate change for terrestrial organisms is climate space modeling (CSM). CSMs use distributional information and climate data to project species ranges into the future, usually under the assumption of no evolution and without adjustment for dispersal differences among species [38–40]. Rapid evolutionary changes on very short timescales and high degrees of variation in dispersal ability under climate change have been observed across a wide range of organisms [34,39,41], therefore CSMs are unlikely to generate accurate forecasts of climate change response. The fossil record, which encompasses many disparate environments and climates, might serve as record of a natural experiment by which ecological hypotheses can be tested in the temporal dimension. Fossil collections are a rich historical record of response to various climatic events that can be incorporated into predictive models, and mammals, in particular, are an excellent group for testing the generality of ecological hypotheses because they have an extensive Cenozoic fossil record. However, studies of extinct organisms have focused largely on richness [12,22,23,42,43] or morphology [44],

with limited focus on community composition [20,22]. Because changes in biological communities are not always associated with changes in richness, spatiotemporal patterns of community composition may be better indicators of climate change response [13,18].

We propose that integrating the study of fossil, modern, and projected spatiotemporal patterns of community composition i) allows for the testing of ecological principles in the temporal dimension, ii) provides the most complete picture of diversity responses to climate change, and iii) enables evaluation of the performance of commonly employed CSMs. Our approach of combining the study of fossil, modern, and projected diversity patterns provides novel insights into the ecological and evolutionary processes that drive continental patterns of biodiversity in space and time.

Methods

Data collection and preparation

We downloaded occurrences for modern North American mammals from NatureServe Canada. The extant mammal dataset included 744 species after the exclusion of a small number of unreadable or corrupted files [45]. We restricted our study of fossil mammals to the late Eocene through Pleistocene, thus avoiding the confounding effects of the early Paleogene mammal radiation. We partitioned the fossil mammal occurrence data by North American Land Mammal Age (NALMA) subdivisions because they delineate relatively temporally stable community assemblages and allowed us to obtain a nearly continuous sequence of mammal community change without large intervening gaps. Using NALMA subdivisions leads to time averaging of mammal communities and to differences in sampling (i.e. intensity, geographic coverage etc.) among time periods. However, we use a statistical approach to reduce these biases, described below. We based the dates for all NALMA subdivisions on Woodburne (2004). Further, we combined data for the entire Clarendonian and excluded for the Whitneyan, late Late Hemphillian, and early Chadronian due to poor sampling (Table 1).

We downloaded fossil mammal occurrence data for the Eocene, Oligocene, Pliocene, and Pleistocene from the the Paleobiology Database using the Fossilworks Gateway (fossilworks.org) in July and August, 2012, using the group name 'mammalia' and the following parameters: time intervals = Cenozoic, region = North America, paleoenvironment = terrestrial (primary contributor: John Alroy; literature sources summarized in Appendix S1). We downloaded Miocene mammal occurrence data from the Miocene Mammal Mapping Project in March 2011 [46] using the NALMA subdivision as our search criterion. For all analyses, with the exception of the Miocene, we used paleolatitudes and paleolongitudes. We chose to use MIOMAP for the Miocene data because it is the most complete Miocene dataset. However, MIOMAP does not provide paleo-coordinates. Fortunately, there are only small differences between modern and Miocene latitudes for the downloaded localities. We removed all taxa with equivocal species identifications (e.g. *Equus* sp.) unless they were the only occurrence for a genus. We assumed all occurrences of open nomenclature (e.g. *Equus* cf. *simplicidens*) were correct identifications.

We did not use latitudinal grids for fossil or extant mammals as in previous studies of latitudinal richness gradients [1,47] because our study is focused on community composition. We therefore do not need to clump localities by spatial proximity to employ rarefaction methods. In addition, the uneven spatial distribution of fossil localities makes the use of a grid method impractical. Instead,

Table 1. Summary of sampled North American Land Mammal Age (NALMA) subdivisions.

Epoch	NALMA subdivision	Age Range (Ma)	Midpoint Age (M)	Number of species	Number of fossil localities	Area (km^2)
Pleistocene	Rancholabrean	0.25–0.011	0.1305	222	180	176615.9
Pliocene	Irvingtonian II	0.85–0.25	0.55	189	94	144745.5
Pliocene	Irvingtonian I	1.72–0.85	1.285	102	37	60361.4
Pliocene	Blancan V	2.5–1.72	2.11	165	130	125042.6
Pliocene	Blancan III	4.1–2.5	3.3	183	163	122839.5
Pliocene	Blancan I	4.9–4.1	4.5	85	66	140433.4
Miocene	Early late Hemphillian	6.7–5.9	6.3	68	46	20108.2
Miocene	Late early Hemphillian	7.5–6.7	7.1	63	55	29446.7
Miocene	Early early Hemphillian	9–7.5	8.25	65	47	31455.8
Miocene	Clarendonian	12.5–9	10.75	104	90	36139.8
Miocene	Late Barstovian	14.8–12.5	13.6	195	194	33789.1
Miocene	Early Barstovian	15.9–14.8	15.5	150	168	51753.3
Miocene	Late Hemingfordian	17.5–15.9	16.7	100	83	25478.4
Miocene	Early Hemingfordian	18.8–17.5	18.15	107	105	45531.3
Miocene	Late late Arikareean	19.5–18.8	19.15	108	123	38307.2
Oligocene/Miocene	Early late Arikareean	23.8–19.5	21.65	71	67	37892.2
Oligocene	Late early Arikareean	27.9–23.8	25.85	95	65	20927.8
Oligocene	Early early Arikareean	30–27.9	28.95	116	124	15382.3
Oligocene	Late Orellan	33.1–32	32.55	38	36	17725.7
Oligocene	Early Orellan	33.7–33.1	33.4	88	130	5579.8
Eocene	Middle Chadronian	35.7–34.7	35.3	88	37	10349.7

we created taxon-by-locality occurrence matrices for extant and fossil mammals at the species taxonomic level excluding *Homo sapiens* [20,22]. In all cases, taxa and localities with fewer than two occurrences were removed from the dataset. Final numbers of localities and species are summarized in Table 1.

To make direct comparisons with modern mammals, we created occurrence matrices for extant mammals by pseudo fossil localities, which were generated using an iterative procedure in R with the maptools, sp, gpclib, ggplot2, rgeos, and MASS packages [48–54] (contact corresponding author for R code). To generate pseudo fossil localities and to ensure that we created pseudo fossil localities with the same spatial distributions as the fossil localities, we fit frequency distributions (normal, gamma, or β) to fossil localities for each NALMA subdivision (Fig. S1). We then generated point samples based on the frequency distributions and the number of fossil localities from which we created occurrence matrices (taxon-by-pseudo locality), repeating the procedure 100 times for each NALMA sub-age for a total of 2100 occurrence matrices. Fossil localities do not record the entire community and so show reduced richness compared to the actual communities (however, note that time averaging also increases richness at fossil localities). Further, most fossil localities, unless intensively screen washed, are biased against small species. Therefore, we also intentionally tested for the effects of sampling bias by removing 25%, 50%, and 75% of species from the extant mammal occurrence matrices for a total of 6300 occurrence matrices. Further, we tested for the effects of body mass bias by 25%, 50%, and 75% of species smaller than 5 kg for a total of 6300 occurrence matrices.

Climate space models

To create climate space models, we sampled the ranges of extant North and South American mammals at a series of 5066 points corresponding to a 1° grid (which we only used to project mammal occurrences under climate change models, but not to calculate biodiversity). Due to the focus on North America, we omitted any species with southern hemisphere ranges that did not cross the equator (n = 602; Table S2). We also excluded rare species (present in <20 cells) for which accurate species distribution models could not be generated (n = 361), leaving 706 species for the climate change projections. We extracted mean annual and winter (December, January, February) temperature and mean annual precipitation data from Climate Wizard (www.climatewizard.org) for the period of 1951–2006 and the following SRES scenarios and time periods: B1 2050s, A1b 2050s, A1b 2080s, A2 2050s, and A2 2080s [55] (Table S1). Each of these projections is based on an ensemble of 16 global circulation models [56]. However, to ensure that we sampled a range of potential warming, we also extracted the ensemble lowest B1 2050s projection (hereafter "B1 2050s low") and the ensemble highest A2 2080s projection (hereafter "A2 2080s high"). This gave a range of warming in North America from 1.49°C (B1 2050s low) to 6.78°C (A2 2080s high, see Table S1 for the full range).

We modeled species' ranges with the BIOMOD package in R using generalized linear models, generalized boosted models, classification tree analysis, artificial neural networks, surface range envelopes, flexible discriminant analysis, multiple adaptive regression splines, and random forests [57] (contact corresponding author for R code). We then used these models to make consensus forecasts for each of the projections described above, as well as current climate to evaluate the performance of the models. We tested model performance using area under the receiver operating curve (AUC), true skill statistic (TSS), and proportion correct classification (PCC, Fig. S2). Species and generic presences were determined across the $1°$ latitude-longitude grid to give presence or absence in each location at each time and SRES scenario.

Using the projections described above, we created pseudo localities, as before. From this, we created occurrence matrices as described above. We repeated this process 100 times for each projection for a total of 16,800 occurrence matrices.

Latitudinal turnover gradients (LTGs) and β diversity

We calculated β diversity as the change in mammalian communities across the North American landscape using multivariate dispersion and the Jaccard index for each NALMA sub-age, for modern mammals, and for the climate projections [58]. We calculated Euclidean distances from the centroid for localities using the R package vegan [59]. Larger distances from the centroid indicate greater spatial community turnover and thus higher β diversity. We did not regress the Jaccard index values against distance, as has been used for modern species [2] because we have found such an approach to be highly influenced by species-area relationships.

To estimate ancient, modern, and projected LTG strength for North American mammals, we calculated the amount of community change with latitude using detrended correspondence analysis (DCA; an ordination technique) in the vegan R package [59]. We used explained variance (R^2; how much of the variation in community change is explained by latitude) as a measure of LTG strength [13]. High values of explained variance indicate strong LTGs [60]. We did not compute latitudinal richness gradients because sampling bias (e.g. loss of taxa, body mass bias) is too great (Fraser, D. unpub.).

Sampling bias control

Although we have chosen methods that minimize the effects of sampling bias, we still used multiple methods to control for the non-independence of β diversity from the number of localities, the geographic area sampled, and the number of sampled taxa. We used three approaches. Firstly, we used a re-sampling approach wherein we sub-sampled (without replacement) each NALMA $100×$ using a standardized number of localities (thirty) and limited to localities occurring between $30°$ and $50°$ North latitude. We also re-sampled the extant mammal ranges under various conditions of bias (taxonomic bias through the removal of 25%, 50%, 75% of taxa and body mass bias where we removed 25%, 50%, and 75% of species with a body mass lower than 5 kg) as above to test for direct causality of sampling bias. We also used a method of detrending whereby we regressed LTG strength and β diversity against statistically significant sampling bias metrics and further analyzed the residuals from the model. Finally, we used multivariate linear models to simultaneously account for the model variance explained by sampling and biological phenomena. The last multivariate method is similar to [61] and [62] (also addressed in [63]) who combine the predictive properties of models of biodiversity change and taphonomic bias.

Correlation with climate

We tested for correlations of β diversity and LTG strength with stable oxygen isotopes from benthic foraminifera ($\delta^{18}O$ ‰) [64,65], mean annual precipitation estimated from paleosols [66], number of localities, sampling area (km^2), number of species, latitudinal range (degrees), and length of the sampled interval (Ma) of the fossil localities using generalized least squares and using an autocorrelation structure of order one (corAR1) to account for

temporal autocorrelation in R [67,68]. Best fit models were selected using automated model selection in the MuMIn R package [69] and the Akaike Information Criterion (ΔAIC).

Results

Fossil mammal β diversity showed considerable variation with the warmest intervals (late Eocene, mid-late Oligocene, mid Miocene, and mid Pliocene), but showing generally higher β diversity than with cooler intervals (early Oligocene, late Miocene) (Fig. 1C). The best fit model includes mean annual precipitation (MAP squared), length of the NALMA subdivision, and number of taxa, which together accounts for 67% of model variance (Table 2). β diversity is statistically significant for all three predictors ($p<0.05$). Residual β diversity is significantly explained by MAP only (Table 2; Fig. 2B). Re-sampling did not alleviate the effects of sampling bias; re-sampled β diversity is significantly explained by MAP-squared, number of taxa, and NALMA subdivision length (Table 2). The remainder of the manuscript will discuss the results from the analyses of raw and residual β diversity only.

Mammalian latitudinal turnover gradients (LTGs) are weak prior to the late Miocene (Fig. 1D). Raw LTG strength (i.e. not detrended) peaks during late Miocene (Hemphillian) and late Pleistocene (Rancholabrean) (Fig. 1D). The best fit model includes mean annual precipitation (MAP) [66], number of taxa, area (km^2) and an the interaction of area and the number of taxa, which explains 47% of the model variance (Table 2; Fig. 2C). LTG strength of late Cenozoic mammal species is statistically significantly explained by all four metrics ($p<0.001$; Table 2). Residual LTG strength is significantly explained only by MAP ($p<0.05$; Table 2; Fig. 2D). As above, re-sampling did not alleviate the effects of sampling bias on LTG strength (Table 2). In other words, even accounting for variables that describe potential sources of bias, a climatic variable (MAP) still explains a significant proportion of the variance.

β diversity is much lower for extant mammals than for extinct mammals (Fig. 3A). LTG strength for extant mammals is also greater than for early to mid Cenozoic fossil mammals, but similar to the values for the late Miocene and Pleistocene (Fig. 3B). Extant mammal β diversity shows a slight decrease under incomplete sampling and a slight increase under body-mass–bias sampling (Fig. 3A), but the change is much smaller than observed for fossil mammals. LTG strength does not appear to be significantly affected by the sample size reduction.

Our forecast models (which showed a strong fit to modern mammalian distributions, see Fig. S2A–C) show a slight increase in β diversity for extant mammals (Fig. 3C), but no substantial change in LTG strength compared to the present (Fig. 3D).

Discussion

Spatiotemporal patterns of β diversity remain poorly studied despite being potentially very useful in conservation biology [17,18,70] and linkage to well-studied biogeographic phenomena such as latitudinal richness gradients [4]. Using an extensive analysis of past and present mammalian communities, we demonstrate that, over the past 36 Ma, spatiotemporal patterns of mammal community composition have varied by orders of magnitude in North America. Specifically, Cenozoic spatial turnover of mammal communities is explained by continental mean annual precipitation (MAP) (Fig. 2A–B), broadly supporting predictions drawn from published studies of modern terrestrial organisms [2,70,71] and our predictions outlined above.

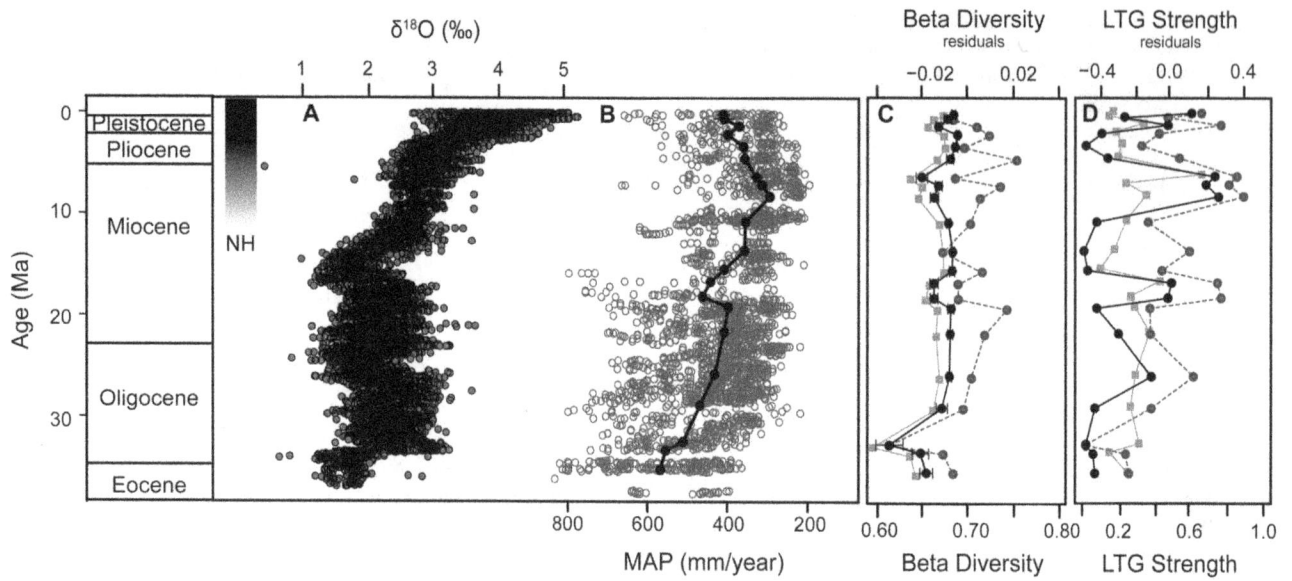

Figure 1. Mid to late Cenozoic trends of (A) $\delta^{18}O$ (‰) from benthic foraminifera (Zachos et al. 2008), (B) mean annual precipitation estimated from paleosols (Retallack, 2007), (C) β diversity of North American mammal species measured using multivariate dispersion (average distance from the centroid), and (D) strength of latitudinal turnover gradients (LTGs) measured as gradient strength for North American fossil mammals. Black lines are raw values, gray lines are residuals from significant sampling bias predictors, and gray dashed lines are re-sampled. Standard errors for re-sampled data are too small to display.

Contemporary ecological theory predicts that mammal diversity either declines monotonically with productivity or shows a unimodal pattern, declining with both low and high productivity [1,2,70,72]. Further, stronger latitudinal diversity gradients are associated with cooler, less productive environments [71] and steeper latitudinal climate gradients [1,70]. Both sets of predictions assume that changes in climate, productivity, and seasonality influence rates of origination and extinction [72,73], niche breadths [74], as well as the carrying capacity of the ecosystem

[75], all factors that change the spatial turnover of terrestrial faunas [70]. Specifically, terrestrial organisms in low latitude, high productivity environments show low rates of speciation and extinction [73], high β diversity [2,76], and weak or absent latitudinal diversity gradients [71]. In contrast, high latitude organisms show high rates of speciation and extinction [73], low β diversity [2,76], and strong latitudinal diversity gradients [71]. Evolutionary history also plays a role in determining rates of spatial community turnover. Modern tropical organisms show

Table 2. Results of best fit generalized least squares models relating β diversity and latitudinal turnover gradient (LTG) strength to mean annual precipitation from paleosols (Retallack, 2007), $\delta^{18}O$ (‰) from benthic forams (mm/year; Zachos et al. 2001; 2008), length of North American Land Mammal Age subdivision, number of taxa sampled, sampling area (km^2), and number of fossil localities.

Dependent Variable	Parameters of Best Fit Model	Variance explained by model (%)	t value	p
Beta Diversity	Mean annual precipitation (quadratic)	66.51	−3.25	0.005
	Length of NALMA subdivision		2.43	0.027
	Number of taxa		5.30	<0.001
Beta Diversity Residuals	Mean annual precipitation (quadratic)	26.48	−3.50	0.002
Beta Diversity Re-sampled	Mean annual precipitation (quadratic)	66.04	−2.39	0.029
	Length of NALMA subdivision		2.51	0.023
	Number of taxa		5.47	<0.001
Latitudinal Turnover Gradient Strength (LTGs)	Mean annual precipitation (quadratic)	46.76	−5.65	<0.001
	Area		−4.62	<0.001
	Number of taxa		−4.36	<0.001
	Area : Number of taxa		4.85	<0.001
LTG Residuals	Mean annual precipitation (linear)	37.48	−3.79	0.001
LTG Re-sampled	Number of taxa	28.59	−2.55	0.020

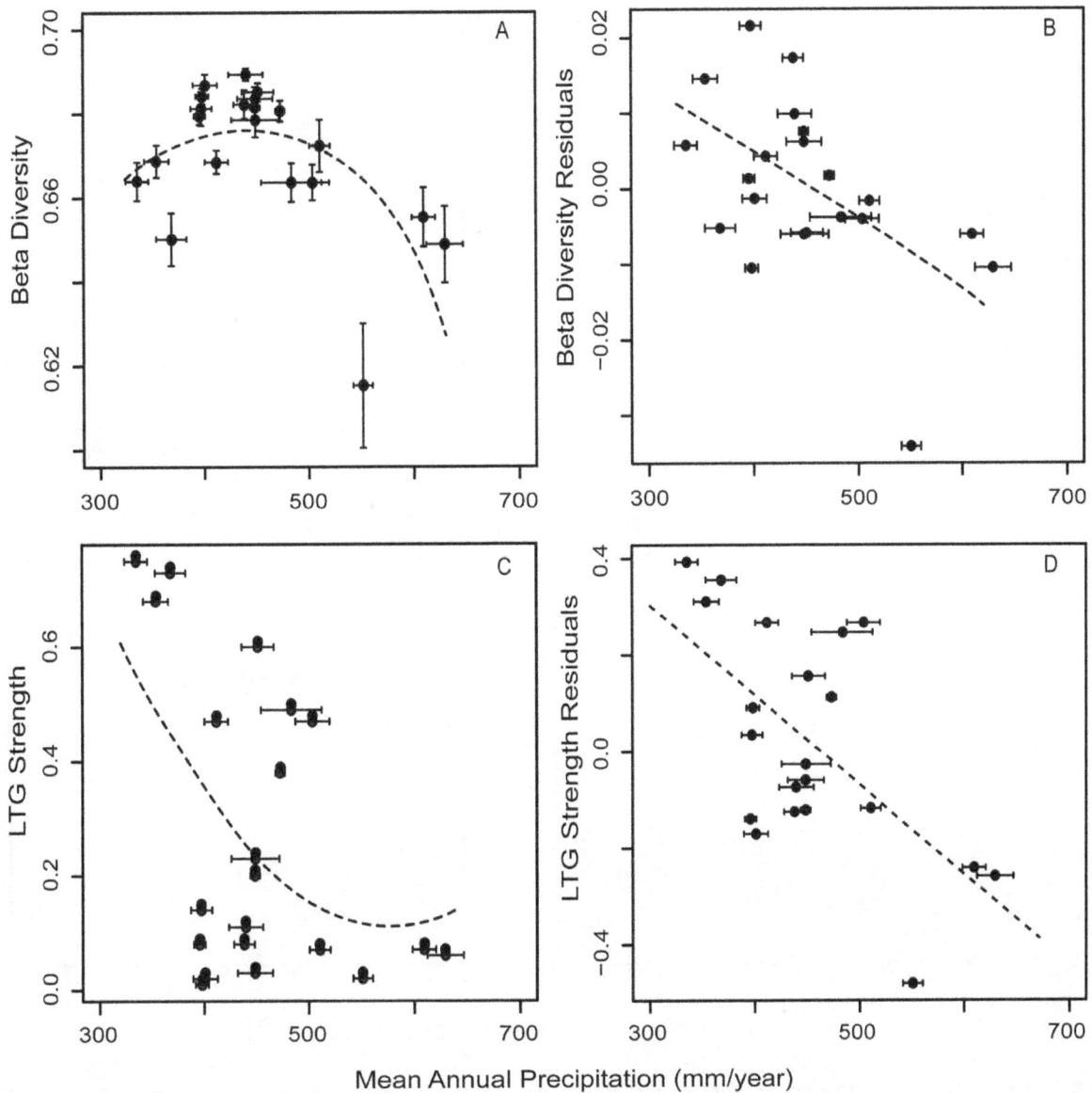

Figure 2. Relationship of mean annual precipitation estimated from paleosols (Retallack, 2007) with North American fossil mammal (A) raw β diversity ($R^2 = 0.43$), (B) residual beta diversity ($R^2 = 0.26$) and (C) raw latitudinal turnover gradient (LTG) strength ($R^2 = 0.25$), and (D) residual LTG strength ($R^2 = 0.37$).

faster turnover than their temperate counterparts regardless of the rate of environmental change [70]. Spatial and, by extension, temporal patterns of β diversity are the result of a mosaic of ecological and evolutionary processes.

Cenozoic fossil mammal β diversity peaked at intermediate values of mean annual precipitation and declined under both drier and wetter conditions (MAP; ~450 mm per year; Fig. 2B), showing a similar shape to latitudinal diversity curves for modern mammals [71]. Mammal β diversity was similarly lowest during periods of relative cooling, including the early Oligocene and late Miocene, coincident with declining atmospheric CO_2 [77–80] and, in the latter case, the expansion of ice sheets in the Northern Hemisphere [27,36], strengthening of thermohaline circulation [27,37,81–84], and transition from C_3 to C_4 dominated ecosystems at middle latitudes [66,85,86]. Declining β diversity during the late Miocene is also coincident with increased maximum body

mass [87], an ecologically relevant characteristic linked to lower ecosystem energy [88,89]. Water is a key component in photosynthesis and therefore net primary productivity (NPP) and MAP are correlated at a global scale, showing an asymptotic relationship [90]. Our results therefore suggest that putatively lower energy ecosystems (e.g. early Oligocene, late Miocene) supported more spatially homogenous mammal faunas than putatively higher energy ecosystems (e.g. late Eocene, mid Miocene, mid Pliocene). Temporal changes in fossil mammal β diversity (this study) are therefore conceptually similar to spatial patterns observed in extant mammals.

Early Oligocene mammals had lower β diversity than expected based on MAP (Fig. 1C; Fig. 2A). The early Oligocene is associated with rapid global cooling [64] and expansion of open grassy ecosystems [91], which may have resulted in lower ecosystem energy. However, our taxonomic sample is the poorest

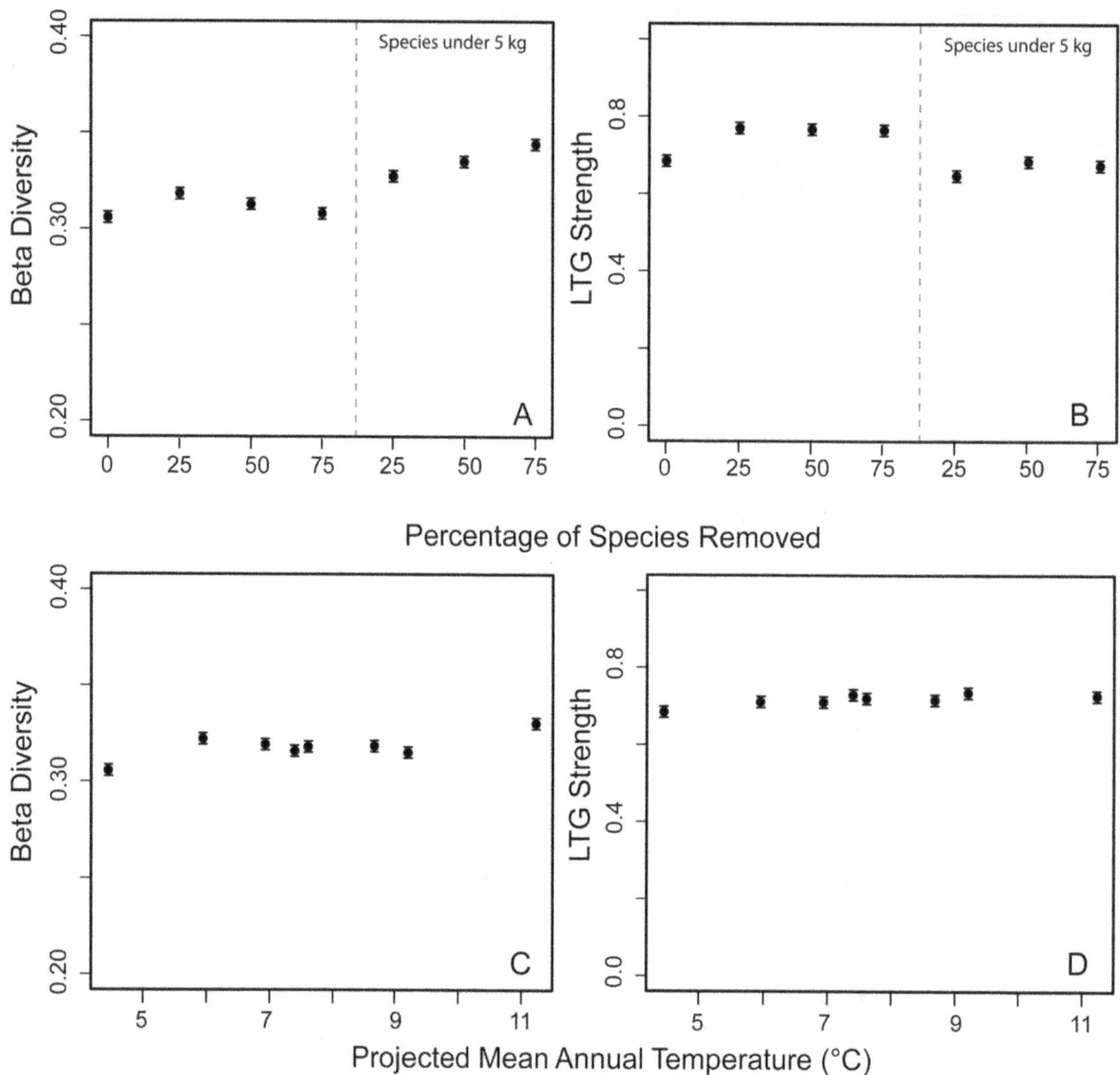

Figure 3. (A) β diversity (distance from centroid) and (B) latitudinal turnover gradients (LTG) strength of extant North American mammals under incomplete taxonomic sampling (removal of 25, 50, and 75% of species in sample) and body mass bias (removal of 25, 50, 75% of species smaller than 5 kg) and (C) β diversity (distance from centroid) and (D) latitudinal turnover gradients (LTG) strength of extant North American mammals under several International Panel on Climate Change scenarios (Special Reports on Emissions Scenarios).

for the early Oligocene; number of taxa is a significant predictor of fossil mammal β diversity (Table 2), suggesting some variation in preservation of species among NALMA subdivisions. Rarefied diversity also shows little change from the late Eocene to the early Oligocene [10]. However, our incomplete sampling trials show that removing even 75% of species reduces β diversity by a negligible amount (Fig. 3A), suggesting that at least some (but not all) of the observed decline in early Oligocene β diversity may have been climatically driven.

The magnitude of the latitudinal turnover gradient (LTG) for fossil mammals shows a temporally cyclic pattern that increases in amplitude during the late Cenozoic as well as a general trend toward stronger LTGs (Fig. 1D), coincident with the formation of ice on Svalbard at ~15 Ma and perennial Arctic sea ice at ~14 Ma, declining atmospheric CO_2 [37], and declining terrestrial MAP (Fig. 2B). Specifically, LTGs are strongest when precipitation is lowest (putatively lower productivity environments)

and weakest at when precipitation is highest (putatively high productivity environments; Fig. 2B), similar to modern mammals that show weak or absent latitudinal diversity gradients in the tropics and strong diversity gradients at mid to high latitudes [71]. Climate gradients are steeper at mid to high latitudes in North America due to the albedo of high latitude glaciation. Northern glaciation is an important means by which solar radiation is reflected from high latitudes, resulting in cool, low productivity Arctic environments [92,93]. Mammal communities are sorted along a latitudinal axis according to their climatic tolerances and the process of abiotic filtering, whereby taxa meet the limits of their environmental tolerances and are excluded from communities farther north [94]. Although late Miocene sea and land ice thickness and extent were reduced compared to the modern, increasing northern albedo and strengthening of thermohaline circulation are coincident with that strengthening of mammal

LTGs during the late Miocene (25–60% stronger than for any preceding NALMA; Fig. 1D) [27,81–84].

At first glance, the Pliocene appears to be anomalous because the magnitude of the mammalian LTG declines dramatically (60–70% reduction in the magnitude of the LTG; Fig. 1D). However, evidence from fossil deposits on Ellesmere Island show that approximately 3.5 Ma the Pliocene Arctic was ~14–22°C warmer than present [83,95,96] with an associated reduced volume of Arctic sea ice [27,82]. Pliocene Arctic warming is similarly coincident with reduced richness gradients of marine zooplankton [81]. The Pliocene might therefore be the "exception" that proves the rule.

Under modern global warming, Arctic winter temperatures have increased at a greater rate than at southern latitudes [97]. Long-term projections suggest boosts in high latitude net primary productivity due to increasing nitrogen fertilization and increases in mean annual precipitation of 100–150 mm per year or 5–20% at middle to high latitudes [98]. From our analyses of fossil North American mammals and published studies of beta diversity [18], we therefore expect weakened climate gradients and thus weakened LTGs due to northward range shifting, and, in the long-term, declining β diversity under the influence of modern anthropogenic climate change. β diversity decline may be facilitated by the homogenization of communities due to any of the following (note the lack of mutual exclusivity): i) extinction of species with small geographic ranges and replacement with wide-ranging species, ii) evolution toward larger range sizes within species, and, iii) invasion by wide-ranging species even without the extinction of residents [18]. However, our climate space models that are based on SRES scenarios corresponding to absolute mean annual temperatures of 4.4–11.2°C (averaged across North and South America) did not show changes in mammal LTGs or β diversity (Fig. 3C–D). We suggest that climate space models (CSMs) are unlikely to accurately forecast the outcomes of anthropogenic climate change for modern mammals because current CSM algorithms do not incorporate microevolutionary, macroevolutionary, or ecological processes, such as niche shifts, niche creation, and differences in dispersal abilities that are inherent in the response of animals to climate change. However, even on modern ecological timescales, rapid evolutionary changes and niche shifts have been observed in native and invasive populations [41], and this local adaptation complicates the prediction of range shifts. On longer timescales, taxa adapt to new climates and the processes of speciation and extinction help form new terrestrial communities. Without the explicit inclusion of evolutionary parameters and historical data for the taxa of interest, we are unlikely to accurately predict long-term changes in terrestrial biodiversity patterns.

We have shown here that macroecological patterns of North American mammal community composition varied considerably over the past 35 million years in response to changes in global climate change and Arctic glaciation (Fig. 1C–D). Furthermore, our comparison of fossil evidence with climate-space forecast models (CSMs) suggests that CSMs (in which species are modeled to simply track climate variables) may distort the degree of community composition change we should expect in the future. A unifying ecological theory relating diversity to climate must address both the spatial and temporal dimensions of diversity, as well as both richness and community composition. However, studies of organismal richness are far more common than studies of community composition (β diversity), despite the importance of the latter in conservation and their vast potential for contributing to our understanding of the processes underlying modern biodiversity. Studying the community composition of fossil animals represents a new frontier in paleontological research with potential to truly inform modern conservation.

Supporting Information

Figure S1　Maps of North America showing the distribution of fossil localities for all sampled North American Land Mammal Age subdivisions.

Figure S2　Model fit statistics for climate space models of extant North American mammals. Model performance was tested using area under the operating curve (A; AUC), the true skill statistics (B; TSS), and the proportion of correct classification (C).

Table S1　Summary of Special Emissions Report Scenarios (SERs) to which we fit climate models for extant mammalian species.

Table S2　List of mammalian taxa included and excluded from the species distribution models.

Appendix S1　Sources for the majority of mammal occurrence data downloaded from the Fossilworks database.

Acknowledgments

We thank John P. Hunter for a thorough review of this paper. Further, we thank John Alroy for his substantial contributions to the fossil data used in this analysis, accessed via his Fossilworks website, and his detailed review of the paper. We would also like to thank two anonymous reviewers, D. Currie, M. Clementz, M. Churchill, R. Haupt, J. Hoffmann, and E. Lightner for reviewing earlier versions of this manuscript, as well as L. Fahrig and S. Kim for constructive comments on this project.

Author Contributions

Conceived and designed the experiments: DF CH NR. Performed the experiments: DF. Analyzed the data: DF. Contributed reagents/materials/analysis tools: CH RG. Contributed to the writing of the manuscript: DF. Manuscript copyediting: CH RG NR.

References

1. Hawkins BA, Field R, Cornell HV, Currie DJ, Guegan JF, et al. (2003) Energy, water, and broad-scale geographic patterns of species richness. Ecology 84: 3105–3117.

2. Qian H, Badgley C, Fox DL (2009) The latitudinal gradient of beta diversity in relation to climate and topography for mammals in North America. Global Ecology and Biogeography 18: 111–122.

3. Condit R, Pitman N, Leigh EG Jr, Chave J, Terborgh J, et al. (2002) Beta-diversity in tropical forest trees. Science 295: 666–669.

4. Baselga A, Lobo JM, Svenning JC, Aragón P, Araújo MB (2012) Dispersal ability modulates the strength of the latitudinal richness gradient in European beetles. Global Ecology and Biogeography 21: 1106–1113.

5. Engle VD, Summers JK (1999) Latitudinal gradients in benthic community composition in Western Atlantic estuaries. Journal of Biogeography 26: 1007–1023.

6. Condamine FL, Sperling FAH, Wahlberg N, Rasplus JY, Kergoat GJ (2012) What causes latitudinal gradients in species diversity? Evolutionary processes and ecological constraints on swallowtail biodiversity. Ecology Letters 15: 267–277.

7. Currie DJ, Fritz JT (1993) Global patterns of animal abundance and species energy use. Oikos 67: 56–68.

8. Currie DJ, Francis AP, Kerr JT (1999) Some general propositions about the study of spatial patterns of species richness. Ecoscience 6: 392–399.

9. Mittelbach GG, Schemske DW, Cornell HV, Allen AP, Brown JM, et al. (2007) Evolution and the latitudinal diversity gradient: speciation, extinction and biogeography. Ecology Letters 10: 315–331.

10. Figueirido B, Janis CM, Pérez-Claros JA, Renzi MD, Palmqvist P (2012) Cenozoic climate change influences mammalian evolutionary dynamics. Proceedings of the National Academy of Sciences USA 109: 722–727.

11. Sepkoski JJ (1998) Rates of speciation in the fossil record. Philosophical Transactions of the Royal Society of London B 353: 315–326.

12. Mayhew PJ, Bell MA, Benton TG, McGowan AJ (2012) Biodiversity tracks temperature over time. Proceedings of the National Academy of Sciences USA 109: 15141–15145.

13. Kent R, Bar-Massada A, Carmel Y (2011) Multiscale analyses of mammal species composition-environment relationship in the contiguous USA. PLoS One 6: e25440.

14. Legendre P, Borcard D, Peres-Neto PR (2005) Analyzing beta diversity: partitioning the spatial variation of community composition data. Ecological Monographs 75: 435–450.

15. Whittaker RJ, Willis KJ, Field R (2001) Scale and species richness: towards a general, hierarchical theory of species diversity. Journal of Biogeography 28: 453–470.

16. Soininen J (2010) Species turnover along abiotic and biotic gradients: patterns in space equal patterns in time? BioScience 60: 433–439.

17. Hassall C, Hollinshead J, Hull A (2012) Temporal dynamics of aquatic communities and implications for pond conservation. Biodiversity and Conservation 21: 829–852.

18. Dornelas M, Gotelli NJ, McGill B, Shimadzu H, Moyes F, et al. (2014) Assemblage time series reveal biodiversity change but not systematic loss. Science 344: 296–299.

19. Janis CM, Damuth J, Theodor JM (2000) Miocene ungulates and terrestrial primary productivity: where have all the browsers gone? Proceedings of the National Academy of Sciences USA 97: 7899–7904.

20. Atwater AL, Davis EB (2011) Topographic and climate change differentially drive Pliocene and Pleistocene mammalian beta diversity of the Great Basin and Great Plains provinces of North America. Evolutionary Ecology Research 13: 833–850.

21. Finarelli JA, Badgley C (2010) Diversity dynamics of Miocene mammals in relation to the history of tectonism and climate. Proceedings of the Royal Society of London, Series B 277: 2721–2726.

22. Davis EB (2005) Mammalian beta diversity in the Great Basin, western USA: palaeontological data suggest deep origin of modern macroecological structure. Global Ecology and Biogeography 14: 479–490.

23. Barnosky AD, Hadly EA, Bell CJ (2003) Mammalian response to global warming on varied temporal scales. Journal of Mammalogy 84: 354–368.

24. Barnosky AD (2005) Effects of Quanternary climatic change on speciation in mammals. Journal of Mammalian Evolution 12: 247–264.

25. Eberle J, Fricke H, Humphrey J (2009) Lower-latitude mammals as year-round residents in Eocene Arctic forests. Geology 37: 499–502.

26. Eberle JJ, Fricke HC, Humphrey JD, Hackett L, Newbrey MG, et al. (2010) Seasonal variability in Arctic temperatures during early Eocene time. Earth and Planetary Science Letters 296: 481–486.

27. Polyak L, Alley RB, Andrews JT, Brigham-Grette J, Cronin TM, et al. (2010) History of sea ice in the Arctic. Quaternary Science Reviews 29: 1757–1778.

28. Clementz MT, Sewall JO (2011) Latitudinal gradients in greenhouse seawater $\delta 18O$: evidence from Eocene sirenian tooth enamel. Science 332: 455–458.

29. Micheels A, Bruch A, Mosbrugger V (2009) Miocene climate modelling sensitivity experiments for different CO_2 concentrations. Palaeontologia Electronica 12: 5A.

30. Post E, Forchhammer MC, Bret-Harte MS, Callaghan TV, Christensen TR, et al. (2009) Ecological dynamics across the Arctic associated with recent climate change. Science 325: 1355–1358.

31. Primack RB, Ibáñez I, Higuchi H, Lee SD, Miller-Rushing AJ, et al. (2009) Spatial and interspecific variability in phenological responses to warming temperatures. Biological Conservation 142: 2569–2577.

32. Bradshaw WE, Holzapfel CM (2006) Evolutionary response to rapid climate change. Science 312: 1477–1478.

33. Parmesan C (2006) Ecological and evolutionary responses to recent climate change. Annual Review of Ecology and Systematics 37: 637–639.

34. Chen IC, Hill JK, Ohlemüller R, Roy DB, Thomas CD (2011) Rapid range shifts of species associated with high levels of climate warming. Science 333: 1024–1026.

35. Parmesan C, Yohe G (2003) A globally coherent fingerprint of climate change impacts across natural systems. Nature 421: 37–42.

36. Foster GL, Lunt DJ, Parrish RR (2009) Mountain uplift and the threshold for sustained Northern Hemisphere glaciation. Climate of the past discussions 5: 2439–2464.

37. Foster GL, Lear CH, Rae JWB (2012) The evolution of pCO_2, ice volume and climate during the middle Miocene. Earth and Planetary Science Letters 341–344: 243–254.

38. Lawler JJ, White D, Neilson RP, Blaustein AR (2006) Predicting climate-induced range shifts: model differences and model reliability. Global Change Biology 12: 1568–1584.

39. Hoffmann AA, Sgró CM (2011) Climate change and evolutionary adaptation. Nature 470: 479–485.

40. Thuiller W, Münkemüller T, Lavergne S, Mouillot D, Mouquet N, et al. (2013) A road map for integrating eco-evolutionary processes into biodiversity models. Ecology Letters 16: 94–105.

41. Lavergne S, Mouquet N, Thuiller W, Ronce O (2010) Biodiversity and climate change: integrating evolutionary and ecological responses of species and communities. Annual Review of Ecology, Evolution, and Systematics 41: 321–350.

42. Sepkoski JJ (1997) Biodiversity: past, present, and future. Journal of Paleontology 71: 533–539.

43. Rose PJ, Fox DL, Marcot J, Badgley C (2011) Flat latitudinal gradient in Paleocene mammal richness suggests decoupling of climate and biodiversity. Geology 39: 163–166.

44. Secord R, Bloch JI, Chester SGB, Boyer DM, Wood AR, et al. (2012) Evolution of the earliest horses driven by climate change in the Paleocene-Eocene thermal maximum. Science 335: 959–962.

45. Patterson BD, Ceballos G, Sechrest W, Tognelli MF, Brooks T, et al. (2007) Digital distribution maps of the mammals of the Western Hemisphere, version 3.0. NatureServe, Arlington, Virginia, USA.

46. Carrasco MA, Kraatz BP, Davis EB, Barnosky AD (2005) Miocene mammal mapping project (MIOMAP). University of California Museum of Paleontology.

47. McCoy ED, Connor EF (1980) Latitudinal gradients in the species diversity of North American mammals. Evolution 34: 193–203.

48. Lewin-Koh NJ, Bivand R (2008) maptools package version 0.8–16.

49. Pebesma EJ, Bivand RS (2005) Classes and methods for spatial data in R. R News 5.

50. Bivand RS, Pebesma EJ, Gomez-Rubio V (2008) Applied spatial data analysis with R. New York: Springer.

51. Peng RD (2007) The gpclib package version 1.5–5.

52. Wickham H (2009) ggplot2: elegant graphics for data analysis. New York: Springer.

53. Bivand R, Rundel C (2012) rgeos: interface to geometry engine version 0.2–16.

54. Venables WN, Ripley BD (2002) Modern and applied statistics with S. New York: Springer.

55. Nakicenovic N, Swart R (2000) Emissions scenarios: a special report of Working Group III of the Intergovernmental Panel on Climate Change. Cambridge: Cambridge University Press.

56. Girvetz EH, Zganjar C, Raber GT, Maurer EP, Kareiva P, et al. (2009) Applied climate-change analysis: the climate wizard tool. PLoS One 4: e8320.

57. Thuiller W, Georges D, Engler R (2012) BIOMOD: Ensemble platform for species distribution modeling. Ecography 32: 369–373.

58. Anderson MJ, Ellingsen KE, McArdle BH (2006) Multivariate dispersion as a measure of beta diversity. Ecology Letters 9: 683–693.

59. Oksanen J, Blanchet FG, Roeland Kindt PL, Minchin PR, O'Hara RB, et al. (2012) Package vegan version 2.0–7.

60. Tuomisto H, Ruokolainen K (2006) Analyzing and explaining beta diversity? understanding the targets of different methods of analysis. Ecology 87: 2697–2708.

61. Benson RBJ, Mannion PD (2012) Multi-variate models are essential for understanding vertebrate diversification in deep time. Biology Letters 8: 127–130.

62. Mannion PD, Upchurch P, Carrano MT, Barrett PM (2011) Testing the effect of the rock record on diversity: a multidisciplinary approach to elucidating the generic richness of sauropodomorph dinosaurs through time. Biological Reviews 86: 157–181.

63. Benton MJ, Dunhill AM, Lolyd GT, Marx FG (2011) Assessing the quality of the fossil record: insights from vertebrates. In: A. J McGowan and A. B Smith, editors. Comparing the geological and fossil records: implications for biodiversity studies. London: Geological Society of London. 63–94.

64. Zachos JC, Dickens GR, Zeebe RE (2008) An early Cenozoic perspective on greenhouse warming and carbon-cycle dynamics. Nature 451: 279–283.

65. Zachos J, Pagani M, Sloan L, Thomas E, Billups K (2001) Trends, rhythms, and aberrations in global climate 65 Ma to present. Science 292: 686–693.

66. Retallack GJ (2007) Cenozoic paleoclimate on land in North America. Journal of Geology 115: 271–294.

67. Development core team R (2012) R: A language and environment for statistical computing. Vienna, Austria: Foundation for Statistical Computing.

68. Dornelas M, Magurran AE, Buckland ST, Chao A, Chazdon RL, et al. (2013) Quantifying temporal change in biodiversity: challenges and opportunities. Proceedings of the Royal Society B 280: 1–10.

69. Bartoń K (2013) Multi-model inference package 'MuMIn' version 1.10.0 (http://cran.r-project.org/web/packages/MuMIn/MuMIn.pdf).

70. Buckley LB, Jetz W (2008) Linking global turnover of species and environments. Proceedings of the National Academy of Sciences USA 105: 17836–17841.

71. Currie DJ (1991) Energy and large-scale patterns of animal- and plant-species richness. American Naturalist 137: 27–49.

72. VanderMeulen MA, Hudson AJ, Scheiner SM (2001) Three evolutionary hypotheses for the hump-shaped productivity–diversity curve. Evolutionary Ecology Research 3: 379–392.

73. Weir JT, Schluter D (2007) The latitudinal gradient in recent speciation and extinction rates of birds and mammals. Science 315: 1574–1576.

74. Vázquez DP, Stevens RD (2004) The latitudinal gradient in niche breadth: concepts and evidence. American Naturalist 164: E1–E19.

75. Buckley LB, Davies J, Ackerly DD, Kraft NJB, Harrison SP, et al. (2010) Phylogeny, niche conservatism and the latitudinal diversity gradient in mammals. Proceedings of the Royal Society, Series B 277: 2121–2138.

76. Qian H, Xiao M (2012) Global patterns of the beta diversity energy relationship in terrestrial vertebrates. Acta Oecologica 39: 67–71.

77. Franks PJ, Beerling DJ (2009) Maximum leaf conductance driven by CO_2 effects on stomatal size and density over geologic time. Proceedings of the National Academy of Sciences USA 106: 10343–10347.

78. DeConto RM, Pollard D, Wilson PA, Pälike H, Lear CH, et al. (2008) Thresholds for Cenozoic bipolar glaciation. Nature 455: 652–657.

79. Tripati AK, Roberts CD, Eagle RA (2009) Coupling of CO_2 and ice sheet stability over major climate transitions of the last 20 million years. Science 326: 1394–1397.

80. Zhang YG, Pagani M, Liu Z, Bohaty SM, DeConto R (2013) A 40-million-year history of atmospheric CO_2. Philosophical Transactions of the Royal Society, Series A 371: 1–20.

81. Yasuhara M, Hunt G, Dowsett HJ, Robinson MM, Stoll DK (2012) Latitudinal species diversity gradient of marine zooplankton for the last three million years. Ecology Letters 15: 1174–1179.

82. Haywood AM, Valdes PJ, Sellwood BW, Kaplan JO, Dowsett HJ (2001) Modelling middle Pliocene warm climates of the USA. Palaeontologia Electronica 4: 1–21.

83. Ballantyne AP, Greenwood DR, Damsté JSS, Csank AZ, Eberle JJ, et al. (2010) Significantly warmer Arctic surface temperatures during the Pliocene indicated by multiple independent proxies. Geology 38: 603–606.

84. Ballantyne AP, Rybczynski N, Baker PA, Harington CR, White D (2006) Pliocene Arctic temperature constraints from the growth rings and isotopic composition of fossil larch. Palaeogeography, Palaeoclimatology, Palaeoecology 242: 188–200.

85. Fox DL, Honey JG, Martin RA, Peláez-Campomanes P (2012) Pedogenic carbonate stable isotope record of environmental change during the Neogene in the southern Great Plains, southwest Kansas, USA: Oxygen isotopes and paleoclimate during the evolution of C_4-dominated grasslands. Geological Society of America Bulletin 124: 431–443.

86. Strömberg CAE, McInerney FA (2011) The Neogene transition from C_3 to C_4 grasslands in North America: assemblage analysis of fossil phytoliths. Paleobiology 37: 50–71.

87. Smith FA, Boyer AG, Brown JH, Costa DP, Dayan T, et al. (2010) The evolution of maximum body size of terrestrial mammals. Science 330: 1216–1219.

88. Freckleton RP, Harvey PH, Pagel M (2003) Bergmann's rule and body size in mammals. American Naturalist 161: 821–825.

89. Blackburn TM, Gaston KJ, Loder N (1999) Geographic gradients in body size: a clarification of Bergmann's rule. Diversity & Distributions 5: 165–174.

90. Del Grosso S, Parton W, Stohlgren T, Zheng D, Bachelet D, et al. (2008) Global potential net primary production predicted from vegetation class, precipitation, and temperature. Ecology 89: 2117–2126.

91. Jacobs BF, Kingston JD, Jacobs LL (1999) The origin of grass-dominated ecosystems. Annals of Missouri Botanical Garden 86: 590–643.

92. Alexeev VA, Langen PL, Bates JR (2005) Polar amplification of surface warming on an aquaplanet in "ghost forcing" experiments without sea ice feedbacks. Climate Dynamics 24: 655–665.

93. Holland MM, Bitz CM (2003) Polar amplification of climate change in coupled models. Climate Dynamics 21: 221–232.

94. Soininen J, McDonald R, Hillebrand H (2007) The distance decay of similarity in ecological communities. Ecography 30: 3–12.

95. Csank AZ, Tripati AK, Patterson WP, Eagle RA, Rybczynski N, et al. (2011) Estimates of Arctic land surface temperatures during the early Pliocene from two novel proxies. Earth and Planetary Science Letters 304: 291–299.

96. Rybczynski N, Gosse JC, Harington CR, Wogelius RA, Hidy AJ, et al. (2013) Mid-Pliocene warm-period deposits in the High Arctic yield insight into camel evolution. Nature Communications 4: 1–9.

97. Kaplan JO, Bigelow NH, Prentice IC, Harrison SP, Bartlein PJ, et al. (2003) Climate change and Arctic ecosystems: 2. Modeling, paleodata-model comparisons, and future projections. Journal of Geophysical Research 108: 1–17.

98. Oechel WC, Vourlitis GL (1994) The effects of climate change on land-atmosphere feedbacks in arctic tundra regions. Trends in Ecology and Evolution 9: 324–329.

13

Opportunistic Feeding Strategy for the Earliest Old World Hypsodont Equids: Evidence from Stable Isotope and Dental Wear Proxies

Thomas Tütken[1]*, Thomas M. Kaiser[2], Torsten Vennemann[3], Gildas Merceron[4]

1 Steinmann-Institut für Geologie, Mineralogie und Paläontologie, University of Bonn, Bonn, Germany, 2 University Hamburg, Biozentrum Grindel, Hamburg, Germany, 3 Institut de Géochimie, Université de Lausanne, Lausanne, Switzerland, 4 iPHEP UMR 7262 CNRS, Université de Poitiers, Poitiers, France

Abstract

Background: The equid *Hippotherium primigenium*, with moderately hypsodont cheek teeth, rapidly dispersed through Eurasia in the early late Miocene. This dispersal of hipparions into the Old World represents a major faunal event during the Neogene. The reasons for this fast dispersal of *H. primigenium* within Europe are still unclear. Based on its hypsodonty, a high specialization in grazing is assumed although the feeding ecology of the earliest European hipparionines within a pure C$_3$ plant ecosystem remains to be investigated.

Methodology/Principal Findings: A multi-proxy approach, combining carbon and oxygen isotopes from enamel as well as dental meso- and microwear analyses of cheek teeth, was used to characterize the diet of the earliest European *H. primigenium* populations from four early Late Miocene localities in Germany (Eppelsheim, Höwenegg), Switzerland (Charmoille), and France (Soblay). Enamel δ^{13}C values indicate a pure C$_3$ plant diet with small (<1.4‰) seasonal variations for all four *H. primigenium* populations. Dental wear and carbon isotope compositions are compatible with dietary differences. Except for the Höwenegg hipparionines, dental microwear data indicate a browse-dominated diet. By contrast, the tooth mesowear patterns of all populations range from low to high abrasion suggesting a wide spectrum of food resources.

Conclusions/Significance: Combined dental wear and stable isotope analysis enables refined palaeodietary reconstructions in C$_3$ ecosystems. Different *H. primigenium* populations in Europe had a large spectrum of feeding habits with a high browsing component. The combination of specialized phenotypes such as hypsodont cheek teeth with a wide spectrum of diet illustrates a new example of the Liem's paradox. This dietary flexibility associated with the capability to exploit abrasive food such as grasses probably contributed to the rapid dispersal of hipparionines from North America into Eurasia and the fast replacement of the brachydont equid *Anchitherium* by the hypsodont *H. primigenium* in Europe.

Editor: Peter Stuart Ungar, University of Arkansas, United States of America

Funding: This study was partly financed by the Swiss National Science Foundation (SNF) grant 200021-100530/1 to TV and the Deutsche Forschungsgemeinschaft Emmy Noether-Program TU 148/2 to TT and KA-1525 4-1/4-2 and KA-1525 9-1, KA 1525 8-1 to TMK and is publication no. 44 of the DFG Research Unit 771 "Function and enhanced efficiency in the mammalian dentition–phylogenetic and ontogenetic impact on the masticatory apparatus". The study was further funded by a research grant from the Fyssen Foundation (Paris, France) to GM. The funders had no role in study design, data collection and analysis, decision to publish, or preparation of the manuscript.

Competing Interests: The authors have declared that no competing interests exist.

* E-mail: tuetken@uni-bonn.de

Introduction

The radiation of tridactyl equids took place from the Middle Miocene onwards in North America, a few million years before their dispersal throughout Old World continents [1,2]. This is one of the most popular textbook examples of evolution taught in palaeontology [3]. The expansion of grasslands in North America has initially been considered *the* environmental factor driving the evolution of high-crowned molars (hypsodonty) as an adaptation to abrasive diets (i.e. silica-rich grasses) among Miocene hipparionines, as well as many other grazing mammals [2,4]. Abrasive material responsible for most of the tooth wear in large mammalian herbivores may have been plant silica (phytoliths) and/or mineral dust (grit) on the vegetation. The importance of grit versus phytoliths as abrasives for dental wear is, however, still

controversial [5,6,7,8,9]. Open, arid (grassland) habitats clearly contain more abrasive food compared to more humid, forested habitats. Hypsodonty is thus considered as an adaptation to a more abrasive diet from grasslands because grasses contain more abrasive biogenic silica in the form of phytoliths (and/or grit) compared to leaves [10,11,12]. Recent studies question the causal link between grasslands and hypsodonty [9] and suggest a much more complex interplay of various factors [13,14]. Indeed, based on phytolith analyses, Strömberg [14] conclude that grasslands were already more widespread in the Early Miocene prior to the increase of hypsodonty among many mammal taxa. Moreover, previous investigations suggest various ecological habits among Mio-Pliocene hypsodont hipparionines from North America, covering the entire spectrum between grazing and browsing

[12,15]. Thus characterizations of ungulate palaeo-diets solely based on crown height often do not reflect the full dietary breadth of a taxon. Selective pressures for crown height may have been weak in North American Equidae during prolonged periods of their evolution [12]. However, a stronger selection for the evolution of high-crowned dentitions occurred during the Early Miocene shortly before the first appearance of Equinae, the horse subfamily in which hypsodonty evolved [12] and that later migrated to Europe.

Neogene Old World equids are all derived from a North American common ancestor that dispersed into Eurasia across Beringia during times of glacio-eustatic sea level fall [2]. The arrival and subsequent radiation of hypsodont hipparionine horses of the *Cormohipparion* clade at the base of the Late Miocene is one of the most important dispersal events of the Eurasian Neogene. The first occurrence of high-crowned tridactyl hipparionine horses in western Eurasia ('*Hipparion*' Datum) has recently been pushed back to the early Late Miocene (MN 9, early Vallesian, ~11.2 Ma) with discovery of *Hippotherium primigenium* in Atzelsdorf, Austria [16]. From this locality, isolated hipparionine molars were found in fluvio-lacustrine deposits in the palaeo-Danube delta, sandwiched between marine strata of two transgressions of Lake Pannon; the upper transgression being dated to between 11.2 and 11.1 Ma [16,17]. The arrival of hipparionine horses in the eastern Mediterranean region at around 12–11 Ma was traditionally thought to mark the simultaneous westward expansion of savanna vegetation across the Old World. However, in Anatolia open landscapes with C_3 grasslands were actually widespread from the Middle Miocene, prior the first occurrence of hipparionines in the Old World [18]. Indeed, C_4 monocotyledons did not radiate across Western Eurasia during the Neogene [19,20].

This study aims to investigate the feeding ecology of the earliest hipparionines from Europe within a pure C_3 plant ecosystem in order to better understand one of the most significant faunal events during the Tertiary, the dispersal of hipparions into the Old World. Two scenarios can be considered: (1) hipparionines, because of their hypsodont cheek teeth, exploited grassy abrasive vegetation in open landscapes to spread into the Old World or, (2) their dispersal illustrates a new example of the Liem's paradox [21,22], inferring that specialized phenotypes (in this case hypsodont cheek teeth) enabled the species to occupy a wide ecological niche (a mixed-feeding or even browsing trait). These two hypotheses are addressed by analyzing combined dental wear (micro−/mesowear) and stable isotope (C and O) analyses of teeth from four populations of hipparionines that are amongst the earliest hipparionines in Europe.

The four localities, Eppelsheim (EP), Höwenegg (HO), Charmoille (CH), and Soblay (SOB), belong to the Late Miocene (MN9–MN10). The first two sites are situated in southern Germany, Charmoille is located in Switzerland and Soblay near Lyon in eastern France. In the latter locality two hipparionine taxa may have co-habited [23]. However, the large equid *Hippotherium primigenium* is always dominant in the faunal assemblage. Bernor et al. [24] thus assemble most of the Vallesian large hipparionines from Europe into a taxonomic unit defined as the "*Hippotherium primigenium*" complex.

Four different dietary proxies are examined: molar mesowear and microwear patterns in combination with carbon and oxygen isotope compositions of enamel carbonate of the early European *H. primigenium*. Stable isotope compositions of tooth enamel combined with dental micro- and mesowear analyses provide complementary information regarding dietary intake, habitat, and niche partitioning. These methods evaluate fundamentally different chemical (isotopes) and mechanical (dental wear) food

properties. Although these dietary proxies are not completely taxon-independent due to some influences of the animals (digestive) physiology and masticatory food processing, this multi-proxy approach allows a refined dietary reconstruction for hypsodont horses. It contributes to a better understanding of the dietary flexibility and will thus help us better understand the successful rapid dispersal of *H. primigenium* into the Old World. Dental wear analysis will help to resolve the grazer-browser dichotomy while carbon and oxygen isotope analyses will yield complementary information regarding habitat and food properties.

Fossil Sites with Early Hippotherium Analyzed in this Study

The Eppelsheim locality (EP) belongs to the oldest deposits of the Miocene Rhine River, exposed at many places in the Rhine-Hesse area of Germany (Fig. 1). These fluvial sediments of the Eppelsheim Formation [25,26], traditionally known as "Dinotheriensande", have yielded many localities mainly placed stratigraphically within the Vallesian Land Mammal Age in the lower part of MN 9, the age of which is approximately 11.5 to 9.5 Ma [27,28,29]. However, in addition to Late Miocene (Vallesian) taxa the Dinotheriensand Fauna also contains early and late Middle Miocene mammal faunas [30]. Here, the focus is on Eppelsheim, situated 30 km south of the city of Mainz, which has provided the richest assemblage of mammalian remains among the Dinotheriensande complex [31,32]. In addition, this assemblage contains the type-species sample of *Hippotherium primigenium* (MEYER, 1829), which is presently housed at the Forschungsinstitut Senckenberg (Frankfurt).

The Höwenegg locality (HO), located in southern Germany (Fig. 1), belongs to the lacustrine deposits of the Höwenegg-Formation, a sequence of light grey marl layers alternating with reddish-brown layers of tuffaceous mudflows. The Höwenegg-Formation was deposited in a lake that formed following the

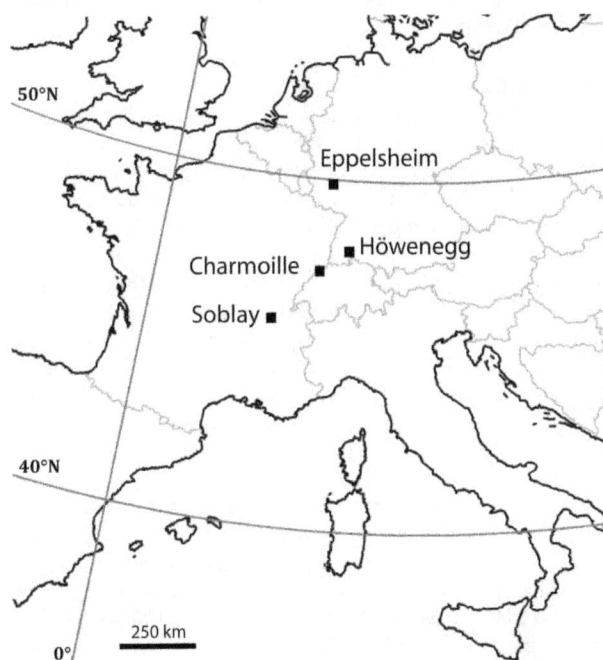

Figure 1. Geographic location of the investigated Late Miocene sampling sites.

eruption of hornblende-bearing pyroclastics during the Late Miocene. The Höwenegg-Formation was dated to 10.3 ± 0.19 Ma by a single hornblende crystal using $^{40}Ar/^{39}Ar$ from the hornblende tuff of the Höwenegg sequence [33,34]. The early MN 9 Höwenegg locality is important due to this radiometric date for the 'Hipparion' Datum and because it contains completely articulated skeletons of *H. primigenium*.

The Late Miocene deposits in Soblay (SOB, Ain, France, Fig. 1) are composed of a several meter-thick sequence of lignites alternating with marls, which overly Upper Jurassic calcareous marine sediments. All fossil remains belong to the second lignite unit, which contains a rich mammalian fauna including 67 species of large and small mammals [23]. Based on the latter, the fauna is biostratigraphically dated to MN 10.

The Charmoille site (CH) is a sand pit close to the small town of Charmoille in the canton of Jura in northern Switzerland (Fig. 1). The faunal assemblage of Charmoille suggests that the locality belongs to the MN9 [35]. Indeed, the presence of *Hippotherium primigenium* together with two rhinocerotids *Aceratherium* cf. *incisivum* and *Dicerorhinus sansaniensis* support an early Vallesian age [36]. (See Text S1 for more detailed site and palaeoenvironmental descriptions).

Combined Stable Isotope and Dental Wear Analyses - a Multi-proxy Dietary Approach

Hypsodont horse teeth grow over a period of several years and record in their enamel dietary, climatic and environmental information. Teeth of modern and fossil hypsodont horses have enamel growth rates of 35 to 40 mm/year and total enamel mineralization takes about 1 to 2.8 years depending on tooth type (molars usually take 1 to 1.5 years) [37,38]. Similar rates of tooth mineralization can be assumed for the slightly less hypsodont Miocene *Hippotherium primigenium*. Due to a short turnover time of body water (about 14 days [37]) and its dissolved inorganic carbon pool, the carbon and oxygen isotopic composition of incrementally growing teeth of hypsodont herbivores record a time-series of the isotopic composition of dietary intake and seasonality [37,39,40,41,42].

Stable Carbon Isotopes

The carbon isotope composition ($\delta^{13}C$) of animal tissues reflects that of the ingested food [43,44] and is mainly related to the proportions of isotopically distinct C_3 and C_4 plants from dietary intake [44,45]. C_3 plants, which include trees, most shrubs and many cold-season, temperate latitudes and high altitude grasses, discriminate strongly against the heavy ^{13}C isotope and therefore have lower $\delta^{13}C$ values than C_4 plants, ranging from -34 to -22‰, with an average of -27‰±3 [46]. C_4 plants include mostly warm-season grasses that discriminate less against the heavy ^{13}C isotope and therefore have higher, less negative $\delta^{13}C$ values of between -17 and -9‰, with an average of -13‰±2 [46]. Carbon ingested with diet and incorporated into the structural carbonate of the enamel apatite is enriched in ^{13}C by about $+14$‰ relative to the plants ingested by large non-ruminant herbivorous mammals such as horses, depending on the digestive physiology and rate of methane production of the animal [47,48]. Therefore, skeletal apatite of extant animals with a pure C_3 plant diet have average $\delta^{13}C$ values of about -13‰ while animals feeding on a pure C_4 plant diet have an average $\delta^{13}C$ value of about $+1$‰ [49,50]. The large C_4 grass-dominated grassland ecosystems of savannas known today first evolved globally in the late Miocene [20,51]. However, this is in contrast with data from Europe where C_3 plants dominated the vegetation during the Neogene, as indicated by the observed low $\delta^{13}C$ values in fossil mammal teeth of herbivores

[19,52]. In a C_3 plant-dominated ecosystem C_3 grazers and C_3 browsers cannot be easily distinguished by carbon isotope analysis of most tissues. However, the enamel carbon isotopic composition in C_3 ecosystems enables the niche partitioning and habitat differences of herbivorous mammals to be determined [53,54,55,56]. This is due to the variability of $\delta^{13}C$ values even within the C_3 plant groups, due to variations in light and water availability as well as position in the forest canopy or habitat temperature, depending on latitude and altitude [57,58,59].

Stable Oxygen Isotopes

Mammalian bioapatite records body water and hence ingested meteoric water $\delta^{18}O$ values. The body water $\delta^{18}O$ value of obligate drinkers, such as most large mammals including equids, is linearly related to that of the drinking water [60,61,62]. The $\delta^{18}O_{H2O}$ values of meteoric water vary within ecosystems due to changes in air temperature and/or amount of precipitation or evaporation [63,64]. These meteoric water $\delta^{18}O_{H2O}$ differences can be used to infer climatic conditions such as ambient air temperature and aridity as well as animal drinking behaviour [60,61,65,66]. Herbivorous mammals derive their water from three sources: (1) surface water, (2) water from food, and (3) metabolic water from food processing, specifically during oxidation of carbohydrates [61,67]. Thus physiological, environmental, and behavioural factors can influence enamel $\delta^{18}O$ values, particularly the water dependency of the animal [61,65,68]. Enamel $\delta^{18}O$ values of obligate drinking mammals are dependent on meteoric $\delta^{18}O$ values whereas drought-tolerant animals usually have less negative $\delta^{18}O$ values because they obtain proportionally more water from ^{18}O-enriched food sources such as leaves, fruits or seeds [61,69]. Browsing taxa that ingest a higher proportion of ^{18}O-enriched water with their food often have higher relative ^{18}O values compared to sympatric grazing taxa [68,70]. Therefore, $\delta^{18}O$ values allow us to draw inferences regarding habitat properties, feeding ecology, drinking behaviour and humidity [65,69,70,71]. Seasonality is recorded as $\delta^{18}O$ amplitude changes in high crowned horse teeth which allows the evaluation of climatic changes [37,72,73]. Combined oxygen and carbon isotope analyses on hypsodont teeth also allow for the reconstruction of seasonal changes of ingested water and diet [42,72,74].

Dental Meso- and Microwear

Mechanical abrasion caused by dietary intake leaves its traces at the enamel surface [7,75,76]. Thus micro- and mesowear investigations of fossil herbivore teeth provide additional information about the type of consumed plant material and the palaeohabitat [77,78,79,80,81,82,83,84,85,86].

Dental facet development on the molar surfaces of living herbivorous ungulates appears to be strongly tied to their feeding styles [86]. These mesowear patterns reflect the long-term (several months to years) diet and can be used to infer broad dietary habits in extinct ungulates [12,82,85]. Dental mesowear reflects the degree of attritional and abrasive wear on the molar occlusal surface. Attritional wear is due to tooth-on-tooth contact and results in high crown relief and sharp cusp apices. Abrasive wear, on the other hand, is due to the alteration of enamel tissue by food items during mastication. In contrast to attritional wear, abrasion obliterates dental facets resulting in lower crown relief and rounder apices on cheek teeth [86].

Dental microwear patterns reflect a short-term (a few days to weeks) dietary signal of the physical properties of the last food items consumed by an individual [7,76,79]. Ungulates, whose main food resources are graminoids (including grasses and grass-like plants, such as sedges), have higher densities of scratches (elongated microwear scars) than pits (short and round microwear scars) [87]. This intense scratching observed in grazing ungulates is due to the abrasiveness of monocotyledons. The cell walls of monocotyledons contain a high concentration of silica phytoliths [88,89], the abrasiveness of which is considered as an adaptive response to herbivory [90]. In contrast to monocotyledons, dicotyledons have fewer silica phytoliths, such that browsing ungulates have a lower ratio of scratches to pits compared to grazers. Beyond the grazer/browser dichotomy, the dental microwear method has been used to detect more subtle feeding preferences such as browsers whose diets contain fruits and seeds, or mixed-feeders that switch from grazing to browsing on a daily basis [91,92].

It is worth noting here that two recent studies based on scratch tests and hardness measurements conclude that silica phytoliths do not appear to influence the enamel enough to scratch it, but identify grit as the major factor driving dental abrasion [8,9]. In contrast, empirical dental microwear data from mammals indicates that variations in food properties play a major role in controlling the microwear patterns. These studies include investigations of mammal populations from the polyspecific scale [93,94] to the monospecific scale [79,95], as well as from controlled feeding studies [7,76] and abrasion experiments with teeth mounted on a tribometer [75]. For example, reindeer feeding on ground lichens in tundra settings or camelids browsing in arid and dusty habitats differ from other ungulates by a higher amount of pits on the enamel surface [5,91]. Thus ingestion of mineral particles does not cause a "grazing" dental microwear pattern with many scratches for these browsing species, demonstrating that the grit contribution to dental wear is minor compared to abrasive plant matter.

Results

Carbon and Oxygen Isotope Compositions of the Hipparion Teeth

Average $\delta^{13}C$ values of the carbonate in the enamel bioapatite from 22 *Hippotherium* teeth of the four Late Miocene localities EP, CH, HO, and SOB range from -14.6 to $-10.2‰$ (Table 1). These values are typical for a pure C_3 plant diet in all analyzed *Hippotherium* individuals. However, there are significant inter-site differences: EP and CH specimens have more negative mean $\delta^{13}C$ values ranging from -13.8 to $-13.6‰$, whereas HO and SOB both have approximately 2‰ less negative $\delta^{13}C$ values of around $-11.5‰$ (Fig. 2, Table 1). These between-site differences in $\delta^{13}C$ are significant (Kruskal-Wallis test: H (3, N = 22) = 14.395; $p = 0.002$).

Intra-tooth $\delta^{13}C$ variability of all 10 serially sampled premolars and molars from EP, CH and HO is nearly 6‰ (range: -15.2 to $-9.5‰$), but is small within a single tooth $0.9 \pm 0.3‰$ (range: 0.5 to 1.4‰, Tables S1–S3). The teeth do not have clear seasonal trends in $\delta^{13}C$ values (Fig. 3). Inter-tooth variability of mean enamel $\delta^{13}C$ values in a single site is higher than the seasonal variation within a single tooth, with a range between 1.9 and 2.5‰, except for SOB which has a lower value of 0.5‰. The total range of $\delta^{13}C$ values for all 22 teeth is 4.4‰. The $\delta^{13}C$ values for the teeth of the HO and SOB populations are less negative than those of the other two *H. primigenium* populations from EP and CH (Fig. 2). Thus the HO and SOB hipparionines have ingested more ^{13}C-enriched C_3 plants compared to those of the latter localities.

The mean $\delta^{18}O_{CO3}$ values of the enamel carbonate of the 22 *H. primigenium* teeth analyzed from the four localities range from -9.1 to $-4.5‰$ (Table 1). These between-population differences in $\delta^{18}O_{CO3}$ are significant (Kruskal-Wallis test: H (3,

N = 22) = 10.624; $p = 0.013$). The intra-tooth $\delta^{18}O$ variation of all of the 10 serially sampled premolars and molars has a similar total range from -10.0 to $-4.3‰$ (Tables S1–S3). Intra-tooth variations of $\delta^{18}O_{CO3}$ values are small and range from 0.6 to 2.3‰ with an average of $1.7 \pm 0.7‰$. One tooth from EP (PW 10153) has a much higher variability oϕ 4.0‰, which is likely a result of diagenetic alteration in the lower, less mineralized crown part of this unerupted M2 that also gave the highest $\delta^{18}O_{CO3}$ values (Fig. 3). Some of the teeth show a seasonal trend in $\delta^{18}O$ values but most intra-tooth $\delta^{18}O$ patterns are relatively flat (Fig. 3). The HO and SOB hipparionines, which have the highest enamel $\delta^{13}C$ values, also have the highest $\delta^{18}O_{CO3}$ values (Fig. 2), hence they ingested water more enriched in ^{18}O than that of the other localities.

The seven molars analyzed from Eppelsheim have the lowest mean $\delta^{13}C$ values of -14.6 to $-12.4‰$ and the intra-tooth $\delta^{13}C$ variation of the two serially sampled M2 have a range from 0.5 to 1.3‰. Mean $\delta^{18}O_{CO3}$ values range from -9.0 to $-5.8‰$ while the intra-tooth variation of the oxygen isotope composition is the largest of all four investigated localities ranging from 0.5 to 3.2‰ (Table 1, Fig. 3).

The eight teeth from Höwenegg (P4, P3, M1 and M3 and four other molars) have the highest $\delta^{13}C$ values of all teeth analyzed from all the localities, with a range from -12.7 to $-10.2‰$ (Fig. 2), while the intra-tooth variation of the four serially sampled teeth is low, from 0.3 to 1.1‰. The HO population has 1.5 to 2.5‰ less negative $\delta^{13}C$ values compared to the EP and CH populations but values similar to SOB. $\delta^{18}O_{CO3}$ values for all teeth range from $-6.4‰$ to $-4.5‰$ while the intra-tooth variation is in the range of 1.3 to 2‰ (Table 1, Fig. 3).

The four serially sampled teeth from Charmoille (P3, P4, M2 and M3) have low mean $\delta^{13}C$ values of -14.6 to $-12.6‰$ similar to those of EP and the intra-tooth variation of the $\delta^{13}C$ values ranges from 0.6 to 1.4‰. Mean $\delta^{18}O_{CO3}$ values for all teeth has a range from -9.1 to $-7.3‰$ and the intra-tooth variation is 1.6 to 2.3‰ (Table 1, Fig. 3).

For Soblay only three molars were analyzed in bulk but no serial intra-tooth sampling was performed. $\delta^{13}C$ values have only a narrow range from -11.7 to $-11.2‰$ and $\delta^{18}O_{CO3}$ values range from -6.4 to $-4.8‰$. Teeth from SOB have less negative $\delta^{13}C$ and $\delta^{18}O_{CO3}$ values than the two EP and CH populations, similar to those of HO teeth.

Dental Micro- and Mesowear Signatures

For microwear analysis, specimens of *Hippotherium primigenium* from the four investigated localities were integrated into a model constructed using present day grazing and browsing ungulates through a discriminant analysis (DA). An overall misclassification for extant species is about 11.54% (5.88% for browsers and 20.49% for grazers; Table S4). As expected, the sample of present-day browsers significantly differs from grazers in DA coordinates (*t*-test, $p < 0.001$). Table S4 provides the coefficients of linear discriminant plus the correlation between discriminating variables and discriminant function.

Table 2 and Figure 4 display the classification of extinct hipparionines within this model. Hipparionines from the four Late Miocene localities investigated were not grazers but rather had mixed feeding habits with a strong dominance of browsing. Only one specimen out of 31 from the EP population has a dental microwear pattern similar to extant grazers, most of the remainder having similarities with living browsers. CH and SOB molar microwear patterns also share browsing characteristics. This contrasts with the HO sample for which 5 out of 7 individuals are classified as grazers, whereas the other two fit clearly with

Table 1. Mean enamel $\delta^{13}C$ and $\delta^{18}O$ values of *H. primigenium* teeth.

Specimen no.	Tooth	Locality	n	$\delta^{18}O_c$ VPDB [‰]	σ	$\delta^{13}C_c$ VPDB [‰]	σ	CO_3 [wt. %]
PW 2000/10153	M2	EP	9	−9.0	0.20	−14.2	0.14	5.6
PW 2000/10171	M2	EP	14	−8.2	1.34	−14.6	0.30	6.1
PW 2000/10072	M2	EP	1	−8.7	0.1	−14.5	0.06	4.4
NMM (FZ EQ EP 1)	M	EP	1	−5.8	0.03	−12.4	0.04	5.1
NMM (FZ EQ EP 2)	M	EP	1	−6.6	0.04	−13.0	0.04	4.4
NMM (FZ EQ EP 3)	M	EP	1	−6.7	0.03	−13.7	0.02	5.4
NMM (FZ EQ EP 4)	M	EP	1	−7.2	0.08	−14.1	0.03	5.1
Mean value EP			**7**	**−7.5**	**1.1**	**−13.8**	**0.7**	**5.2**
HW164/55	P4	HO	7	−6.4	0.44	−11.0	0.11	4.7
HW153/55	P3	HO	1	−4.8	0.11	−11.1	0.10	4.7
HW755/59	M1	HO	7	−8.4	0.71	−12.2	0.19	5.4
HMLD Hip. III 53	M	HO	1	−5.9	0.09	−12.5	0.05	4.1
UMz P 1913	M	HO	1	−4.6	0.05	−12.7	0.06	4.4
SMNK Höw 06/117	M	HO	15	−6.0	0.42	−11.5	0.32	4.1
SMNK Höw 06/102	M	HO	1	−5.9	0.07	−11.5	0.07	4.4
SMNK Höw 06/55	M3	HO	15	−4.5	0.64	−10.2	0.29	4.8
Mean value HO			**8**	**−5.4**	**0.7**	**−11.5**	**0.8**	**4.6**
CM 600	P4	CH	15	−7.8	0.58	−14.5	0.44	5.3
CM 274	M3	CH	15	−8.4	0.62	−13.7	0.23	5.6
CM 574	M2	CH	19	−9.1	0.49	−13.4	0.16	5.8
CM 457	P3	CH	13	−7.3	0.84	−12.6	0.32	5.0
Mean value CH			**4**	**−8.2**	**0.7**	**−13.6**	**0.7**	**5.4**
FSL SOB−1	M3	SOB	1	−5.6	0.08	−11.2	0.10	6.4
FSL SOB−2	M3	SOB	1	−6.4	0.09	−11.7	0.07	6.1
FSL SOB−3	M3	SOB	1	−4.8	0.10	−11.4	0.08	5.1
Mean value SOB			**3**	**−5.6**	**0.7**	**−11.5**	**0.2**	**5.9**

extant browsers. The differences in the distribution between the four populations along the discriminant analysis are valid (Kruskal-Wallis test: H (3, N = 51) = 9.975; p = 0.018).

Mesowear data indicate a wide range of values from high attrition-dominated mesowear signatures (score = 0.22) to high-abrasion dominated mesowear scores (score = 2.11; Table 3 and Fig. 4). The most attrition-dominated wear signature is found in the HO population of *H. primigenium*, which plots in the central portion of the browser spectrum. In contrast, the SOB sample has the highest abrasion-dominated signature and therefore plots within the range of extant grazing species. The CH sample plots close to the SOB sample with a molar mesowear score of 1.94. The EP population plots within the mixed feeder range sandwiched between the spectrum for browsers and grazers (Fig. 4). Note that the EP score also falls within the range of fruit-eaters.

Discussion

Feeding Ecology of the Earliest European Hipparionines

All *H. primigenium* teeth of Charmoille, Soblay, Eppelsheim and Höwenegg have low enamel $\delta^{13}C$ values in the range expected for C3 feeders. Thus C4 plants were not a significant part of the diet of *Hippotherium*, confirming that C4 grasses were absent in Europe during the Late Miocene [96,97]. Whether *H. primigenium* were C3 browsers or C3 grazers can not be inferred solely from their carbon isotopes. The small intra-tooth variability of enamel $\delta^{13}C$ values of ≤1.4‰ indicates only small seasonal $\delta^{13}C$ variations within the C3 plants ingested by *Hippotherium*. Intra-site differences of mean enamel $\delta^{13}C$ values are somewhat larger (2 to 2.5‰). These differences in enamel $\delta^{13}C$ values of the investigated *H. primigenium* teeth thus reflect $\delta^{13}C$ variations of the dietary C3 plants (either C3 dicots or C3 monocots) due to seasonal and/or habitat differences. The enamel $\delta^{13}C$ values of 60% of the *H. primigenium* teeth are 1 to 3‰ lower than the mean value of around −11.5‰ expected for herbivores feeding on C3 plants during the Miocene [98].

Low enamel $\delta^{13}C$ values ≤ −13‰ for many of the *H. primigenium* teeth, especially those from EP and CH, suggest feeding was predominantly in forested ecosystems. This is compatible with the palaeoenvironmental reconstructions for these settings based on palaeofloral remains (see Text S1 for details). Forest-dwelling browsers are, due to canopy related effects, expected to have lower $\delta^{13}C$ values compared to more open country or grassland browsers or grazers [55,99]. However, *H. primigenium* was clearly

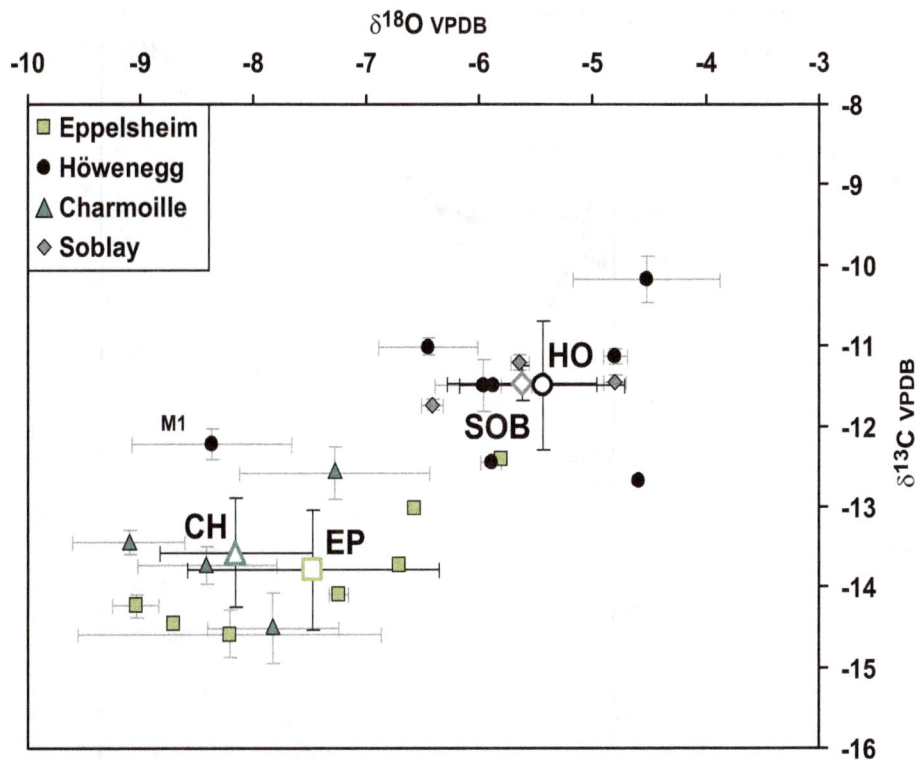

Figure 2. $\delta^{13}C$ and $\delta^{18}O$ values of all 21 *H. primigenium* teeth from the four different localities Eppelsheim (EP), Höwenegg (HO), Charmoille (CH) and Soblay (SOB). Note that the M1 from HO was excluded for the calculation of the HO mean value. For serially sampled teeth, average values and standard deviations are given. Mean values and standard deviations of each locality are plotted as open symbols.

not feeding in a dense evergreen tropical forest with a very strong canopy effect, otherwise much more negative $\delta^{13}C$ enamel values would be expected [55].

In contrast, *H. primigenium* teeth from the HO and SOB populations have around 2‰ higher $\delta^{13}C$ enamel values when compared to those of EP and CH (Fig. 2). This is probably not related to diagenetic alteration as enamel and dentin of the same teeth for the HO sample have distinct differences in their $\delta^{13}C$ values of about 5‰ (Fig. S1). Thus the dentin has higher $\delta^{13}C$ values, that is, values similar to those of the embedding sediment while the enamel has likely preserved original values. Therefore, the higher enamel $\delta^{13}C$ values of the HO and SOB specimens indicate the ingestion of more ^{13}C enriched C_3 plants. This may reflect palaeoenvironmental effects (e.g., water stress, habitat openness) on the plants eaten. An alternative interpretation would be that the HO and SOB *H. primigenium* populations fed predominantly on different plants or plant parts (e.g., more woody plant remains or fruits) with less negative $\delta^{13}C$ values. The HO and SOB specimens also have higher enamel $\delta^{18}O$ values (Fig. 2) compared to the teeth of EP and CH, supporting the suggestion of environmental effects on the ingested plants, and possibly more intense water stress in the HO and SOB palaeoenvironments than in the vicinities of EP and CH. However, for SOB the palaeontological evidence indicates a humid and forested environment in the vicinity [23] and, therefore, does not support water stress. Interestingly, the dental mesowear patterns of the SOB population are consistent with the dental wear pattern of extant grazers (Fig. 4), which would not be expected for a forest dweller. Conversely, more than half of the Soblay specimens (7 out of 10) had molar microwear patterns similar to the browsing species (Table 2 and Figure 4). Thus the SOB *H. primigenium* grazed and

browsed. This suggests that more open and drier areas with grass existed in the vicinity of the humid forested environment and that both settings were likely used for foraging by the SOB population. Predominant grazing best explains the higher $\delta^{13}C$ and $\delta^{18}O$ values and dental mesowear signatures of the SOB teeth. Feeding on canopy-derived fruits and leaves that have higher $\delta^{13}C$ values than subcanopy plants could be an alternative explanation for higher enamel $\delta^{13}C$ values in frugivorous forest dwelling taxa compared to subcanopy browsers [55]. As fruits and seeds tend also to be enriched in ^{18}O relative to the source water [100], frugivores have higher enamel $\delta^{18}O$ values. However, the abrasion dominated mesowear patterns of SOB do not comply with extensive frugivory (Fig. 4).

The HO population has similar $\delta^{13}C$ and $\delta^{18}O$ values compared to those of the SOB (Fig. 2). In contrast to the SOB population, the dental mesowear of the HO population resembles dietary reference taxa of present-day browsers (Fig. 4). However, molar microwear analysis detects some grazing habits for this population. This clearly points to a large dietary flexibility in *Hippotherium* on a meal-by-meal basis and thus its ability to succeed in different environments as a mixed feeder with opportunistic preferences for either browsing or grazing, depending on resource availability.

The low mean $\delta^{13}C$ values of ≤ -13.6‰ for EP and CH (compared with SOB and HO) are compatible with browsing on C_3 plants in a forested environment. The dental mesowear signature of the EP population supports mixed feeding habits, whereas, the dental mesowear analysis on the CH sample indicates a diet richer in grasses. However, note that such an abrasion-dominated mesowear signal could also be due to high fruit consumption. The hypothesis of grazing traits is challenged by the

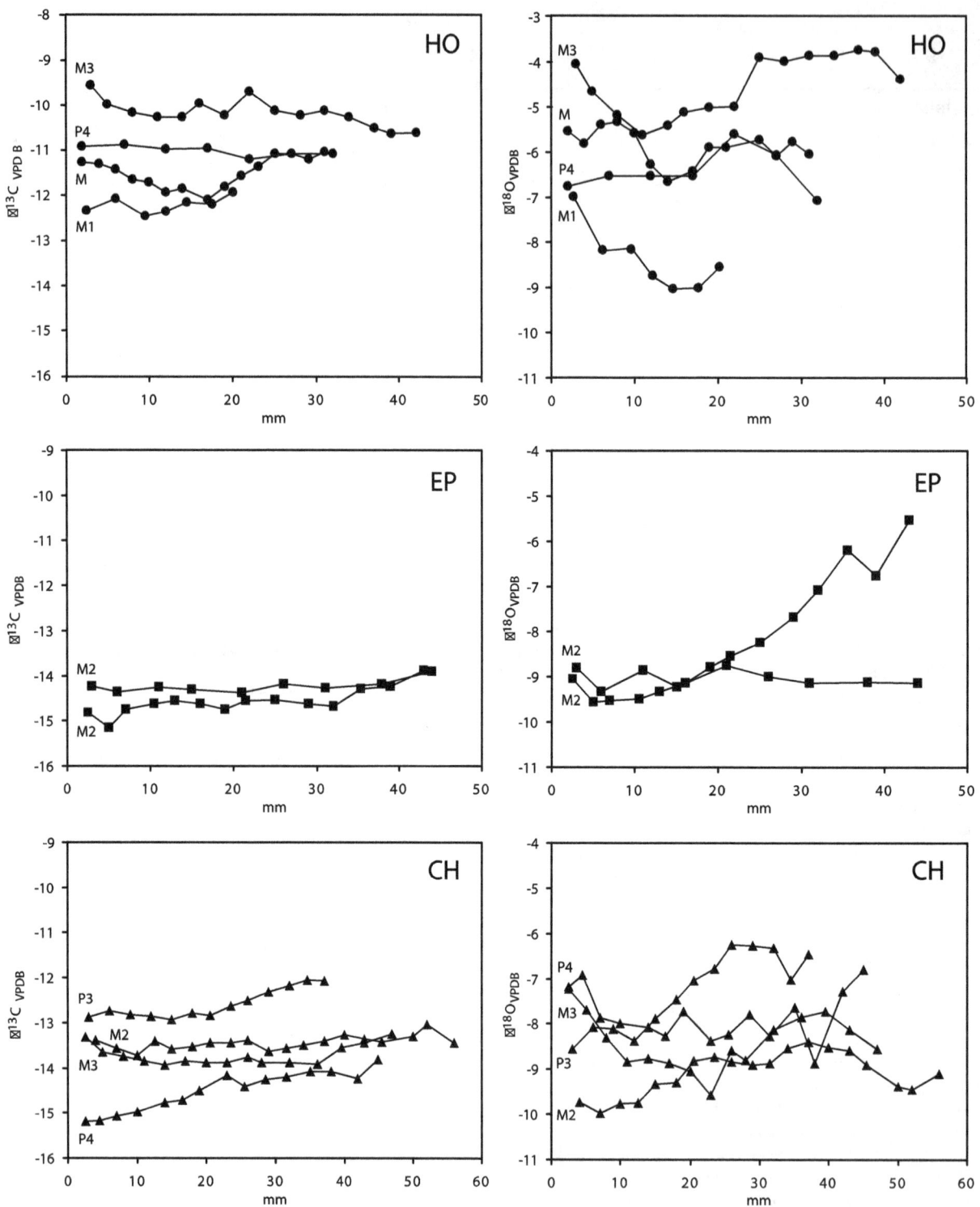

Figure 3. Intra-tooth variation of the enamel δ¹³C and δ¹⁸O values in 10 serially sampled teeth of *Hippotherium primigenium* molar and premolar teeth from Charmoille (CH), Eppelsheim (EP), and Höwenegg (HO). Samples are plotted against the distance from the crown in mm.

dental microwear results of these two populations. In fact, only one individual out of 19 from EP and two of 15 from CH can be considered as grazers; the remaining individuals all plot in the browsers group (Table 2 and Figure 4). These individuals obviously did not graze a considerable amount during the weeks or even months before their death.

Altogether the data support opportunistic and mixed feeding habits for *H. primigenium*. Indeed, long-term (molar mesowear patterns and bulk stable isotope compositions) and short-term

Table 2. Number of individuals n mean m and standard deviation *sd* of the three main dental micro-wear variables (Number of scratches Ns, pits Np and percentage of pits Pp) for the four investigated populations (and sub-groups defined through the discriminant analysis) of *Hippotherium primigenium*.

				Ns		Np		Pp	
				m	sd	m	sd	m	sd
Charmoille	**Complete population**		15	22.4	10.1	41.9	14.5	64.9%	12.5%
	*Specimens classified in sub-groups:**	B95	8	18.0	6.5	48.0	16.8	71.9%	10.0%
		B50	5	21.8	5.4	35.8	9.1	61.9%	6.9%
		G50	1	49.0	–	33.0	–	40.2%	–
		G95	1	34.0	–	33.0	–	49.3%	–
Eppelsheim	**Complete population**		19	14.8	7.8	53.6	17.9	76.4%	15.5%
	Specimens classified in sub-groups:	B95	14	13.7	4.4	59.1	12.2	80.6%	7.6%
		B50	4	12.5	7.7	41.8	24.9	71.5%	22.8%
		G50	1	40	–	23	–	36.5%	–
		G95	0	–	–	–	–	–	–
Höwenegg	**Complete population**		7	29.1	9.2	31.7	28.8	44.5%	26.8%
	Specimens classified in sub-groups:	B95	2	20.0	12.7	70.0	21.2	77.0%	16.3%
		B50	0	–	–	–	–	–	–
		G50	3	30.7	6.1	23.3	4.7	43.2%	0.6%
		G95	2	36.0	2.8	6.0	4.2	14.1%	9.6%
Soblay	**Complete population**		10	28.1	8.3	50.8	17.7	63.0%	12.4%
	Specimens classified in sub-groups:	B95	5	23.0	6.8	64.8	8.2	74.0%	5.1%
		B50	3	35.3	7.8	43.0	12.3	54.6%	1.6%
		G50	2	30.0	4.2	27.5	3.5	47.8%	0.3%
		G95	0	–	–	–	–	–	–

*Each (browsing and grazing) diet category is divided in two sub categories depending on the probability associated with the classification. Specimens that plot with either browsers (with p>0.95; B95) or grazers (with p>0.95; G95) browse and graze, respectively, the few weeks preceding the death. Two other subcategories are defined as browsers with 0.50>p>0.95 (B50) and as grazers with 0.50>p>0.95 (G50).

(molar microwear patterns and serial intra-tooth stable isotope compositions) signals do not always corroborate each other and vary from one population to another and even within a single population. Given the different food properties and timescales represented by the various diet proxies this is not surprising. However, the data clearly demonstrate the potential of combined dental wear and stable isotope analysis to obtain refined and complementary dietary reconstructions of herbivores. This multi-proxy approach reveals dietary flexibility and enables us to gain new insights into the palaeoecology of extinct taxa and their feeding behaviour.

The Habitat of the Earliest European Hipparionines

Differences in vegetation density and/or water stress may explain some of the dietary differences between the *H. primigenium* populations. For instance, the EP habitat was part of the Rhine River stream system in a flat and well-watered landscape. Water stress thus likely played a much less pronounced role at EP. Similarly, this may have been the case for HO as it was situated near a lake in the Hegau volcanic field on a tuff-dominated bedrock and well-drained limestone bedrock. Given the higher enamel $\delta^{13}C$ and $\delta^{18}O$ values from HO, and also SOB, compared to EP and CH, the HO and SOB populations may represent open landscape populations, which in turn would be expected to have had a more grass-dominated dietary regime. This is in agreement with the microwear results from HO teeth, but not for those from SOB where there is robust evidence for wet and forested habitat

[23]. Note, however, that the SOB population has the most abrasion-dominated mesowear pattern of all four populations, probably indicating an overall grass-dominated diet which is in agreement with a more open habitat. In turn, the EP and CH habitats would be considered to have been more closed, and therefore less likely to provide access to grass and other abrasive food items. However, tooth mesowear patterns indicate that EP and CH hipparionines included abrasive components in their diet (grasses, grass-like plants). *Hippotherium primigenium* was thus generally a flexible and opportunistic meal-by-meal mixed feeder highly engaged in browsing, grazing and occasional frugivory and therefore was predominantly a forest dweller during the Late Miocene of Western Europe.

General Implications for Palaeodiet and Palaeoecology of Hippotherium Primigenium

Despite the correlation of crown height with ecological factors relating to diet in modern herbivorous mammals, the correlation cannot prove that crown height is always a reliable predictor of diet in terms of browsing versus grazing [101]. Mihlbachler & Solounias [101] reported a less direct coevolution of diet and crown height in the fossil record of North American Merycoidodontidae compared to previous studies. The question of evolutionary trajectories which may also be responsible for an increase in crown height has been raised a number of times over the last decade, and a reappraisal of the question may now be required.

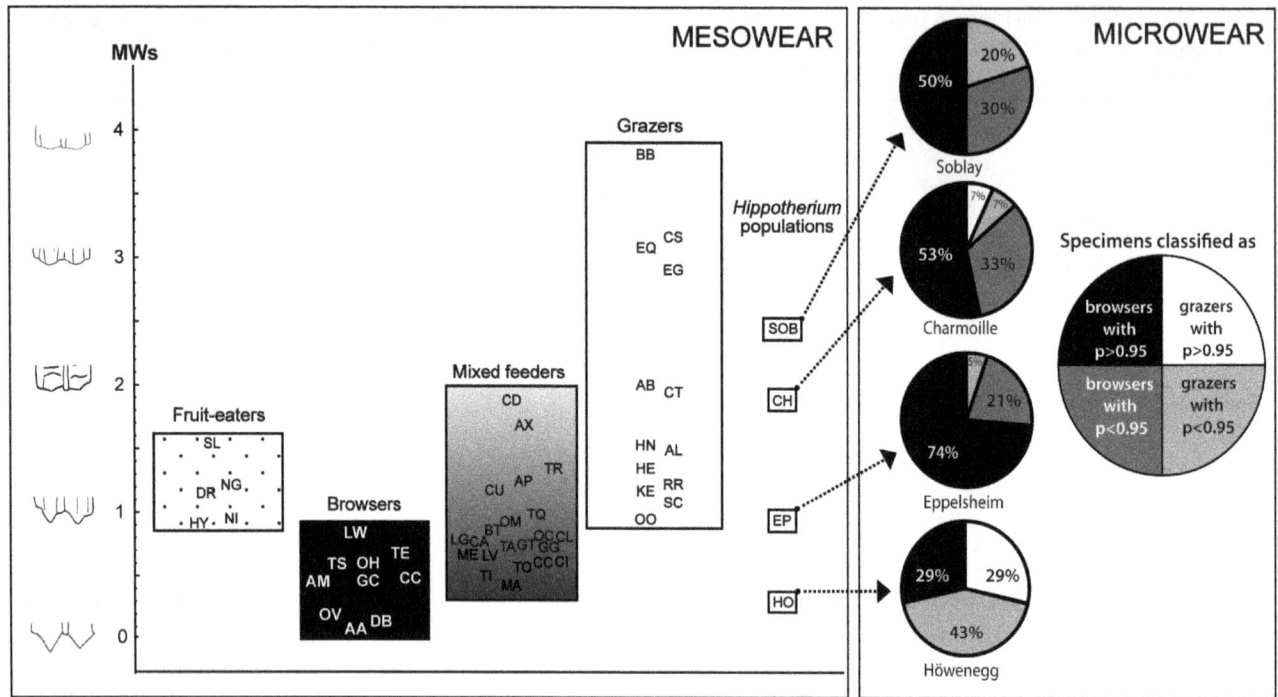

Figure 4. Molar mesowear and microwear patterns of the four populations of *Hippotherium primigenium.* Comparative extant species for microwear analysis are shown in Table S4 and in Table S5 for mesowear scoring.

One could be derived from the current data regarding the broad dietary flexibility of *H. primigenium.*

Because of their hypsodonty, '*Hipparions*' were widely considered as grazers. However, there is another biological benefit of hypsodonty beyond the adaptation to an abrasive (grass) diet; a greater dietary flexibility. The present data indicate that *Hippotherium* had a variable C_3 plant-based diet ranging from browsing to grazing and was thus a flexible and opportunistic mixed-feeder. At around 11 Ma, when hypsodont hipparions arrived in Eurasia, C_4 plants, which comprise the majority of abrasive grasses in extant open environments, were absent or just a minor component in the global flora and did not become widespread until 8 to 6 Ma [19,20]. This indicates that hipparions must have inhabited a Europe with an absence of C_4 grasslands

during the early Late Miocene (MN9) and could only have met their dietary needs with the C_3 environment, which was probably forested. *Hippotherium* would have predominantly fed on available leafy plants such as trees and shrubs within these forested landscapes of Europe but could also switch to a more abrasive grass or even a fruit-based diet. This ecologic flexibility enabled hypsodont *Hippotherium* to feed both on abrasive and less abrasive food items, depending on their availability. This was advantageous for survival in environments with changing seasonality as would have occurred following the Mid-Vallesian crisis, with drier and more seasonal climates prevailing in Europe in the Late Miocene [102,103].

Furthermore, opportunistic mixed-feeding may have facilitated the dispersal of *H. primigenium* via Asia due to any potential climatic (arid areas) and geographic barriers being easier to cope with due to its flexible diet. The North American ancestors of the earliest Eurasian hipparions from the *Cormohipparion* clade are best known from the Early Clarendonian (ca 12 Ma) quarries in Texas and Nebraska [2]. However, the population from which the first Eurasian individuals derived must have existed in Alaska, which would have been a relatively cool and humid place compared to the North American interior [104]. Conversely, preliminary results indicate that habitat conditions were quite dry as far south as central China in the earlier part of the Late Miocene [105]. Fortelius et al. [104] postulate a westward migratory route of earliest Old World hipparions, and consider "traditional" migratory dispersal routes unlikely because of aridity. However, the dietary signatures of this study of the first three-toed horses in Europe are indicative of a variety of dietary sources. Indeed, *Hippotherium primigenium* was likely able to graze as well browse, much in the same way as present-day wild asses. This dietary flexibility would explain its rapid dispersal into Eurasia despite climatic and environmental dispersal barriers, and thus was not necessarily linked to the expansion of grasslands for grazing. The

Table 3. Frequency of hipparionine specimens in the different mesowear categories (in percentage) and mesowear scores.

Populations	HS*	HR	LS	LR	LB	score
Höwenegg	80.00	20.00	0.00	0.00	0.00	0.22
Eppelsheim	18.75	79.17	0.00	2.08	0.00	0.86
Charmoille	9.09	39.39	6.06	39.39	6.06	1.94
Soblay	0.00	47.06	0.00	47.06	5.88	2.11

*HS = Combination of high relief H and sharp cusp S (score = 0),
HR = combination of high relief H and round cusp R (score = 1),
LS = combination of low relief L and sharp cusp S (score = 2), LR = combination of low relief L and round cusp R (score = 3), LB = combination of low relief L and blunt cusp B (score = 4). Average score per population is computed by the mean of the average score per individual in case that several teeth represent an individual.

hypsodonty of *H. primigenium* can thus be considered an example of Liem's paradox in which a specialized phenotype (high crowned cheek teeth) enabled a more flexible mixed-feeding diet.

Given how wide the dietary niche of *H. primigenium* appears to be, its hypsodont dentition does not restrict it to a habitat of abrasive diets. This is also compatible with the reasoning that fallback foods may have more selective force in natural selection than preferred foods [106]. The hypsodont adaptation would allow *H. primigenium* to survive in a complex variety of landscapes, which did not necessarily provide the food items that limit hypsodont herbivores to abrasive diets. *Hippotherium* is thus another example that hypsodonty, although originally an adaptation to abrasive food, can enable a species to become a generalist feeder of both high- and low-abrasion diets, similarly to camelids [107], rhinocerotids [108], and other equids [12]. Finally, hypsodonty was advantageous for *H. primigenium* to expand its habitats and food resources in the increasingly drier and more open Late Miocene landscapes of Europe [109].

The appearance of *Hippotherium* in Europe as a hypsodont, opportunistic mixed feeder put competitive pressure on the brachydont equid *Anchitherium aurelianense* which was feeding in the same ecological niche [110]. This may have contributed to the decline and extinction of *Anchitherium* in the Old World following the Mid-Vallesian crisis, with only a short period of coexistence in the upper Vallesian (MN9) [111,112]. The number of fossil sites with co-occuring *H. primigenium* and *A. aurelianense* specimens are thus limited, hampering a direct diet comparison of sympatric individuals. A rapid replacement of *Anchitherium* by *Hippotherium* seems quite plausible due to the latter having a high ecological tolerance; because of its hypsodonty it was capable of feeding on abrasive diets. Although some *Anchitherium* populations in arid areas such as Spain also have an adaptive increase in crown height, those in humid central European areas (i.e. Germany) did not, and they remained brachydont [113]. Overall, the hypsodont *H. primigenium* was able to live and feed in a variety of habitats and therefore was a very successful migrant species.

Conclusions

To better understand the Late Miocene dispersal of the earliest high-crowned hipparionines in the Old World two hypotheses were investigated: (1) hypsodont cheek teeth facilitated the exploitation of grassy abrasive vegetation in evolving open grassland landscapes and, (2) hypsodont cheek teeth enabled *H. primigenium* to occupy a broad dietary niche. Combined dental meso- and microwear together with enamel oxygen and carbon stable isotope analyses on four populations of hipparionines in Europe, amongst the earliest in the Old World, support mixed feeding habits for *H. primigenium* with a strong reliance on leafy plants rather than on grazing habits. This high-crowned equid was thus adapted to use a wide range of dietary sources and had a large dietary flexibility. Feeding changed partly on a meal-by-meal basis, but was likely also affected by local and seasonal availability of vegetational resources in the natural food supply. This dietary flexibility was likely a key factor for the rapid dispersal of *H. primigenium* from North America into Eurasia, given the different environmental settings in this region.

With the broad feeding habits of *H. primigenium*, its hypsodont dentition may no longer be regarded as a constraint for forage on abrasive items in open landscapes. Instead, it would rather enable opportunistic feeding from grazing to pure browsing for long periods of time; even when no less abrasive diets were available. The hypsodonty of its cheek teeth would allow *H. primigenium* to survive in a variety of landscape types which did not necessarily

provide the food items that would limit hypsodont herbivores to abrasive diets. Thus *Hippotherium* is an example of the Liem's paradox, hypothesizing that specialized phenotypes (here hypsodonty) enable more ecological flexibility in a taxon and the occupation of a broader dietary niche.

The arrival of this ecologically flexible equid must have imposed severe competition on brachydont anchitheres that occupied a variable browsing or mixed feeding niche in the Late Miocene of Europe [110]. The higher permanent dietary flexibility of *H. primigenium* might thus be the cause for the rapid replacement of the low-crowned equid *Anchitherium aurelianense* in Europe during the Late Miocene (MN9–MN10).

Materials and Methods

We thank the Forschungsinstitut und Naturmuseum Senckenberg (Frankfurt), the Staatliches Museum für Naturkunde (Karlsruhe), and the Landesmuseum Rheinland-Pfalz (Mainz), the Collections d'Histoire Naturelle de l'Université de Lyon (A. Prieur) and the Naturhistorisches Museum (Basel, L. Costeur) for access to the fossil specimens investigated in this study. All the specimens were loaned in the aforementioned museum collections and returned after analysis.

Carbon and Oxygen Isotope Analysis of Tooth Enamel

Serial and bulk enamel samples of isolated cheek teeth of *Hippotherium primigenium* from Eppelsheim (EP; n = 7), Höwenegg (HO; n = 8), and Charmoille (CH; n = 4) were analysed for their carbon and oxygen isotopic composition of carbonate in the apatite (Table 1). If possible, only teeth such as P4, M2 and M3, which formed after weaning, were analyzed. Bulk enamel samples of isolated cheek teeth of *Hippotherium primigenium* from Soblay (SOB; n = 3) were also analysed for their carbon and oxygen isotope composition of carbonate in the apatite (see Table 1 for details). For isotope analyses teeth with the least possible wear were selected to sample the complete tooth length and thus longest growth period possible. Due to limited sample availability, teeth of different jaw positions were sampled for the isotopic measurements (Table 1). The tooth cement was removed and the enamel surface mechanically cleaned prior to enamel sampling. Bulk enamel samples were recovered by drilling a line parallel to the growth axis of the tooth along the entire length from the crown to the root. Serial enamel samples were hand-drilled with a Proxxon drill perpendicular to the growth axis on a non-occlusal tooth surface. Six teeth, two per locality for HO, EP, and CH, were serially sampled in 3-mm intervals perpendicular to the growth axis to investigate the seasonal intra-tooth variation of the carbon and oxygen isotopic composition (Table S1–S3).

The enamel powder was pretreated according to Koch et al. [114]. 10 mg enamel powder were soaked for 24 hours with 2% NaOCl and 1 M calcium-acetate buffer solution for 24 hours in a powder/solution ratio of 0.04 g/ml to remove organic substances and diagenetic carbonates, respectively, prior to analysis of the carbon ($\delta^{13}C$) and oxygen ($\delta^{18}O_{CO3}$) isotopic composition of the carbonate in the apatite. About 2 mg pretreated enamel powder was reacted with 100% H_3PO_4 for 90 minutes at 70°C using a ThermoFinnigan Gasbench II [115]. Carbon and oxygen isotope ratios of the generated CO_2 were measured in continuous flow mode on a Finnigan Delta Plus XL isotope ratio gas mass spectrometer at the University of Lausanne and Tübingen. For this reaction the same acid fractionation factor as between calcite and CO_2, was assumed to be applicable. The measured carbon and oxygen isotopic compositions were normalized to the in-house Carrara marble calcite standard that has been calibrated against the international NBS-19 calcite standard. The isotope composition of tooth enamel apatite is reported in the usual δ-notation in per mil (‰) relative to the known isotope reference standard VPDB with $\delta^{13}C$ or $\delta^{18}O$ (‰) = [($R_{sample}/R_{standard}$) −1]×1000, where R_{sample} and $R_{standard}$ are the $^{13}C/^{12}C$ and $^{18}O/^{16}O$ ratios in the sample and standard, respectively. Precision for the carbon ($\delta^{13}C$) and oxygen ($\delta^{18}O$) isotopic composition of carbonate in the apatite is better than ±0.1‰ and ±0.15‰, respectively. The NBS 120c Florida phosphate rock standard, also pre-treated after Koch et al. [114], gave values of $\delta^{13}C_{VPDB}$ = −6.23±0.09‰ and $\delta^{18}O_{VPDB}$ = −2.23±0.10‰ (n = 9).

Microwear Analysis

After initial examination, many specimens were excluded from analysis because they were physically altered during transportation or compaction of sediments [116]. Since microwear patterns may vary from mesial to distal teeth, the analysis is preferentially restricted to upper and lower M1s and M2s. The dental microwear patterns of fossil taxa were compared to those of extant grazing and browsing ungulates including artiodactyls and perissodactyls (Tables S4 and Figure 4). All microwear signatures of this study were only determined by one experienced observer (G. Merceron) to avoid any potential inter-observer error [117].

Regarding dental microwear analysis, several protocols were developed to quantify occlusal wear patterns at a microscopic scale. Differences concern casting procedures applied and the data acquisition. In this study the protocol of Merceron et al. [118] is employed because it combines the high quality of light stereo-microscopy with reliable analyses of high-resolution digitized images. To assess the feeding preferences the hipparionines were compared with extant species using discriminant analysis (DA) (Tables S4 and Figure 4). This multivariate statistical technique first evaluates the ability of the microwear variables to discriminate grazing species from the browsers. The mere pure grazing and browsing present-day ungulates are here considered as model for the DA; no intermediate feeders are considered. The dental microwear pattern of individuals engaged in both grazing and browsing is similar either to that of browsers or to that of grazers depending on their last few meals. A set of 9 variables (Ns, Np, Ls, Pp, Tot, Nws, Nfs, Nlp, Nsp) was considered for running the DA [92]. Because the normality and homogeneity of variance are not guaranteed, the variables were log-transformed. DA requires the very same assumptions as the analysis of variance. Then, the extinct hipparionine specimens are classified into the browsing or the grazing kernel according to the model set up with living species (Figure 2 and Table S4). As suggested by DeGusta and Vrba [119], a threshold probability here set up at 5% is used to distinguish the significant predictions from more questionable reliability.

Mesowear Analysis

Dental facet development on the molar surfaces of herbivorous ungulates appears to be strongly tied to feeding styles [86,120]. Dental mesowear reflects the degree of attritive and abrasive wear on the molar and premolar occlusal surface. Attritive wear is due to the tooth/tooth contact and results in high crown relief and sharp cusp apices. Abrasive wear is due to tooth/food contacts, obliterates dental facets and results in lower crown relief and more round or blunted cusp apices. Initially dental mesowear analysis was based only on the second upper molars [86], however, Kaiser and Solounias [121] extended the analytical model to include three more maxillary tooth positions (P4, M1, and M3) in hypsodont equids. The mesowear method treats ungulate tooth mesowear as two variables: occlusal relief and cusp shape.

Occlusal relief (OR) is classified as high (H) or low (L), depending on how high the cusps rise above the valley between them. The second mesowear variable, cusp shape, includes 3 scored attributes: sharp (S), round (R) and blunt (B) according to the degree of facet development. In addition to established mesowear convention, a combined mesowear score was computed from each population similar [110,122]. The convention used, however differs from that used by Semprebon and Rivals [123] by inclusion of a score for the combination of low reliefs and sharp cusps. This combination is frequently found especially in grazers, but was not accommodated by Semprebon and Rivals [123]. A combination of high relief and sharp cusps was assigned a score of "0", a combination of high relief and round cusp was assigned a score of "1", a combination of low relief and sharp cusp was assigned a score of "2", a combination of low relief and round cusp was assigned a score of "3" and a combination of low relief and blunt cusps was assigned a score of "4". In this convention, a score of 0 represents the most attrition-dominated mesowear signature, while a score of 4 would represent the most abrasion-dominated signature. Individual scores were averaged and a mean score was calculated for each species (Fig. 3). Scores thus indicate the over all abrasiveness of the diet a species has to cope with. As comparative dataset, we consider the medium-size ungulates (artiodactyls and perissodactyls) with sample size clustered in broad diet categories (Figure 4, Table S5).

Supporting Information

Figure S1 $\delta^{13}C$ and $\delta^{18}O$ values of enamel and dentin H. primigenium teeth from the Höwenegg locality as well as on sediment sample from the main fossil-bearing layer. Note the large difference in $\delta^{13}C$ values indicating a diagenetic alteration of the dentine while enamel still has values typical for C_3 feeders.

Table S1   C and O isotopic compositions of the serial sampled Eppelsheim H. primigenium teeth.

Table S2   C and O isotopic compositions of the Höwe-negg H. primigenium teeth.

Table S3   C and O isotopic compositions of the Charmoille H. primigenium teeth.

Table S4   Summary statistics of the main dental microwear variables for the browsing and grazing ungulates that composed the comparative dataset.

Table S5   List of extant species used for mesowear analysis.

Text S1   Detailed site description of the fossil localites.

Acknowledgments

We thank the Forschungsinstitut und Naturmuseum Senckenberg (Frankfurt), the Staatliches Museum für Naturkunde (Karlsruhe), and the Landesmuseum Rheinland-Pfalz (Mainz), the Collections d'Histoire Naturelle de l'Université de Lyon (A. Prieur) and the Naturhistorisches Museum Basel (L. Costeur) for access to the fossil specimens investigated in this study. In particular, we wish to warmly thank Dr. Jens Lorenz Franzen for allowing us to include the hipparion skeletal material from the excavations 1996–2000 of the Forschungsinstitut Senckenberg at Eppelsheim into this study. Furthermore we would like to thank Dr. Burkhard Engesser who provided the hipparionine teeth from Charmoille for the stable isotope analyses. Finally we thank Matt Mihlbachler and one anonymous reviewer for their constructive comments that helped to improve the manuscript significantly. Last but not least Jo Hellawell is acknowledged for proof reading the English.

Author Contributions

Conceived and designed the experiments: TT GM TMK. Performed the experiments: TT GM TMK. Analyzed the data: TT GM TMK. Contributed reagents/materials/analysis tools: TV. Wrote the paper: TT GM TMK.

References

1. Simpson GG (1951) Horses. Oxford: Oxford University Press. 323 p.
2. MacFadden BJ (1992) Fossil Horses: Systematics, Paleobiology, and Evolution of the Family Equidae. Cambridge: Cambridge University Press. 369 p.
3. MacFadden BJ (2005) Fossil horses - evidence for evolution Science 307: 1728–1730.
4. Janis CM (2008) An evolutionary history of browsing and grazing ungulates. Ecological Studies 195: 21–45.
5. Merceron G, Blondel C, Brunet M, Sen S, Solounias N, et al. (2004) The late Miocene paleoenvironment of Afghanistan as inferred from dental microwear in artiodactyls. Palaeogeography, Palaeoclimatology, Palaeoecology 207: 143–163.
6. Merceron G, Schulz E, Kordos L, Kaiser TM (2007) Paleoenvironment of *Dryopithecus brancoi* at Rudabánya, Hungary: evidence from dental meso- and micro-wear analyses of large vegetarian mammals. Journal of Human Evolution 53: 331–349.
7. Schulz E, Piotrowski V, Clauss M, Mau M, Merceron G, et al. (2013) Dietary abrasiveness determines variability in microwear and dental surface texture in rabbits. PLoS One 8: e56167.
8. Lucas PW, Omar R, Al-Fadhalah K, Almusallam AS, Henry AG, et al. (2013) Mechanisms and causes of wear in tooth enamel: implications for hominin diets. Journal of Royal Society Interface 10.
9. Sanson GD, Kerr SA, Gross KA (2007) Do silica phytoliths really wear mammalian teeth? Journal of Archaeological Science 34: 526–531.
10. Hummel J, Findeisen E, Südekum K-H, Ruf I, Kaiser TM, et al. (2011) Another one bites the dust: faecal silica levels in large herbivores correlate with hypsodonty. Proceedings of the Royal Society of London B 278: 1742–1747.
11. Damuth J, Janis C (2011) On the relationship between hypsodonty and feeding ecology in ungulate mammals, and its utility in palaeoecology. Biological Reviews 86: 733–758.
12. Mihlbachler MC, Rivals F, Solounias N, Semprebon GM (2011) Dietary change and evolution of horses in North America. Science 331: 1178–1181.
13. Strömberg CAE (2002) The origin and spread of grass dominated ecosystems in the late Tertiary of North America: preliminary results concerning the evolution of hypsodonty. Palaeogeography, Palaeoclimatology, Palaeoecology 177: 59–75.
14. Strömberg CAE (2006) Evolution of hypsodonty in equids: testing a hypothesis of adaptation. Paleobiology 32: 236–258.
15. MacFadden BJ, Solounias N, Cerling TE (1999) Ancient diets, ecology, and extinction of 5-million-year-old horses from Florida. Science 283: 824–827.
16. Woodburne MO (2009) The early Vallesian vertebrates of Atzelsdorf (Late Miocene, Austria). 9. *Hippotherium* (Mammalia, Equidae). Annalen des Naturhistorischen Museums in Wien, Serie A, 111: 585–604.
17. Harzhauser M (2009) The early Vallesian vertebrates of Atzelsdorf (Late Miocene, Austria). 2. Geology. Annalen des Naturhistorischen Museums in Wien, Serie A, 111: 479–488.
18. Strömberg CAE, Werdelin L, Friis EM, Sarac G (2007) The spread of grass-dominated habitats in Turkey and surrounding areas during the Cenozoic: phytolith evidence. Palaeogeography, Palaeoclimatology, Palaeoecology 250: 18–49.
19. Cerling TE, Wang Y, Quade J (1993) Expansion of C4 ecosystems as an indicator of global ecological change in the late Miocene. Nature 361: 344–345.
20. Cerling TE, Harris JR, MacFadden BJ, Leakey MG, Quade J, et al. (1997) Global vegetation change through the Miocene/Pliocene boundary. Nature 389: 153–158.
21. Liem KF (1980) Adaptive significance of intraspecific and interspecific differences in the feeding repertoires of cichlid fishes. American Zoologist 20: 295–314.
22. Robinson BW, Wilson DS (1998) Optimal foraging, specialization, and a solution to Liem's paradox. American Naturalist 151: 223–235.
23. Ménouret B, Mein P (2008) Les vertébrés du Miocène supérieur de Soblay (Ain, France). Documents du Laboratoire de Géologie de Lyon 165: 1–97.
24. Bernor RL, Koufos GD, Woodburne MO, Fortelius M (1996) The evolutionary history and biochronology of European and Southwest Asian Late Miocene and Pliocene Hipparionine Horses. In: Bernor RL, Fahlbusch V, Mittmann H-W, editors. The Evolution of Western Eurasian Neogene Mammal Faunas. New York: Columbia University Press. pp. 307–338.
25. Grimm MC (2005) Beiträge zur Lithostratigraphie des Paläogens und Neogens im Oberrheingebiet (Oberrheingraben, Mainzer Becken, Hanauer Becken). Geologisches Jahrbuch Hessen 132: 79–112.
26. Franzen JL (2011) Eppelsheim Formation. In: Grimm KI, editor. Stratigraphie von Deutschland IX Tertiär, Teil 1: Oberrheingraben und benachbarte

Tertiärgebiete. Hannover: Deutsche Gesellschaft für Geowissenschaften. pp. 184–187.
27. Woodburne MO, Bernor RL, Swisher CCI (1996) An appraisal of the stratigraphic and phylogenetic bases for the "Hipparion Datum" in the Old World. In: Bernor RL, Fahlbusch V, Mittmann HW, editors. The Evolution of Western Eurasian Neogene Mammal Faunas. New York: Columbia University Press. pp. 124–136.
28. Steininger FF (1999) Chronostratigraphy, Geochronology and Biochronology of the Miocene "European Land Mammal Mega-Zones (ELMMZ)" and the Miocene "Mammal-Zones (MN-Zones)". In: Rössner GE, Heissig K, editors. Land Mammals of Europe. München: Verlag Friedrich Pfeil. pp. 9–24.
29. Andrews PA, Bernor RL (1999) Vicariance Biogeography and Paleoecology of Eurasian Miocene hominoid primates. In: Agusti J, Rook L, Andrews P, editors. The Evolution of Neogene Terrestrial Ecosystems in Europe. Cambridge: Cambridge University Press, pp. 454–488.
30. Böhme M, Aiglstorfer M, Uhl D, Kullmer O (2012) The Antiquity of the Rhine River: Stratigraphic Coverage of the Dinotheriensande (Eppelsheim Formation) of the Mainz Basin (Germany). PLoS ONE 7: e36817.
31. Franzen JL, Fejfar O, Storch G, Wilde V, editors (2003) Eppelsheim 2000 - new discoveries at a classic locality. Rotterdam: Deinsea. 217–234 p.
32. Franzen JL (2000) Auf dem Grunde des Urrheins – Ausgrabungen bei Eppelsheim. Natur und Museum 130: 169–180.
33. Swisher CC (1996) New $^{40}Ar/^{39}Ar$ dates and their contribution toward a revised chronology for the late Miocene of Europe and West Asia. In: Bernor RL, Fahlbusch V, Mittmann H-W, editors. The evolution of western Eurasian Neogene mammal faunas. New York: Columbia University Press. pp. 64–77.
34. Munk W, Bernor RL, Heizmann EPJ, Mittmann HW (2007) Excavations at the Late Miocene MN9 (10.3 Ma) locality of Höwenegg (Hegau), southwest Germany, 2004–2006. Carolinea 65: 5–13.
35. Kälin D (1997) Litho- und Biostratigraphie der mittel- bis obermiozänen Bois de Raube-Formation (Nordwestschweiz). Eclogae Geologicae Helvetiae 90: 97–114.
36. Becker D (2003) Paléoécologie et paléoclimats de la Molasse du Jura (Oligo-Miocène): apport des Rhinocerotoidea (Mammalia) et des minéraux argileux. GeoFocus 9: 327.
37. Sharp ZD, Cerling TE (1998) Fossil isotope records of seasonal climate and ecology: Straight from the horse's mouth. Geology 26: 219–222.
38. Hoppe KA, Stover SM, Pascoe JR, Amundson R (2004) Tooth enamel biomineralization in extant horses: implications for isotopic microsampling. Palaeogeography Palaeoclimatology Palaeoecology 203: 299–311.
39. Bryant JD, Froelich PN, Showers WJ, Genna BJ (1996) Biologic and climatic signals in the oxygen isotopic composition of Eocene-Oligocene equid enamel phosphate. Palaeogeography, Palaeoclimatology, Palaeoecology 126: 75–89.
40. Balasse M (2002) Reconstructing dietary and environmental history from enamel isotopic analysis: time resolution of intra-tooth sequential sampling. International Journal of Osteoarchaeology 12: 155–165.
41. Fricke HC, O'Neil JR (1996) Inter- and intra-tooth variation in the oxygen isotope composition of mammalian tooth enamel phosphate: implications for palaeoclimatological and palaeobiological research. Palaeogeography, Palaeoclimatology, Palaeoecology 126: 91–99.
42. Nelson SV (2007) Isotopic reconstructions of habitat change surrounding the extinction of *Sivapithecus*, a Miocene hominoid, in the Siwalik Group of Pakistan. Palaeogeography, Palaeoclimatology, Palaeoecology 243: 204–222.
43. Tieszen LL, Fagre T (1993) Effect of diet quality and composition on the isotopic composition of respiratory CO₂, bone collagen, bioapatite, and soft tissues. In: Lambert JB, Grupe G, editors. Prehistoric Human Bone: Archaeology at the Molecular Level. Berlin: Springer-Verlag. pp. 121–155.
44. DeNiro MJ, Epstein S (1978) Influence of diet on the distribution of carbon isotopes in animals. Geochimica et Cosmochimica Acta 42: 495–506.
45. Cerling TE, Harris JM, MacFadden BJ (1997) Carbon isotopes, diets of North American equids, and the evolution of North American C4 grasslands. In: Griffiths H, Robinson D, van Gardingen P, editors. Stable isotopes and the integration of biological, ecological, and geochemical processes. Oxford: Bios Scientific Publishers. pp. 363–379.
46. Deines P (1980) The isotopic composition of reduced organic carbon. In: Fritz P, Fontes C, editors. Handbook of Environmental Geochemistry. New York: Elsevier. pp. 239–406.
47. Cerling TE, Harris JM (1999) Carbon isotope fractionation between diet and bioapatite in ungulate mammals and implications for ecological and paleoecological studies Oecologia 120: 347–363.

48. Passey BH, Robinson TF, Ayliffe LK, Cerling TE, Sponheimer M, et al. (2005) Carbon isotope fractionation between diet, breath CO_2, and bioapatite in different mammals. Journal of Archaeological Science 32: 1459–1470.

49. Sullivan CH, Krueger HW (1981) Carbon isotope analysis of separate chemical phases in modern and fossil bone. Nature 301: 177–178.

50. Lee-Thorp JA, van der Merwe NJ (1987) Carbon isotope analysis of fossil bone apatite. South African Journal of Science 83: 712–715.

51. Quade J, Cerling TE, Barry JC, Morgan ME, Pilbeam DR, et al. (1992) A 16-Ma record of paleodiet using carbon and oxygen isotopes in fossil teeth from Pakistan. Chemical Geology 94: 183–192.

52. Mateu Andrés I (1993) A revised list of the European C4 plants. Photosynthetica 26: 323–331.

53. Drucker D, Bocherens H, Bridault A, Billiou D (2003) Carbon and nitrogen isotopic composition of red deer (Cervus elaphus) collagen as a tool for tracking palaeoenvironmental change during the Late Glacial and Early Holocene in the northern Jura (France). Palaeogeography, Palaeoclimatology, Palaeoecology 195: 375–388.

54. Feranec RS, MacFadden BJ (2006) Isotopic discrimination of resource partitioning among ungulates in C_3-dominated communities from the Miocene of Florida and California. Paleobiology 32: 191–205.

55. Cerling TE, Hart JA, Hart TB (2004) Stable isotope ecology in the Ituri Forest. Oecologia 138: 5–12.

56. Tütken T, Vennemann T (2009) Stable isotope ecology of Miocene large mammals from Sandelzhausen, Germany. Paläontologische Zeitschrift 83: 207–226.

57. Heaton THE (1999) Spatial, species, and temporal variations in the $^{13}C/^{12}C$ ratios of C_3 plants: implications for palaeodiet studies. Journal of Archaeological Science 26: 637–649.

58. Kohn MJ (2010) Carbon isotope compositions of terrestrial C3 plants as indicators of (paleo)ecology and (paleo)climate. Proceedings of the National Academy of Sciences of the United States of America 107: 19691–19695.

59. Diefendorf AF, Mueller KE, Wing SL, Koch PL, Freeman KH (2010) Global patterns in leaf ^{13}C discrimination and implications for studies of past and future climate. Proceedings of the National Academy of Sciences of the United States of America 107: 5738–5743.

60. Longinelli A (1984) Oxygen isotopes in mammal bone phosphate: A new tool for palaeohydrological and palaeoclimatological research? Geochimica et Cosmochimica Acta 48: 385–390.

61. Kohn MJ (1996) Predicting animal $\delta^{18}O$: Accounting for diet and physiological adaptation. Geochimica et Cosmochimica Acta 60: 4811–4829.

62. Huertas AD, Iacumin P, Stenni B, Chillon BS, Longinelli A (1995) Oxygen isotope variations of phosphate in mammalian bone and tooth enamel. Geochimica et Cosmochimica Acta 59: 4299–4305.

63. Dansgaard W (1964) Stable isotopes in precipitation. Tellus 16: 436–468.

64. Rozanski K, Araguás-Araguás L, Gonfiantini R (1993) Isotopic patterns in modern global precipitation. Geophysical Monograph 78: 1–36.

65. Levin NE, Cerling TE, Passey BH, Harris JM, Ehleringer JR (2006) A stable isotope aridity index for terrestrial environments. Proceedings of the National Academy of Sciences of the United States of America 103: 11201–11205.

66. Tütken T, Vennemann TW, Janz H, Heizmann EPJ (2006) Palaeoenvironment and palaeoclimate of the Middle Miocene lake in the Steinheim basin, SW Germany: A reconstruction from C, O, and Sr isotopes of fossil remains. Palaeogeography, Palaeoclimatology, Palaeoecology 241: 457–491.

67. Bryant JD, Froelich PN (1995) A model of oxygen isotope fractionation in body water of large mammals. Geochimica et Cosmochimica Acta 60: 4523–4537.

68. Kohn MJ, Schoeninger MJ, Valley JW (1996) Herbivore tooth oxygen isotope compositions: Effects of diet and physiology. Geochimica et Cosmochimica Acta 60: 3889–3896.

69. Ayliffe LK, Chivas AR (1990) Oxygen isotope composition of the bone phosphate of Australian kangaroos: Potential as a palaeoenvironmental recorder. Geochimica et Cosmochimica Acta 54: 2603–2609.

70. Sponheimer M, Lee-Thorp JA (1999) Oxygen isotopes in enamel carbonate and their ecological significance. Journal of Archaeological Science 26: 723–728.

71. Kohn MJ, Schoeninger MJ, Valley JW (1998) Variability in oxygen isotope compositions of herbivore teeth: reflections of seasonality or developmental physiology? Chemical Geology 152: 97–112.

72. Nelson SV (2005) Paleoseasonality inferred from equid teeth and intra-tooth isotopic variability. Palaeogeography, Palaeoclimatology, Palaeoecology 222: 122–144.

73. van Dam JA, Reichart GJ (2009) Oxygen and carbon isotope signatures in late Neogene horse teeth from Spain and application as temperature and seasonality proxies. Palaeogeography, Palaeoclimatology, Palaeoecology 274: 64–81.

74. Balasse M, Ambrose SH, Smith AB, Price D (2002) The seasonal mobility model for prehistoric herders in the south-western Cape of South Africa assessed by isotopic analysis of sheep tooth enamel. Journal of Archaeological Science 29: 917–932.

75. Gügel IL, Grupe G, Kunzelmann K-H (2001) Simulation of dental microwear: Characteristic traces by opal phytoliths give clues to ancient human dietary behavior. American Journal of Physical Anthropology 114: 124–138.

76. Teaford MF, Oyen OJ (1989) In vivo and in vitro turnover in dental microwear. American Journal of Physical Anthropology 80: 447–460.

77. Scott RS, Teaford MF, Ungar PS (2012) Dental microwear texture and anthropoid diets. American Journal of Physical Anthropology 147: 551–579.

78. Scott JR (2012) Dental microwear texture analysis of extant African Bovidae. Mammalia 76: 157–174.

79. Merceron G, Escarguel G, Angibault J-M, Verheyden-Tixier H (2010) Can dental microwear textures record dietary inter-individual dietary variations? PLoS ONE 5(3): e9542.

80. Ungar PS, Scott RS, Scott JR, Teaford MF (2008) Dental microwear analysis: historical perspectives and new approaches In: Irish JD, Nelson GC, editors. Volume on Dental Anthropology. Cambridge: Cambridge University. pp. 389–425.

81. Ungar PS, Merceron G, Scott RS (2007) Dental microwear texture analysis of Varswater bovids and Early Pliocene paleoenvironments of Langebaanweg, Western Cape Province, South Africa. Journal of Mammalian Evolution 14: 163–181.

82. Blondel C, Merceron G, Andossa L, Mackaye HT, Vignaud P, et al. (2010) - (Chad) and early hominid habitats in Central Africa. Palaeogeography, Palaeoclimatology, Palaeoecology 292: 184–191.

83. Rivals F, Mihlbachler MC, Solounias N (2007) Effect on the ontogenetic-age distribution in fossil and modern samples on the interpretation of ungulate paleodiets using the mesowear method. Journal of Vertebrate Paleontology 27: 763–767.

84. Franz-Odendaal T, Solounias N (2004) Comparative dietary evaluations of an extinct giraffid (Sivatherium hendeyi) (Mammalia, Giraffidae, Sivatheriinae) from Langebaanweg, South Africa (Early Pliocene). Geodiversitas 26: 675–685.

85. Kaiser TM (2003) The dietary regimes of two contemporary populations of Hippotherium primigenuim (Perissodactyla, Equidae) from the Vallesian (Upper Miocene) of Southern Germany. Palaeogeography, Palaeoclimatology, Palaeoecology 198: 381–402.

86. Fortelius M, Solounias N (2000) Functional characterization of ungulate molars using the abrasion-attrition wear gradient: A new method for reconstructing paleodiets. American Museum Novitates 3301: 1–36.

87. Solounias N, Teaford MF, Walker A (1988) Interpreting the diet of extinct ruminants: the case of a non-browsing giraffid. Paleobiology 14: 287–300.

88. Kaufman PB, Dayanandan P, Franklin CI (1985) Structure and function of silica bodies in the epidermal system of grass bodies. Annals of Botany 55: 487–507.

89. Lanning FC, Eleuterius LN (1989) Silica deposition in some C_3 and C_4 species of grasses, sedges, and composites in the USA. Annals of Botany 63: 395–410.

90. Mac Naughton SJ, Tarrants JL, Mac Naughton MM, Davis RH (1985) Silica as a defense against herbivory and a growth promotor in African grasses. Ecology 66: 528–535.

91. Solounias N, Semprebon G (2002) Advances in the reconstruction of ungulates ecomorphology with application to early fossil equids. American Museum Novitates 3366: 1–49.

92. Merceron G, Kaiser TM, Kostopoulos DS, Schulz E (2010) Ruminant diet and the Miocene extinction of European great apes. Proceedings of the Royal Society B 277: 3105–3112.

93. Walker A, Hoeck HN, Perez L (1978) Microwear of mammalian teeth as an indicator of diet. Science 201: 908–910.

94. Ramdarshan A, Alloing-Séguier T, Merceron G, Marivaux L (2012) Ecological niche partitioning in a modern South American primate community: implications for extinct species. PLoS One 6: e27392.

95. Teaford MF, Robinson JG (1989) Seasonal or ecological differences in diet and molar microwear in Cebus nigrivittatus. American Journal of Physical Anthropology 80: 391–401.

96. Ehleringer JR, Cerling TE, Helliker BR (1997) C_4 photosynthesis, atmospheric CO_2, and climate. Oecologia 112: 285–299.

97. Quade J, Cerling TE, Andrews P, Alpagut A (1995) Paleodietary reconstruction of Miocene faunas at Pasalar, Turkey using stable carbon and oxygen isotopes of fossil tooth enamel. Journal of Human Evolution 28: 373–384.

98. Passey BH, Cerling TE, Perkins ME, Voorhies MR, Harris JM, et al. (2002) Environmental change in the great plains: An isotopic record from fossil horses. The Journal of Geology 110: 123–140.

99. Van der Merwe NJ, Medina E (1989) Photosynthesis and $^{13}C/^{12}C$ ratios in Amazonian rain forests. Geochimica et Cosmochimica Acta 53: 1091–1094.

100. Yakir D (1997) Oxygen-18 of leaf water: a crossroad for plant associated isotopic signals. In: Griffith H, editor. Stable isotopes: integration of biological, ecological and geochemical processes. Oxford: BIOS. pp. 147–168.

101. Mihlbachler MC, Solounias N (2006) Coevolution of tooth crown height and diet in Oreodonts (Merycoidodontidae, Artiodactyla) examined with phylogenetically independant contrast. Journal of Mammalian Evolution 13: 11–36.

102. de Bonis L, Bouvrain G, Geraads D, Koufos GD (1992) Multivariate study of the late Cenozoic mammalian faunal compositions and paleoecology. Paleontologia i Evolució 24–25: 93–101.

103. Fortelius M, Eronen J, Liu L, Pushkina D, Tesakov A, et al. (2006) Late Miocene and Pliocene large land mammals and climatic changes in Eurasia. Palaeogeography, Palaeoclimatology, Palaeoecology 238: 219–227.

104. Fortelius M, Eronen J, Liu L, Pushkina D, Tesakov A, et al. (2003) Continental-scale hypsodonty patterns, climatic paleobiogeography and dispersal of Eurasian Neogene large mammal herbivores. In: Reumer JWF, Wessels W, editors. Distribution and Migration of Tertiary Mammals in Eurasia: DEINSEA. pp. 1–11.

105. Zhang Z-Q, Gentry AW, Kaakinen A, Liu L-P, Lunkka JP, et al. (2002) Land mammal faunal sequence in the late Miocene of China: new evidence from Lantian, Shaanxi province. Vertebrata Palasiatica 40: 166–176.

106. Laden G, Wrangham R (2005) The rise of the hominids as an adaptative shift in fallback foods: Plant underground strorage organs (USOs) and australopith origins. Journal of Human Evolution 49: 482–498.

107. Feranec RS (2003) Stable isotopes, hypsodonty, and the paleodiet of *Hemiauchenia* (Mammalia: Camelidae): a morphological specialization creating ecological generalization. Paleobiology 29: 230–242.

108. Kahlke R-D, García N, Kostopoulos DS, Lacombat F, Lister AM, et al. (2011) Western Palaearctic palaeoenvironmental conditions during the Early and early Middle Pleistocene inferred from large mammal comunities, and implications for hominin dispersal in Europe. Quaternary Science Reviews 30: 1368–1395.

109. Jernvall J, Fortelius M (2002) Common mammals drive the evolutionary increase of hypsodonty in the Neogene. Nature 417: 538–540.

110. Kaiser TM (2009) *Anchitherium aurelianense* (Equidae, Mammalia)–a brachydont "dirty browser" in the community of herbivorous large mammals from Sandelzhausen (Miocene, Germany). Paläontologische Zeitschrift 83: 131–140.

111. Daxner-Höck G, Bernor RL (2009) The early Vallesian vertebrates of Atzelsdorf (Late Miocene, Austria) 8. *Anchitherium*, Suidae, and Castoridae (Mammalia). Annalen des Naturhistorischen Museums in Wien, Serie A, 111: 557–584.

112. Sondaar PY (1974) The Hipparion of the Rhone valley. Geobios 7: 289–306.

113. Eronen JT, Evans AR, Fortelius M, Jernvall J (2010) The impact of regional climate on the evolution of mammals: a case study using fossil horses. Evolution 64: 398–408.

114. Koch PL, Tuross N, Fogel ML (1997) The effects of sample treatment and diagenesis on the isotopic integrity of carbonate in biogenic hydroxylapatite Journal of Archaeological Science 24: 417–429.

115. Spötl C, Vennemann TW (2003) Continuous-flow isotope ratio mass spectrometric analysis of carbonate minerals. Rapid Communication in Mass Spectrometry 17: 1004–1006.

116. King T, Andrews P, Boz B (1999) Effect of taphonomic processes on dental microwear. American Journal of Physical Anthropology 108: 359–373.

117. Mihlbachler MC, Brian BL, Caldera-Siu A, Chan D, Lee R (2012) Error rates and observer bias in dental microwear analysis using light microscopy. Palaeontologia Electronica 15: 1–22.

118. Merceron G, Blondel C, de Bonis L, Koufos GD, Viriot L (2005) A new dental microwear analysis: application to extant Primates and *Ouranopithecus macedoniensis* (Late Miocene of Greece). Palaios 20: 551–561.

119. DeGusta D, Vrba E (2005) Methods for inferring paleohabitats from discrete traits of the bovid postcranial skeleton. Journal of Archaeological Science 32: 1115–1123.

120. Kaiser TM, Bernor R, Fortelius M, Scott R (2000) Ecological diversity in the Neogene genus *Hippotherium* (Perissodactyla, Equidae) from the late Miocene of Central Europe. Journal of Vertebrate Paleontology 20 Supplement: 51A.

121. Kaiser TM, Fortelius M (2003) Differential mesowear in occluding upper and lower molars: Opening mesowear analysis for lower molars and premolars in hypsodont horses. Journal of Morphology 258: 63–83.

122. Kaiser TM (2011) Feeding ecology and niche partitioning of the Laetoli ungulate faunas. In: Harrison T, editor. Paleontology and Geology of Laetoli: Human Evolution in Context: Volume 2: Fossil Hominins and the Associated Fauna (Vertebrate Paleobiology and Paleoanthropology). Springer. pp. 329–354.

123. Semprebon GM, Rivals F (2007) Was grass more prevalent in the pronghorn past? An assessment of the dietary adaptations of Miocene to Recent Antilocapridae (Mammalia: Artiodactyla). Palaeogeography, Palaeoclimatology, Palaeoecology 253: 332–347.

14

Skull Ecomorphology of Megaherbivorous Dinosaurs from the Dinosaur Park Formation (Upper Campanian) of Alberta, Canada

Jordan C. Mallon[1][*][¤]**, Jason S. Anderson**[2]

1 Department of Biological Sciences, University of Calgary, Calgary, Alberta, Canada, 2 Department of Comparative Biology & Experimental Medicine, University of Calgary, Calgary, Alberta, Canada

Abstract

Megaherbivorous dinosaur coexistence on the Late Cretaceous island continent of Laramidia has long puzzled researchers, owing to the mystery of how so many large herbivores (6–8 sympatric species, in many instances) could coexist on such a small (4–7 million km²) landmass. Various explanations have been put forth, one of which–dietary niche partitioning–forms the focus of this study. Here, we apply traditional morphometric methods to the skulls of megaherbivorous dinosaurs from the Dinosaur Park Formation (upper Campanian) of Alberta to infer the ecomorphology of these animals and to test the niche partitioning hypothesis. We find evidence for niche partitioning not only among contemporaneous ankylosaurs, ceratopsids, and hadrosaurids, but also within these clades at the family and subfamily levels. Consubfamilial ceratopsids and hadrosaurids differ insignificantly in their inferred ecomorphologies, which may explain why they rarely overlap stratigraphically: interspecific competition prevented their coexistence.

Editor: Richard J. Butler, Ludwig-Maximilians-Universität München, Germany

Funding: Funding to JCM was provided by a Natural Sciences and Engineering Research Council Alexander Graham Bell Canada Graduate Scholarship, Alberta Innovates Technology Futures graduate student scholarship, Jurassic Foundation grant and a Queen Elizabeth II Graduate Scholarship. The funders had no role in study design, data collection and analysis, decision to publish, or preparation of the manuscript.

Competing Interests: The authors have declared that no competing interests exist.

* E-mail: jmallon@mus-nature.ca

¤ Current address: Palaeobiology, Canadian Museum of Nature, Ottawa, Ontario, Canada

Introduction

Megaherbivore Diversity on Laramidia

Megaherbivorous dinosaur diversity on the Late Cretaceous island continent of Laramidia [1] was exceptionally high (particularly during the late Campanian [2–4]), and various metabolic, demographic, and biogeographic considerations about these animals (reviewed in Mallon et al. [5]) have caused many to wonder how so many large herbivores could coexist on such a small landmass. Two main hypotheses have traditionally been given in response to this question. One is that plant resources on Laramidia were not limiting, due to dinosaurian bradymetabolism [2,6,7], elevated Late Cretaceous primary productivity [7,8], and/ or predation pressure [5]. Alternatively, Laramidian plant resources may have been limiting, and megaherbivorous dinosaur coexistence was achieved via dietary niche partitioning [3,9,10]. This hypothesis has received little attention in the literature and is the focus of the present study.

The upper Campanian Dinosaur Park Formation (DPF) of Alberta is the uppermost unit of the Belly River Group, and comprises alluvial, estuarine, and paralic facies [11,12]. We chose the megaherbivore assemblage of the DPF as a study model for three reasons: (1) the fossil record of the DPF is exceptionally rich [13]; (2) it preserves the same suite of megaherbivorous dinosaur taxa present in most time-contemporaneous strata elsewhere in western North America (e.g., ankylosaurids, nodosaurids, centro-

saurines, chasmosaurines, hadrosaurines, and lambeosaurines; Figure 1); (3) its biostratigraphy is well understood [14–17] so that constituent taxa can be compared in biologically meaningful ways. These same considerations have influenced use of the DPF as a model in other studies of palaeoecology and taphonomy [13,17,18].

Herbivore Ecomorphology

If natural selection has acted to allow vertebrate herbivores to forage optimally [19,20] on different plants or plant parts, those herbivores should exhibit a variety of skull morphologies that correspond to variation in the plants they eat. The relationship between an organism and its environment ('synerg' sensu Bock and von Wahlert [21]) comprises the interaction between the biological role of some feature of that organism and the selection force acting upon it by the environment. The ecological morphology (ecomorphology) of an organism is therefore a reflection of the environment in which its parent population evolved [22]. This relationship is imperfect, owing to redundancy in the form-function complex [23] and to the confounding effects of phylogenetic inertia [24]. Nonetheless, a considerable body of work has demonstrated a fundamental relationship between herbivore skull morphology and the physical properties of the plants on which they feed (e.g., [25–33]).

With these principles in mind, numerous authors have suggested that dietary niche partitioning among the megaherbivorous

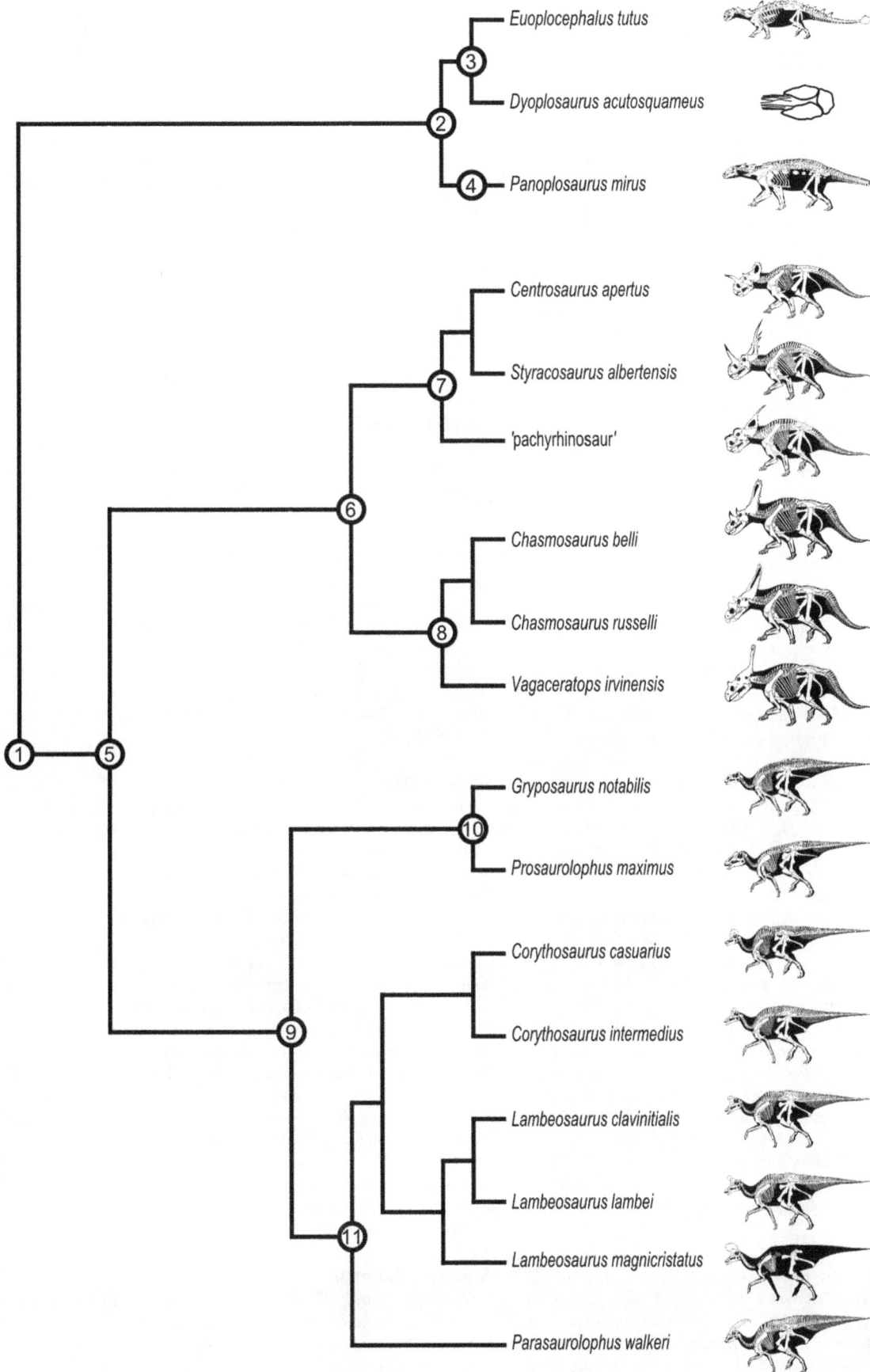

Figure 1. Phylogenetic relationships of megaherbivorous dinosaurs from the Dinosaur Park Formation. Suprageneric taxonomy: 1, Genasauria; 2, Ankylosauria; 3, Ankylosauridae; 4, Nodosauridae; 5, Cerapoda; 6, Ceratopsidae; 7, Centrosaurinae; 8, Chasmosaurinae; 9, Hadrosauridae; 10, Hadrosaurinae; 11, Lambeosaurinae. After Butler et al. [149], Prieto-Márquez [166], Sampson et al. [167], and Thompson et al. [168]. Skeletal drawings (not to scale) by G. S. Paul (used with permission).

dinosaurs from the DPF might have been facilitated via differential skull morphology [34–46]; however, this hypothesis has never been tested systematically. The present study seeks to test the hypothesis that the long-term coexistence of these animals was facilitated by dietary niche partitioning, with special focus given to inferring the ecomorphologies of their skulls.

Institutional Abbreviations

All specimens were studied with permission from the following institutions: AMNH, American Museum of Natural History, New York; CMN, Canadian Museum of Nature, Ottawa; FMNH, Field Museum of Natural History, Chicago; NHMUK, Natural History Museum, London; ROM, Royal Ontario Museum, Toronto; TMM, Texas Memorial Museum, Austin; TMP, Royal Tyrrell Museum of Palaeontology, Drumheller, Alberta; UALVP, University of Alberta Laboratory of Vertebrate Palaeontology, Edmonton; USNM, National Museum of Natural History, Washington, D. C.; YPM, Yale Peabody Museum, New Haven.

Materials and Methods

Hypotheses

Ricklefs and Miles [47] noted that, in ecological communities shaped by the forces of competition, species overlap in morphospace tends to be minimized, reflecting the different niche requirements of the constituent species. The corollary of this is that, in communities where niche partitioning plays a negligible role, the morphological overlap of species is unconstrained.

In view of these considerations, our null hypothesis is that Late Cretaceous plant resources of Laramidia were not limiting, and that the coexistence of the megaherbivorous dinosaurs from the DPF was not facilitated by niche partitioning. In this case, we would expect to find significant overlap in morphospace, particularly between closely related species, reflecting the similar dietary niche requirements of the megaherbivores.

Our alternative hypothesis is that plant resources were limiting, and that the coexistence of the megaherbivores was facilitated by niche partitioning. If true, we would expect that species overlap in morphospace should be minimized, reflecting their different dietary niche requirements.

Ricklefs and Miles ([47]: p. 30) also emphasized that meaningful interpretations of morphospace require "a judicious selection of morphological variables reflecting a priori biomechanical function". The morphometric model used here was therefore conceived in light of biomechanical design analyses and form-function correlations [48–50] observed in living herbivores. Using a combination of ordination and statistical methods (described below), we sought to quantify the degree to which megaherbivores from the DPF overlap in these morphological parameters.

Morphometrics

The ecomorphological model employed here comprises 12 linear measurements of the skull (Figure 2; Table 1), selected because of their perceived ability to reflect such aspects as plant quality, mechanical properties, and growth habit. The choice of variables stemmed from a literature pertaining to a variety of vertebrates, including lizards, turtles, birds, ungulates, macropodids, and primates. Only specimens preserving more than half

of the measurements were included in the analysis to reduce the confounding effects of missing data. The total dataset (Table S1), encompassing nearly all suitable material available from the DPF, comprised 82 specimens spanning 12 megaherbivorous dinosaur genera from the clades Ankylosauria, Ceratopsidae, and Hadrosauridae, all from the DPF. The ankylosaurs *Dyoplosaurus* [51,52] and *Scolosaurus* [53,54], and the ceratopsid *Spinops* [55] are absent from the dataset because they lack suitable skull material (the exact provenance of *Spinops* and *Scolosaurus* are also uncertain, and may be situated in the underlying Oldman Formation). Juvenile specimens, identified by their small size and undeveloped cranial ornamentation (e.g., [16,36]), are excluded because body size tends to be an ecologically discriminating factor [56], and their inclusion would only serve to obscure the results with respect to the question of interspecific dietary niche partitioning. Moreover, juvenile specimens are not available for all species, and selective inclusion of these specimens for some species and not others would further confound the results. We took measurements to the nearest mm with dial callipers or with a tailor's measuring tape, where appropriate. When one side of the skull was damaged or more poorly preserved than the other, we measured only the best-preserved side; otherwise, we averaged bilateral measurements to yield a single value. In many instances, the data were multivariate non-normal, which is not ideal for use with many ordination methods [57]. We therefore log-transformed the data to produce linear relationships between variables with log-normal distributions [58], which we verified using an omnibus test for multivariate normality [59].

Missing Data

As is common in palaeontology (e.g., [43]), missing data is an issue because it hinders the implementation of otherwise helpful ordination procedures used to aid the interpretation of morphometric data. Traditionally, numerous imputation methods have been applied in morphometric studies, but many of these are inadvisable [60,61]. For example, deletion methods (e.g., listwise deletion, pairwise deletion) tend to decrease statistical power, bias parameter estimates, and lead to mathematically inconsistent matrices that are not positive definite. Similarly, many substitution methods (e.g., substitution of means, prediction by regression) lead to underestimated variances and spurious statistical significance. We therefore used the principal-component method of imputation, which accurately estimates missing values and does not suffer from the aforementioned shortcomings [60]. It works by substituting the column mean and iteratively running principal component analysis (PCA) to improve the estimates until convergence is reached. PCA allows the projection of a multivariate dataset down to a few orthogonal dimensions of maximal variance (principal components or PCs) to simplify interpretation of the data distribution [62].

Statistical Comparisons

We drew statistical comparisons between taxonomic samples using those imputed PCA scores accounting for a significant majority (>95%) of the total variance. However, we relaxed this constraint where appropriate. We made comparisons in a hierarchical fashion at coarse (family/suborder), medium (subfamily/family), and fine (genus) taxonomic scales. We did not

A

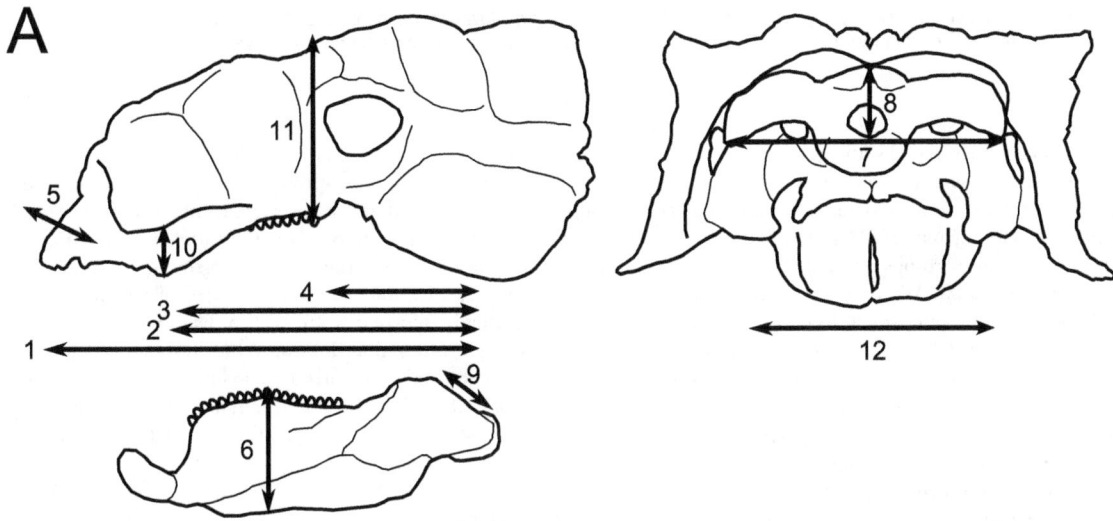

B

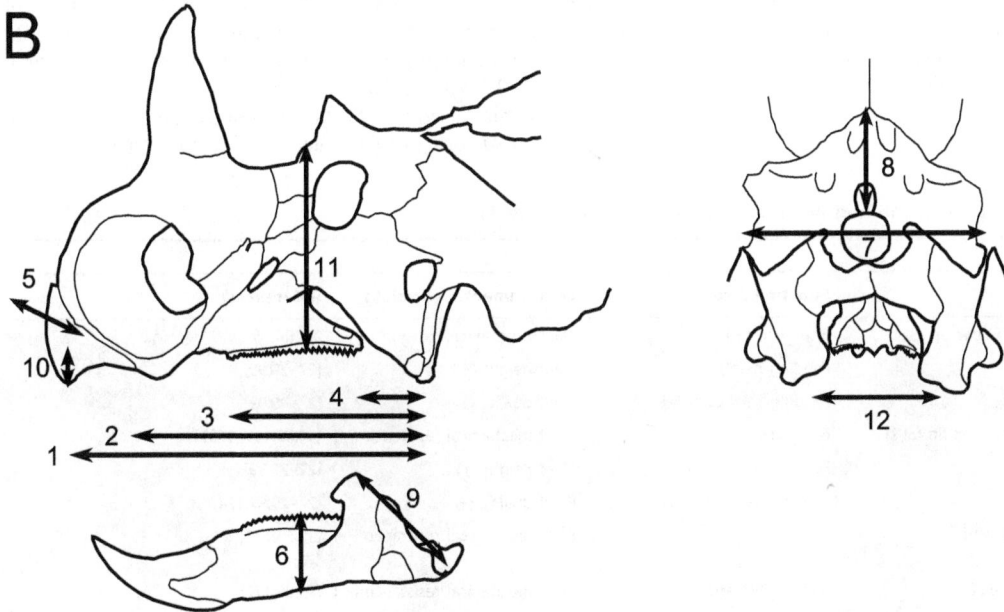

C

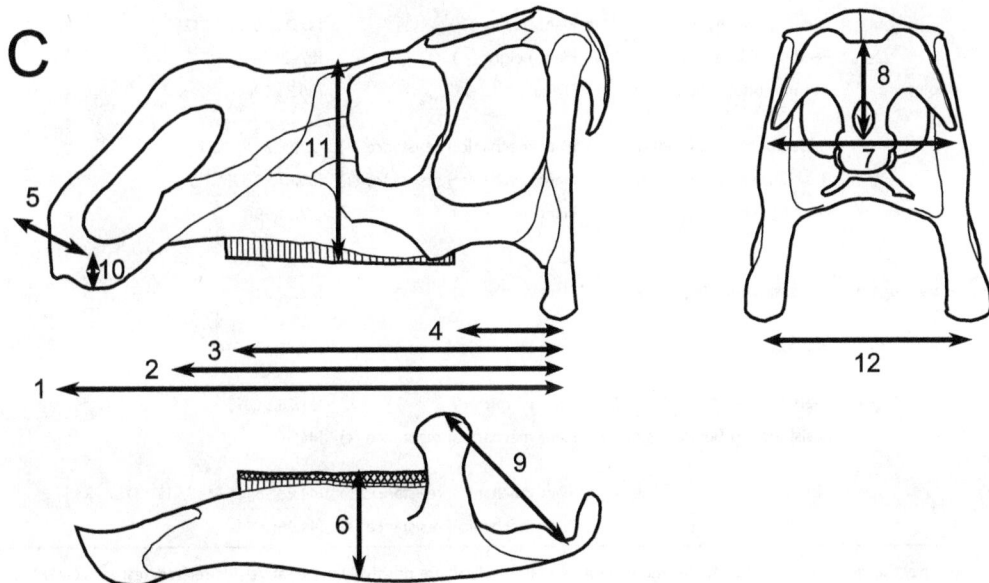

Figure 2. Linear measurements used in this study (compare with Table 1). A, ankylosaur skull in left lateral (left) and caudal (right) views; B, ceratopsid skull in left lateral (left) and caudal (right) views; C, hadrosaurid skull in left lateral (left) and caudal (right) views.

consider the species level because sample size was generally too low at this resolution to permit meaningful statistical comparisons. We used non-parametric multivariate analysis of variance (NPMANOVA) as a statistical test because samples were typically quite small (n <30) and non-normal. NPMANOVA tests for differences between two or more groups of multivariate data, based on any distance measure [63]. We used the Mahalanobis distance measure [64] because it is better suited to non-spherically symmetric data than the traditional Euclidean distance measure. In NPMANOVA, significance is estimated by permutation across groups, which we performed using 10,000 replicates.

We likewise conducted post-hoc pairwise comparisons using NPMANOVA with Bonferroni correction. Bonferroni correction was designed to counteract the problem of multiple comparisons, whereby the probability of committing a type I error increases with the number of simultaneous comparisons being made [58]. This problem is rectified by multiplying the p-value by the number of pairwise comparisons, effectively lowering the significance level. However, because Bonferroni correction provides little power and is probably too conservative [58,65], we also report uncorrected probabilities for interpretation.

We examined those variables that best distinguish the samples using discriminant function analysis (DFA) of the imputed PCA scores. DFA is an ordination procedure whereby two or more groups of multivariate data are projected onto a reduced set of dimensions in a way that maximizes the ratio of between-group variance to within-group variance. For N groups, there are N-1 discriminant axes of diminishing importance, of which only the first few are usually informative [62]. DFA, like PCA, returns both a series of eigenvalues that indicates the amount of variation explained by each axis, and a set of loadings that denotes the importance of each variable as a discriminator along each axis. We performed all statistical and ordination procedures using the software program PAST 2.12 [57].

Time-averaging

Because the DPF does not represent a single assemblage of contemporaneous organisms, time-averaging is an issue. This has the effect of masking palaeoecological patterns that are otherwise distinguishable only at fine temporal resolutions [66]. For this reason, we minimized the effects of time-averaging by making the above comparisons within each of the two most inclusive

Table 1. Form-function complex of the herbivore skull (compare with Figure 1).

Variable	Functional correlate	Environmental correlate	References
1. Distance from jaw joint to rostral beak tip (SL1)	Bite force (−)	Plant mechanical resistance (−)	[44,85,169–173]
	Feeding height (−)	Plant height (−)	[27–29,99]
	Feeding selectivity (+)	Plant quality (+)	[27–29,99]
2. Distance from jaw joint to caudal beak tip (SL2)	Bite force (−)	Plant mechanical resistance (−)	[44,85,169–173]
	Feeding height (+)	Plant height (+)	[27–29,99]
	Feeding selectivity (+)	Plant quality (+)	[27–29,99,174]
3. Distance from jaw joint to mesial end of tooth row (SL3)	Bite force (−)	Plant mechanical resistance (−)	[35,85,172]
4. Distance from jaw joint to distal end of tooth row (SL4)	Bite performance (−)	Plant mechanical resistance (−)	[35,85,172]
5. Maximum beak width (BW)	Feeding selectivity (−)	Plant quality (−)	[25,26,28–31,99,175]
	Feeding height (−)	Plant height (−)	[27–29,99]
6. Mandible depth (MD), measured at midpoint of tooth row	Accommodate cheek teeth (+)	Dietary grit (+)	[28,31,99]
	Adductor muscle insertion (+)	Plant mechanical resistance (+)	[27,28,31,81,83,175]
	Resistance to bending stress (+)	Plant mechanical resistance (+)	[80,81,98,99]
7. Paroccipital process breadth (PPB), measured as the sum of the lengths of the left and right paroccipital processes	Feeding height (−)	Plant height (−)	[27,31,99]
8. Occiput height (OH), measured from ventral edge of foramen magnum to dorsal edge of occiput	Feeding height (−)	Plant height (−)	[27,28,31]
9. Distance from jaw joint to coronoid process apex (JCP)	Bite force (+)	Plant mechanical resistance (+)	[35,85,172,176]
10. Depression of snout below occlusal plane (SP)	Feeding height (−)	Plant height (−)	[27,28,31,99]
11. Cranial height (CH), measured from base of tooth row to dorsal surface of orbit	Resistance to bending stress (+)	Plant mechanical resistance (+)	[44]
	Bite force (+)	Plant mechanical resistance (+)	[44,88,96,] [172,173,177,178,]
12. Distance between quadrates	Bite force (+)	Plant mechanical resistance (+)	[44,96]

Plus (+) and minus (−) symbols indicate whether the variable and its functional and environmental correlates are positively or negatively correlated, respectively, when all other variables are held constant.

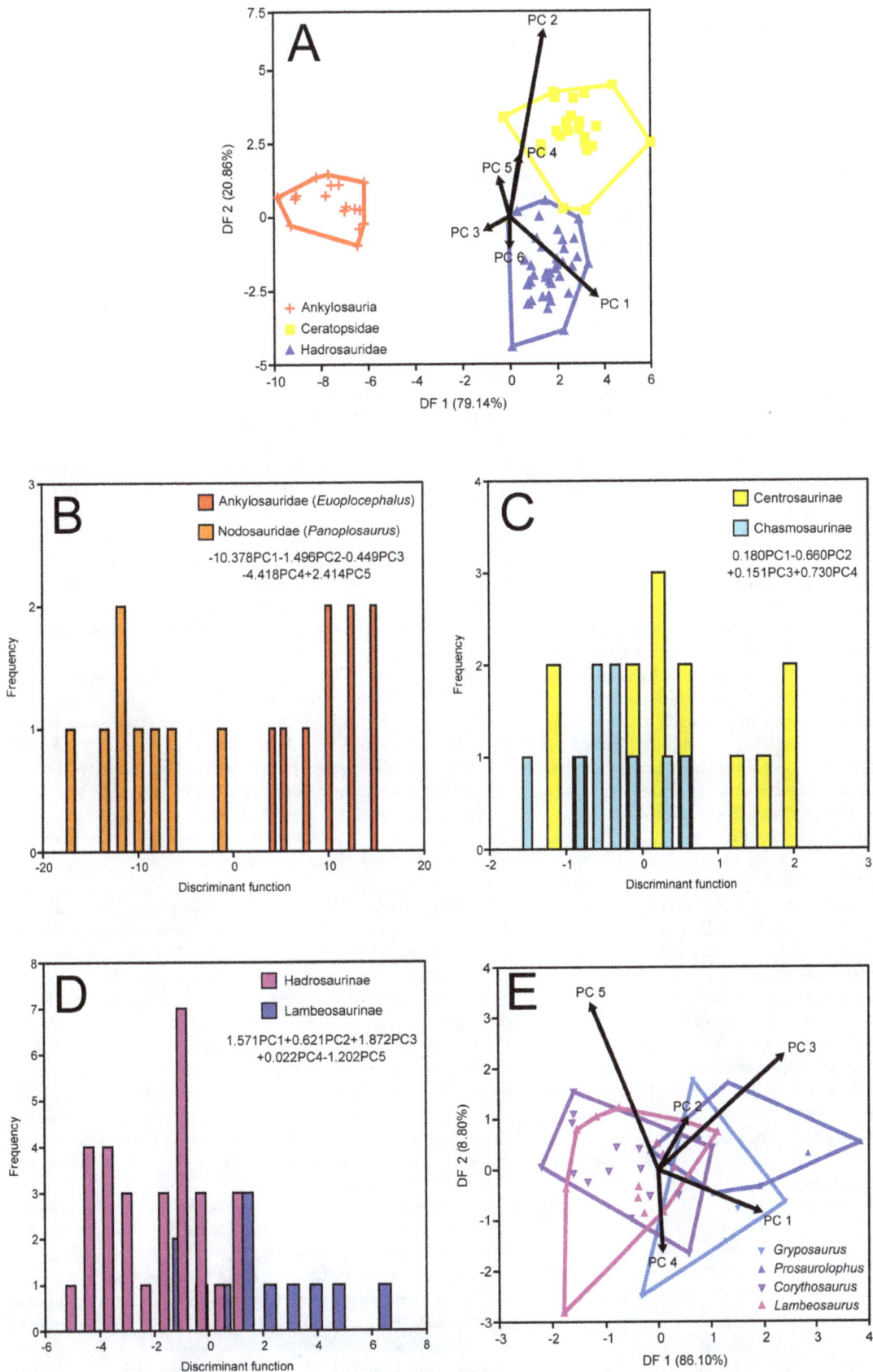

Figure 3. Time-averaged DFAs. A, coarse-scale analysis; B, ankylosaur family analysis; C, ceratopsid subfamily analysis; D, hadrosaurid subfamily analysis; E, hadrosaurid genus analysis.

Table 2. NPMANOVA results for the time-averaged coarse scale (suborder/family) taxonomic comparisons (10,000 permutations).

	Ankylosauria (n = 17)	Ceratopsidae (n = 23)	Hadrosauridae (n = 42)
Ankylosauria (n = 17)		1.00×10^{-4}	1.00×10^{-4}
Ceratopsidae (n = 23)	0.0003		1.00×10^{-4}
Hadrosauridae (n = 42)	0.0003	0.0003	

Bonferroni corrected p-values shown in lower left triangle; uncorrected p-values shown in upper right triangle. Significant results reported in bold.
Total analysis: N = 82, F = 16.18, $p = 1.0 \times 10^{-4}$.

Megaherbivore Assemblage Zones (MAZs) identified by Mallon et al. [17]. To summarize, MAZ-1 encompasses the lower 28 m of the DPF, and MAZ-2 encompasses intervals from 29–52 m. Although this time-constrained approach theoretically increases the probability of recovering differences that would otherwise be masked by the effects of time-averaging, there is a trade-off in that sample size (and hence statistical power) is reduced considerably. Also, this approach does not completely remove the effects of time-averaging because the MAZs are themselves time-averaged over a period of approximately 600 Ka [17].

Caveats

Dinosaurs almost certainly inhabited environments quite unlike those of the present. For example, although grasses were in existence during the Late Cretaceous [67], they did not form the extensive grasslands observed today [68]. Angiosperm-dominated forests also were quite rare. Instead, angiosperms likely took the form of herbs and shrubs growing in open and disturbed habitats [69–75]. As a result, conifer forests are thought to have been sparser than at present, with sunlight penetrating fully through to the ground [76]. It is therefore reasonable to ask whether modern herbivore ecomorphology should serve to inform interpretations of dinosaur palaeoecology. Two observations are offered in response to this concern. First, many modern plant genera are, in fact, known from fossil deposits of the Late Cretaceous Western Interior, including the DPF [77,78]. It is therefore likely that many dinosaurs did consume plants similar to those alive today, despite the fact that the environments in which they lived were different in many other respects from those existing presently. Second, the independent acquisition of certain traits in response to different, but mechanically similar, plants suggests that herbivore morphology is at least partially decoupled from phylogeny and likely adheres to certain general functional principles. For example, granivores of all types [79–83] repeatedly evolve short skulls, deep jaws, rostrally displaced jaw adductors, and durophagous dentitions (when present); the taxonomic identity of either the granivore or the seed in question is irrelevant. Therefore, the approach taken here, whereby various aspects of palaeodiet are inferred from form-function correlations, is warranted.

Results

The supporting ordination data (eigenvalues and variable loadings) for the results presented below are given in tables S2–S27 of Information S1.

Time-averaged Approach

NPMANOVA of the first six PCs reveals significant differences among the most inclusive clades (N = 82, F = 16.18, $p < 0.0001$). Posthoc pairwise comparisons show that Ankylosauria, Ceratopsidae, and Hadrosauridae each differ significantly from one another (Table 2). The corresponding DFA (Figure 3A) yields a 97.56% successful classification rate. The first discriminant function (DF 1) accounts for 79.14% of the total between-group variance. Ankylosaurs score negatively on this axis, whereas ceratopsids and hadrosaurids score positively. Ceratopsids place slightly more distally on DF 1 than hadrosaurids. PC 1 loads strongly and positively on DF 1, indicating that ankylosaurs differ from ceratopsids and hadrosaurids in having smaller skulls, which are relatively broader transversely, and relatively shorter tooth rows and deeper mandibles. DF 2 accounts for the remaining between-group variance. This axis best separates ceratopsids from hadrosaurids, with ankylosaurs falling in between. PC 2 loads strongly and positively on this axis, indicating that hadrosaurids possess transversely narrower paroccipital processes and snouts with a strong ventral deflection, followed sequentially by ankylosaurs and ceratopsids.

Ankylosauria. The ankylosaur families Ankylosauridae and Nodosauridae (represented by *Euoplocephalus* and *Panoplosaurus*, respectively) are significantly different from one another, as revealed by NPMANOVA of the first five PCs (N = 17, F = 3.095. $p < 0.0001$). The corresponding DFA perfectly discriminates *Euoplocephalus* and *Panoplosaurus* (Figure 3B). The separation is most strongly influenced by PC 1, which loads negatively on the discriminant axis. Thus, *Panoplosaurus* mainly differs from *Euoplocephalus* in having a greater offset between the jaw joint and coronoid apex.

Ceratopsidae. NPMANOVA of the ceratopsid subfamilies Centrosaurinae and Chasmosaurinae (chiefly represented by *Centrosaurus* and *Chasmosaurus*, respectively), using the first four PCs, produces no significant difference (N = 23, F = 1.022, $p = 0.424$). The p-value decreases if all 12 PCs are included in the comparison ($p = 0.077$), but otherwise remains insignificant. DFA of the first four PCs yields a 73.91% successful classification rate (Figure 3C). PC 4 loads most strongly and positively on the discriminant axis, indicating that centrosaurines generally have taller crania with slightly more distally extended tooth rows than chasmosaurines (but not significantly so). More comprehensive genus-level comparisons within subfamilies are not possible due to sample size limitations.

Hadrosauridae. NPMANOVA of the first five PCs yields a significant difference between the hadrosaurid subfamilies Hadrosaurinae and Lambeosaurinae (N = 42, F = 4.19, $p < 0.001$). DFA yields an 88.10% successful classification rate, with lambeosaurines scoring more negatively on the discriminant axis, and hadrosaurines scoring more positively (Figure 3D). In order of decreasing magnitude, the two subfamilies are best discriminated by PCs 3 and 1, both of which load positively on the discriminant axis. Thus, hadrosaurines primarily differ from lambeosaurines in having larger skulls (PC 1) that are transversely narrower, and with less ventrally deflected beaks (PC 3).

We subjected the hadrosaurines *Gryposaurus* and *Prosaurolophus*, and the lambeosaurines *Corythosaurus* and *Lambeosaurus*, to a genus-level NPMANOVA of the first five PCs (we excluded *Parasaurolophus* due to a lack of sufficient material). We recovered significant differences among the genera (N = 41, F = 1.804, p<0.05). Posthoc pairwise comparisons reveal that the differences occur between lambeosaurines and hadrosaurines; there are no significant differences within these two subfamilies (Table 3). DFA of the first five PCs yields a 56.10% successful classification rate. DF 1 captures 86.10% of the total between-group variance, and DF 2 captures 8.80%. Generally, lambeosaurine genera score more negatively along DF 1, whereas hadrosaurine genera score more positively (Figure 3E). There is poor separation along DF 2. Examination of the loadings reveals that PCs 1 and 3 both load strongly and positively on DF 1, which unsurprisingly mirror the shape changes captured by the hadrosaurine-lambeosaurine analysis above.

MAZ-1

NPMANOVA of the first five PCs recovers significant differences among ankylosaurs, ceratopsids, and hadrosaurids (N = 40, F = 10.33, p<0.0001). Posthoc pairwise comparisons demonstrate that each of these clades differs significantly from the other (Table 4). The corresponding DFA yields a 100% successful classification rate. The ordination results (Figure 4A) correspond to those of the time-averaged analysis, such that all three clades occupy similar areas of morphospace; however, there is better separation of all taxa–particularly ceratopsids and hadrosaurids–probably a reflection of the overall smaller sample size. Although the discriminant axes capture a similar amount of between-group variation as the time-averaged DFA, their loadings differ slightly. In the MAZ-1 analysis, PCs 1 and 2 load subequally on the first axis. This appears to be a consequence of the increased group separation along DF 1. PC 1 captures a similar signal to that reported for the time-averaged analysis, separating ankylosaurs from ceratopsids and hadrosaurids on the basis of skull length and breadth, tooth row length, and mandible depth. Conversely, PC 2 appears to reflect the fact that ceratopsids possess broader paroccipital processes and less ventrally deflected snouts than hadrosaurids. Evidently, some of the signal captured by DF 2 in the original, time-averaged DFA has 'leaked' over onto DF 1 in this analysis. PC 1 loads strongly and negatively on DF 2, whereas PC 2 loads strongly and positively. The morphological signal captured by DF 2 is similar to that captured by DF 1; the different loadings between the first two DF axes simply reflect the different relative positions of the taxa along those axes.

Ankylosauria. We did not conduct statistical comparisons of ankylosaurs due to sample size limitations.

Ceratopsidae. Centrosaurines and chasmosaurines (represented solely by *Centrosaurus* and *Chasmosaurus*, respectively) cannot be distinguished from one another using NPMANOVA of the first four PCs (N = 12, F = 1.799, p>0.05), but the two taxa are significantly different with the inclusion of PC 5 (N = 12, F = 1.92, p<0.05). DFA using the first five PCs yields a 100% successful classification rate, but given the particularly small chasmosaurine sample (n = 4), this result may be artificially inflated. Chasmosaurines score negatively on the discriminant axis, and centrosaurines score positively (Figure 4B). PC 1 loads most strongly and negatively on the discriminant axis, indicating that chasmosaurines generally possess a transversely wider occipital region of the skull. PC 4 also loads strongly and positively on the axis, reflecting the fact that centrosaurines possess a relatively wider beak and dorsoventrally deeper skulls.

Hadrosauridae. Hadrosaurines and lambeosaurines are significantly different (N = 22, F = 2.586, p<0.01), as revealed by NPMANOVA of the first six PCs. DFA results in a 95.45% successful classification rate. Lambeosaurines generally score negatively on the discriminant axis, whereas hadrosaurines score positively (Figure 4C). PC 1 loads most strongly and positively on the discriminant axis, indicating that hadrosaurines possess larger skulls with slightly wider occipital regions than lambeosaurines. PCs 3 and 5 also load strongly and positively on the discriminant axis, but their signals are more difficult to interpret because their loadings sometimes conflict. Both PCs reveal that hadrosaurines have relatively narrower snouts than lambeosaurines.

We included *Gryposaurus*, *Corythosaurus*, and *Lambeosaurus* in a genus level comparison. No significant differences are recovered among the genera (N = 21, F = 1.437, p>0.05), but the posthoc pairwise comparisons do support the contention that hadrosaurines and lambeosaurines tend to be most different (Table 5). DFA yields a 71.43% successful classification rate. The ordination and loadings correspond to those of the subfamily comparisons (Figure 4D).

MAZ-2

The MAZ-2 analyses do not include ankylosaurs due to sample size limitations [17]. NPMANOVA of the first three PCs yields a highly significant difference between ceratopsids and hadrosaurids (N = 16, F = 5.434, p = 0.001). Both taxa are perfectly discriminated using DFA. Hadrosaurids score negatively on the discriminant axis, whereas ceratopsids score positively (Figure 5A). This separation is most influenced by PC 1, which correlates negatively with the discriminant axis. Thus, hadrosaurids are distinguished from ceratopsids primarily by their ventrally deflected rostra and transversely narrow skulls.

Ceratopsidae. We did not conduct statistical comparisons of ceratopsids due to sample size limitations.

Hadrosauridae. Hadrosaurines and lambeosaurines (represented by *Prosaurolophus* and *Lambeosaurus*, respectively) are significantly different, as revealed by NPMANOVA of the first four PCs

Table 3. NPMANOVA results for the time-averaged hadrosaurid genus comparisons (10,000 permutations).

	Gryposaurus (n = 5)	*Prosaurolophus* (n = 7)	*Corythosaurus* (n = 15)	*Lambeosaurus* (n = 15)
Gryposaurus (n = 5)		0.7053	0.08309	**0.05389**
Prosaurolophus (n = 7)	1		**0.0049**	**0.0286**
Corythosaurus (n = 15)	0.4986	**0.0294**		0.8046
Lambeosaurus (n = 15)	0.3234	0.1716	1	

Bonferroni corrected p-values shown in lower left triangle; uncorrected p-values shown in upper right triangle. Significant results reported in bold.
Total analysis: N = 41, F = 1.804, p = 0.0245.

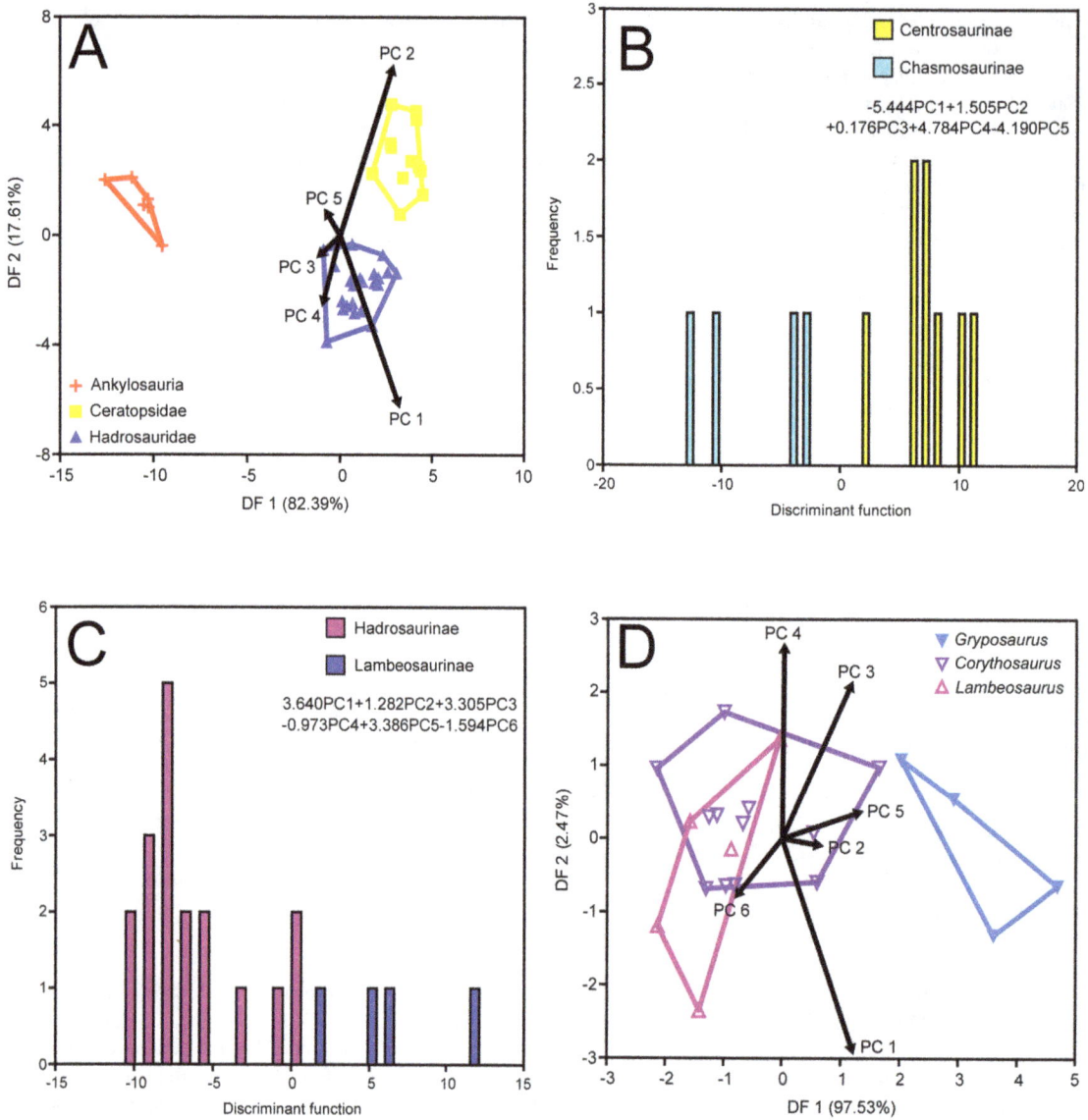

Figure 4. Time-constrained MAZ-1 DFAs. A, coarse-scale analysis; B, ceratopsid subfamily analysis; C, hadrosaurid subfamily analysis; D, hadrosaurid genus analysis.

($N = 11$, $F = 2.026$, $p<0.05$). DFA results in a 90.91% successful classification rate. Lambeosaurines score more negatively on the discriminant axis, and hadrosaurines score more positively (Figure 5B). PC 1 loads most strongly and positively on this axis, reflecting the larger skull size of *Prosaurolophus* relative to *Lambeosaurus*.

Discussion

Palaeodietary Implications

Ankylosauria. The inferred ankylosaur ecomorph is characterized by a small, proportionally wide skull, with a relatively deep mandible and short tooth row. The snout is ventrally deflected, but

Table 4. NPMANOVA results for the MAZ-1 coarse scale (suborder/family) taxonomic comparisons (10,000 permutations).

	Ankylosauria (n = 6)	Ceratopsidae (n = 12)	Hadrosauridae (n = 22)
Ankylosauria (n = 6)		**1.00×10^{-4}**	**1.00×10^{-4}**
Ceratopsidae (n = 12)	**0.0003**		**1.00×10^{-4}**
Hadrosauridae (n = 22)	**0.0003**	**0.0003**	

Bonferroni corrected *p*-values shown in lower left triangle; uncorrected *p*-values shown in upper right triangle. Significant results reported in bold.
Total analysis: N = 40, F = 10.33, $p = 1.0\times10^{-4}$.

Table 5. NPMANOVA results for the MAZ-1 hadrosaurid genus comparisons (10,000 permutations).

	Gryposaurus (n = 4)	Corythosaurus (n = 12)	Lambeosaurus (n = 6)
Gryposaurus (n = 4)		**0.0171**	0.195
Corythosaurus (n = 12)	**0.05129**		0.9403
Lambeosaurus (n = 6)	0.5849	1	

Bonferroni corrected *p*-values shown in lower left triangle; uncorrected *p*-values shown in upper right triangle. Significant results reported in bold.
Total analysis: N = 21, F = 1.437, *p* = 0.1211.

not as strongly as in hadrosaurids. First-hand examination of ankylosaur specimens reveals that the depth of the mandible is exaggerated by the dorsal bowing of the tooth row (Figure 2A). Ankylosaurs are not strongly distinguished from either ceratopsids or hadrosaurids based on the distance between the jaw joint and coronoid process apex. This is somewhat surprising because both ceratopsids and hadrosaurids possess elevated coronoid processes and depressed jaw joints that are otherwise not developed to the same degree in ankylosaurs [84–86]. It may be that the ankylosaur coronoid apex is more rostrally displaced than in ceratopsids and hadrosaurids, resulting in subequal measurements of this variable. Further work illuminating the differences in jaw mechanics between these taxa is in progress.

Because jaw adductor muscle mass–and by extension, bite force–generally scales positively with skull size (e.g., [87–91]), it is likely that ankylosaurs possessed a weaker bite than the larger ceratopsids and hadrosaurids. Likewise, the rostral placement of the tooth row relative to the coronoid process in ankylosaurs means that they did not possess as powerful a bite as the other two taxa, in which the tooth row extends caudal to the coronoid process, resulting in increased leverage of the distal tooth row [84–86]. Other evidence cited in favour of a relatively weak bite in ankylosaurs is the presence of small, phyliform teeth with peg-like roots [92–94], and simple jaw musculature [95].

Nonetheless, the ankylosaur skull exhibits other features thought to correlate with either high bite forces or repetitive masticatory movements–both adaptations for comminuting resistant plant matter. For example, the proportionally great transverse breadth of the skull may have accommodated larger jaw adductor muscles.

This explanation was offered by Herrel et al. [96] to account for the fact that finches with relatively wide skulls also possess the highest bite forces. Henderson [44] likewise used beam theory to show that wider skulls are able to resist high torsional stresses incurred by elevated bite forces. Furthermore, the curved tooth row of ankylosaurs is reminiscent of that of grazing macropodoids, the function of which Sanson [97] surmised was to concentrate bite forces in response to a tough diet. The secondary increase in the depth of the mandible likewise would have served to withstand repeated bending forces associated with mastication, preventing bone fatigue [98,99]. Finally, Vickaryous et al. [100] cited the presence of an ossified secondary palate in ankylosaurs as evidence that their skulls were adapted to resisting strain resulting from complex jaw movements used in the comminution of tough plants.

Besides reconstructed feeding envelopes [5], several other morphological characters attest to the low-browsing habit of ankylosaurs. First is the broad, ventrally-deflected snout, which is otherwise observed most frequently among grazing bovids [99]. The relatively great breadth of the snout undoubtedly enables these mammals to feed more efficiently on fibrous, low-growing grasses [25–32]. However, the purpose of the ventral deflection of the snout in these bovids is not yet fully understood. It may serve to bring the cropping mechanism (incisors) closer to the ground, similar to what has been proposed for marine grazing dugongs [101], but this is speculation. It may also reflect the fact that grazers tend to have faces more strongly flexed on the basicranium than browsers [27,28,31], but the reason for this correlation is likewise unknown.

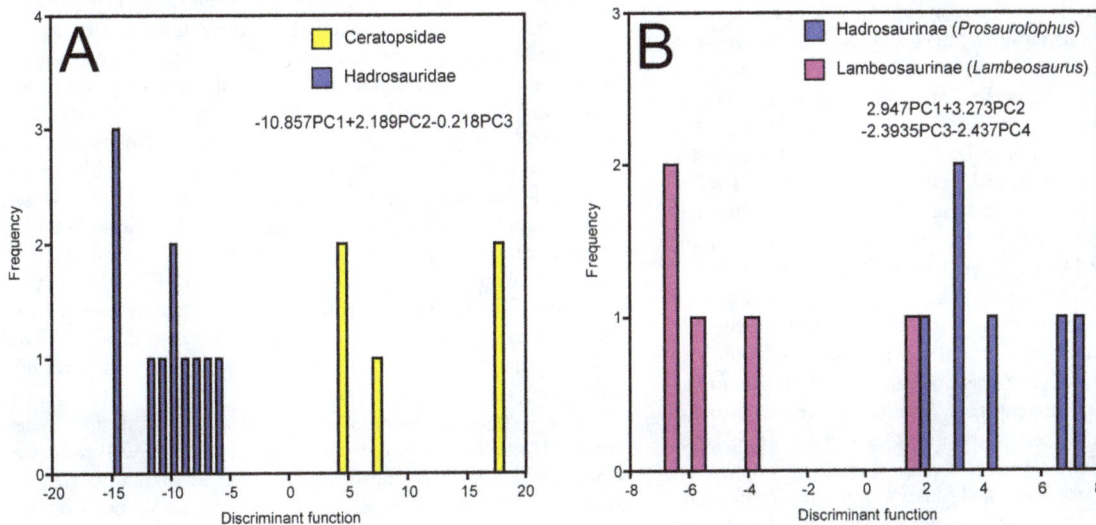

Figure 5. Time-constrained MAZ-2 DFAs. A, coarse-scale analysis; B, hadrosaurid analysis.

Second is the relatively great transverse breadth of the paroccipital processes, which is also common among grazing bovids. Spencer [99] suggested that this may reflect the fact that grazers tend to use sharp head movements for cropping forage, effected by the nuchal musculature, whereas browsers rely more on their lips and tongue. Perhaps ankylosaurs also relied on head movements to sever plant food, but no corresponding study on head mobility in these animals has been conducted to date. Challenging this hypothesis is the observation of Maryańska [102] that ankylosaurs possessed a well-developed hyoid apparatus and entoglossal process, which would have supported a long and mobile tongue. Ankylosaurs may have used such a tongue in the cropping of vegetation.

Therefore, while it is likely that they consumed soft, pulpy plant tissues (e.g., fruits [93,103]), ankylosaurs probably subsisted on tough leaves that required more thorough mastication as well [104]. This interpretation is corroborated by circumstantial evidence in the form of a cololite associated with a Lower Cretaceous ankylosaurid from Australia [105]. The fossil comprises angiosperm fruits or endocarps, small seeds, possible fern sporangia, and abundant vascular tissue (probably leaves). The plant material exhibits signs of having been comminuted by the jaws [105]; however, the finding of gastroliths associated with a specimen of *Panoplosaurus mirus* (ROM 1215 [106]) suggests that additional food processing occurred in the gizzard. If so, then it is likely that the ankylosaur skull ecomorph does not accurately reflect the associated palaeodiet. Nonetheless, there is some doubt about whether the gastroliths truly pertain to the specimen in question, as neither the field notes nor the original description [92] mention the existence of gizzard stones (K. Seymour, pers. comm., 2011).

Ankylosaur families differ in their mandibular morphologies such that nodosaurids possess a relatively greater offset between the jaw joint and coronoid apex than ankylosaurids. This suggests that the mechanical advantage of the nodosaurid mandible was elevated relative to that of ankylosaurids (due to the increased length of the applied force moment arm), resulting in a more powerful bite. Supporting this interpretation, Carpenter [106] and Vickaryous [107] reported on the existence of dorsoventrally deep (fused) vomers with a distally dilated process among nodosaurids, which may have served to further dissipate stress associated with either elevated bite forces or repetitive masticatory movements. Therefore, it seems likely that nodosaurids subsisted on harder or tougher plants than ankylosaurids, necessitating a more powerful bite and cranial structures associated with stress distribution.

Perhaps surprisingly, the contention of Carpenter [38–41] that ankylosaurids and nodosaurids differ appreciably in relative beak width is not well-supported here. In the time-averaged ankylosaur analysis above, the separation of the two families along PC 1 is due in part to the relatively wider beak of ankylosaurids, but this variable loads comparatively weakly on the axis, and its signal is otherwise contradicted by loadings on other PCs. It is possible that relative beak width did not prove to be a stronger discriminator of ankylosaurids and nodosaurids because: (1) it was overwhelmed by other, stronger loading variables; (2) it was not captured by the first PCs considered here; or (3) it was not captured at all due to the confounding effects of missing data. Additional research into the specific question of ankylosaur beak width variation is in progress.

Ceratopsidae. The inferred ceratopsid ecomorph is characterized by a particularly large and narrow skull, distally-elongate tooth row, and rostrally projecting snout. Although the relative transverse width of the paroccipital processes is most developed in ankylosaurs, the separation of ceratopsids from hadrosaurids along

DF 2 of the time-averaged analysis suggests that the former taxon is characterized by slightly wider paroccipital processes as well.

Two features in particular attest to the especially powerful bite of ceratopsids. The first is overall skull size, which is the largest of any of the forms from the DPF. The second is the distal extension of the tooth row beyond the apex of the coronoid process. Ostrom [35,85] demonstrated that this morphology equates to a shift in the behaviour of the jaw mechanism, from a class 3 to a class 1 lever, because the relative lengths of the applied and resistance force moment arms are switched. Therefore, the ceratopsid jaw mechanism appears to have been more efficient than that of ankylosaurs. The elevation of the coronoid process and concomitant depression of the jaw joint would have further served to enhance the leverage of the ceratopsid mandible [35,85].

The transversely wide paroccipital processes of ceratopsids–although not as developed as in ankylosaurs–may correlate with low browsing. On the other hand, it may reflect the development of the nuchal musculature in support of the large parietosquamosal frill. Paradoxically, although ceratopsids appear to have been restricted to feeding below one metre from the ground [5], the cropping mechanism is not ventrally deflected as in mammalian grazers [99]. This might be attributable to the great mobility of the head, which could have pivoted easily about the spherical occipital condyle to bring the beak near to the ground [108].

Bearing these points in mind, ceratopsids can be characterized as low-level browsers that probably sustained themselves on mechanically resistant vegetation requiring high bite forces. Mechanical resistance comprises various physical properties such as strength, toughness, and 'hardness' (a general term that encompasses the properties of plasticity and stiffness). The bladed dentition of ceratopsids [34,35,85,109–111] almost certainly was not suitable for processing particularly strong or hard plant types, which require a durophagous dentition [82]. Therefore, it is likely that ceratopsids specialized on tough plant parts that resisted crack propagation, such as low-growing, woody browse. The 'weedy' angiosperms of the Late Cretaceous, which grew most commonly in coastal plain settings [112] alongside ceratopsids [113], may have provided an abundant and renewable food resource for these animals [114–116]. The interpretation of ceratopsids as woody browse specialists might help to explain the narrowness of their beaks, which would have restricted them to selective foraging, but more work in this area is required.

NPMANOVA indicates that centrosaurines and chasmosaurines probably differ in their skull proportions, but low sample size generally impedes the interpretation of the results. For example, whereas the two subfamilies differ primarily according to cranial depth and distal tooth row extension in the time-averaged comparison (where the probabilities are not quite significant), their differences are better attributed to the transverse width of the occipital region in the MAZ-1 comparison (where the probabilities are significant). It is possible that morphological disparity between centrosaurines and chasmosaurines may truly manifest itself differently within MAZ-1, but it is also likely that the smaller samples in this assemblage zone do not adequately capture the true ecological signal therein, and, in fact, artificially inflate statistical significance by reducing taxonomic overlap in morphospace [117].

There is some evidence, however, that ceratopsid subfamilies differ at least partly according to cranial depth in MAZ-1, as in the time-averaged analysis. This might be taken as tentative support for the finding of Henderson [44] that centrosaurines possess taller crania than chasmosaurines, making them more resistant to bending and torsional stresses. These differences were said to have facilitated niche partitioning between the two subfamilies, as centrosaurines presumably would have been capable of subsisting on a more resistant plant diet than sympatric chasmosaurines. Nonetheless, although Henderson [44] was careful to account for the confounding effects of taphonomic distortion, he considered only a single specimen per species, and therefore did not account for intraspecific variation. This omission is likely to have introduced some systematic bias into the results because numerous studies have shown that individual ceratopsid species actually vary quite widely, even when ontogenetic effects are accounted for [34,109,118–126]. For example, long-faced *Centrosaurus apertus* have been described (e.g., *Ce.* "*longirostris*" [127]), as well as short-faced *Chasmosaurus belli* (e.g., *Ch.* "*brevirostris*" [34]). It is therefore necessary that statistical approaches be taken to account for the significance of this variation.

Hadrosauridae. The inferred hadrosaurid ecomorph is characterized by a relatively large, narrow skull (though not as large as in ceratopsids), distally extended tooth row, and ventrally deflected snout. Hadrosaurids do not differ appreciably from ceratopsids in either the offset between the jaw joint and coronoid process apex or the distal extension of the tooth row, so it is likely that both jaw systems shared a similar mechanical advantage. However, the smaller size of the hadrosaurid skull suggests that these animals possessed a slightly weaker bite than ceratopsids.

Reconstruction of the hadrosaurid feeding envelope suggests that these animals could browse at heights up to 4 m above ground level [5], but it is otherwise unclear which height they browsed at most regularly. Unfortunately, those morphological features of the skull that correlate with feeding height do not clarify the matter. For example, the paroccipital processes are relatively narrow, a condition common among high-level browsers. However, the ventral deflection of the snout is most commonly observed among low-level grazers. Relative beak width is intermediate between that of ankylosaurs and ceratopsids, and does not otherwise provide convincing evidence for browse height. This unique combination of morphological characters might therefore indicate that hadrosaurids were equally comfortable browsing at both high and low levels. It is not difficult to imagine these animals feeding low in the herb layer, occasionally rearing up to feed bipedally among the surrounding shrubs when a herd of low-browsing ceratopsids passed through the area [5].

The strong jaws and large feeding heights of hadrosaurids suggests that these animals could subsist on a variety of plant types, and, as the largest members of the DPF megaherbivore assemblage, hadrosaurids likely possessed correspondingly large niche breadths [128]. Their dentition was probably capable of both crushing and shearing functions [43,46], and could therefore process both tough and hard foodstuffs, encompassing a variety of browse types. Circumstantial evidence in favour of this hypothesis comes by way of fossil gut contents (enterolites), associated with various hadrosaurids, that contain conifer and angiosperm twigs and stems, bark, seeds, and leaves [129–131], although some have also cautioned that these materials may have been washed into the gut cavity post-mortem [130,132]. Chin and Gill [133] and Chin [134] also reported on hadrosaurid coprolites containing an abundance of conifer wood, which cannot have been derived allochthonously.

Hadrosaurid subfamilies differ most noticeably in the development of cranial crests [135–137], but from the perspective of inferred dietary ecomorphology, hadrosaurine skulls are consistently larger than those of lambeosaurines. Hutchinson [138] noted that, in cases where two closely-related species, occupying the same position on the food-web, coexist, the skull of the larger form usually exceeds that of the smaller form in length by a ratio of ~1.3, a figure that has come to be known as the Hutchinsonian

ratio [139]. Although there is some question as to the actual statistical validity of this ratio [140–144], it is generally thought that these size differences are what allow closely related species to specialize on different foodstuffs, thereby circumventing interspecific competition. Interestingly, the ratio of mean skull length (measured as the distance from the jaw joint to the premaxillary apex) between hadrosaurines (754 mm) and lambeosaurines (612 mm) is 1.23, which is close to the figure of 1.3 noted by Hutchinson [138]. This difference in size could mean that the larger hadrosaurines incorporated less digestible plant matter in their diet than lambeosaurines. Although we found no other morphological characters to corroborate this hypothesis, it has been noted that hadrosaurines tend to exhibit more steeply inclined tooth facets than lambeosaurines, equating to a higher capacity for shearing in the former taxon [43,46,145]. This, in turn, would allow hadrosaurines to more effectively rend tougher, more fibrous plant tissues [82].

Numerous authors [37,42,43,45] have argued that hadrosaurines generally possess relatively wider, squarer beaks than lambeosaurines, attributing this distinction to differences in their feeding ecologies. Unfortunately, previous attempts to quantify variation in hadrosaurid beak shape have not controlled for time-averaging, and have instead grouped forms spanning much of the Late Cretaceous. Thus, genera such as the late Campanian *Lambeosaurus* were compared alongside the late Maastrichtian *Edmontosaurus*, although the two were separated in time by ~10 Ma. The time-constrained approach taken here suggests that sympatric hadrosaurines and lambeosaurines did not always differ in beak shape. Overlap in beak shape was noted by both Carrano et al. [43] and Whitlock [45], but the palaeoecological implications of this were not addressed. In fact, the DFA results suggest that, if anything, hadrosaurines possessed relatively narrower beaks than lambeosaurines within the boundaries of the DPF, but a posthoc Mann-Whitney U test reveals that the differences in arcsine-transformed relative beak width are not quite significant at $p < 0.05$ (N = 33, U = 64, $p = 0.079$).

Chapman and Brett-Surman [146] also noted, on the basis of geometric morphometrics, that lambeosaurines have more ventrally deflected beaks than hadrosaurines. There is some support for this position, particularly in light of the results for the time-averaged analysis. This might be taken as evidence that lambeosaurines habitually fed closer to the ground than hadrosaurines. If so, this interpretation runs contrary to that of Carrano et al. [43], who suggested, on the basis of beak, tooth, and hindlimb morphology, that hadrosaurines foraged near to the ground in open habitats, whereas lambeosaurines foraged in closed habitats.

Finally, the results of Dodson's [36] morphometric investigation of lambeosaurine skulls are supported here. His survey comprised 48 variables measured over 36 specimens, and the data were examined using bivariate and multivariate approaches. Dodson ([36]: p. 50) noted that "the differences among the five [morphological patterns among lambeosaurines] relate not to structures that have apparent significance in the differential utilization of trophic resources necessary for the coexistence of closely related species of large animals. Instead, they are confined to several parameters of the bony crest." This finding was especially surprising to Dodson because it was then assumed that the DPF represented a single 'snapshot' in time, and that all dinosaurs from the DPF were contemporaneous. This hypothesis has since been falsified, and the temporal overlap of the lambeosaurines minimized [15,17,147,148].

Evolutionary Palaeoecology

Of the 12 genera considered here–six of which typically coexisted at any given time [17]–five or six distinct ecomorphs are recovered using skull morphometrics. Ankylosaurs, ceratopsids, and hadrosaurids are each characterized by unique morphologies, the distinguishing characteristics being concentrated in the absolute size of the skull, and in its relative width, degree of ventral deflection of the snout, distal extension of the tooth row, and depth of the mandible. Ceratopsids and hadrosaurids are more alike than ankylosaurs, probably a reflection of the more recent common ancestry of the first two taxa (Figure 1, [149]). This implies that ankylosaurs were least likely to compete with the other megaherbivores from the DPF simply as a result of their more distant phylogenetic relatedness. Corroborating this hypothesis, the convergent evolution of dental batteries in ceratopsids and hadrosaurids suggests adaptation to the comminution of similar plant food (but see [114]). Ankylosaurids and nodosaurids are themselves distinguished primarily by differences in the construction of the mandible, and hadrosaurines and lambeosaurines differ mainly in skull size. Due to sample size limitations, it is difficult to determine whether centrosaurines and chasmosaurines differ at all, but there is reasonable evidence to suggest that they did. How they differ is also not immediately obvious, and the signal may change depending on whether time-averaging is minimized. Nonetheless, conventional knowledge that the two subfamilies differ according to cranial depth is tentatively confirmed. Therefore, with the above considerations in mind, the contention that dietary niche partitioning supported the rich megaherbivore diversity of the DPF is confirmed by this study (Figure 6).

There is no evidence for dietary niche partitioning among genera belonging to the same subfamily. For example, all hadrosaurine genera occupy nearly identical regions of morphospace, as do all genera within Lambeosaurinae, Centrosaurinae, and Chasmosaurinae. This is hardly surprising in light of the fact that consubfamilial taxa rarely lived in sympatry [17]. In fact, it may be precisely because of the morphological similarity of such taxa that they were unable to coexist, owing to the effects of competitive exclusion. In those rare instances where such taxa do overlap in time [17], their coexistence is either short-lived (e.g., ~214 Ka for *Corythosaurus* and *Lambeosaurus*) or involves rare or transient forms (e.g., *Parasaurolophus*). The apparent rarity of the ankylosaurid *Dyoplosaurus* in MAZ-1 of the DPF, where *Euoplocephalus* is most common [17], also fits this pattern. These stable biostratigraphic patterns might be taken as evidence that the megaherbivorous dinosaur assemblage of the DPF was structured by the effects of competition, rather than having assembled via stochastic processes [150]. The rarity of sympatric consubfamilial taxa suggests that niche space was saturated and could not accommodate the addition of new species without the concomitant loss of already established competitors. This hypothesis further predicts that taxonomic overlap in morphospace should remain negligible with the recovery of additional fossils, and that the assemblage should adhere to certain 'assembly rules' [151] whereby the addition of new species results in an increase in total morphospace (niche) volume, an increase in morphological (niche) specificity, or the localized extinction of competitors [47]. Where taxonomic overlap in morphospace does occur, it should involve only rare taxa that would not have posed serious competition to more established members of the assemblage.

The temporal stability of the morphological patterns identified here (compare Figures 3–5), spanning ~1.5 Ma, supports the contention of Brinkman et al. [152] that fossils from the upper Campanian Belly River Group (which includes the DPF) constitute a chronofauna. This is a term introduced by Olson ([153]: p. 181) to refer to "a geographically restricted, natural assemblage of interacting animal populations that has maintained its basic structure over a geologically significant period of time." Thus, while individual species appear and disappear with time, the ecological relationships of the chronofauna remain stable. Olson [153] accounted for this stability with reference to environmental stasis, but the megaherbivore chronofauna of the DPF appears to have been rather impervious to environmental change, as the formation itself records the gradual transgression of the Western Interior Seaway [11,12]. Perhaps the change was slow enough that the megaherbivore chronofauna could adapt accordingly, but there is no evidence to date that species turnover in the DPF responded to environmental change [17]. The apparent displacement of ankylosaurs from the upper limits of the DPF [17,113,154] may indicate a reshaping of the megaherbivore chronofauna in response to the encroaching sea, with ankylosaur congeners appearing in younger sediments of the Horseshoe Canyon Formation subsequent to the regression of the sea [52]. Alternatively, DiMichele et al. [155] have also invoked evolved mutualisms, historical contingency, and the 'law of large numbers' to account for ecological stasis in fossil assemblages.

Figure 6. Depiction of dietary niche partitioning among megaherbivorous dinosaurs from the DPF (MAZ-2). Left to right: *Chasmosaurus belli, Lambeosaurus lambei, Styracosaurus albertensis, Euoplocephalus tutus, Prosaurolophus maximus, Panoplosaurus mirus.* A herd of *S. albertensis* looms in the background. Image courtesy of J.T. Csotonyi.

Conclusions

In their appraisal of research into dinosaur feeding behaviour, Barrett and Rayfield ([156]: p. 218) recently lamented the lack of studies attempting to "place feeding within more holistic evolutionary or ecological frameworks". The present study is an attempt to address this concern by examining dinosaur feeding in a geographically and temporally constrained manner, thereby approximating true ecological relationships. The implementation of statistical procedures also allows for more robust comparisons and provides a means by which to gauge the palaeoecological significance of variation.

This study supports the hypothesis that the great standing crop megaherbivore diversity of the DPF (and by extrapolation, much of Laramidia) is largely attributable to dietary niche partitioning. Coexisting ankylosaurids and nodosaurids, hadrosaurines and lambeosaurines, and probably centrosaurines and chasmosaurines, differ significantly in their morphologies, and likely differed in their food preferences as a result. The interpretation that niche partitioning facilitated megaherbivore coexistence in the DPF can be tested further by examining the response of the assemblage structure to the appearance of new species as new fossil discoveries are made, and by examining other ecological proxies including dental microwear and stable isotopes. The inferred ecological relationships appear to have been stable over the ~1.5 million year span of the DPF, as revealed by time-constrained analyses of morphological patterns. This stability is characteristic of chronofaunas [153].

Variation in such aspects as body size, beak breadth, jaw mechanics, and tooth morphology are commonly cited as evidence for dinosaur feeding ecology [9,37–46,69,70,157–164], but this study identifies several other morphological variables that may help to reveal subtle differences in dinosaur palaeoecology. Chief among these are variables relating to the development of the nuchal musculature and the ventral deflection of the beak. Unfortunately, the functional significance of these and other morphological variables is only poorly understood, emphasizing the need for further detailed analyses of herbivore functional morphology and the development of general functional principles [165].

Supporting Information

Information S1 Supporting ordination data (eigenvalues and variable loadings) for the results of this study.

Table S1 Raw data used in this study.

Acknowledgments

We thank P. Dodson, K. Ruckstuhl, M. Ryan, J. Theodor, and F. Therrien for reviewing early drafts of the manuscript. A. Farke and an anonymous reviewer also provided constructive feedback following our submission. Ø. Hammer offered valuable technical assistance. C. Mehling (AMNH); W. Simpson (FMNH); P. Barrett and S. Chapman (NHMUK); M. Currie, A. McDonald, and K. Shepherd (CMN); B. Iwama, and K. Seymour (ROM); P. Owen (TMM); T. Courtenay, J. Gardner, F. Hammer, G. Housego, B. Strilisky, and J. Wilke (TMP); P. Currie, A. Locock, and R. Holmes (UALVP); M. Carrano and M. Brett-Surman (USNM); and D. Brinkman (YPM) gave access to specimens in their care. V. Arbour, N. Campione, D. Evans, R. Holmes and A. Murray, and S. Maidment provided accommodations over the course of this study.

Author Contributions

Conceived and designed the experiments: JCM. Performed the experiments: JCM. Analyzed the data: JCM. Contributed reagents/materials/analysis tools: JCM JSA. Wrote the paper: JCM JSA.

References

1. Archibald JD (1996) Dinosaur extinction and the end of an era: What the fossils say. New York: Columbia University Press. 237 p.
2. Lehman TM (1997) Late Campanian dinosaur biogeography in the Western Interior of North America. In: Wolberg DL, Stump E, Rosenberg GD, editors. Dinofest International: Proceedings of a symposium sponsored by Arizona State University. Philadelphia: Academy of Natural Sciences. 223–240.
3. Lehman TM (2001) Late Cretaceous dinosaur provinciality. In: Tanke DH, Carpenter K, editors. Mesozoic vertebrate life. Bloomington: Indiana University Press. 310–328.
4. Gates TA, Prieto-Márquez A, Zanno LE (2012) Mountain building triggered Late Cretaceous North American megaherbivore dinosaur radiation. PLoS ONE 7(8): e42135. doi:10.1371/journal.pone.0042135.
5. Mallon JC, Evans DC, Ryan MJ, Anderson JS (2013) Feeding height stratification among the herbivorous dinosaurs from the Dinosaur Park Formation (upper Campanian) of Alberta, Canada. BMC Ecology 13: 14. doi:10.1186/1472-6785-13-14.
6. Farlow JO, Dodson P, Chinsamy A (1995) Dinosaur biology. Annu Rev Ecol Syst 26: 445–471.
7. Sampson SD (2009) Dinosaur odyssey: Fossil threads in the web of life. Berkeley: University of California Press. 332 p.
8. Ostrom JH (1964) A reconsideration of the paleoecology of hadrosaurian dinosaurs. Am J Sci 262: 975–997.
9. Coe MJ, Dilcher DL, Farlow JO, Jarzen DM, Russell DA (1987) Dinosaurs and land plants. In Friis EM, Chaloner WG, Crane PR, editors. The origins of angiosperms and their biological consequences. Cambridge: Cambridge University Press. 225–258.
10. Sander PM, Gee CT, Hummel J, Clauss M (2010) Mesozoic plants and dinosaur herbivory. In: Gee CT, editor. Plants in Mesozoic time: Morphological innovations, phylogeny, ecosystems. Bloomington: Indiana University Press. 331–359.
11. Eberth DA, Hamblin AP (1993) Tectonic, stratigraphic, and sedimentologic significance of a regional discontinuity in the upper Judith River Group (Belly River wedge) of southern Alberta, Saskatchewan, and northern Montana. Can J Earth Sci 30: 174–200.
12. Eberth DA (2005) The geology. In Currie PJ, Koppelhus EB, editors. Dinosaur Provincial Park: A spectacular ancient ecosystem revealed. Bloomington: Indiana University Press. 54–82.
13. Currie PJ, Koppelhus EB, editors. (2005) Dinosaur Provincial Park: A spectacular ancient ecosystem revealed. Bloomington: Indiana University Press. 648 p.
14. Ryan MJ (2003) Taxonomy, systematics and evolution of centrosaurine ceratopsids of the Campanian Western Interior Basin of North America. Unpublished PhD dissertation. Calgary: University of Calgary. 578 p.
15. Ryan MJ, Evans DC (2005) Ornithischian dinosaurs. In: Currie PJ, Koppelhus EB, editors. Dinosaur Provincial Park: A spectacular ancient ecosystem revealed. Bloomington: Indiana University Press. 312–348.
16. Evans DC (2007) Ontogeny and evolution of lambeosaurine dinosaurs (Ornithischia: Hadrosauridae). Unpublished PhD dissertation. Toronto: University of Toronto. 497 pp.
17. Mallon JC, Evans DC, Ryan MJ, Anderson JS (2012) Megaherbivorous dinosaur turnover in the Dinosaur Park Formation (upper Campanian) of Alberta, Canada. Palaeogeogr Palaeoclimatol Palaeoecol 350–352: 124–138.
18. Brown CM, Evans DC, Campione NE, O'Brien LJ, Eberth DA (2013) Evidence for taphonomic size bias in the Dinosaur Park Formation (Campanian, Alberta), a model Mesozoic terrestrial alluvial-paralic system. Palaeogeogr Palaeoclimatol Palaeoecol 372: 108–122.
19. MacArthur RH, Pianka ER (1966) On optimal use of a patchy environment. Am Nat 100: 603–609.
20. Charnov EL (1976) Optimal foraging, the marginal value theorem. Theor Popul Biol 9: 129–136.
21. Bock WJ, von Wahlert G (1965) Adaptation and the form-function complex. Evolution 19: 269–299.
22. Wainwright PC, Reilly SM, editors. (1994) Ecological morphology: Integrative organismal biology. Chicago: University of Chicago Press. 376 p.
23. Lauder GV (1995) On the inference of function from structure. In: Thomason JJ, editor. Functional morphology in vertebrate paleontology. Cambridge: Cambridge University Press. 1–18.
24. Losos JB, Miles DB (1994) Adaptation, constraint, and the comparative method: Phylogenetic issues and methods. In: Wainwright PC, Reilly SM, editors. Ecological morphology: Integrative organismal biology. Chicago: University of Chicago Press. 60–98.
25. Gordon IJ, Illius AW (1988) Incisor arcade structure and diet selection in ruminants. Funct Ecol 2: 15–22.

26. Janis CM, Ehrhardt D (1988) Correlation of relative muzzle width and relative incisor width with dietary preference in ungulates. Zool J Linnean Soc 92: 267–284.

27. Janis CM (1990) Correlation of cranial and dental variables with dietary preferences in mammals: A comparison of macropodoids and ungulates. Mem Queensl Mus 28: 349–366.

28. Janis CM (1995) Correlations between craniodental morphology and feeding behavior in ungulates: Reciprocal illumination between living and fossil taxa. In Thomason JJ, editor. Functional morphology in vertebrate paleontology. Cambridge: Cambridge University Press. 76–98.

29. Solounias N, Moelleken SMC (1993) Dietary adaptation of some extinct ruminants determined by premaxillary shape. J Mammal 74: 1059–1071.

30. Dompierre H, Churcher CS (1996) Premaxillary shape as an indicator of the diet of seven extinct late Cenozoic New World camels. J Vert Paleontol 16: 141–148.

31. Mendoza M, Janis CM, Palmqvist P (2002) Characterizing complex craniodental patterns related to feeding behaviour in ungulates: A multivariate approach. J Zool 258: 223–246.

32. Fraser D, Theodor JM (2011) Anterior dentary shape as an indicator of diet in ruminant artiodactyls. J Vert Paleontol 31: 1366–1375.

33. Fraser D, Theodor JM (2011) Comparing ungulate dietary proxies using discriminant function analysis. J Morphol 272: 1513–1526.

34. Lull RS (1933) A revision of the Ceratopsia or horned dinosaurs. Mem Peabody Mus Nat Hist 3: 1–175.

35. Ostrom JH (1966) Functional morphology and evolution of the ceratopsian dinosaurs. Evolution 20: 290–308.

36. Dodson P (1975) Taxonomic implications of relative growth in lambeosaurine hadrosaurs. Syst Biol 24: 37–54.

37. Dodson P (1983) A faunal review of the Judith River (Oldman) Formation, Dinosaur Provincial Park, Alberta. Mosasaur 1: 89–118.

38. Carpenter K (1982) Skeletal and dermal armor reconstructions of *Euoplocephalus tutus* (Ornithischia: Ankylosauridae) from the Late Cretaceous Oldman Formation of Alberta. Can J Earth Sci 19: 689–697.

39. Carpenter K (1997) Ankylosauria. In Currie PJ, Padian K, editors. Encyclopedia of Dinosaurs. San Diego, Academic Press. 16–20.

40. Carpenter K (1997) Ankylosaurs. In Farlow JO, Brett-Surman MK, editors. The complete dinosaur. Bloomington: Indiana University Press. 307–316.

41. Carpenter K (2004) Redescription of *Ankylosaurus magniventris* Brown 1908 (Ankylosauridae) from the Upper Cretaceous of the Western Interior of North America. Can J Earth Sci 41: 961–986.

42. Bakker RT (1986) The dinosaur heresies: New theories unlocking the mystery of the dinosaurs and their extinction. New York: Zebra Books. 481 p.

43. Carrano MT, Janis CM, Sepkoski JJ Jr (1999) Hadrosaurs as ungulate parallels: Lost lifestyles and deficient data. Acta Palaeontol Pol 44: 237–261.

44. Henderson DM (2010) Skull shapes as indicators of niche partitioning by sympatric chasmosaurine and centrosaurine dinosaurs. In Ryan MJ, Chinnery-Allgeier BJ, Eberth DA, editosr. New perspectives on horned dinosaurs: The Royal Tyrrell Museum ceratopsian symposium. Bloomington: Indiana University Press 293–307.

45. Whitlock JA (2011) Inferences of diplodocoid (Sauropoda: Dinosauria) feeding behavior from snout shape and microwear analyses. PLoS ONE 6(4): e18304. doi:10.1371/journal.pone.0018304.

46. Erickson GM, Krick BA, Hamilton M, Bourne GR, Norell MA, et al. (2012) Complex dental structure and wear biomechanics in hadrosaurid dinosaurs. Science 338: 98–101.

47. Ricklefs RE, Miles DB (1994) Ecological and evolutionary inference from morphology: An ecological perspective. In: Wainwright PC, Reilly SM, editors. Ecologisl morphology: Integrative organismal biology. Chicago: University of Chicago Press. 13–41.

48. Rudwick MJS (1964) The inference of function from structure in fossils. Brit J Philos Sci 15: 27–40.

49. Radinsky LB (1987) The evolution of vertebrate design. Chicago: University of Chicago Press. 188 p.

50. Shockey BJ, Croft DA, Anaya F (2007) Analysis of function in the absence of extant functional homologues: A case study using mesotheriid notoungulates (Mammalia). Paleobiology 33: 227–247.

51. Parks WA (1924) *Dyoplosaurus acutosquameus*, a new genus and species of armoured dinosaur; and notes on a skeleton of *Prosaurolophus maximus*. Univ Toronto Stud Geol Ser 18: 1–35.

52. Arbour VM, Burns ME, Sissons RL (2009) A redescription of the ankylosaurid dinosaur *Dyoplosaurus acutosquameus* Parks, 1924 (Ornithischia: Ankylosauria) and a revision of the genus. J Vert Paleontol 29: 1117–1135.

53. Nopcsa BF (1928) Palaeontological notes on reptiles. Geol Hungarica Ser Palaeontol 1: 1–84.

54. Penkalski P, Blows WT (2013) *Scolosaurus cutleri* (Ornithischia: Ankylosauria) from the Upper Cretaceous Dinosaur Park Formation of Alberta, Canada. Can J Earth Sci 50: 171–182.

55. Farke AA, Ryan MJ, Barrett PM, Tanke DH, Braman DR, et al. (2011) A new centrosaurine from the Late Cretaceous of Alberta, Canada, and the evolution of parietal ornamentation in horned dinosaurs. Acta Palaeontol Pol 56: 691–702.

56. Peters RH (1983) The ecological implications of body size. Cambridge: Cambridge University Press. 329 p.

57. Hammer Ø, Harper DAT, Ryan PD (2001) PAST: paleontological statistics software package for education and data analysis. Palaeontol Electron 4(1): 9. Available: http://palaeo-electronica.org/2001_1/past/issue1_01.htm. Accessed 8 April 2013.

58. Sokal RR, Rohlf FJ (1995) Biometry (third edition). New York: W. H. Freeman and Company. 887 p.

59. Doornik JA, Hansen H (2008) An omnibus test for univariate and multivariate normality. Oxford B Econ Stat 70: 927–939.

60. Strauss R E, Atanassov NMN, Alves J (2003) Evaluation of the principal-component and expectation-maximization methods for estimating missing data in morphometric studies. J Vert Paleontol 23: 284–296.

61. Brown CM, Arbour JH, Jackson DA (2012) Testing of the effect of missing data estimation and distribution in morphometric multivariate data analyses. Syst Biol 61: 941–954.

62. Hammer Ø, Harper D (2006) Paleontological data analysis. Malden: Blackwell Publishing. 351 p.

63. Anderson MJ (2001) A new method for non-parametric multivariate analysis of variance. Austral Ecol 26: 32–46.

64. Mahalanobis PC (1936) On the generalised distance in statistics. Proc Nat Inst Sci India 2: 49–55.

65. Nakagawa S (2004) A farewell to Bonferroni: The problems of low statistical power and publication bias. Behav Ecol 15: 1044–1045.

66. Behrensmeyer AK, Hook RW (1992) Paleoenvironmental contexts and taphonomic modes. In Behrensmeyer AK, Damuth JD, DiMichele WA, Potts R, Sues H-D, Wing SL, editors. Terrestrial ecosystems through time: Evolutionary paleoecology of terrestrial plants and animals. Chicago: University of Chicago Press. 15–136.

67. Prasad V, Strömberg CAE, Alimohammadian H, Sahni A (2005) Dinosaur coprolites and the early evolution of grasses and grazers. Science 310: 1177–1180.

68. Taggart RE, Cross AT (1997) The relationship between land plant diversity and productivity and patterns of dinosaur herbivory. In: Wolberg DL, Stump E, Rosenberg GD, editors. Dinofest International: Proceedings of a symposium sponsored by Arizona State University. Philadelphia: Academy of Natural Sciences. 403–416.

69. Bakker RT (1978) Dinosaur feeding behaviour and the origin of flowering plants. Nature 274: 661–663.

70. Béland P Russell DA (1978) Paleoecology of Dinosaur Provincial Park (Cretaceous), Alberta, interpreted from the distribution of articulated vertebrate remains. Can J Earth Sci 15: 1012–1024.

71. Krassilov VA (1981) Changes of Mesozoic vegetation and the extinction of dinosaurs. Palaeogeogr Palaeoclimatol Palaeoecol 34: 207–224.

72. Retallack GJ, Dilcher DL (1986) Cretaceous angiosperm invasion of North America. Cretaceous Res 7: 227–252.

73. Crane PR (1987) Vegetational consequences of the angiosperm diversification. In Friis EM, Chaloner WG, Crane PR, editors. The origins of angiosperms and their biological consequences. Cambridge: Cambridge University Press. 107–144.

74. Wing SL, Tiffney BH (1987) The reciprocal interaction of angiosperm evolution and tetrapod herbivory. Rev Palaeobot Palynol 50: 179–210.

75. Tiffney BH (1992) The role of vertebrate herbivory in the evolution of land plants. Palaeobotanist 41: 87–97.

76. Wolfe JA, Upchurch GR Jr (1987) North American nonmarine climates and vegetation during the Late Cretaceous. Palaeogeogr Palaeoclimatol Palaeoecol 61: 33–77.

77. Braman DR, Koppelhus EB (2005) Campanian palynomorphs. In Currie PJ, Koppelhus ED, editors. Dinosaur Provincial Park: A spectacular ancient ecosystem revealed. Bloomington: Indiana University Press. 101–130.

78. Koppelhus EB (2005) Paleobotany. In Currie PJ, Koppelhus EB, editors. Dinosaur Provincial Park: A spectacular ancient ecosystem revealed. Bloomington: Indiana University Press. 131–138.

79. Bowman RI (1961) Morphological differentiation and adaptation in the Galápagos finches. Univ Calif Publ Zool 58: 1–326.

80. Bouvier M, Hylander WL (1981) Effect of bone strain on cortical bone structure in macaques (*Macaca mulatta*). J Morphol 167: 1–12.

81. Anapol F, Lee S (1994) Morphological adaptation to diet in platyrrhine primates. Am J Phys Anthropol 94: 239–261.

82. Lucas PW (2004) Dental functional morphology: How teeth work. Cambridge: Cambridge University Press. 355 p.

83. Nogueira MR, Monteiro LR, Peracchi AL, de Araújo AFB (2005) Ecomorphological analysis of the masticatory apparatus in the seed-eating bats, genus *Chiroderma* (Chiroptera: Phyllostomidae). J Zool 266: 355–364.

84. Ostrom JH (1961) Cranial morphology of the hadrosaurian dinosaurs of North America. Bull Am Mus Nat Hist 122: 39–186.

85. Ostrom JH (1964) A functional analysis of jaw mechanics in the dinosaur *Triceratops*. Postilla 88: 1–35.

86. Tanoue K, Grandstaff BS, You H-L, Dodson P (2009) Jaw mechanics in basal Ceratopsia (Ornithischia, Dinosauria). Anat Rec 292: 1352–1369.

87. Kiltie RA (1982) Bite force as a basis for niche differentiation between rain forest peccaries (*Tayassu tajacu* and *T. pecari*). Biotropica 14: 188–195.

88. Herrel A, de Grauw E, Lemos-Espinal JA (2001) Head shape and bite performance in xenosaurid lizards. J Exp Zool 290: 101–107.

89. Verwaijen D, Van Damme R, Herrel A (2002) Relationships between head size, bite force, prey handling efficiency and diet in two sympatric lacertid lizards. Funct Ecol 16: 842–850.

90. Meers MB (2002) Maximum bite force and prey size of *Tyrannosaurus rex* and their relationships to the inference of feeding behaviour. Hist Biol 16: 1–12.

91. Erickson GM, Gignac PM, Steppan SJ, Lappin AK, Vliet KA, et al. (2012) Insights into the ecology and evolutionary success of crocodilians revealed through bite-force and tooth-pressure experimentation. PLoS ONE 7(3):e31781. doi:10.1371/journal.pone.0031781.

92. Russell LS (1940) *Edmontonia rugosidens* (Gilmore), an armoured dinosaur from the Belly River Series of Alberta. Univ Toronto Stud Geol Ser 43: 1–28.

93. Weishampel DB (1984) Interactions between Mesozoic plants and vertebrates: Fructifications and seed predation. N Jb Geol Paläont Abh 167: 224–250.

94. Galton PM (1986) Herbivorous adaptations of Late Triassic and Early Jurassic dinosaurs. In Padian K, editor. The beginning of the age of dinosaurs: Faunal change across the Triassic-Jurassic boundary. Cambridge: Cambridge University Press. 203–221.

95. Haas G (1969) On the jaw musculature of ankylosaurs. Am Mus Novit 2399: 1–11.

96. Herrel A, Podos J, Huber SK, Hendry AP (2005) Evolution of bite force in Darwin's finches: A key role for head width. J Evol Biol 18: 669–675.

97. Sanson GD (1989) Morphological adaptations of teeth to diets and feeding in the Macropodoidea. In: Grigg G, Jarman P, Hume I, editors. Kangaroos, wallabies and rat-kangaroos. New South Wales: Surrey Beatty & Sons Pty Limited. 151–168.

98. Hylander WL (1979) The functional significance of primate mandibular form. J Morphol 160: 223–239.

99. Spencer LM (1995) Morphological correlates of dietary resource partitioning in the Belly River Bovidae. J Mammal 76: 448–471.

100. Vickaryous MK, Maryańska T, Weishampel DB (2004) Ankylosauria. In: Weishampel DB, Dodson P, Osmólska H, editors. The Dinosauria (second edition). Berkeley: University of California Press. 363–392.

101. Marshall CD, Maeda H, Iwata M, Furuta M, Asano S, et al. (2003) Orofacial morphology and feeding behaviour of the dugong, Amazonian, West African and Antillean manatees (Mammalia: Sirenia): functional morphology of the muscular-vibrissal complex. J Zool 259: 245–260.

102. Maryańska T (1977) Ankylosauridae (Dinosauria) from Mongolia. Palaeontol Pol 37: 85–151.

103. Mustoe GE (2007) Coevolution of cycads and dinosaurs. Cycad Newsl 30: 6–9.

104. Rybczynski N, Vickaryous MK (2001) Evidence of complex jaw movement in the Late Cretaceous ankylosaurid *Euoplocephalus tutus* (Dinosauria: Thyreophora). In: Carpenter K, editor. The armored dinosaurs. Bloomington: Indiana University Press. 299–317.

105. Molnar RE, Clifford HT (2001) An ankylosaurian cololite from the Lower Cretaceous of Queensland, Australia. In: Carpenter K, editor. The armored dinosaurs. Bloomington: Indiana University Press. 399–412.

106. Carpenter K (1990) Ankylosaur systematics: example using *Panoplosaurus* and *Edmontonia* (Ankylosauria: Nodosauridae). In: Carpenter K, Currie PJ, editors. Dinosaur systematics: Approaches and perspectives. Cambridge: Cambridge University Press. 281–298.

107. Vickaryous MK (2006) New information on the cranial anatomy of *Edmontonia rugosidens* Gilmore, a Late Cretaceous nodosaurid dinosaur from Dinosaur Provincial Park, Alberta. J Vert Paleontol 26: 1011–1013.

108. Tait J, Brown B (1928) How the Ceratopsia carried and used their head. Trans R Soc Can Ser 3 22: 13–23.

109. Hatcher JB, Marsh OC, Lull RS (1907) The Ceratopsia. Monogr U S Geol Survey 49: 1–300.

110. Lull RS (1908) The cranial musculature and the origin of the frill in the ceratopsian dinosaurs. Am J Sci 25: 387–399.

111. Varriale FJ (2011) Dental microwear and the evolution of mastication in the ceratopsian dinosaurs. Unpublished PhD dissertation. Baltimore: Johns Hopkins University. 470 p.

112. Upchurch GR Jr, Wolfe JA (1993) Cretaceous vegetation of the Western Interior and adjacent regions of North America. Geol Assoc Can Spec Pap 39: 243–281.

113. Brinkman DB, Ryan MJ, Eberth DA (1998) The paleogeographic and stratigraphic distribution of ceratopsids (Ornithischia) in the upper Judith River Group of western Canada. Palaios 13: 160–169.

114. Dodson P (1993) Comparative craniology of the Ceratopsia. Am J Sci 293A: 200–234.

115. Dodson P (1996) The horned dinosaurs: A natural history. Princeton: Princeton University Press. 346 p.

116. Dodson P (1997) Neoceratopsia. In Currie PJ, Padian K, editors. Encyclopedia of Dinosaurs. San Diego: Academic Press. 473–478.

117. Strauss RE (2010) Discriminating groups of organisms. In: Elewa AMT, editor. Morphometrics for nonmorphometricians. Berlin: Springer-Verlag. 73–91.

118. Dodson P (1990) On the status of the ceratopsids *Monoclonius* and *Centrosaurus*. In Carpenter K, Currie PJ, editors. Dinosaur systematics: Approaches and perspectives. Cambridge: Cambridge University Press. 231–243.

119. Lehman TM (1990) The ceratopsian subfamily Chasmosaurinae: Sexual dimorphism and systematics. In Carpenter K, Currie PJ, editors. Dinosaur systematics: Approaches and perspectives. Cambridge: Cambridge University Press. 211–229.

120. Ostrom JH, Wellnhofer P (1986) The Munich specimen of *Triceratops* with a revision of the genus. Zitteliana 14: 111–158.

121. Ostrom JH, Wellnhofer P (1990) *Triceratops*: An example of flawed systematics. In: Carpenter K, Currie PJ, editors. Dinosaur systematics: Approaches and perspectives. Cambridge: Cambridge University Press. 245–254.

122. Godfrey SJ, Holmes R (1995) Cranial morphology and systematics of *Chasmosaurus* (Dinosauria: Ceratopsidae) from the Upper Cretaceous of western Canada. J Vert Paleontol 15: 726–742.

123. Forster CA (1996) Species resolution in *Triceratops*: Cladistic and morphometric approaches. J Vert Paleontol 16: 259–270.

124. Farke AA (2006) Cranial osteology and phylogenetic relationships of the chasmosaurine ceratopsid *Torosaurus latus*. In Carpenter K, editor. Horns and beaks: Ceratopsian and ornithopod dinosaurs. Bloomington: Indiana University Press. 235–257.

125. Currie PJ, Langston W Jr., Tanke DH (2008) A new species of *Pachyrhinosaurus* (Dinosauria, Ceratopsidae) from the Upper Cretaceous of Alberta, Canada. In Currie PJ, Langston W Jr., Tanke DH, editors. A new horned dinosaur from an Upper Cretaceous bone bed in Alberta. Ottawa: NRC Research Press. 1–108.

126. Mallon JC, Holmes R, Eberth DA, Ryan MJ, Anderson JS (2011) Variation in the skull of *Anchiceratops* (Dinosauria, Ceratopsidae) from the Horseshoe Canyon Formation (Upper Cretaceous) of Alberta. J Vert Paleontol 31: 1047–1071.

127. Sternberg CM (1940) Ceratopsidae from Alberta. J Paleontol 14: 468–480.

128. Owen-Smith RN (1988) Megaherbivores: The influence of very large body size on ecology. Cambridge: Cambridge University Press. 369 p.

129. Kräusel R (1922) Die Nahrung von *Trachodon*: Paläont Z 4: 80.

130. Currie PJ, Koppelhus EB, Muhammad AF (1995) Stomach contents of a hadrosaur from the Dinosaur Park Formation (Campanian, Upper Cretaceous) of Alberta, Canada. In Sun A, Wang Y, editors. Sixth symposium on Mesozoic terrestrial ecosystems and biota, short papers. Beijing: China Ocean Press 111–114.

131. Tweet JS, Chin K, Braman DR, Murphy NL (2008) Probable gut contents within a specimen of *Brachylophosaurus canadensis* (Dinosauria: Hadrosauridae) from the Upper Cretaceous Judith River Formation of Montana. Palaios 23: 624–635.

132. Abel O (1922) Diskussion zu den Vorträgen R. Kräusel und F. Versluys. Paläont. Z 4: 87.

133. Chin K, Gill BD (1996) Dinosaurs, dung beetles, and conifers: participants in a Cretaceous food web. Palaios 11: 280–285.

134. Chin K (2007) The paleobiological implications of herbivorous dinosaur coprolites from the Upper Cretaceous Two Medicine Formation of Montana: Why eat wood? Palaios 22: 554–566.

135. Lull RS, Wright NE (1942) Hadrosaurian dinosaurs of North America. Geol Soc Am Spec Pap 40: 1–242.

136. Hopson JA (1975) The evolution of cranial display structures in hadrosaurian dinosaurs. Paleobiology 1: 21–43.

137. Horner JR, Weishampel DB, Forster CA (2004) Hadrosauridae. In Weishampel DB, Dodson P, Osmólska H, editors. The Dinosauria (second edition). Berkeley: University of California Press. 438–463.

138. Hutchinson GE (1959) Homage to Santa Rosalia or why are there so many kinds of animals? Am Nat 93: 145–159.

139. Lewin R (1983) Santa Rosalia was a goat. Science 221: 636–639.

140. Grant PR (1972) Convergent divergent character displacement. Biol J Linnean Soc 4: 39–68.

141. Horn HS, May RM (1977) Limits to similarity among coexisting competitors. Nature 270: 660–661.

142. Grant PR, Abbott I (1980) Interspecific competition, null hypotheses and island biogeography. Evolution 34: 332–341.

143. Schoener TW (1984) Size differences among sympatric, bird-eating hawks: A worldwide survey. In: Strong DR, Simberloff D, Abele LG, Thistle AB, editors. Ecological communities: Conceptual issues and the evidence. Princeton: Princeton University Press. 254–281.

144. Pianka ER (1994) Comparative ecology of *Varanus* in the Great Victoria Desert. Aust J Ecol 19: 395–408.

145. Weishampel DB (1984) Evolution of jaw mechanics in ornithopod dinosaurs. Adv Anat Embryol Cell Biol 87: 1–109.

146. Chapman RE, Brett-Surman MK (1990) Morphometric observations on hadrosaurid ornithopods. In Carpenter K, Currie PJ, editors. Dinosaur systematics: Approaches and perspectives. Cambridge: Cambridge University Press. 163–177.

147. Sternberg CM (1950) Steveville–West of the Fourth Meridian, Alberta. Geological Survey of Canada Topographic Map 969A. Ottawa: Geological Survey of Canada.

148. Currie PJ, Russell DA (2005) The geographic and stratigraphic distribution of articulated and associated dinosaur remains. In Currie PJ, Koppelhus EB, editors. Dinosaur Provincial Park: A spectacular ancient ecosystem revealed. Bloomington: Indiana University Press. 537–569.

149. Butler RJ, Upchurch P, Norman DB (2008) The phylogeny of the ornithischian dinosaurs. J Syst Palaeontol 6: 1–40.

150. Hubbell SP (2001) The unified neutral theory of species abundance and diversity. Princeton: Princeton University Press. 448 p.

151. Diamond JM (1975) Assembly of species communities. Cody ML, Diamond JM, editors. Ecology and evolution of communities. Cambridge: Harvard University Press. 342–444.

152. Brinkman DB, Russell AP, Eberth DA, Peng J (2004) Vertebrate palaeocommunities of the lower Judith River Group (Campanian) of southeastern Alberta, Canada, as interpreted from vertebrate microfossil assemblages. Palaeogeogr Palaeoclimatol Palaeoecol 213: 295–313.

153. Olson EC (1952) The evolution of a Permian vertebrate chronofauna. Evolution 6: 181–196.

154. Brinkman DB (1990) Paleoecology of the Judith River Formation (Campanian) of Dinosaur Provincial Park, Alberta, Canada: evidence from vertebrate microfossil localities. Palaeogeogr Palaeoclimatol Palaeoecol 78: 37–54.

155. DiMichele WA, Behrensmeyer AK, Olszewski TD, Labandeira CC, Pandolfi JM, Wing SL, Bobe R (2004) Long-term stasis in ecological assemblages: Evidence from the fossil record. Annu Rev Ecol Evol Syst 35: 285–322.

156. Barrett PM, Rayfield EJ (2006) Ecological and evolutionary implications of dinosaur feeding behaviour. Trends Ecol Evol 21: 217–224.

157. Weishampel DB, Norman DB (1989) Vertebrate herbivory in the Mesozoic; jaws, plants, and evolutionary metrics. Geol Soc Am Spec Pap 238: 87–100.

158. Fiorillo AR (1991) Dental microwear on the teeth of *Camarasaurus* and *Diplodocus*: implications for sauropod paleoecology. Kielan-Jaworowska Z, Heintz N, Nakrem HA, editors. Fifth symposium on Mesozoic terrestrial ecosystems and biota. Oslo: Paleontologisk Museum, University of Oslo. 23–24.

159. Fiorillo AR (1998) Dental microwear patterns of the sauropod dinosaurs *Camarasaurus* and *Diplodocus*: Evidence for resource partitioning in the Late Jurassic of North America. Hist Biol 13: 1–16.

160. Fiorillo AR (2008) Lack of variability in feeding patterns of the sauropod dinosaurs *Diplodocus* and *Camarasaurus* (Late Jurassic, western USA) with respect to climate as indicated by tooth wear features. Sankey JT, Baszio S, editors. Vertebrate microfossil assemblages: Their role in paleoecology and paleobiogeography. Bloomington: Indiana University Press. 104–116.

161. Calvo JO (1994) Jaw mechanisms in sauropod dinosaurs. Gaia 10: 183–194.

162. Stevens KA, Parrish JM (1999) Neck posture and feeding habits of two Jurassic sauropod dinosaurs. Science 284: 798–800.

163. Stevens KA, Parrish JM (2005) Digital reconstructions of sauropod dinosaurs and implications for feeding. Currie Rogers KA, Wilson JA, editors. The sauropods: Evolution and paleobiology. Berkeley: University of California Press. 178–200.

164. Upchurch P, Barrett PM (2000) The evolution of sauropod feeding mechanisms. Sues HD, editor. Evolution of herbivory in terrestrial vertebrates. Cambridge: Cambridge University Press. 79–122.

165. Bock WJ (1977) Toward an ecological morphology. Vogelwarte 29: 127–135.

166. Prieto-Márquez A (2010) Global phylogeny of Hadrosauridae (Dinosauria: Ornithopoda) using parsimony and Bayesian methods. Zool J Linn Soc 159: 435–502.

167. Sampson SD, Loewen MA, Farke AA, Roberts EM, Forster CA, et al. (2010) New horned dinosaurs from Utah provide evidence for intracontinental dinosaur endemism. PLoS One 5(9): e12292. doi:10.1371/journal.pone.0012292.

168. Thompson RS, Parish JC, Maidment SCR, Barrett PM (2012) Phylogeny of the ankylosaurian dinosaurs (Ornithischia: Thyreophora). J Syst Palaeontol 10: 301–312.

169. Bowman RI (1961) Morphological differentiation and adaptation in the Galápagos finches. Univ Calif Publ Zool 58: 1–326.

170. Greaves WS (1991) The orientation of the force of the jaw muscles and the length of the mandible in mammals. Zool J Linn Soc 102: 367–374.

171. King G (1996) Reptiles and herbivory. London: Chapman and Hall. 160 p.

172. Metzger KA, Herrel A (2005) Correlations between lizard cranial shape and diet: A quantitative, phylogenetically informed analysis. Biol J Linn Soc 86: 433–466.

173. Stayton CT (2005) Morphological evolution of the lizard skull: A geometric morphometric survey. J Morphol 263: 47–59.

174. Greaves WS (1978) The jaw lever system in ungulates: A new model. J Zool 184: 271–285.

175. Solounias N, Dawson-Saunders B (1988) Dietary adaptations and paleoecology of the Late Miocene ruminants from Pikermi and Samos in Greece. Palaeogeogr Palaeoclimatol Palaeoecol 65: 149–172.

176. Greaves WS (1974) Functional implications of mammalian jaw joint position. Forma et Functio 7: 363–376.

177. Herrel A, Van Damme R, Vanhooydonck B, De Vree F (2001) The implications of bite performance for diet in two species of lacertid lizards. Can J Zool 79: 662–670.

178. Herrel A, O'Reilly JC, Richmond AM (2002) Evolution of bite performance in turtles. J Evol Biol 15: 1083–1094.

15

Virtual Reconstruction and Prey Size Preference in the Mid Cenozoic Thylacinid, *Nimbacinus dicksoni* (Thylacinidae, Marsupialia)

Marie R. G. Attard[1,2]*, William C. H. Parr[1], Laura A. B. Wilson[1], Michael Archer[3], Suzanne J. Hand[3], Tracey L. Rogers[1], Stephen Wroe[2]

1 Evolution and Ecology Research Centre, School of Biological, Earth and Environmental Sciences, University of New South Wales, Sydney, New South Wales, Australia, **2** Function, Evolution and Anatomy Research laboratory, Zoology, School of Environmental and Rural Sciences, University of New England, New South Wales, Australia, **3** Evolution of Earth and Life Sciences Research Centre, School of Biological, Earth and Environmental Sciences, University of New South Wales, Sydney, New South Wales, Australia

Abstract

Thylacinidae is an extinct family of Australian and New Guinean marsupial carnivores, comprising 12 known species, the oldest of which are late Oligocene (~24 Ma) in age. Except for the recently extinct thylacine (*Thylacinus cynocephalus*), most are known from fragmentary craniodental material only, limiting the scope of biomechanical and ecological studies. However, a particularly well-preserved skull of the fossil species *Nimbacinus dicksoni*, has been recovered from middle Miocene (~16-11.6 Ma) deposits in the Riversleigh World Heritage Area, northwestern Queensland. Here, we ask whether *N. dicksoni* was more similar to its recently extinct relative or to several large living marsupials in a key aspect of feeding ecology, i.e., was *N. dicksoni* a relatively small or large prey specialist. To address this question we have digitally reconstructed its skull and applied three-dimensional Finite Element Analysis to compare its mechanical performance with that of three extant marsupial carnivores and *T. cynocephalus*. Under loadings adjusted for differences in size that simulated forces generated by both jaw closing musculature and struggling prey, we found that stress distributions and magnitudes in the skull of *N. dicksoni* were more similar to those of the living spotted-tailed quoll (*Dasyurus maculatus*) than to its recently extinct relative. Considering the Finite Element Analysis results and dental morphology, we predict that *N. dicksoni* likely occupied a broadly similar ecological niche to that of *D. maculatus*, and was likely capable of hunting vertebrate prey that may have exceeded its own body mass.

Editor: Kornelius Kupczik, Friedrich-Schiller-University Jena, Germany

Funding: This research was funded by Australian Research Council grants to S. Wroe (DP0666374 and DP0987985), and to M. Archer, and S. J. Hand (LP100200486 and DP1094569). M. Attard is supported by the Postgraduate Writing and Skills Transfer Award sponsored by the Evolution and Ecology Research Centre, University of New South Wales. L. Wilson is supported by the Swiss National Science Foundation (PBZHP3_141470). The funders had no role in study design, data collection and analysis, decision to publish, or preparation of the manuscript.

Competing Interests: The authors have declared that no competing interests exist.

* E-mail: marie.attard@une.edu.au

Introduction

Thylacinids first appear in the Australian fossil record during the late Oligocene (~24 Ma) and include the largest representatives of the Dasyuromorphia, i.e., families Thylacinidae, Dasyuridae and Myrmecobiidae [1–8]. A wide range of feeding ecologies are known within the order. They include omnivores, insectivores, small prey specialists, hypercarnivores and osteophageous species [9]. Variation in the dentition, skull shape and body size (~1–60 kg) of thylacinids suggests considerable trophic diversity within the family [10,11]. In addition to the recently extinct thylacine or Tasmanian 'tiger' (*Thylacinus cynocephalus*), eleven extinct species of thylacinid have been described [4,12–18]. Up to five species may have co-existed in the Riversleigh World Heritage Area, northwestern Queensland between the late Oligocene (~24 Ma) to middle Miocene (16-11.6 Ma) [12,16]. See Table S1 for the temporal and geographic distribution of all thylacinid species.

The Riversleigh thylacinids inhabited forests [19,20]. These regions were also occupied by an assortment of other carnivorous/omnivorous taxa, including 'giant' carnivorous rat-kangaroos (*Ekaltadeta* spp.), crocodiles (Mekosuchinae, e.g. *Baru darrowi* and *Trilophosuchus rackhami*), flightless dromornithid birds, marsupial lions (Thylacoleonidae), bandicoots (Peramelemorphia), dasyurids (Dasyuridae), pythons (Pythonidae), madtsoiid snakes and the world's oldest known venomous snakes [21–23]. Subsequent drying of the Australian continent from the late Miocene (11.6–5.3 Ma) led to the gradual replacement of forest environments with open woodlands, shrublands and grasslands [19–21]. These changes appear to broadly correlate with declining thylacinid diversity [24].

To date, interpretations of the ecology and feeding behavior of fossil thylacinids have been largely qualitative. This is, at least in part, because most extinct species are known only from jaw fragments and teeth. The near-complete skull of *Nimbacinus dicksoni* [4,14,18], a medium-sized thylacinid, provides an opportunity to more fully investigate feeding ecology in an extinct thylacinid.

Nimbacinus dicksoni was approximately 5 kg in body mass [10]. Fossils of *N. dicksoni* have been recovered from Oligocene-Miocene

(~24–5.3 Ma) deposits in the Riversleigh World Heritage Area, northwestern Queensland and Bullock Creek, Northern Territory [12,14,16,18]. Its dentition is less specialized than that of the species of *Thylacinus*, but broadly similar to that of the living dasyurid, the spotted-tailed quoll (*Dasyurus maculatus*) in the arrangement and geometry of molar shearing crests typically associated with carnivory [18,25].

To date conflicting evidence has been presented regarding the body size of prey *N. dicksoni* may have hunted. Predictions of bite force adjusted for body mass, based on application of 2D beam theory, have suggested that *N. dicksoni* may have taken relatively large prey, as does the slightly smaller *D. maculatus* [26]. However, shape analysis of the cranium has suggested that the species may have been restricted to smaller prey and/or included a higher proportion of invertebrate food in its diet [11].

The loads imposed on an animal during prey acquisition and feeding play an important role in the evolution of its skull morphology [27]. Testing hypotheses regarding the relationship between the form and function of skulls from extinct species requires an understanding of this relationship in living animals [28]. A comparative biomechanics approach involving living analogues has increasingly been applied to predict the feeding ecology and predatory behavior of extinct species [29–34]. To gain further insight in the feeding ecology of *N. dicksoni*, here we perform a biomechanical analysis of the skull of *N. dicksoni* to predict its mechanical behavior in response to loads simulating the capture and processing of prey.

Finite Element Analysis (FEA) is a computer modeling approach now commonly used by biologists and paleontologists to examine and compare mechanical performance in biological structures in comparative contexts [35–41]. In FEA, continuous structures, such as the skull, are divided into discrete, finite numbers of elements, allowing the prediction of mechanical behavior for complex geometric shapes. The structure is analyzed in the form of a matrix algebra problem that is solved with the aid of a computer [42].

Studies of feeding ecology for thylacinids have primarily focused on the most recently extinct member of the family, *T. cynocephalus*, which survived in Tasmania until 1936 [43]. Our understanding of the ecology of *T. cynocephalus* is chiefly based on morphological comparisons and 2D beam theory [11,26,44,45], as well as anecdotal accounts of their behavior in the wild [46,47]. Elbow joint morphology of *T. cynocephalus* evidently most closely resembles that of extant ambush predators, a compromise between efficient distance locomotion and the ability to manipulate and grapple with prey [48]. Three-dimensional biomechanical modeling of the skull of *T. cynocephalus*, extant dasyurids and an introduced Australian predator (*Canis lupus dingo*) have suggested potential limitations in prey body size [32,35].

Morphological and biomechanical comparisons including sympatric native predators in Tasmania indicate that the diet of *T. cynocephalus* may have overlapped considerably with the two largest extant marsupial carnivores, the Tasmanian devil (*Sarcophilus harrisii*) and spotted-tailed quoll (*D. maculatus*) [35,44,49]. These species represented the three largest marsupial carnivores in Tasmania at the time of European settlement.

Sarcophilus harrisii is the largest living marsupial carnivore (mean adult male weight 8.7 kg, female 6.1 kg) [50]. They are the only specialized scavengers among living marsupials, filling a broadly similar ecological niche to that of osteophagous hyenas [44]. They are also opportunistic hunters and are known to prey on mammals that may exceed their body mass [51,52]. Relative to its body size, its predicted bite force is greater than that of any other extant mammal studied to date [26].

Quolls are represented by four extant species in Australia and two in New Guinea [53]. The largest quoll, *Dasyurus maculatus* (maximum weight 7 kg) has a broad diet mainly consisting of mammals and insects, but will occasionally feed on birds and reptiles [54,55]. Larger prey species constitute a higher proportion of the diet of adult male *D. maculatus*, while females and immature *D. maculatus* more frequently feed on smaller bodied mammals and invertebrates [56,57]. The northern quoll (*Dasyurus hallucatus*) is the smallest and most arboreal of the four Australian quolls, weighing up to 1.2 kg [58,59]. Although primarily insectivorous, this active hunter can feed on a variety of foods: fruits, small mammals, birds, reptiles, frogs and carrion [60–62].

In this study we aim to determine whether *N. dicksoni* was capable of killing large prey relative to their body size, or was restricted to catching relatively small bodied species. By digital reconstruction of its skull and applied 3D FEA, we compare its mechanical performance with that of three extant marsupial carnivores; *S. harrisii*, *D. maculatus* and *D. hallucatus*. We use previously applied scaling procedures [31] to account for differences in body mass, allowing for comparison of results between species. We predict that *N. dicksoni* will show similar distributions and magnitudes of craniomandibular stress to that of *D. maculatus* due to similarities in their dental morphology and size [18]. We also include *T. cynocephalus* to establish whether the biomechanical performance of *N. dicksoni* more closely resembles that of this larger, more derived thylacinid than dasyurids. A general assessment of the phylogenetic relationships of dasyuromophians, including taxa examined in this study is shown in Figure S1. We test if the biomechanical patterns and the inferred feeding aspects in *T. cynocephalus* in a recent study [35] are derived. We hypothesize that the relatively long rostrum of *T. cynocephalus* will result in higher stresses in the skull during biting and prey procurement than *N. dicksoni* and other dasyuromorphians.

Materials and Methods

Specimens

The skull of *Nimbacinus dicksoni* (QMF36357) was recovered from AL90 Site on the Gag Plateau of the Riversleigh World Heritage area in northwestern Queensland. Precise locality details can be provided on application to the Queensland Museum. This specimen was collected under permits issued by Queensland Department of Environment and Heritage and Environment Australia, and registered in the paleontological collections of the Queensland Museum, Brisbane, Australia. All necessary permits were obtained for the described study, which complied with all relevant regulations.

The biomechanical performance of the *Nimbacinus dicksoni* skull was compared with that of four dasyuromorphian species covering a range of craniodental morphologies and feeding ecologies. These comprized three extant dasyurids [*Dasyurus hallucatus* TMM M-6921; *D. maculatus* UNSW Z20; *Sarcophilus harrisii* AM10756] and one thylacinid (*Thylacinus cynocephalus* AM1821). Institutional abbreviations are QMF (Queensland Museum Fossil, Queensland), TMM (Texas Memorial Museum, Austin), UNSW (University of New South Wales, Sydney) and AM (Australian Museum, Sydney).

We generated 3D finite element models (FEMs) of each skull on the basis of computed tomography X-ray (CT) scan data. Digimorph (University of Texas; http://www.digimorph.org) was the source of CT data of a *D. hallucatus* skull (0.0784 mm slice thickness, 0.0784 mm inter-slice distance). Permission to use Digimorph derived CT scan data was granted by Dr. Timothy Rowe, Project Director of Digimorph. Other skulls were scanned in a Toshiba Aquillon 16 scanner (ToshibaMedical Systems Corporation, Otawara, Tachigi, Japan) at the Mater Hospital, Newcastle, NSW (1 mm slice thickness, 0.8 mm inter-slice distance, 240 mm field of view). The Australian Museum, Queensland Museum and University of New South Wales granted the loan of specimens to obtain CT data for this study.

We used the same specimen CT scans of *T. cynocephalus*, *D. maculatus* and *S. harrisii* as examined by Attard et al. [35], but constructed the FEMs again using a higher resolution mesh than that used by Attard et al. [35]. More specifically, the 3D surface meshes which formed the bases of the FEMs were computed using the High Quality as opposed to Medium Quality option in Mimics (ver. 13.2). The generated surface meshes were then converted to FEMs in STRAND7 (ver. 2.4) following previously established protocols [31,35,63]. Additionally, 3D objects of *T. cynocephalus* and all dasyurid skulls were exported as separate stl files for the creation of interactive 3D pdf documents that can be viewed using Adobe Reader (Figure S2, S3, S4, S5).

Digital Reconstruction of *Nimbacinus dicksoni*

For detailed descriptions of *N. dicksoni* see Muirhead and Archer [14] and Wroe and Musser [18]. Previous analysis has suggested that, within the family, the dentition of *N. dicksoni* is less derived than the recent *Thylacinus cynocephalus* for at least 12 features, but that relatively few cranial specializations in *T. cynocephalus* distinguish the two species. These two taxa share at least three cranial features not present in the most generalized thylacinid known from significant cranial material, the late Oligocene *Badjcinus turnbulli* [14,18].

The skull of *N. dicksoni* is well preserved, although some regions are absent or damaged. Specifically, some damage/deformation is present at the postorbital processes, frontal, maxillary and nasal bones, which are compressed dorsoventrally (Figure 1). These damaged regions were reconstructed according to the morphology of surrounding bone regions once the damaged areas had been isolated and deleted [64]. Regions of bone that showed only minor damage were smoothed to create a coherent surface mesh for later solid meshing.

The right and left dentaries were largely intact but missing the superior regions of the coronoid processes, the temporomandibular joints (TMJ), condyles and angular processes (Figure 1). The anterior of the mandible is broken, separating both dentaries. We used the right dentary as a basis for reconstruction because its dentition was more complete, with only the incisors missing (Figure 1). We used a surface mesh of the right dentary of *D. maculatus* to reconstruct posterior regions of the right dentary of *N. dicksoni*. *Dasyurus maculatus* was chosen as its mandible was most similar in shape to that of *N. dicksoni* [11], thereby minimizing the extent of warping needed (and see below).

Reconstruction involved scaling the dentary of *D. maculatus* to the same size as that of *N. dicksoni* on the basis of skull length (condylo-basal). The missing posterior region of the *N. dicksoni* specimen was then isolated on the *D. maculatus* specimen and the mesh fitted to the existing structure in the mesh of *N. dicksoni* using Iterative Closest Point (ICP) registration. ICP is an algorithm that revises the transformation needed to minimize the distance between the points of two partially overlapping meshes. This process re-oriented the *D. maculatus* dentary in accordance with the morphology of the *N. dicksoni* dentary [65]. The anterior region of the *D. maculatus* dentary was deleted and the posterior region 'warped' so that overlapping regions of the coronoid process and angular process from the *D. maculatus* mesh matched the existing morphology of *N. dicksoni*. The manual warping method was used as much of the target (fossil) morphology was missing, making it

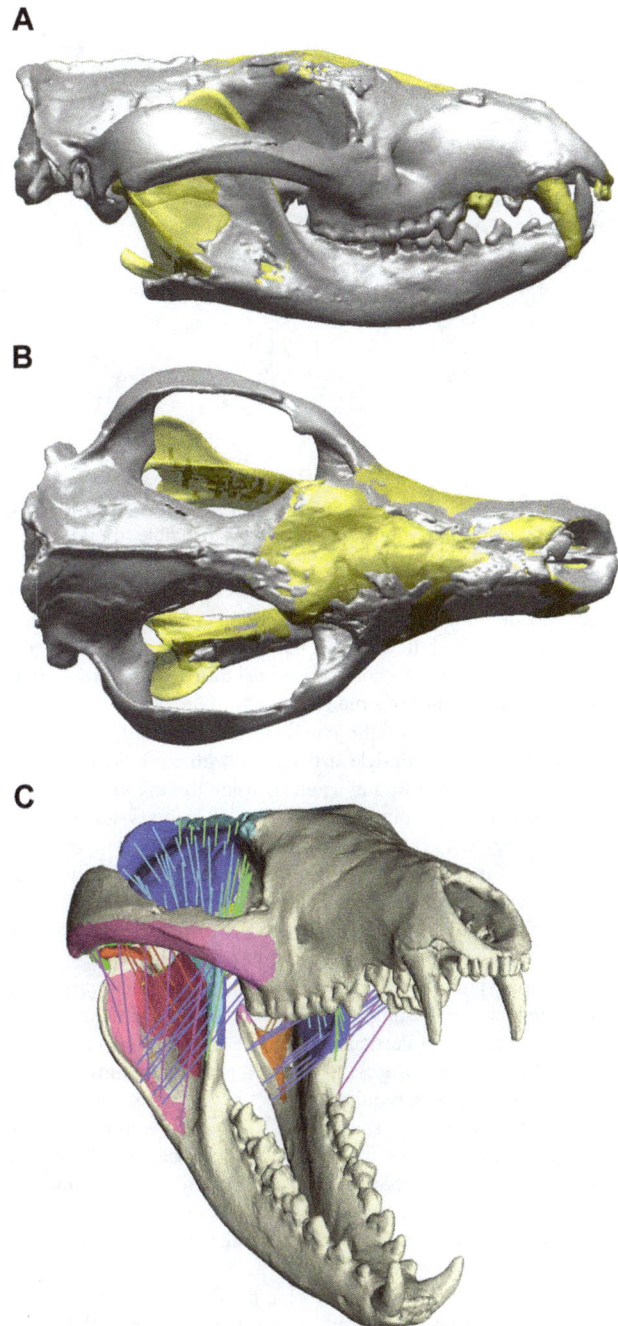

Figure 1. Digital reconstruction of *Nimbacinus dicksoni*. Original (grey) and reconstructed 3D (yellow) in (A) lateral view and (B) dorsal view. (C) Pre-processed Finite Element model of *N. dicksoni*, showing jaw musculature represented by trusses.

impossible to apply homologous landmarks on both complete (*D. maculatus*) and incomplete fossil (*N. dicksoni*) specimens. Procedures to warp overlapping skull regions followed established protocols used by Oldfield et al. and Parr et al. [38,66]. Manual warping works by establishing a grid of control points around the complete model (note that at this stage the incomplete fossil model has been ICP registered with the scaled complete model by matching the orientations of the regions of the jaw that are present in both models). These control points are then manipulated so that the surface morphology of the complete model matches that of the

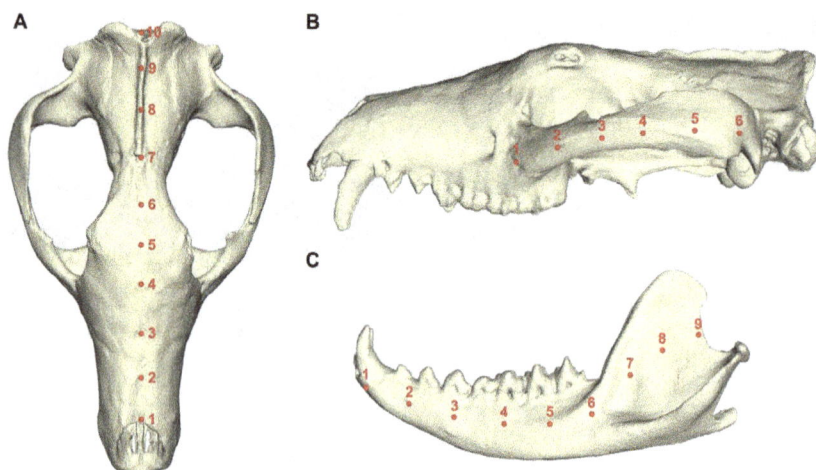

Figure 2. Position of nodes selected on each model to measure von Mises stress. Nodes were selected at equidistant points along the (A) mid-sagittal plane, (B) zygomatic arch and (C) mandible to measure the distribution of von Mises stress for each loading case.

target (fossil). This is another variation of Template Mesh Deformation [66], but with the template points being the grid control points around the complete model rather than homologous anatomical points on both models.

Similarly, the TMJ of the complete (*D. maculatus*) model was warped so that the condyle articulated with and fitted the *N. dicksoni* cotyle of the cranium, again by using the manual template mesh deformation warping method. The left dentary was created by mirroring the reconstructed right dentary. These were positioned so that the condyles articulated with the cranium, the outer surfaces of the lower molars made contact with the inner surface of the upper molars, and the tips of the lower canines aligned with their 'sockets' in the cranium (see Figure 1).

It is important to note that the shape of the warp was determined by the existing regions of the *N. dicksoni* dentary; the need for the condyle to articulate with the cotyle to form the TMJ and for the coronoid process to fit between the cranium and the zygomatic arch. These requirements act as restraints on the warp such that the shape of the starting mesh (*D. maculatus* in this case) is not important in the sense that the warping process would always end with a similarly shaped posterior region of the mandible regardless of which taxon was used. We reiterate that *D. maculatus* was used because it was the most similar in shape [11] and therefore required less 'warping'.

The *N. dicksoni* cranium was missing the following teeth: left I1-4, right I1, 3-4, both right and left C1 and right LDP2. The existing I2 and LDP2 on *N. dicksoni* were mirrored. All incisors were missing from the mandible. Incisors from *D. maculatus* were isolated, scaled and fitted into the empty tooth sockets on *N. dicksoni*. Figure 1 displays the completed reconstruction of *N. dicksoni*.

Finite Element Models

The assembly of FEMs largely follows previously published procedures [30,31,35]. As the skull of *N. dicksoni* was not fully preserved, we were unable to assign multiple material properties to the digital reconstruction without introducing additional assumptions. Consequently, as in most FEA incorporating fossil material [30,39,67], all FEMs were homogeneous and assigned a single material property for cortical bone ($E = 13.7$ GPa, $v = 0.3$, where E is Young's modulus of elasticity and v is Poisson's ratio) [68] to enable direct comparisons between species. Poisson's ratio and

Young's modulus are fundamental metrics in the comparison of stress or deformation for any material when strained elastically, including homogeneous materials [69]. Young's modulus is a measure of stiffness in the material, whereas Poisson's ratio is the negative ratio between transverse strain and longitudinal strain in an elastic material subjected to uniaxial stress [70].

Each homogeneous model was comprized of four-noded tetrahedral elements or 'bricks' (tet4). FEMs for QMF36357, AM1821, AM10756, UNSW Z20 and TMM M-6921 were comprized of 1564048, 1429714, 1799292, 1402103 and 1956942 bricks respectively. Tet4 models are theoretically less accurate than models comprized of tet10 elements. However, the models used in this study are large and any difference in accuracy between results from tet4 versus tet10 models will diminish as the number of elements is increased. Comparable analyses comparing tet4 and tet10 based models much smaller than those used here (<252000 elements) found differences of <10% [27].

Modeling Masticatory Muscle Forces

Jaw elevators were modeled as seven muscle subdivisions: *temporalis superficialis, temporalis profundus, masseter superficialis, masseter profundus, zygomatico mandibularis, pterygoideus internus* and *pterygoideus externus* [32]. Proportions used for each jaw muscle division were based on muscle mass proportions from a dissected Virginia opossum (*Didelphis virginiana*) [71]. Muscle forces were predicted on the basis of maximum cross-sectional areas (CSA) using the 'dry skull' method [72]. To improve the accuracy of our CSA measurements, we used our FEMs to record the co-ordinates of ~100 nodes at the perimeter of each muscle cross sectional area [73]. The FEM was moved to the correct orientation described by Thomason [70] to select nodes outlining the CSA. The node co-ordinates were then plotted in plane geometry software, GEUP 5 (version 5.0.3) and connected to form a multi-sided polygon. The area of the polygon was measured to estimate the CSA of each major jaw closing muscle. To minimize the incidence of artefacts at bite points and muscle origin and insertion areas, surface regions at these sites were tessellated using a network of stiff beam elements [74].

Restraints, Loading Conditions and Scaling

Dasyurids frequently use a penetrating canine bite to kill prey [44,75–78] which involves the application of a bending load [79].

Table 1. Predicted body mass and masticatory muscle forces for modeled dasyuromorphians.

Species	Predicted body mass (kg)	Temporalis muscle force (N)	Masseteric muscle force (N)	Total muscle force (N)
Dasyurus hallucatus	0.78	67.60	55.89	123.49
Dasyurus maculatus	2.88	211.01	178.67	389.67
Nimbacinus dicksoni	5.25	282.38	368.33	650.71
Sarcophilus harrisii	14.20	300.46	384.73	685.19
Thylacinus cynocephalus	32.49	706.64	843.21	1,549.85

Predicted body mass (kg) calculated using the regression equation for dasyuromorphians provided by Myers [76] based on lower molar row length. Temporalis and masseteric muscle forces (N) were calculated based on cross-sectional area [67].

We simulated bilateral canine biting (intrinsic load) and four extrinsic loads to simulate loads generated by struggling prey (axial twist, lateral shake, pullback and dorsoventral) for all models using protocols described by Attard et al. [35] and following McHenry et al. [31]. Extrinsic loads were modeled without applying bite forces so as to clearly reveal the different influences of each separate loading [31]. A gape angle of 35° was applied in all linear static load cases.

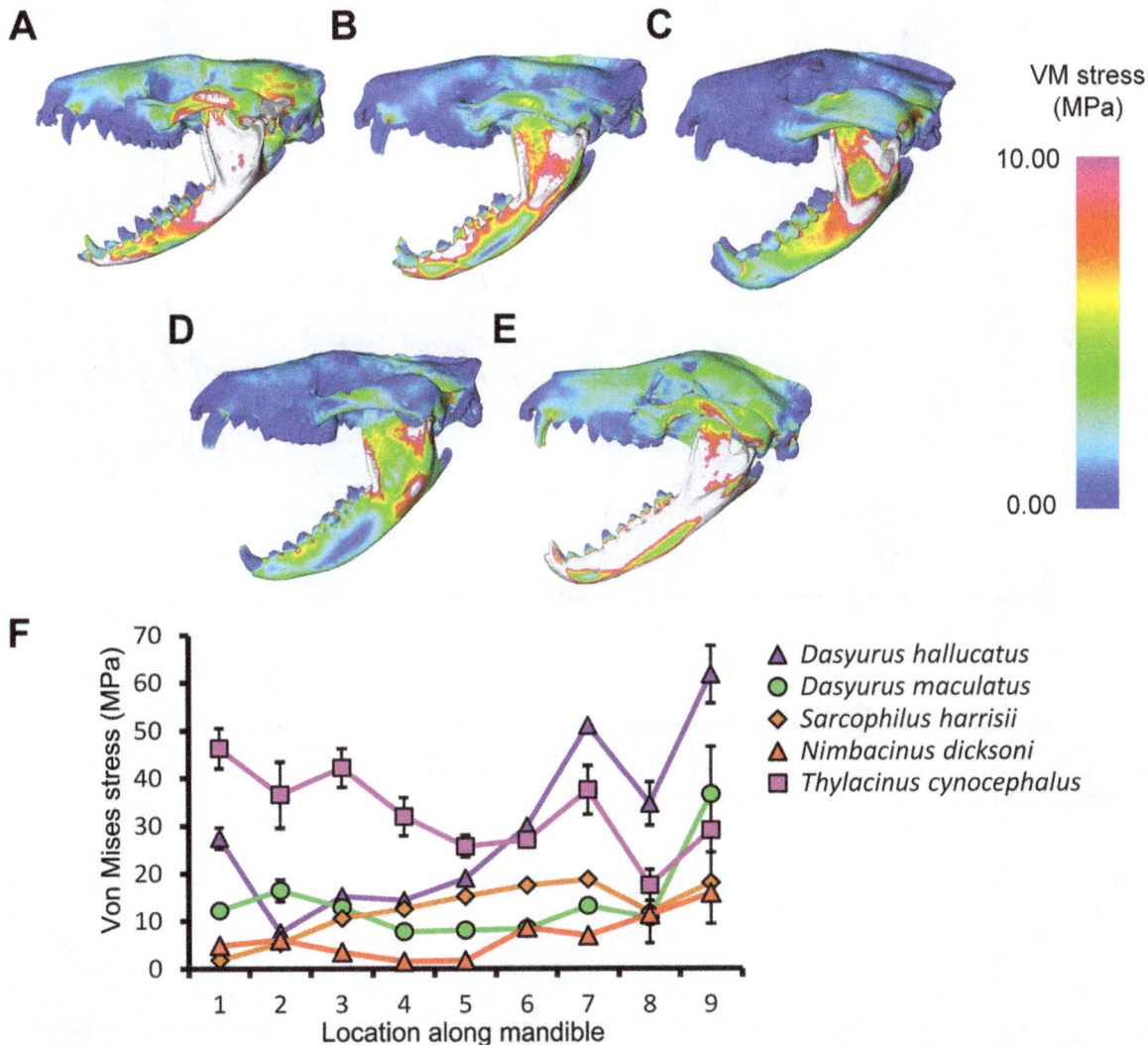

Figure 3. Von Mises stress under a bilateral canine bite in lateral view. The models are subjected to a load applied to both canines, with bite force scaled based on theoretical body mass. Species modeled were (A) *Dasyurus hallucatus*, (B) *Dasyurus maculatus*, (C) *Sarcophilus harrisii*, (D) *Nimbacinus dicksoni* and (E) *Thylacinus cynocephalus*. White colored regions of the skull represent VM stress above 10 MPa. (F) Distribution of von Mises stress was measured from anterior to posterior along the mandible.

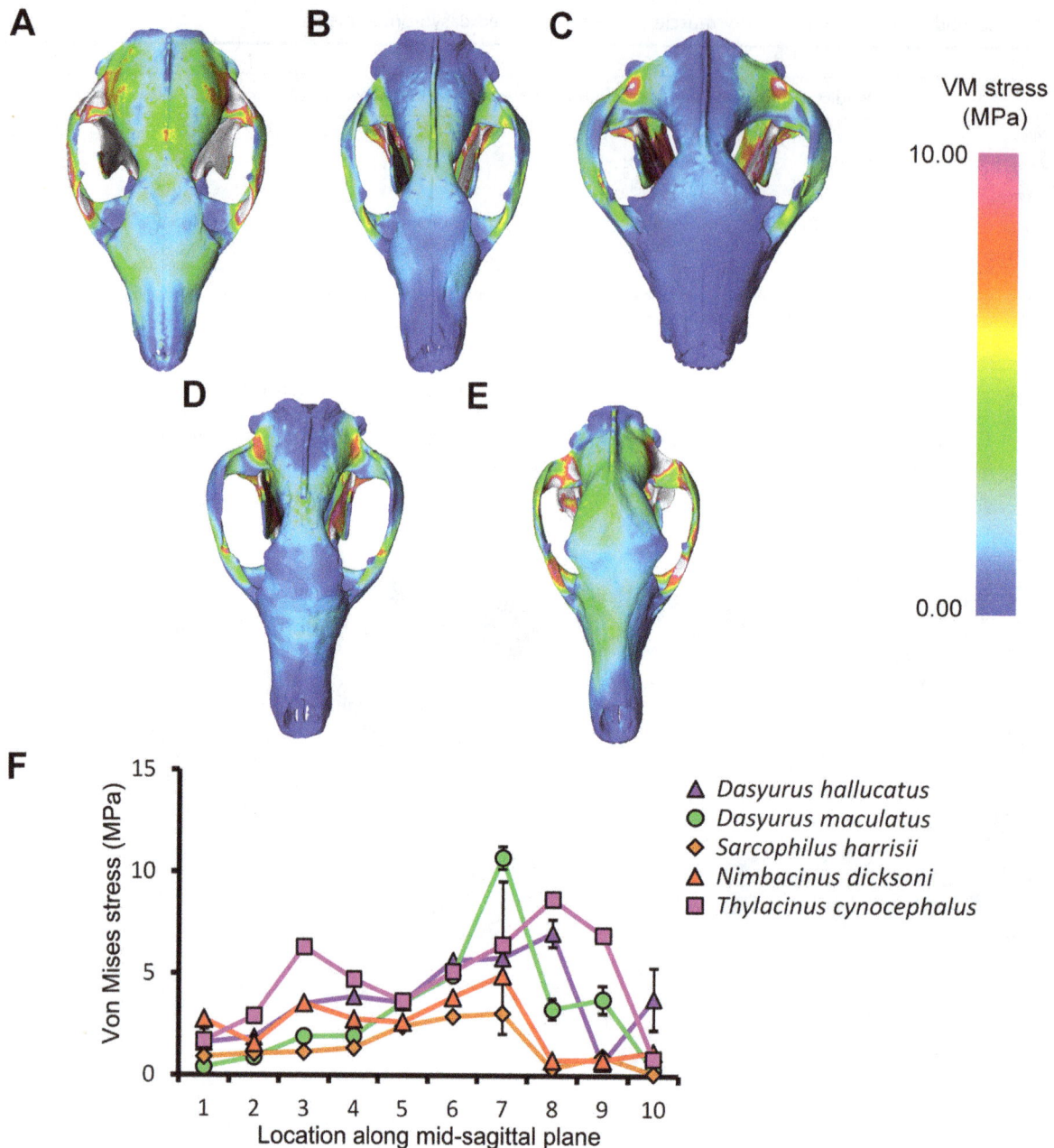

Figure 4. Von Mises stress under a bilateral canine bite in dorsal view. The models are subjected to a load applied to both canines, with bite force scaled based on theoretical body mass. Species modeled were (A) *Dasyurus hallucatus*, (B) *Dasyurus maculatus*, (C) *Sarcophilus harrisii*, (D) *Nimbacinus dicksoni* and (E) *Thylacinus cynocephalus*. White colored regions of the skull represent VM stress above 10 MPa. (F) Distribution of von Mises stress was measured from anterior to posterior along the mid-sagittal plane.

A considerable size range exists between specimens considered in the present study. The relationship between bite force and body mass is negatively allometric [26,80]. To account for differences in body mass, a second series of load cases were solved following the scaling procedures of McHenry et al. [31]. Here, for each model, an estimate of bite force was made based on regression of body mass to predicted bite force for dasyuromorphians [$z = 0.6998$ (log y)+1.8735, where and y = mass (g) and z = bite force at canines (N)] [26], with body mass for each specimen predicted using the equation based on lower molar row length [log $y = -1.075 + 3.209$(log x), where x = lower molar length (mm), and y = mass (g)] as presented by Myers [81]. Muscle forces were then scaled for each specimen to achieve bite forces predicted on the basis of body mass. FEMs were solved using these scaled muscle forces. Prediction of bite force based on body mass using the regression equation provided in Wroe et al. [26] is close to that which would be expected following a 2/3 power relationship, whereby muscle force is proportional to area while body mass is proportional to volume [82]. The maximum bite force measured in Newtons (N) was also estimated for intrinsic loads (Table S2) using FEMs with un-scaled, specimen-specific estimated muscle forces (Table S3). Three dimensional approaches are likely to be more accurate than 2D based approaches [83].

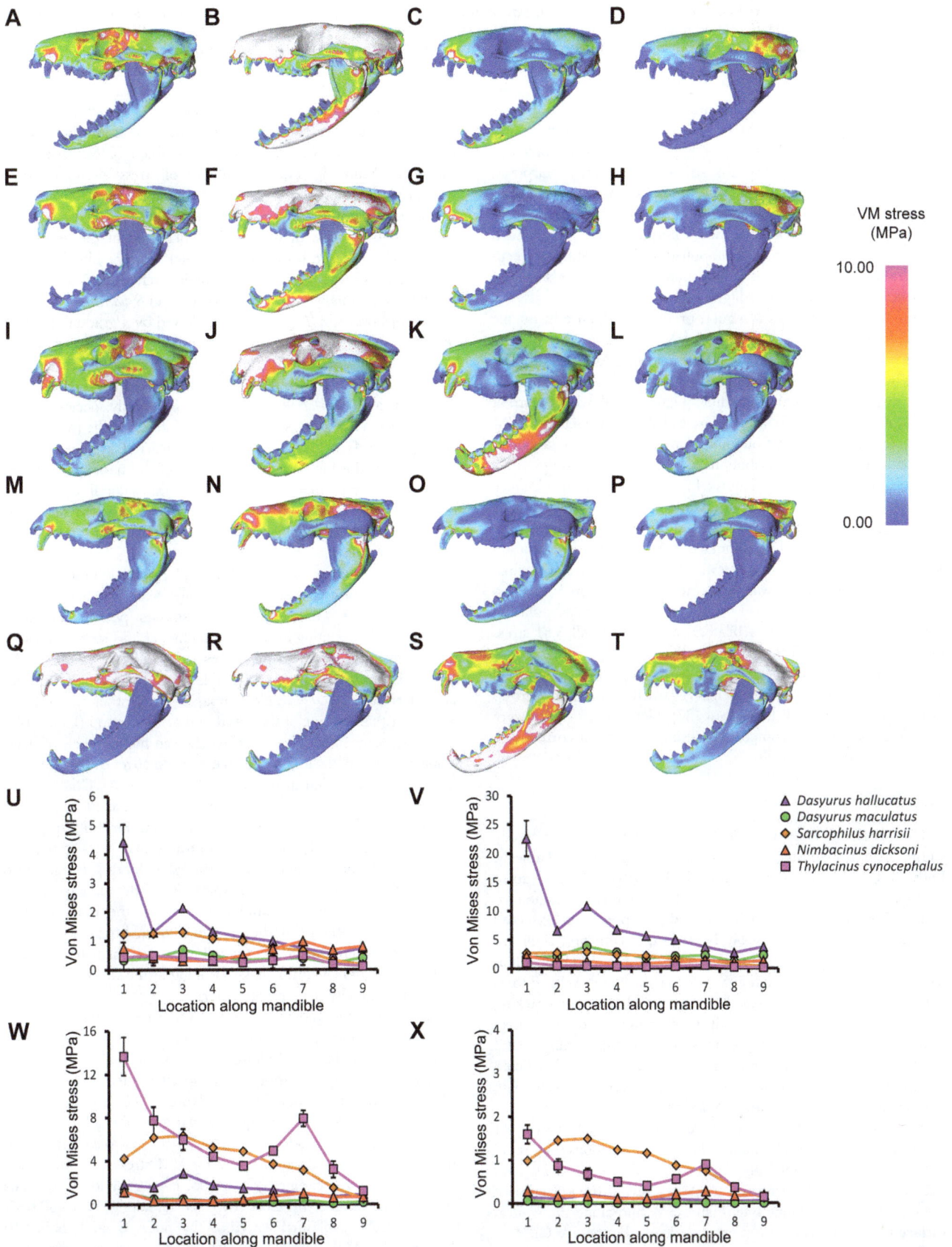

Figure 5. Von Mises stress under extrinsic loads in lateral view. The models are subjected to various loads applied to the canines, including a (A, E, I, M, Q) lateral shake, (B, F, J, N, R) axial twist, (C, G, K, O, S) pullback and (D, H, L, P, T) dorsoventral. The force applied was equivalent to 100 times the animal's estimated body mass for an axial twist, and 10 times the animal's estimated body mass for a lateral shake, pullback and dorsoventral

shake. Species compared were (A–D) *Dasyurus hallucatus*, (E–H) *Dasyurus maculatus*, (I–L) *Sarcophilus harrisii*, (M–P) *Nimbacinus dicksoni* and (Q–T) *Thylacinus cynocephalus*. White colored regions of the skull represent VM stress above 10 MPa. Distribution of von Mises (VM) stress was measured from anterior to posterior along the mandible for a (U) lateral shake, (V) axial twist, (W) pullback and (X) dorsoventral.

A H-frame connecting the canines of the upper and lower jaws was used to apply extrinsic forces, with forces applied at the center of the frame [32,35]. The force (N) applied to extrinsic loads was an arbitrary figure, applied for strictly comparative purposes, equivalent to 100 times the animal's estimated body mass for an axial twist, and 10 times the animal's estimated body mass for a lateral shake, pullback and dorsoventral shake [81]. Each simulation in which forces are applied with the anterior teeth (canines) restrained is a test for the hypothesis that stress will be highest for species with the longest rostrum.

Von Mises (VM) stress is a good predictor of failure in ductile materials such as bone [84,85] and VM stress is used here as a metric for comparison between models following Attard et al. [35]. Nodes were selected at equidistant points along the mid-sagittal plane, zygomatic arch and mandible (Figure 2) following Attard et al. [35] and at each node values were calculated by averaging VM stress recorded in the surrounding elements to assess changes in stress magnitudes and distributions under different loadings.

Principal component analysis (PCA) was used to visualize differences between species in average VM stress values for equidistant nodes along the mid-sagittal plane (N = 10). PCA is an ordination technique that summarizes the maximum variation among a set of variables on few, uncorrelated axes (principal components) [86]. PCA was performed separately on VM stress values for each extrinsic load (axial twist, lateral shake, pullback and dorsoventral) and for a bilateral canine bite. All VM stress values were log transformed prior to PCA.

Differences in VM stress values between species were compared using a Kruskal-Wallis test, which is regarded as a multiple-group extension of the Man-Whitney test [87]. Significance values were corrected for multiple comparisons using Bonferroni corrections, as a conservative approach.

Results

The predicted body mass (kg) of each species was generally within the expected range for each of the extant species (Table 1). Body mass estimates ranged from 0.78 kg for *D. hallucatus*, up to 32.49 kg for *T. cynocephalus*. However, the body mass estimated for *S. harrisii* of 14.20 kg was slightly above the upper limit observed for males (13 kg) [88], possibly because the teeth and skull are relatively large in this species. The robust craniodental morphology and relatively large teeth in *S. harrisii* are probably related to its habitual osteophagy, as has been observed in bone-cracking carnivorans [89]. To obtain body mass estimates for these taxa using simple or multiple regressions adjusted from cranidoental variables may lead to overestimates of body mass. Predicted maximum muscle forces for *N. dicksoni* (651 N) were relatively high, approaching those of the larger *S. harrisii* (685 N) (Table 1).

Thylacinus cynocephalus displayed comparatively high levels of VM stress in the cranium and mandible for most simulations (Figure 3–6, S3). This is consistent with results of Attard et.al. [35]. *Dasyurus hallucatus* showed relatively high levels of stress in the posterior of the mandible for a canine bite (Figure 3A), and along the ventral surface of the ramus for most extrinsic loads (Figure 5A–C).

The regions of highest stress along the dentary of *N. dicksoni* were located at the coronoid fossa and condylar process (Figure 3D). These regions of peak stress may be in part artifacts of reconstruction. Otherwise the dentary of *N. dicksoni* revealed

similar stress patterns for a bilateral bite to *D. maculatus* (Figure 3). The distribution of stress for *N. dicksoni* in the cranium in response to a bilateral bite was intermediate between *S. harrisii* and *D. maculatus* (Figure 4). The magnitudes of stress along the mid-sagittal plane of *N. dicksoni* were slightly higher than for *S. harrisii* and lower than for *D. maculatus* (Figure 4F).

The highest stress in the cranium occurred at the zygomatic arch for all species in response to a bilateral canine bite (Figure 4). Von Mises stress along the zygomatic arch during a bilateral canine bite gradually increased posteriorly in *S. harrisii*, while stress peaked at node 3 in *T. cynocephalus* followed by a gradual decrease posteriorly (Figure S6A). The three other species displayed two peaks in stress along the zygomatic arch during a bilateral canine bite; one at the middle and the other at the posterior region of the zygomatic arch. *Thylacinus cynocephalus* was the only species to show two distinct peaks in stress for a bilateral bite along the mid-sagittal crest (Figure 4F). These stress points occurred at the temporal ridge and at the most narrowed region of the nasal (Figure 4E). Von Mises stress measured along the mid-sagittal crest for a bilateral bite revealed one point of peak stress halfway along the frontal of *D. maculatus*, *S. harrisii* and *N. dicksoni* and at the temporal ridge for *D. hallucatus* (Figure 4).

Stress was quite evenly distributed along the dentaries for all species in response to lateral shaking and axial twisting, with the exception of *D. hallucatus*, wherein stresses peaked anteriorly (Figure 5U–V) resulting in significantly higher VM stress values for that species compared to all others ($\chi^2 = 32.87$, $P<0.0001$). An axial twist resulted in much higher levels of stress along the mid-sagittal crest for *D. hallucatus* compared to all other species, and peaked at the anterior of the nasal and at the frontal (Figure 6B). *Sarcophilus harrisii* and *T. cynocephalus* showed higher levels of stress along the mandible for a pullback and dorsoventral shake than other species included in this study (Figure 5W–X). Comparisons of mandible VM stress values revealed significant differences between species after Bonferroni correction for both pullback ($\chi^2 = 33.28$, $P<0.001$) and dorsoventral shake ($\chi^2 = 35.61$, $P< 0.0001$), with the exception of *T. cynocephalus* and *S. harrisii*. Two points of peak stress were apparent along the dentary for *T. cynocephalus* in these two simulations; one at the most anterior point, and the second at the coronoid fossa. Stress distribution along the dentary of *S. harrisii* followed a similar trend for a pullback and dorsoventral shake; peaking at the ramus inferior to M1 then gradually decreasing posteriorly.

PCA results for mid-sagittal node VM stress values (Figure 7) showed that a high proportion of variance could be explained in all cases by two Principal Component (PC) axes (>85%). These plots provide an appreciation of interspecific differences across all 10 mid-sagittal nodes and bite simulations. PCA results indicate that the main axes of interspecific variance for all bites were explained by either nodes 1 and/or 7–10. *Thylacinus cynocephalus* and *S. harrisii* differed significantly for VM stress values under a bilateral bite at the canines ($\chi^2 = 12.95$, $P = 0.04$) and PCA results indicated separation of those two species along PC1 (60.6%), which largely explained variance in node 8 and node 10 (Figure 7A). PC1 for a bilateral canine bite revealed close similarities between *N. dicksoni* and *D. maculatus*, whereas PC2 (28.9%) clearly separated *N. dicksoni* from *D. maculatus* and reflected differences in node 7 (as seen in Figure 4).

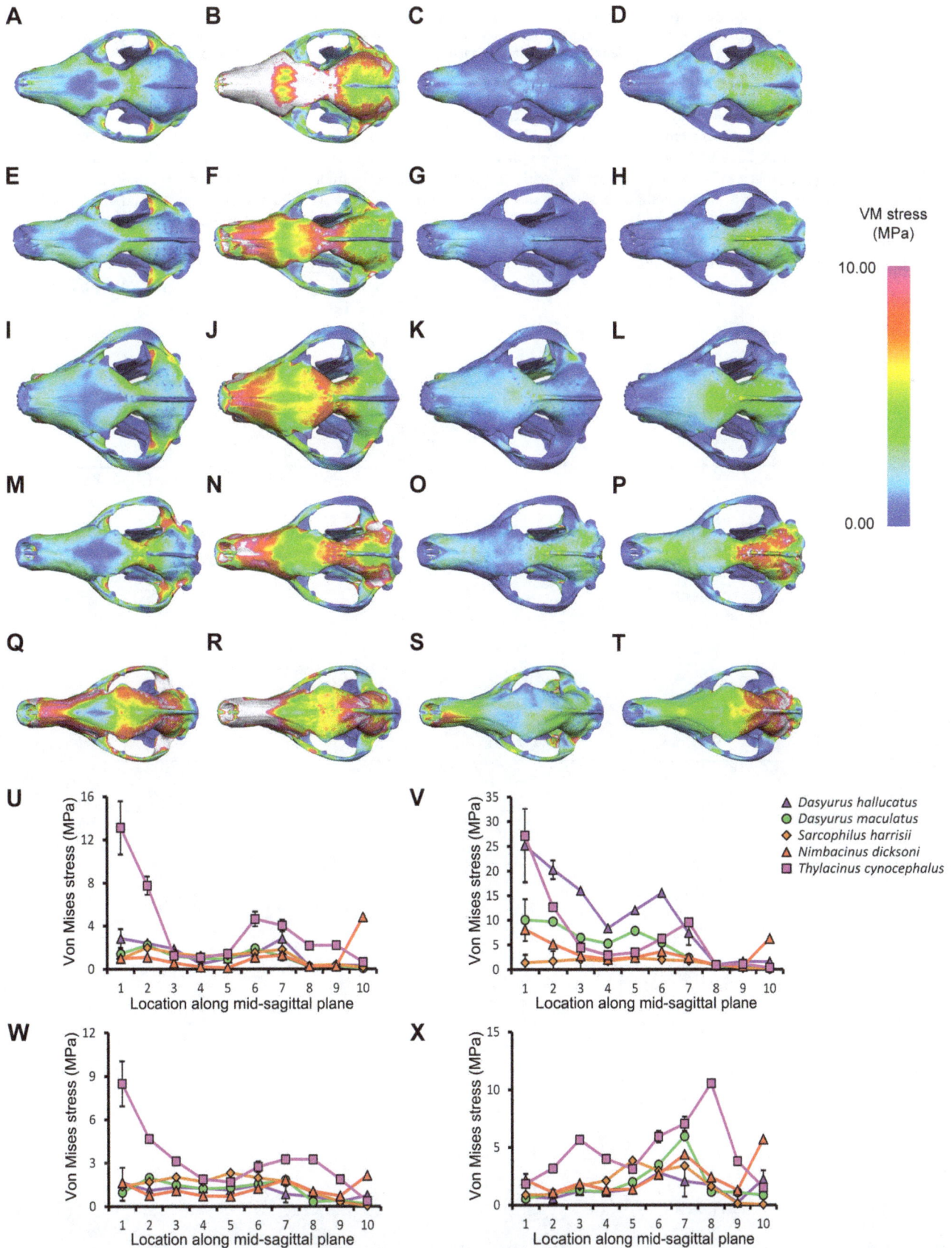

Figure 6. Von Mises stress under extrinsic loads in dorsal view. The models are subjected to various loads applied to the canines, including a (A, E, I, M, Q) lateral shake, (B, F, J, N, R) axial twist, (C, G, K, O, S) pullback and (D, H, L, P, T) dorsoventral. The force applied was equivalent to 100 times the animal's estimated body mass for an axial twist, and 10 times the animal's estimated body mass for a lateral shake, pullback and dorsoventral

shake. Species compared were (A–D) *Dasyurus hallucatus*, (E–H) *Dasyurus maculatus*, (I–L) *Sarcophilus harrisii*, (M–P) *Nimbacinus dicksoni* and (Q–T) *Thylacinus cynocephalus*. White colored regions of the skull represent VM stress above 10 MPa. Distribution of von Mises (VM) stress was measured from anterior to posterior along the mid-sagittal plane for a (U) lateral shake, (V) axial twist, (W) pullback and (X) dorsoventral.

For a lateral shake, PC1 (51.0%) explained change at nodes 1 and 10 and separated *T. cynocephalus* from *N. dicksoni* (Figure 7B), as also seen in Figure 6U. PC2 (42.0%) for a lateral shake revealed differences among all species for nodes 7 and 10, with *T. cynocephalus* closely resembling *N. dicksoni* compared to other species (Figure 7B). Interspecific differences were not significant for a lateral shake after Bonferroni correction, and before correction those distinguished *T. cynocephalus* from *N. dicksoni*, *D. maculatus* and *S. harrisii* ($\chi^2 = 9.27$, $P = 0.01$–0.03). PCA for an axial twist

(Figure 7C) showed that the extremes of PC1 (66.1%) were delimited by *S. harrisii* and *D. hallucatus*, and the remaining taxa were located in between those two species. PC1 mainly accounted for differences among species for node 1 values (as seen also in Figure 6V), which were high for *D. hallucatus* and low for *S. harrisii*, whereas PC2 (26.1%) largely summarized node 10 values. Pairwise comparisons of VM stress values were not significant for an axial twist, however before Bonferroni correction, differences between *S. harrisii* and *D. hallucatus* ($\chi^2 = 10.73$, $P = 0.03$), *N. dicksoni*

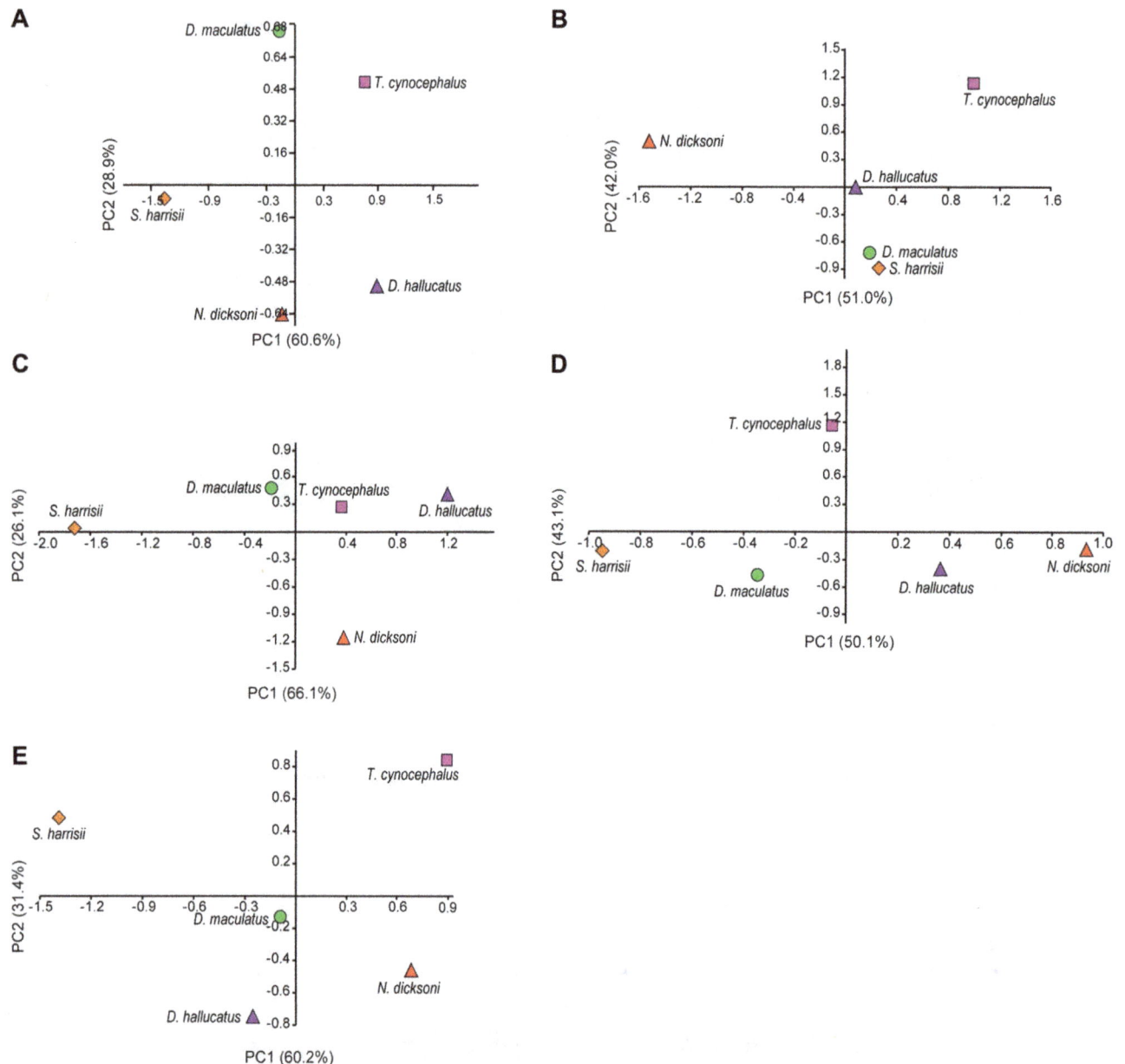

Figure 7. Principal components one and two of von Mises stress along mid-sagittal plane. Results of principal components analysis to compare stress among species along mid-sagittal nodes for each loading case, including (A) bilateral canine bite, (B) lateral shake, (C) axial twist, (D) pullback and (E) dorsoventral. Key to symbols: (pink square) *Thylacinus cynocephalus*; (red triangle) *Nimbicinus dicksoni*; (orange diamond) *Sarcophilus harrisii*; (green circle) *Dasyurus maculatus*; (purple triangle) *Dasyurus hallucatus*.

($\chi^2 = 10.73$, $P = 0.01$) and *T. cynocephalus* ($\chi^2 = 10.73$, $P = 0.04$) could be distinguished from one another. For a pullback bite, PC1 (50.1%) mainly explained interspecific differences for node 10, and PC2 (43.1%) explained change at nodes 1 and 8 (Figure 7D). VM stress values for *T. cynocephalus* were different from those for *D. hallucatus* ($\chi^2 = 14.37$, $P = 0.02$) and *D. maculatus* ($\chi^2 = 14.37$, $P = 0.04$) for a pullback bite. For the dorsoventral shake, PC1 explained 60.2% of variance and reflected differences between *T. cynocephalus* and *S. harrisii*, whereas values across nodes were more similar among the remaining three species, located toward the middle of PC1 (Figure 7E). PC2 (31.4%) separated *D. hallucatus* from *T. cynocephalus*, and pairwise comparisons revealed node values to be different between those two species ($\chi^2 = 14.78$, $P = 0.01$).

Discussion

Differences in biomechanical performance between the three extant dasyurids included in this study appear consistent with their respective known feeding behaviors. *Dasyurus hallucatus* showed comparatively higher levels of stress in most simulations than *S. harrisii* and *D. maculatus*. *Dasyurus hallucatus* eats invertebrates and other relatively small prey [60–62], which may not require adaptation to sustain the full range of extrinsic loads simulated here. This species shows particularly high VM stress in axial twisting, especially in contrast to *S. harrisii* However, it performs relatively well under pull-back loading, which may be linked to a capacity for pulling invertebrates from the ground. Observational studies on wild *D. hallucatus* will be required to confirm the functional role of their skull in prey acquisition. Future work on the comparative musculoskeletal anatomy and collection of in vivo or ex vivo biomechanical data of the extant species would likely improve the predictive power of current bite force and muscle force estimations. Overall consistencies found between known prey size and biomechanical performance for extant dasyuromorphians underscore the potential value of projections based on comparative FEA for extinct/fossil taxa.

Our comparative biomechanical modeling of dasyuromorphian skulls suggests considerable differences in predatory behaviors between the two thylacinids considered here. Our 3D based results indicate that the Oligocene to Miocene *N. dicksoni* had a high bite force for its size, comparable to that of extant dasyurids known to take relatively large prey, *D. maculatus* and *S. harrisii* [48,60]. In light of similar levels of 'carnassialization' (development of relatively long, high amplitude vertical shearing crests) in the cheektooth dentition with *D. maculatus*, and a lack of obvious dental specialization consistent with regular bone-cracking, our results suggest a predominantly carnivorous diet for *N. dicksoni* that may have included relatively large prey. *Dasyurus maculatus* are opportunistic hunters, varying their diet in response to environmental disturbances and short-term fluctuations in prey abundance [54,56]. They will prey on vertebrate species up to and sometimes exceeding their own body mass. Prey includes bandicoots, smaller dasyurids, possums, smaller macropodoids, snakes, lizards, birds and frogs, as well as invertebrates. Potential prey for a fox-sized thylacinid living in the closed forest communities of Riversleigh likely included many small to medium-size birds, frogs, lizards and snakes, as well as a wide range of marsupials, including bandicoots (peramelemorphians), dasyurids (dasyuromorphians), kangaroos (macropodoids), thingodontans (yalkiparidontians), marsupial moles (notoryctemorphians) and wombats (vombatoids) [12].

Although our FEA results for *N. dicksoni* suggest a capacity to kill prey approaching or exceeding its own body mass, its prey range may have been limited by competition with sympatric carnivores. The extent of niche overlap and competition within this ancient, medium-large sized carnivore community may have been partially alleviated by occupying different habitats and specializing in different hunting strategies. The recovery of a near complete skeleton of *N. dicksoni* [14] will provide further information on the locomotion and predatory behavior based on postcranial material; for example, was *N. dicksoni* as arboreal as the extant *D. maculatus*?

Differences in mechanical performance suggest that *T. cynocephalus* is unusual relative to other dasyuromorphians, including, *N. dicksoni*, as indicated by distinctly higher VM stresses than all other species in response to each loading case. *Thylacinus cynocephalus*, in contrast to *N. dicksoni*, has completely lost the metaconid on the lower molars and has a proportionally much larger postmetacrista on the upper molars. On the basis of traditional beam theory we predicted that taxa with longer rostra would exhibit higher stress [90], as evident in the long-snouted *T. cynocephalus* relative to shorter-snouted dasyuromorphians. Differences between *T. cynocephalus* and other species were also significant for three out of five simulations examined after conservative Bonferroni correction for multiple testing. These results further support the contention by Attard et al. [35] that niche breadth in *T. cynocephalus* may have been more limited and that it likely preyed on relatively small to medium-sized vertebrates such as wallabies, possums and bandicoots.

Although measures of skull performance in response to forces imposed by struggling prey revealed closer similarity between the fossil thylacinid *N. dicksoni* and large extant carnivorous dasyurids, than with *T. cynocephalus*, there were differences. Our reconstruction suggests that the TMJ was more elevated in *N. dicksoni* than in *D. maculatus*, and higher relative to the height of the cheektooth row. The TMJ is a complex joint and is important for occlusion and mastication [91,92]. The position of the TMJ can influence bite strength and muscle activation [93]. The position of the TMJ along the anterior-posterior axis tends to lie closer to the plane of the tooth row in carnivorous taxa [11]. Conclusive determination of the precise position and morphology of the TMJ in *N. dicksoni* must await the discovery of more complete cranial material.

Morphological evidence from past studies further demonstrates diversity within this family. The smallest thylacinid, *Muribacinus gadiyuli*, is believed to have fed on relatively small vertebrates and invertebrates because it lacks a number of dental features present in large prey specialists (e.g., robust protoconids and brachycephalization) such as similarly sized *D. maculatus* [17]. The variety of feeding behaviors among thylacinids may have helped facilitate their co-existence within different ecological niches that were later filled by diversifying carnivorous dasyurids.

Supporting Information

Figure S1 Phylogenetic tree of dasyuromophians investigated in this study. One of several recent assessments of the phylogenetic relationships of dasyuromorphians, including taxa that have been examined in this study (Wroe & Musser 2001).

Figure S2 Interactive 3D pdf showing the digitally segmented cranium and mandible of *Thylacinus cynocephalus*.

Figure S3 Interactive 3D pdf showing the digitally segmented cranium and mandible of *Sarcophilus harrisii*.

Figure S4 Interactive 3D pdf showing the digitally segmented cranium and mandible of *Dasyurus maculatus*.

Figure S5 Interactive 3D pdf showing the digitally segmented cranium and mandible of *Dasyurus hallucatus*.

Figure S6 Von Mises stress along zygomatic arch for all loading cases. Distribution of von Mises (VM) stress was measured from anterior to posterior along the zygomatic arch for a (A) bilateral canine bite, (B) lateral shake, (C) axial twist, (D) pullback and (E) dorsoventral.

Table S1 Temporal and geographic distribution of thylacinid species. Abbreviations: Aust, Australian mainland; E., Early; L., Late; M., Middle; Mio, Miocene; NG, New Guinea; NT, Northern Territory; Oligo, Oligocene; Plio, Pliocene; Qld, Queensland; Tas, Tasmania.

Table S2 Maximum bite forces (N) for un-scaled homogeneous models for a bilateral canine bite.

Table S3 Muscle forces used for each jaw muscle division in un-scaled intrinsic models. Species studied were *Dasyurus hallucatus, Dasyurus maculatus, Sarcophilus harrisii, Nimbacinus dicksoni* and *Thylacinus cynocephalus*. These were calculated using muscle mass proportions from dissected *Didelphis virginiana* (Turnbull 1970). Muscle forces were scaled for a bilateral canine bite by multiplying the muscle force by the ratio between bite force estimated using body mass regressions and maximum bite force estimated from the un-scaled model.

References S1   Supporting Information references.

Acknowledgments

We thank Sandy Ingleby from the Australian museum for providing several comparative specimens and the makers of Digimorph for access to CT scan data.

Author Contributions

Conceived and designed the experiments: MRGA SW. Performed the experiments: MRGA WCHP. Analyzed the data: MRGA LABW WCHP. Contributed reagents/materials/analysis tools: SW. Wrote the paper: MRGA WCHP LABW MA SJH TLR SW.

References

1. Krajewski C, Buckley L, Westerman M (1997) DNA phylogeny of the marsupial wolf resolved. Proc R Soc Lond, Ser B: Biol Sci 264: 911–917.
2. Krajewski C, Driskell AC, Baverstock PR, Braun MJ (1992) Phylogenetic relationships of the thylacine (Mammalia: Thylacinidae) among dasyuroid marsupials: evidence from cytochrome b DNA sequences. Proc R Soc Lond, Ser B: Biol Sci 250: 19–27.
3. Lowenstein JM, Sarich VM, Richardson BJ (1981) Albumin systematics of the extinct mammoth and Tasmanian wolf. Nature 291: 409–411.
4. Muirhead J, Wroe S (1998) A new genus and species, *Badjcinus turnbulli* (Thylacinidae: Marsupialia), from the late Oligocene of Riversleigh, northern Australia, and an investigation of thylacinid phylogeny. J Vert Paleontol 18: 612–626.
5. Sarich V, Lowenstein JM, Richardson BJ (1982) Phylogenetic relationships of the thylacine (*Thylacinus cynocephalus*, Marsupialia) as reflected in comparative serology. In: Archer M, editor. Carnivorous Marsupials. Sydney: Royal Zoological Society of New South Wales. pp. 445–476.
6. Szalay FS (1982) A new appraisal of marsupial phylogeny and classification. In: Archer M, editor. Carnivorous Marsupials. Sydney: Royal Zoological Society of New South Wales. pp. 621–640.
7. Thomas RH, Schaffner W, Wilson AC, Paabo S (1989) DNA phylogeny of the extinct marsupial wolf. Nature 340: 465–467.
8. Wroe S, Archer M (2006) Origins and early radiations of marsupials. In: Merrick JR, Archer M, Hickey GM, Lee MSY, editors. Evolution and Biogeography of Australasian Vertebrates. Sydney: Australian Scientific Publishing. pp. 517–540.
9. Goswami A, Milne N, Wroe S (2011) Biting through constraints: cranial morphology, disparity and convergence across living and fossil carnivorous mammals. Proceedings of the Royal Society B: Biological Sciences 278: 1831–1839.
10. Wroe S (2001) *Maximucinus muirheadae*, gen. et sp. nov. (Thylacinidae: Marsupialia), from the Miocene of Riversleigh, north-western Queensland, with estimates of body weights for fossil thylacinids. Aust J Zool 49: 603–614.
11. Wroe S, Milne N (2007) Convergence and remarkably consistent constraint in the evolution of mammalian carnivore skull shape. Evolution 61: 1251–1260.
12. Archer M, Arena DA, Bassarova M, Beck RMD, Black K, et al. (2006) Current status of species-level representation in faunas from selected fossil localities in the Riversleigh World Heritage Area, northwestern Queensland. Alcheringa 30: 1–17.
13. Muirhead J (1997) Two new thylacines (Marsupialia: Thylacinidae) from early Miocene sediments of Riversleigh, northwestern Queensland and a revision of the family Thylacinidae. Mem Queensl Mus 41: 367–377.
14. Muirhead J, Archer M (1990) *Nimbacinus dicksoni*, a plesiomorphic thylacine (Marsupialia: Thylacinidae) from Tertiary deposits of Queensland and the Northern Territory. Mem Queensl Mus 28: 203–221.
15. Murray PF (1997) *Thylacinus megiriani*, a new species of thylacine (Marsupialia: Thylacinidae) from the Ongeva local fauna of central Australia. Rec S Aust Mus 30: 43–61.
16. Murray PF, Megirian D (2000) Two new genera and three new species of Thylacinidae (Marsupialia) from the Miocene of the Northern Territory, Australia. The Beagle, Records of the Museums and Art Galleries of the Northern Territory 16: 145–162.
17. Wroe S (1996) *Muribacinus gadiyuli*, (Thylacinidae: Marsupialia), a very plesiomorphic thylacinid from the Miocene of Riversleigh, northwestern Queensland, and the problem of paraphyly for the Dasyuridae. J Paleontol 70: 1032–1044.
18. Wroe S, Musser A (2001) The skull of *Nimbacinus dicksoni* (Thylacinidae: Marsupialia). Aust J Zool 49: 487–514.
19. Archer M, Hand S, Godthelp H (1991) Riversleigh: The Story of Animals in Ancient Rainforests of Inland Australia. Sydney: Reed Books. 264 p.
20. Travouillon KJ, Legendre S, Archer M, Hand SJ (2009) Palaeoecological analyses of Riversleigh's Oligo-Miocene sites: implications for Oligo-Miocene climate change in Australia. Palaeogeogr, Palaeoclimatol, Palaeoecol 276: 24–37.
21. Archer M, Brammall J, Field J, Hand SJ, Hook C (2002) The Evolution of Australia: 110 million years of change. Sydney: Australian Museum. 91 p.
22. Wroe S (2002) A review of terrestrial mammalian and reptilian carnivore ecology in Australian fossil faunas and factors influencing their diversity: The myth of reptilian domination and its broader ramifications. Aust J Zool 49: 603–614.
23. Wroe S, Myers TJ, Wells RT, Gillespie A (1999) Estimating the weight of the Pleistocene marsupial lion, *Thylacoleo carnifex* (Thylacoleonidae: Marsupialia): implications for the ecomorphology of a marsupial super-predator and hypotheses of impoverishment of Australian marsupial carnivore faunas. Aust J Zool 47: 489–498.
24. Wroe S (2003) Australian marsupial carnivores: an overview of recent advances in palaeontology. In: Jones M, Dickman C, Archer M, editors. Predators with Pouches: The Biology of Carnivorous Marsupials. Collingwood: CSIRO Publishing. pp. 102–123.
25. Wroe S, Brammall J, Cooke BN (1998) The skull of *Ekaltadeta ima* (Marsupialia, Hypsiprymnodontidae?): an analysis of some marsupial cranial features and a re-investigation of propleopine phylogeny, with notes on the inference of carnivory in mammals. J Paleontol 72: 738–751.
26. Wroe S, McHenry C, Thomason J (2005) Bite club: comparative bite force in big biting mammals and the prediction of predatory behaviour in fossil taxa. Proc R Soc Lond, Ser B: Biol Sci 272: 619–625.
27. Dumont ER, Piccirillo J, Grosse IR (2005) Finite element analysis of biting behavior and bone stress in the facial skeletons of bats. Anat Rec 283: 319–330.
28. Ross CF (2005) Finite element analysis in vertebrate biomechanics. Anat Rec A Discov Mol Cell Evol Biol 283A: 253–258.
29. Tseng ZJ (2009) Cranial function in a late Miocene *Dinocrocuta gigantea* (Mammalia: Carnivora) revealed by comparative finite element analysis. Biol J Linn Soc 96: 51–67.
30. Wroe S, Ferrara TL, McHenry CR, Curnoe D, Chamoli U (2010) The craniomandibular mechanics of being human. Proc R Soc Lond, Ser B: Biol Sci 277: 3579–3586.

31. McHenry CR, Wroe S, Clausen PD, Moreno K, Cunningham E (2007) Supermodeled sabercat, predatory behavior in *Smilodon fatalis* revealed by high-resolution 3D computer simulation. Proc Natl Acad Sci USA 104: 16010–16015.

32. Wroe S, Clausen P, McHenry C, Moreno K, Cunningham E (2007) Computer simulation of feeding behaviour in the thylacine and dingo as a novel test for convergence and niche overlap. Proc R Soc Lond, Ser B: Biol Sci 274: 2819–2828.

33. Young MT, Rayfield EJ, Holliday CM, Witmer LM, Button DJ, et al. (2012) Cranial biomechanics of *Diplodocus* (Dinosauria, Sauropoda): testing hypotheses of feeding behaviour in an extinct megaherbivore. Naturwissenschaften 99: 637–643.

34. Bell PR, Snively E, Shychoski L (2009) A comparison of the jaw mechanics in hadrosaurid and ceratopsid dinosaurs using finite element analysis. Anat Rec (Hoboken) 292: 1338–1351.

35. Attard MRG, Chamoli U, Ferrara TL, Rogers TL, Wroe S (2011) Skull mechanics and implications for feeding behaviour in a large marsupial carnivore guild: the thylacine, Tasmanian devil and spotted-tailed quoll. J Zool 285: 292–300.

36. Chamoli U, Wroe S (2011) Allometry in the distribution of material properties and geometry of the felid skull: Why larger species may need to change and how they may achieve it. J Theor Biol 283: 217–226.

37. Moazen M, Curtis N, Evans SE, O'Higgins P, Fagan MJ (2008) Combined finite element and multibody dynamics analysis of biting in a *Uromastyx hardwickii* lizard skull. J Anat 213: 499–508.

38. Oldfield CC, McHenry CR, Clausen PD, Chamoli U, Parr WCH, et al. (2012) Finite element analysis of ursid cranial mechanics and the prediction of feeding behaviour in the extinct giant *Agriotherium africanum*. J Zool 286: 163–170.

39. Rayfield EJ, Norman DB, Horner CC, Horner JR, Smith PM, et al. (2001) Cranial design and function in a large theropod dinosaur. Nature 409: 1033–1037.

40. Slater GJ, Figueirido B, Louis L, Yang P, Van Valkenburgh B (2010) Biomechanical Consequences of Rapid Evolution in the Polar Bear Lineage. PLoS ONE 5: e13870.

41. Strait DS, Grosse IR, Dechow PC, Smith AL, Wang Q, et al. (2010) The structural rigidity of the cranium of *Australopithecus africanus*: implications for diet, dietary adaptations, and the allometry of feeding biomechanics. Anat Rec (Hoboken) 293: 583–593.

42. Thresher RW, Saito GE (1973) The stress analysis of human teeth. J Biomech 6: 443–449.

43. Paddle R (2000) The Last Tasmanian tiger: the History and Extinction of the Thylacine. Oakleigh, VIC: Cambridge University Press. 273 p.

44. Jones ME, Stoddart DM (1998) Reconstruction of the predatory behaviour of the extinct marsupial thylacine (*Thylacinus cynocephalus*). J Zool 246: 239–246.

45. Jones ME (2003) Convergence in ecomorphology and guild structure among marsupial and placental carnivores. In: Jones ME, Dickman C, Archer M, editors. Predators with Pouches: The Biology of Carnivorous Marsupials. Collingwood: CSIRO Publishing. pp. 285–269.

46. Guiler ER (1985) Thylacine: The tragedy of the Tasmanian tiger. Melbourne: Oxford University Press. 207 p.

47. Bailey C (2003) Tiger Tales: Stories of the Tasmanian Tiger. Sydney: HarperCollins Publishers. 164 p.

48. Figueirido B, Janis CM (2011) The predatory behaviour of the thylacine: Tasmanian tiger or marsupial wolf? Biol Lett 7: 937–940.

49. Jones ME, Barmuta LA (1998) Diet overlap and abundance of sympatric dasyurid carnivores: a hypothesis of competition? J Anim Ecol 67: 410–421.

50. Bradshaw CJA, Brook BW (2005) Disease and the devil: density-dependent epidemiological processes explain historical population fluctuations in the Tasmanian devil. Ecography 28: 181–190.

51. Guiler E (1970) Obsevations on the Tasmanian devil, *Sarcophilus harrisii* (Marsupialia : Dasyuridae) I. Numbers, home, range, movements and food in two populations. Aust J Zool 18: 49–62.

52. Taylor RJ (1986) Notes on the diet of the carnivorous mammals of the upper Henry River region, Western Tasmania. Pap Proc R Soc Tasman 120: 7–10.

53. Groves CP (2005) Order Dasyuromorphia. In: Wilson DE, Reeder DM, editors. Mammal Species of the World: A Taxonomic and Geographic Reference. 3rd ed. Baltimore: Johns Hopkins University Press. pp. 23–37.

54. Edgar R, Belcher C (1995) Spotted-tailed quoll, *Dasyurus maculatus*. In: Strahan R, editor. The Mammals of Australia. Sydney: Reed. pp. 67–69.

55. Glen AS, Dickman CR (2006) Diet of the spotted-tailed quoll (*Dasyurus maculatus*) in eastern Australia: effects of season, sex and size. J Zool 269: 241–248.

56. Jones ME (1997) Character displacement in Australian dasyurid carnivores: size relationships and prey size patterns. Ecology 78: 2569–2587.

57. Dawson JP, Claridge AW, Triggs B, Paull DJ (2007) Diet of a native carnivore, the spotted-tailed quoll (*Dasyurus maculatus*), before and after an intense wildfire. Wildl Res 34: 342–351.

58. Braithwaite RW, Begg RJ (1995) Northern quoll *Dasyurus hallucatus* Gould, 1842. In: Strahan R, editor. The Mammals of Australia: National Photographic Index of Australian Wildlife. Sydney: Reed Books. pp. 65–66.

59. Strahan R (1995) The Mammals of Australia. Sydney: New Holland Publishing Pty Ltd.

60. Belcher CA (1995) Diet of the Tiger quoll (*Dasyurus maculatus*) in East Gippsland, Victoria. Wildl Res 22: 341–357.

61. Oakwood M (1997) The ecology of the northern quoll, *Dasyurus hallucatus* [PhD thesis]. Canberra: Australian National University. 556 p.

62. Pollock AB (1999) Notes on status, distribution and diet of northern quoll *Dasyurus hallucatus* in the Mackay-Bowen area, mideastern Queensland. Aust Zool 31: 388–395.

63. Wroe S (2008) Cranial mechanics compared in extinct marsupial and extant African lions using a finite-element approach. J Zool 274: 332–339.

64. Benazzi S, Bookstein FL, Strait DS, Weber GW (2011) A new OH5 reconstruction with an assessment of its uncertainty. J Hum Evol 61: 75–88.

65. Besl PJ, McKay ND (1992) A method for registration of 3D shapes. IEEE Transactions on Pattern Analysis and Machine Intelligence. pp. 239–256.

66. Parr WCH, Wroe S, Chamoli U, Richards HS, McCurry MR, et al. (2012) Toward integration of geometric morphometrics and computational biomechanics: New methods for 3D virtual reconstruction and quantitative analysis of Finite Element Models. J Theor Biol 301: 1–14.

67. Rayfield EJ (2007) Finite element analysis and understanding the biomechanics and evolution of living and fossil organisms. Annu Rev Earth Planet Sci 35: 541–576.

68. Cook SD, Weinstein AH, Klawitter JJ (1982) A three-dimensional finite element analysis of a porous rooted Co–Cr–Mo alloy dental implant. J Dent Res 61: 25–29.

69. Dumont ER, Grosse IR, Slater GJ (2009) Requirements for comparing the performance of finite element models of biological structures. J Theor Biol 256: 96–103.

70. Greaves G, Greer A, Lakes R, Rouxel T (2011) Poisson's ratio and modern materials. Nat Mater 10: 823–837.

71. Turnbull WD (1970) Mammalian masticatory apparatus. Fieldiana: Geology 18: 149–356.

72. Thomason JJ (1991) Cranial strength in relation to estimated biting forces in some mammals. Can J Zool 69: 2326–2333.

73. Chamoli U (2011) Biomechanics of the felid skulls: A comparative study using finite element approach [Masters thesis]. Sydney: University of New South Wales. 144 p.

74. Clausen P, Wroe S, McHenry C, Moreno K, Bourke J (2008) The vector of jaw muscle force as determined by computer-generated three dimensional simulation: a test of Greaves' model. J Biomech 41: 3184–3188.

75. Jones ME (1995) Guild structure of the large marsupial carnivores in Tasmania [PhD thesis]. Hobart: University of Tasmania. 176 p.

76. Pellis SM, Nelson A (1984) Some aspects of predatory behaviour of the quoll *Dasyurus viverrinus* (Marsupialia: Dasyuridae). Aust Mammal 7: 5–15.

77. Pellis SM, Officer RCE (1987) An analysis of some predatory behaviour patterns in four species of carnivorous marsupials (Dasyuridae), with comparative notes on the Eutherian cat *Felis catus*. Ethology 75: 177.

78. Fleay D (1932) The rare dasyures (native cats). Vic Nat 49: 63–69.

79. Dumont ER, Herrel A (2003) The effect of gape angle and bite point on bite force in bats. J Exp Biol 206: 2117–2123.

80. Christiansen P, Wroe S (2007) Bite forces and evolutionary adaptations to feeding ecology in carnivores. Ecology 88: 347–358.

81. Myers TJ (2001) Marsupial body mass prediction. Aust J Zool 49: 99–118.

82. Wroe S, Chamoli U, Parr WCH, Clausen P, Ridgely R, et al. (2013) Comparative biomechanical Modeling of Metatherian and Placental Saber-Tooths: A Different Kind of Bite for an Extreme Pouched Predator. PLoS ONE 8: e66888.

83. Ellis JL, Thomason JJ, Kebreab E, France J (2008) Calibration of estimated biting forces in domestic canids: comparison of post-mortem and in vivo measurements. J Anat 212: 769–780.

84. Nalla RK, Kinney JH, Ritchie RO (2003) Mechanistic fracture criteria for the failure of human cortical bone. Nat Mater 2: 164–168.

85. Tsafnat N, Wroe S (2010) An experimentally validated micromechanical model of a rat vertebra under compressive loading. J Anat 218: 40–46.

86. Mitteroecker P, Gunz P (2009) Advances in geometric morphometrics. Evol Biol 36: 235–247.

87. Zar JH (1996) Multiple regression and correlation. Biostatistical Analysis 3rd ed Upper Saddle River, NJ: Prentice Hall: 353–360.

88. Owen D, Pemberton D (2005) Tasmanian Devil: A Unique and Threatened Animal. Crows Nest, NSW: Allen and Unwin. 225 p.

89. Figueirido B, Tseng ZJ, Martín-Serra A (2013) Skull shape evolution in durophagous carnivorans. Evolution 67: 1975–1993.

90. Walmsley CW, Smits PD, Quayle MR, McCurry MR, Richards HS, et al. (2013) Why the long face? The mechanics of mandibular symphysis proportions in crocodiles. PLoS ONE 8: e53873.

91. Hylander WL (1979) An experimental analysis of temporomandibular joint reaction force in macaques. Am J Phys Anthropol 51: 433–456.

92. Breul R, Mall G, Landgraf J, Scheck R (1999) Biomechanical analysis of stress distribution in the human temporomandibular-joint. Ann Anat 181: 55–60.

93. Hickman DM, Cramer R (1998) The effect of different condylar positions on masticatory muscle electromyographic activity in humans. Oral Surg Oral Med Oral Pathol Oral Radiol Endod 85: 18–23.

16

Impact of Geography and Climate on the Genetic Differentiation of the Subtropical Pine *Pinus yunnanensis*

Baosheng Wang[1ɘ]**, Jian-Feng Mao**[2ɘ]**, Wei Zhao**[3]**, Xiao-Ru Wang**[1]*

1 Department of Ecology and Environmental Science, Umeå University, Umeå, Sweden, **2** National Engineering Laboratory for Forest Tree Breeding, Key Laboratory for Genetics and Breeding of Forest Trees and Ornamental Plants of Ministry of Education, Beijing Forestry University, Beijing, People's Republic of China, **3** State Key Laboratory of Systematic and Evolutionary Botany, Institute of Botany, Chinese Academy of Sciences, Beijing, People's Republic of China

Abstract

Southwest China is a biodiversity hotspot characterized by complex topography, heterogeneous regional climates and rich flora. The processes and driving factors underlying this hotspot remain to be explicitly tested across taxa to gain a general understanding of the evolution of biodiversity and speciation in the region. In this study, we examined the role played by historically neutral processes, geography and environment in producing the current genetic diversity of the subtropical pine *Pinus yunnanensis*. We used genetic and ecological methods to investigate the patterns of genetic differentiation and ecological niche divergence across the distribution range of this species. We found both continuous genetic differentiation over the majority of its range, and discrete isolated local clusters. The discrete differentiation between two genetic groups in the west and east peripheries is consistent with niche divergence and geographical isolation of these groups. In the central area of the species' range, population structure was shaped mainly by neutral processes and geography rather than by ecological selection. These results show that geographical and environmental factors together created stronger and more discrete genetic differentiation than isolation by distance alone, and illustrate the importance of ecological factors in forming or maintaining genetic divergence across a complex landscape. Our findings differ from other phylogenetic studies that identified the historical drainage system in the region as the primary factor shaping population structure, and highlight the heterogeneous contributions that geography and environment have made to genetic diversity among taxa in southwest China.

Editor: Ting Wang, Wuhan Botanical Garden, Chinese Academy of Sciences, China

Funding: This study was supported by grants from the Natural Science Foundation of China (NSFC 30830010 and 31100158; http://www.nsfc.gov.cn), the National Basic Research Program of China (2009CB119104; http://www.973.gov.cn), and Vetenskapsrådet, Sweden (http://www.vr.se/). The funders had no role in study design, data collection and analysis, decision to publish, or preparation of the manuscript.

Competing Interests: The authors have declared that no competing interests exist.

* E-mail: xiao-ru.wang@emg.umu.se

ɘ These authors contributed equally to this work.

Introduction

Genetic differentiation is strongly influenced by neutral processes and ecological selection. Population divergence engendered by geographical isolation as a consequence of topographical change and recent climatic oscillations are well documented in phylogeographic studies [1,2]. In these situations, physical distance and geographical barriers are major factors limiting gene flow, and populations diverge via genetic drift. In contrast, under ecological selection, migration may occur between populations located in close proximity but adapted to distinct niches. However, the fitness of immigrants or hybrids may be less than that of an existing population in a given environment, and this will limit the potential for genetic exchange [3]. Thus, niche divergence and local adaptation may produce or maintain genetic divergence even if physical barriers to dispersal eventually disappear [4,5,6]. Determining the role of environmental factors in causing genetic differentiation and assessing their importance relative to that of historical isolation, have been challenging [7]. In recent years, the combination of informative molecular markers, spatial statistics

and high resolution geographic information system (GIS) data has made it more feasible to carry out explicit evaluation of environmental influences on the distribution of genetic variation [8,9,10]. This approach offers the opportunity of assessing how specific landscape and environmental features have shaped gene flow between populations and the extent of local adaptation [11]. Correlation between environmental and genetic gradients can often provide initial evidence of the impact of natural selection and local adaptation [12,13,14]. Such information is important for understanding the neutral and selective processes driving divergence and, ultimately, speciation.

Southwest (SW) China is a biodiversity hotspot characterized by complex topography, heterogeneous regional climates and rich flora [15,16]. In particular, Yunnan Province has a climate and ecology distinct from those of the majority of the Eurasian continent, in that much of the region has been free from glacial advances and retreats, creating a region with high biodiversity that has been maintained for millions of years [17,18,19]. Thus, local adaptation and ecological divergence have potentially had sufficient time to influence the pattern of genetic differentiation in many local species. For this reason, the area is particularly attractive for studies on the roles played by macro- and micro-evolutionary processes in the evolution of biodiversity and speciation. The topography of SW China is characterized by a number of large valley systems, e.g. those of the Jinsha (Upper Yangtze), Mekong and Salween Rivers. These deep valleys, together with the high mountains surrounding them, have been identified as strong geographic barriers to dispersal, which have defined the phylogeography of regional flora [20,21]. The geometry and evolution of fluvial systems in this region have been affected to a great extent by tectonic changes in the Tibetan Plateau. During the most recent episodes of uplift of the eastern Tibetan Plateau, which occurred in the Late Miocene-Pliocene, major river drainage systems in SW China were reorganized and reinforced [22]. Species in this region responded uniquely to these landscape changes. In a number of conifers, herbs and shrubs, phylogeographic studies have revealed major landscape effects in which the current mountain and valley systems have acted as natural dispersal barriers [20,23,24], while in some other plants, freshwater fishes and amphibian species, the spatial genetic structure was found to reflect the historical geography of the region rather than the current geography [25,26,27,28,29,30]. These phylogeographic analyses to date have concentrated primarily on the effects of neutral processes on the pattern of genetic variation, and the roles of environmental adaptation and ecology-driven genetic divergence have seldom been examined.

Pinus yunnanensis is a subtropical pine endemic to SW China, which has a continuous distribution in the Yunnan-Guizhou region at elevations ranging from 700–3000 m above sea level across all the major river valleys [31,32]. Climatic conditions vary between regions divided by the mountain chains, and pronounced morphological variations in this pine have been recorded across its range [33,34,35,36]. It has hybridized with another Asian pine, *Pinus tabuliformis*, generating a homoploid hybrid, *Pinus densata* [37,38,39,40]. Mitochondrial (mt) and chloroplast (cp) DNA markers have been used to investigate the direction of hybridization and extent of introgression among these three species [39,41]. Moderate levels of total mtDNA and cpDNA diversity were detected in *P. yunnanensis*, of which 45% and 4%, respectively, resided between populations [39]. In most of these early investigations, *P. yunnanensis* was used as a parental reference to characterize the hybrid nature of *P. densata* and therefore only relevant representative populations were sampled and analysed. The western and south-eastern marginal populations of *P. yunnanensis*, which have distinct ecological and morphological characters, were inadequately represented or not sampled at all in the previous mt- and cpDNA analyses. The species-wide pattern of genetic diversity in *P. yunnanensis* therefore remains unknown. Moreover, the ecological and phylogeographic processes responsible for the current population structure of *P. yunnanensis* have not been explicitly addressed. In this study, we sampled 16 populations throughout the range of *P. yunnanensis* to cover all ecological habitats, especially those on the western and south-eastern peripheries. Both mtDNA and cpDNA variations were analyzed, together with environmental data, in order to assess the influence of ecological and historical factors on genetic divergence in *P. yunnanensis*. We addressed the following questions: How is genetic diversity distributed geographically in *P. yunnanensis*, and does the observed genetic pattern reflect the modern or the historical geography? What is the extent of environmental heterogeneity within the species' range, and could ecological factors have promoted genetic differentiation in this pine?

Materials and Methods

Ethics Statement

Here we state that the sampling of *P. yunnanensis* populations used in our study did not require any specific permission from any authority as it is a dominant forest species in SW China. Thus this study does not involve endangered or protected species.

Population Sampling, Sequencing and Genotyping

We sampled 255 individuals from 16 populations throughout the range of *P. yunnanensis*. The distribution of these populations is illustrated in Fig. 1. The name, location, and sample size of each population are listed in Table 1.

Eleven of the 16 populations (Nos. 1, 3–8 and 11–14) have been characterized for mtDNA and cpDNA variation in a previous study [39]. The data are based on three mtDNA segments (*nad*1 intron 2, *nad*4 intron 3 and *nad*5 intron 1) and five cpDNA microsatellite (cpSSR) loci (Pt45002, Pt71936, Pt87268, PCP1289 and PCP41131) [42,43]. The other five populations (Nos. 2, 9–10 and 15–16) were collected *de novo* for this study. For these additional populations, composite seed samples were collected from more than 100 mature trees in each stand, and then 16 bulked seeds per population were used to grow small seedlings for genotyping. Total DNA was extracted using a Plant Genomic DNA Kit (Tiangen, Beijing, China) according to the manufacturer's instructions. These new samples were characterized with the same set of mt and cp genetic markers used by Wang et al. [39]. For mtDNA, the purified PCR products were sequenced directly using an ABI 3730 automated sequencer (PE Applied Biosystems). For cpDNA, PCR products were resolved using a CEQ8000 capillary sequencer (Beckman-Coulter). Allele identification and genotyping were performed using CEQ8000 Fragment Analysis software (Beckman-Coulter).

Haplotype Network Analysis

Mitochondrial DNA sequences were aligned using Clustal X 1.81 [44], and alignments were further refined manually. Unique mt sequences (mitotypes) for *nad*1, *nad*4, *nad*5 and the combination of these three mtDNA sequences were identified among the sampled individuals. Their relationships were then established by median-joining networks using Network v. 4.6.1.0 [45]. Ten mitotypes (M1–10) identified in a sister species, *P. tabuliformis* [39], were used as an outgroup in the mtDNA network. A complex 27-bp insertion/deletion region was found in each of the three mt segments when aligning the sequences between *P. yunnanensis* and *P. tabuliformis*. Following the approach of Wang et al. [39], we treated distinct sequence types in this 27-bp region as having arisen from different insertion events in order to obtain the most compact network.

For cpDNA data, size scores for the five cpSSR loci in each individual were combined into a 5-locus chloroplast haplotype (chlorotype). Relationships among the chlorotypes were reconstructed using the median-joining model implemented in Network. For simplicity, singletons were excluded from the network analysis.

Genetic Diversity Analyses

All genetic diversity analyses were based on individual genotypes. For both mt- and cpDNA, the observed number of haplotypes and genetic diversity were calculated for each population and for the species. Genetic differentiation among populations and groups of populations were estimated by analysis of molecular variance (AMOVA) [46] with significance tests based on 10 000 permutations. These analyses were performed using Arlequin v. 3.0 [47]. The genetic divergence index (*D*) proposed by

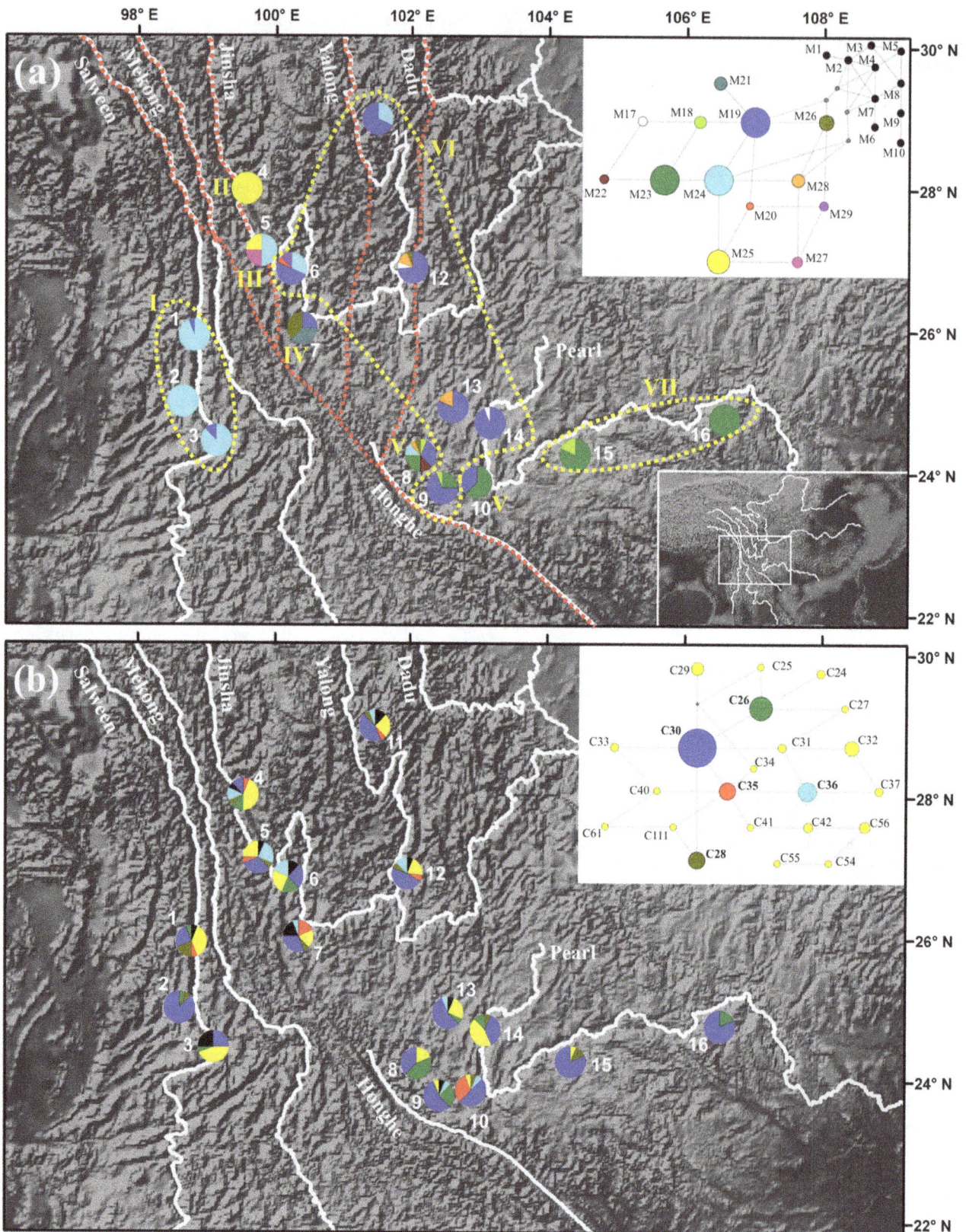

Figure 1. Mitotype (a) and chlorotype (b) composition of the 16 populations of *Pinus yunnanensis.* (a) Pie charts show the proportions of mitotypes in each population. Seven groups (I–VII) defined by mtDNA SAMOVA are shown. The current major rivers in Southwest China are illustrated in white, and the dashed red lines indicate the paleo-drainage routes before the major river reorganization (adapted from Clack et al. [22]). In the mtDNA network, each link represents one mutation step. Circle size is proportional to the frequency of mitotypes over all populations. Mitotype nomenclature follows that in Wang et al. [39]. M1–M10 occurred exclusively in *P. tabuliformis* and are used as an outgroup in this study. M17–M29 were detected in *P. yunnanensis* and are colored individually. (b) Pie charts show the proportions of chlorotypes in each population; singletons are

grouped and shown in black. Relationships among 22 common chlorotypes are shown in the network, in which each link represents one mutation step. Circle size is proportional to the frequency of mitotypes over all populations. Chlorotype nomenclature follows that in Wang et al. [39]. The five most common chlorotypes (frequency >5%; C26, C28, C30, C35 and C36) are indicated in bold, and colored green, brown, dark blue, red and light blue, respectively.

Jost [48] was also calculated using the software package SPADE (available at http://chao.stat.nthu.edu.tw/softwareCE.html). Jost's D provides a measure of actual differentiation of haplotypic frequencies among populations that is mathematically independent of within-population diversity [48].

The population structure was analyzed by comparing two coefficients of population divergence for both mtDNA (G_{ST} and N_{ST}) and cpSSR (G_{ST} and R_{ST}). G_{ST} is based solely on allele frequencies, while N_{ST} (or R_{ST}) takes into account similarities or relatedness among haplotypes. Thus, a significantly higher value for N_{ST} (or R_{ST}) than for G_{ST} implies that closely related haplotypes occur geographically closer to each other than distantly related haplotypes, indicating significant phylogeographic structure. The program Permut & CpSSR v. 2.0 [49] was applied to compare G_{ST} vs. N_{ST} or R_{ST} values using 10 000 random permutations.

To further assess genetic structure in $P.$ $yunnanensis$, the spatial variance in mitotype and chlorotype distributions was analyzed using SAMOVA 1.0 [50]. This program implements a simulated annealing approach to define groups of populations (K) that maximize the proportion of total divergence due to differences between groups of populations (F_{CT}). For each mitotype and chlorotype dataset, K values ranging from 2 to 10 were tested to search for the K that gave the highest F_{CT}. The significance of each F_{CT} was tested by simulating the annealing process 1000 times.

Finally, historical population expansion was assessed by mismatch distributions of both mt- and cpDNA data, using Arlequin. In this analysis, cpSSR data were coded in a binary

format following the method described by Navascués et al. [51]. A total of 10 000 parametric bootstrap replicates was used to generate an expected distribution under a model of sudden demographic expansion [52], and to test the goodness-of-fit of the demographic model.

Ecological Niche Modeling and Partial Mantel Test

We extracted ecological data from Mao & Wang [32] to perform ecological niche modeling for $P.$ $yunnanensis$. These data consist of 148 geo-referenced occurrence records and 14 environmental variables (Table S1). The 148 occurrence points were filtered spatially such that only one point occurred within each 1 km^2 grid cell (the maximum sampling resolution of our environmental data). We then used these locations for inclusion in GIS environmental layers. The 14 environmental variables were first examined for pairwise correlations within the distribution of $P.$ $yunnanensis$. Highly correlated variables could result in over-fitting of niche models and should thus be removed. After evaluation, we retained eight variables with pairwise Pearson correlation coefficients $r<0.70$ for subsequent analyses (Table S1), to minimize over-fitting of niche models and improve the interpretability of niche axes in the multivariate analyses. All selected environmental layers were converted to the same resolution at a grid cell size of 30×30 arc-seconds (1 km^2), and analyzed using the raster package (avaiable at http://raster.r-forge.r-project.org) in R and ArcGIS 9.2 (Environmental Systems Research Institute, Redlands, CA).

We performed a TwoStep clustering analysis in SPSS 13.0 (SPSS, Chicago) to quantitatively assess environmental heteroge-

Table 1. Geographic locations, sample sizes (N), number of haplotypes (n$_h$), and genetic diversity (H$_e$) of the 16 Pinus yunnanensis populations included in this study.

Population	Longitude (E)	Latitude (N)	Altitude (m)	mtDNA N	n$_h$	H$_e$	cpDNA N	n$_h$	H$_e$	Population code in Wang et al. [39]
1 Gongshan	98°49′	25°58′	1616	16	2	0.125	16	9	0.900	49
2 Tengchong	98°39′	25°02′	1580	16	1	0	16	3	0.242	New
3 Baoshan	99°08′	24°28′	1897	16	2	0.233	16	11	0.933	50
4 Zhongdian 1	99°32′	28°09′	3048	16	1	0	16	9	0.917	45
5 Zhongdian 2	100°03′	27°11′	2009	16	3	0.667	16	8	0.858	46
6 Lijiang	100°13′	26°53′	2493	16	4	0.675	16	9	0.900	47
7 Binchuan	100°21′	25°58′	3141	16	3	0.700	16	9	0.883	48
8 Yuxi	102°09′	24°15′	1849	16	7	0.850	16	4	0.692	54
9 Shiping	102°29′	23°43′	1428	16	3	0.492	16	5	0.650	New
10 Jianshui	102°57′	23°50′	2084	16	2	0.500	16	5	0.683	New
11 Jiulong	101°30′	29°00′	3129	16	2	0.458	16	10	0.825	44
12 Miyi	102°01′	26°55′	2047	15	4	0.467	15	8	0.791	51
13 Kunming	102°37′	24°58′	2242	16	2	0.325	16	8	0.700	52
14 Yiliang	103°10′	24°43′	1846	16	2	0.125	16	8	0.850	53
15 Luoping	104°24′	24°17′	1643	16	2	0.325	16	3	0.342	New
16 Leye	106°34′	24°48′	1039	16	1	0	16	2	0.325	New
Total				255	13	0.773	255	39	0.778	

neity within *P. yunnanensis*. This analysis estimates the number of ecotypic clusters within *P. yunnanensis* and their membership based on our 148 occurrence points and 8 environmental variables. Firstly, a sequential clustering approach was implemented to divide records into subclusters by constructing a modified cluster feature (CF) tree. The process scans records one by one and merges them into subclusters based on a distance defined by the log-likelihood decrease. Secondly, subclusters identified in step one were grouped into the desired number of clusters that maximize the Bayesian information criterion (BIC). For each ecotypic cluster pair, the relative contribution of environmental variables to their discrimination was evaluated by discriminant function analysis (DFA) using SPSS, and Wilks' λ was used to test the null hypothesis that the two clusters have identical means for the specific variables.

We also performed principal components analysis (PCA) to further investigate ecological differentiation within *P. yunnanensis*. PCA was applied to scaled data for all eight environmental variables corresponding to 148 *P. yunnanensis* occurrence records, without a prior designation of ecotypic clusters. The relative contribution of each environmental parameter to the formation of niche spaces was then represented in a PCA distance biplot, and the magnitude and statistical significance of niche shifts among the occurrence clouds in the PCA graph were assessed using between-class inertia percentages and 99 Monte-Carlo randomization tests [53]. The PCA was performed and the PCA biplot generated using ade4 [54].

We then followed the procedure and parameter settings described in Mao & Wang [32] to construct the distribution range of each ecotypic cluster. Based on the 148 occurrence data points and 8 environmental variables, we simulated species distribution models (SDMs) via maximum entropy using Maxent 3.3.1 with default settings [55]. The predictive power of each model for the region where it was calibrated was evaluated, with 25% of the occurrence dataset being chosen at random and compared with the model output created with the remaining 75% of the present dataset. Ten thousand background points were sampled to construct a predicted range distribution for each *P. yunnanensis* cluster. Model accuracy was evaluated by assessing the area under the curve (AUC) of the receiver-operating characteristic (ROC) plot [56]. According to Swets' scale [57], predictions are considered poor when AUC values are in the range 0.5–0.7, useful in the range 0.7–0.9, and good when greater than 0.9 (1 is perfect).

We also performed a niche-identity test to examine the null hypothesis that each pair of the ecotypic clusters is distributed in identical environmental space. This test compares the similarity of an ecotypic cluster's actual niches to a distribution of niche similarities, obtained from pairs of pseudoniches based on randomly reshuffled occurrence points of the two clusters. The niche-identity test was performed in ENMTools [58] with 100 pseudoreplicates, and niche overlap between each pair of the ecotypic clusters was assessed by Schoener's D [59] and Warren's I [60] similarity index.

Finally, we applied a niche space-based multivariate test [61] to assess the possibility that the allopatrically-distributed ecotypic clusters occupy similar niches. This test compares background divergence (d_b) with observed niche divergence (d_n) in the PCA-reduced axes, with the null hypothesis $d_b = d_n$ [61]. Niche divergence is supported if $d_b < d_n$ and the observed niche divergence itself (d_n) is significant (according to a t-test), whereas niche conservatism is supported if $d_b > d_n$. For each climatic cluster, the eight environmental variables, longitude, latitude, and altitude were extracted from the occurrence points and from 1000 random

background points within the background region of each ecotype using the packages dismo (available at http://cran.r-project.org/web/packages/dismo) and raster in R. The eight variables were reduced by PCA of the correlation matrix with the ade4 package. Correlations between the reduced PCA axes and the geographical variables (longitude, latitude and altitude) were examined by a nonparametric correlation test implemented in perm [62]. The background area for each cluster was delineated by SDMs from Maxent modeling at a baseline threshold obtained by minimizing the sum of sensitivity and specificity on the test data. In this study, d_n and d_b were computed as the differences between the mean scores of 75% random samples of the occurrence points of the two niches being compared (d_n) and of the 1000 background points of the two compared background habitats (d_b), in the reduced PCA axes. The distributions of d_b and d_n were generated with 1000 resamplings, and the mean of d_n was compared to the 95% confidence interval of d_b to determine its significance. The significance of the observed divergence between two compared niches was determined by a permutation t-test in perm.

To determine whether ecological factors explain genetic differentiation above and beyond differentiation due simply to isolation by distance (IBD), we performed partial Mantel tests on distances between populations. We compared matrices of pairwise genetic distance (F_{ST}) vs. geographic distance and genetic distance (F_{ST}) vs. ecological distance, controlling for ecological distance and geographic distance, respectively. Because the sample sites of all 16 populations were included in the 148 occurrence points for the species used in PCA, we estimated ecological distance by calculating the Euclidean distance between population pairs in a principal components space defined by the first two PC axes. Partial Mantel tests were performed with Arlequin, and 10 000 permutations were used in significance testing.

Results

Distribution of mtDNA and cpDNA Diversity

Sequences of the three selected mtDNA segments were obtained from 255 trees. When the three mtDNA segments were combined, a total of 13 mitotypes (M17– M29) were identified (Table S2). All of them have been reported by Wang et al. [39], and sequences of these mitotypes are available from GenBank accessions HM467712-HM467735. Network analysis showed that the 13 mitotypes were distinctly separated from those of *P. tabuliformis* (M1–M10), which was used as an outgroup, and all neighboring mitotypes differed by only one mutational step (Fig. 1a). A marked geographic pattern of mitotype distribution was observed in *P. yunnanensis*. The three most common mitotypes (M19, M23 and M24) were found in the central, south-eastern and western regions of the *P. yunnanensis* distribution, respectively. The other mitotypes were restricted to local populations at low frequencies, except for M25 which was fixed in population no. 4. In the case of the cp genome, a total of 39 chlorotypes (including 17 singletons) were detected over the five concatenated cpSSR loci (Table S2). After excluding the 17 singletons, network analysis of the 22 chlorotypes revealed a close relationship among them (Fig. 1b). The five most common chlorotypes (frequency >5%; C26, C28, C30, C35 and C36, which are shown in green, brown, dark blue, red and light blue, respectively in Fig. 1b) dominated all populations, with a total frequency of 72%, of which C30 (dark blue) contributed an overall frequency of 45%.

Total mtDNA diversity H_T (0.804) across all populations was much higher than the average within-population diversity H_S (0.371), resulting in strong between-population differentiation ($G_{ST} = 0.538$, $D = 0.688$). In contrast, the H_T value of 0.816 based on chlorotype variations was close to H_S (0.728), and both G_{ST} (0.108) and D (0.209) values were much lower than those for mtDNA (Table 2). AMOVA confirmed these findings, showing that 55.48% of the total diversity was due to population divergence for mtDNA, while only 6.94% in the case of the cpDNA divergence (Table 3). The levels of population differentiation observed in this study for both mt and cpDNA were higher than those previously reported (44.60% and 3.88%, respectively) for *P. yunnanensis* [39]. This increase is mainly caused by the inclusion of two new populations (Nos. 15 and 16), with distinct genetic compositions, from the south-eastern periphery. The contrasting pattern of population differentiation revealed by mtDNA and cpDNA loci in *P. yunnanensis* reflects the different modes of inheritance of the two cytoplasmic genomes. In the genus *Pinus*, variation in the mitochondrial genome represents the gene flow that is mediated by seed, while variation in the chloroplast genome represents gene flow attributable to both seed and pollen [63,64].

Comparisons of mtDNA G_{ST} *vs.* N_{ST} and cpDNA G_{ST} *vs.* R_{ST} indicated that N_{ST} and R_{ST} values were not significantly higher than G_{ST} values in *P. yunnanensis* (Table 2), suggesting a lack of phylogeographic structure in this species. However, mtDNA SAMOVA detected the presence of meaningful phylogeographic grouping, in which seven population groups (I–VII) were identified (Figs. 1 and S1). Groups I, VI and VII each spanned a large geographical area. Group I included three western populations (Nos. 1–3) dominated by mitotype M24, group VI included six central populations (Nos. 6, 9 and 11–14) dominated by M19, and group VII included two south-eastern populations (Nos. 15 and

16) in which M23 predominated. The other four groups (II–V) were each of restricted distribution in the central *P. yunnanensis* area. Groups II, III and IV were each represented by a single population, Nos. 4, 5 and 7, respectively, while group V included two populations, Nos. 8 and 10 (Fig. 1a). Group II was monomorphic for M25, while groups III–V each had multiple mitotypes in comparable proportions. SAMOVA of cpDNA variation failed to reveal any meaningful phylogeographic grouping (Fig. S1).

MtDNA-based mismatch distribution rejected a recent population expansion model at the species level ($P_{(SSD)}<0.01$, Table 2), but supported it at group levels ($P_{(SSD)}>0.05$). This analysis was not performed on group II because of its monomorphism. It was noticeable that the τ value of groups I and VII (3.0) was 2–3 times greater than that of the other groups (1.2–1.8). Based on the relationship τ = 2ut [52], the expansion time (t) is proportional to τ. If the mutation rate (u) is assumed to be constant within a species, the higher τ values of group I and VII would indicate that their expansion predated that of the other groups. A cpDNA-based mismatch distribution test indicated that a model postulating recent population expansion is supported at both the species and the ecotype level.

Niche Differentiation across the Species' Range

TwoStep clustering analysis grouped the 148 occurrence sites into three ecotypic clusters, Py-eco1, Py-eco2 and Py-eco3. These three clusters were geographically structured, and occupied the western, central and south-eastern ranges of *P. yunnanensis*,

Table 2. Average genetic diversity within populations (H_S), total genetic diversity (H_T), three coefficients of population divergence for mtDNA (G_{ST}, N_{ST} and Jost's D) and for cpDNA (G_{ST}, R_{ST} and Jost's D), and mismatch distribution test for *Pinus yunnanensis*.

	No. of populations	H_S (SE)	H_T (SE)	G_{ST} (SE)	N_{ST} or R_{ST} (SE)	Jost's D	Mismatch distribution N	τ	$P_{(SSD)}$	Raggedness index
mtDNA										
Species-wide	16	0.371 (0.068)	0.804 (0.036)	0.538 (0.091)	0.554 (0.091)	0.688 (0.016)	255	1.4	0.007	0.098**
Within groups[†] Group I	3	0.119 (0.067)	0.120 (0.063)	0.003 (0.062)	0.003 (0.062)	0 (0.016)	48	3.0	0.297	0.593
Group II	1	NC	NC	NC	NC	NC	NC	NC	NC	NC
Group III	1	NC	NC	NC	NC	NC	16	1.2	0.572	0.093
Group IV	1	NC	NC	NC	NC	NC	16	1.3	0.491	0.110
Group V	2	NC	NC	NC	NC	NC	32	1.7	0.111	0.124
Group VI	6	0.424 (0.075)	0.467 (0.077)	0.094 (NC)	0.066 (0.022)	0.076 (0.052)	95	1.8	0.518	0.140
Group VII	2	NC	NC	NC	NC	NC	32	3.0	0.184	0.452
Within ecotypes[‡] Py-eco1	3	0.119 (0.067)	0.120 (0.063)	0.003 (0.062)	0.003 (0.062)	0 (0.016)	48	3.0	0.299	0.593
Py-eco2	11	0.478 (0.076)	0.754 (0.074)	0.366 (0.110)	0.423 (0.116)	0.529 (0.038)	95	1.6	0.067	0.077
Py-eco3	2	NC	NC	NC	NC	NC	32	3.0	0.182	0.452
cpDNA Species-wide	16	0.728 (0.058)	0.816 (0.053)	0.108 (0.030)	0.099 (0.026)	0.209 (0.049)	255	0.8	0.598	0.020
Within ecotypes[‡] Py-eco1	3	0.692 (0.225)	0.820 (0.173)	0.157 (0.228)	0.105 (0.262)	0.417 (0.140)	48	10.2	0.670	0.027
Py-eco2	11	0.812 (0.033)	0.870 (0.037)	0.067 (0.018)	0.132 (0.032)	0.138 (0.066)	95	2.0	0.689	0.030
Py-eco3	2	NC	NC	NC	NC	NC	32	3.0	0.617	0.202

SE, standard error; *N*, sample size; τ, expansion parameter; $P_{(SSD)}$, SSD *P*-value; NC, not calculated due to low variation among populations;
**P<0.01;
[†], Grouping follows the division resulting from mtDNA SAMOVA;
[‡], Grouping follows the division resulting from TwoStep niche clustering analysis.

Table 3. Analysis of molecular variance (AMOVA) for mtDNA and cpDNA in *Pinus yunnanensis*.

	Source of variation	d.f.	SS	Variance components	Percentage of variation	F-statistics
mtDNA	Among 16 populations	15	90.003	0.358	55.48**	$F_{ST}=0.55**$
	Within populations	239	68.750	0.288	44.52	
	Total	254	158.753	0.646		
	Among 7 SAMOVA groups	6	85.411	0.413	57.77**	$F_{CT}=0.58**$
	Among populations within groups	9	4.592	0.014	1.96**	$F_{SC}=0.05**$
	Within populations	239	68.750	0.288	40.27**	$F_{ST}=0.60**$
	Total	254	158.753	0.714		
	Among 3 ecotypes	2	40.262	0.267	34.41**	$F_{CT}=0.34**$
	Among populations within ecotypes	13	49.741	0.222	28.58**	$F_{SC}=0.44**$
	Within populations	239	68.750	0.288	37.01**	$F_{ST}=0.63**$
	Total	254	158.753	0.777		
cpDNA	Among 16 populations	15	19.913	0.045	6.94**	$F_{ST}=0.07**$
	within populations	239	144.950	0.606	93.06	
	Total	254	164.863	0.652		
	Among 3 ecotypes	2	5.265	0.025	3.72**	$F_{CT}=0.04*$
	Among populations within ecotypes	13	14.648	0.033	4.92**	$F_{SC}=0.05**$
	Within populations	239	144.950	0.606	91.36**	$F_{ST}=0.09**$
	Total	254	164.863	0.664		

*$P<0.05$;
**$P<0.01$.

respectively (Fig. 3). DFA applied to all eight environmental variables supported the hypothesis that all ecotypic cluster pairs are significantly differentiated (Wilks' λ, $P<0.01$, Table 4). Py-eco1 diverged from Py-eco2 mainly on the basis of wet day frequency (WET), while Py-eco3 diverged from both Py-eco1 and Py-eco2 on the basis of temperature variability (bio3).

PCA of the eight environmental factors identified two components (with eigenvalues >1) that collectively explained 60.5% of the observed variation in the 148 occurrence records, accounting for 31.7% and 28.8% of the total variation, respectively (Fig. 2). The relative contributions of the different environmental variables to PC1 and PC2 are illustrated in the PCA distance biplot. PC1 is closely associated with temperature, soil type and seasonality (e.g. bio3, bio4, WET and SpH), while PC2 is associated mainly with precipitation (e.g. bio12 and bio15). The 148 occurrence sites were divided into three clearly separated environmental spaces in the Cartesian coordinates formed by the first two principal components. The niche centroids diverged strongly between the three clusters with a between-group inertia value of 0.44 ($P=0.001$). This division was in good agreement with that produced by the TwoStep clustering analysis (Fig. 2). The PCA distance biplot shows that the three ecotypic clusters diverged from each other along both PC1 and PC2. According to the reduced dimensionality of the ecological spaces, Py-eco1 occupies a niche with a mild, moist and low-seasonality climate. Py-eco2 is more seasonal than Py-eco1, while Py-eco3 is characterized by a drier climate. Taken together, our results suggest that each of the climatic clusters identified here represents a niche with unique ecological characteristics.

Based on the occurrence records for each climatic cluster, we generated geographic distribution maps projecting the areas in which each cluster might occur (Fig. 3). The niche modeling

accurately predicted the distribution of the three clusters identified by the TwoStep clustering analysis, with all training and test AUC values being greater than 0.99 ($P<0.0001$). The predicted distributions of the three clusters were generally consistent with the geographical ranges of their occurrence points, except that the projection for Py-eco3 extended beyond the observed distribution range of the species (Fig. 3). This inconsistency could be due to the limited numbers of occurrence points from the Py-eco3 region used in the modeling, which may have resulted in overestimation of the range of Py-eco3.

The niche similarity between Py-eco1 and Py-eco3 (Schoener's $D=0.21$ and Warren's $I=0.37$) was the lowest among the three niche pair comparisons (Table 4). A background test conducted by a multivariate method supported significant niche divergence between the three ecotypic clusters. Four axes were identified (each with an eigenvalue >1) that explained more than 85% of the total variation in each of the three pairwise comparisons (Table 5). In all pairwise tests, divergence was detected along all niche axes (i.e. $d_n>d_b$, and d_n is significant), except in the comparisons of PC1 for Py-eco1 vs. Py-eco3 and Py-eco2 vs. Py-eco3 (Table 5).

Effects of Environmental and Geographical Factors on Genetic Differentiation

The ecological niche clusters are broadly congruent with the grouping obtained by mtDNA SAMOVA. Py-eco1 and Py-eco3 correspond to SAMOVA group I and VII, respectively, while Py-eco2 covers populations from all other groups (II–VI) from the central distribution. Hierarchical AMOVA for mtDNA variation showed significant divergence between ecotypic clusters, since 34.41% of the variation occurred among ecotypes (Table 3). In

Table 4. The eight environmental variables (abbreviations in parentheses) used in this study, their contributions in discriminant function analysis (DFA) in pairwise comparisons of three ecotypes, and their similarities assessed based on Schoener's D and Warren's I index.

Environmental variables	Py-eco1 vs. Py-eco2	Py-eco1 vs. Py-eco3	Py-eco2 vs. Py-eco3
Isothermality (bio3)	0.17	**0.55**	**−0.83**
Temperature seasonality (bio4)	−0.06	**−0.27**	**0.41**
Maximum temperature of warmest month (bio5)	−0.06	−0.21	0.28
Annual precipitation (bio12)	**0.45**	0.01	**0.43**
Precipitation seasonality (bio15)	**−0.44**	<0.01	−0.35
Soil organic carbon (SC)	−0.15	0.05	−0.27
Soil pH (SpH)	0.38	0.22	−0.19
Wet day frequency (WET)	**0.81**	**0.29**	0.02
Pairwise comparision			
DFA (Wilks's λ)	0.21**	0.04**	0.21**
Schoener's D	0.42	0.21	0.25
Warren's I	0.73	0.37	0.45

Values corresponding to the three most significant variables are in boldface.
**P<0.01.

contrast, genetic differentiation on the basis of cpDNA was low (3.72%) between ecotypic clusters (Table 3).

For mtDNA, partial Mantel tests across 16 populations detected low but significant correlations between genetic distance and geographic distance ($r_{gen-geo}$ = 0.22, P<0.05; controlling for ecological distance) and between genetic distance and ecological distance ($r_{gen-eco}$ = 0.28, P<0.05; controlling for geographic distance; Table 6). At the ecotype level, this test could be performed only for Py-eco2, due to the low level of polymorphism and limited number of populations in Py-eco1 and 3. In Py-eco2, population genetic distance correlated only with geographic distance ($r_{gen-geo}$ = 0.18, P<0.05; $r_{gen-eco}$ = 0.03, P>0.05). For the same reason, this test could be performed for only one (group VI) of the seven mtDNA SAMOVA groups. Within group VI, the genetic distance correlated with neither geographic nor ecological distance ($r_{gen-geo}$ = −0.12, P>0.05; $r_{gen-eco}$ = 0.20, P>0.05). For cpDNA, the genetic distance correlated with geographic distance only at the species level ($r_{gen-geo}$ = 0.25, P<0.01; $r_{gen-eco}$ = −0.06, P>0.05), and within Py-eco2 ($r_{gen-geo}$ = 0.35, P<0.01; $r_{gen-eco}$ = −0.25, P>0.05; Table 6). These results indicate that both geographic and environmental factors contributed to the pattern of mtDNA variation across the species as a whole, but only geographic distance affected cpDNA relatedness between populations.

Discussion

Phylogeography of *P. yunnanensis*

Located on the south-eastern margin of the Tibetan Plateau, SW China has undergone dramatic geomorphological changes during the most recent uplift of the plateau since the Late Miocene-Pliocene [65,66,67]. Reconstruction of the paleo-landscape of the region suggests that a large-scale dendritic drainage network formed on a regional low-relief landscape [65,66,67]. The initial drainage system was characterized by multiple southward-flowing rivers draining into the South China Sea through the ancient Honghe River (Fig. 1a). This landscape was destroyed by a series of river reversal and capture events and aggressive river incision in response to the uplift of the eastern plateau that was initiated between 13 and 9 million years ago (MYA) [65,67]. The

paleo-Mekong and Salween Rivers separated from the paleo-Honghe River, forming parallel rivers. The modern Mekong River drains southward into the South China Sea independently from the Honghe River, and the modern Salween River drains into the Indian Ocean. The paleo-Honghe River then split into two further systems: the northern branches, the Jinsha, Yalong and Dadu Rivers, connected to become the modern Jinsha River, which radically altered its southward course to an eastward one and now drains into the East China Sea (see [22] and Fig. 1a). The southern section of the paleo-Honghe River, which became disconnected from the upper streams, drains into the South China Sea following the course of the paleo-river. The southeastern margin of the Tibetan Plateau is now characterized by localized gorges, 2–3 km in depth, which major rivers have incised into the regionally elevated, low-relief, relict topography which represents the landscape that existed throughout the eastern margin prior to regional uplift [66,67].

Molecular phylogeographic studies of endemic freshwater fishes and amphibian species in this region have suggested that the river rearrangements facilitated their genetic divergence, with estimates of divergence time falling between the Late Miocene and the Pleistocene [25,26,27,28]. However, our understanding of the impact of landscape changes on the distribution, evolution and genetic structure of plant taxa of the region is limited (but see [29,30]). *Pinus yunnanensis* is a dominant conifer of SW China with a continuous distribution, which probably extended beyond its extant range further into the north before the uplift of the eastern Tibetan Plateau [39]. We are interested to know whether the phylogeographic structure of *P. yunnanensis* has been strongly influenced by landscape changes in the past. Such studies are needed in order to understand the role played by habitat structure in the evolution of biodiversity in regional flora.

Analysis of mtDNA diversity revealed both discrete and continuous spatial structure in *P. yunnanensis*. MtDNA SAMOVA divided the species into 7 groups. Group I populations (Nos. 1–3) are located to the west of the paleo-Honghe River, and well separated from the other groups. This group is dominated by M24, a mitotype also detected in three northern populations (Nos. 5, 6 and 11) at relatively high frequencies. In addition, this

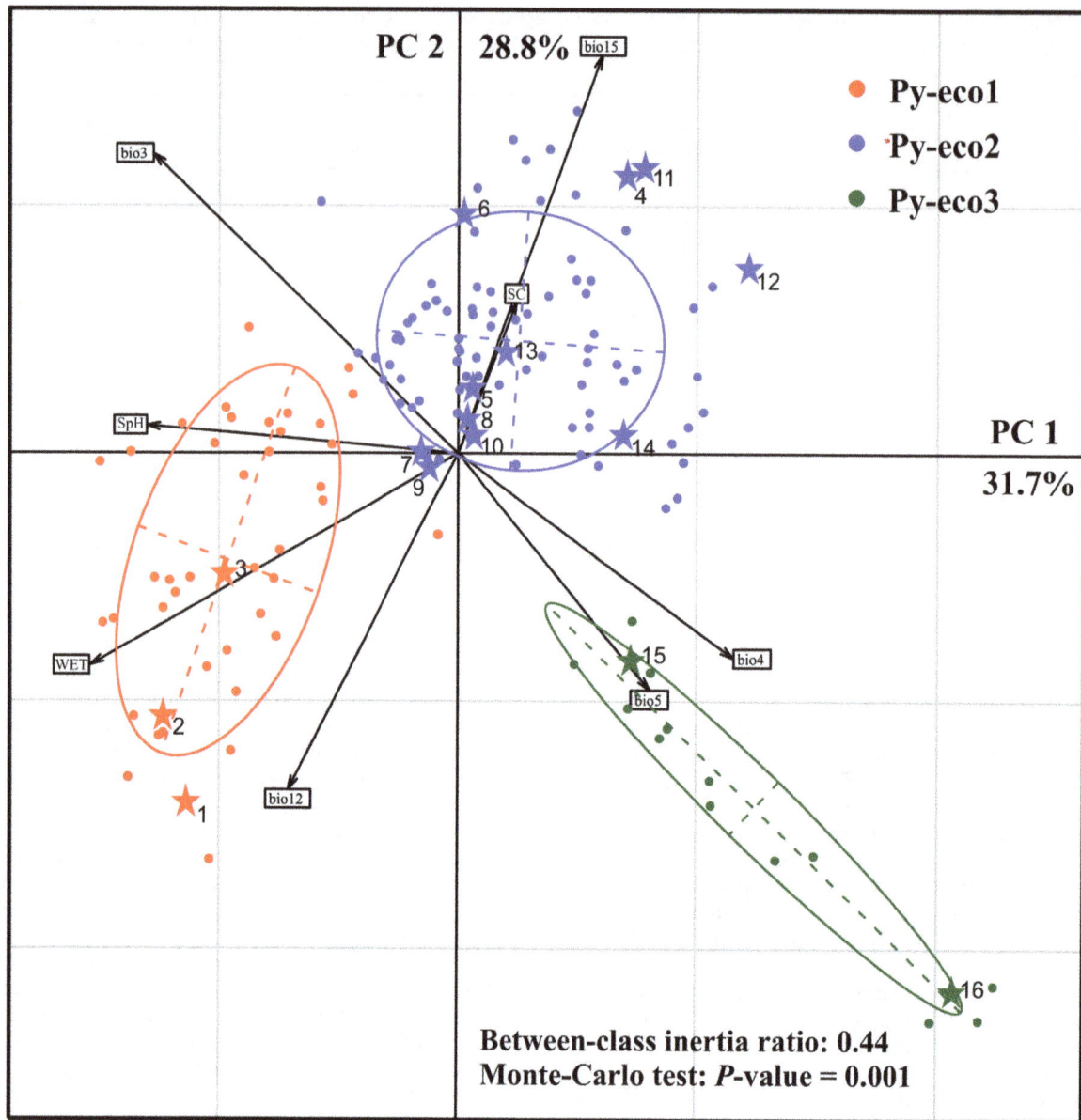

Figure 2. Principal components analysis distance biplot for the 148 *Pinus yunnanensis* **occurrence sites based on eight environmental variables.** Three occurrence clouds in the PCA graph are each outlined with a 1.5 inertia ellipse. The division of the 148 occurrence sites into three ecotypic clusters, Py-eco1, Py-eco2 and Py-eco3, by the TwoStep clustering analysis are shown in red, blue and green, respectively. The 16 populations sampled for genetic data analysis are each denoted by a star and numbered as in Table 1.

mitotype was found in the ancient hybrid zone between *P. yunnanensis* and *P. tabuliformis* located allopatrically north of the current *P. yunnanensis* range [39], suggesting that historically M24 was widespread in the northern region. The sharing of M24 between group I and the northern populations indicates that there was probably a connection between these regions before the uplift of the eastern Tibetan Plateau. The low relief of the paleo-landscape would have facilitated regional population connectivity, and this pattern is still visible in the extant population structure due to the low mutation rate and non-recombinant nature of the mt genome and the long generation time of pine species. During the landscape changes that took place in the Late Miocene, M24 drifted to near fixation in the western periphery, and the region became isolated from seed exchange by the wide Salween and

Mekong Rivers that function as barriers to seed dispersal. Similarly, group VII populations (Nos. 15 and 16), which represent the most south-easterly range of *P. yunnanensis*, are dominated by M23, a mitotype shared with three other southern populations (Nos. 8–10). Although the Pearl River separates group VII from the other populations, connectivity in this southern range is apparent, and M23 has drifted to a high frequency in the south-eastern periphery. It might be argued that the regional fixation of M24 and M23 could be due to introgression from neighboring species. This hypothesis seems unlikely because these two mitotypes were not detected in three other pines of the subgenus *Pinus* found in nearby regions, *Pinus massoniana*, *Pinus kesiya* and *Pinus merkusii* [39].

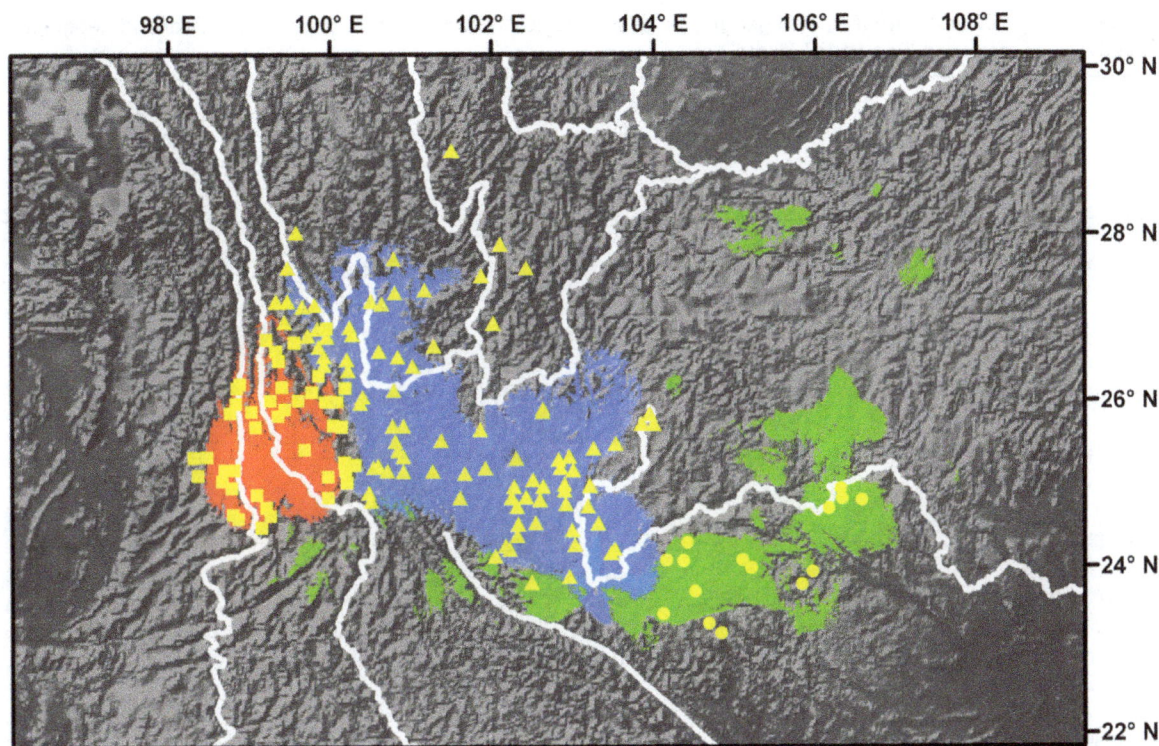

Figure 3. Predicted distribution for three *Pinus yunnanensis* ecotypes. The distribution ranges with probability of occurrence greater than 0.5, 0.5 and 0.6 for Py-eco1, Py-eco2 and Py-eco3 are shown in red, blue and green, respectively. Occurrence points used in the modeling are indicated by yellow squares, triangles and dots for Py-eco1, Py-eco2 and Py-eco3, respectively. The white lines show the current major rivers in southwest China.

Spatial expansion can favor the fixation of low frequency alleles by drift in newly colonized areas [68,69]. The mismatch distribution observed for mtDNA suggested that *P. yunnanensis* was in population equilibrium and had not undergone recent demographic expansion at the species level, but regional expansion was detected for all population groups. In addition, expansion in group I and VII seemed to have occurred earlier than that in other groups. This result suggests that groups I and

Table 5. Divergence on the independent niche axes between niche pairs.

	Py-eco1 vs. Py-eco2				Py-eco1 vs. Py-eco3				Py-eco2 vs. Py-eco3			
	PC1	PC2	PC3	PC4	PC1	PC2	PC3	PC4	PC1	PC2	PC3	PC4
d_n	**2.22** D	**1.80** D	**0.26** D	**0.43** D	**2.21** C	**1.90** D	**0.60** D	**0.67** D	**2.47** C	**0.55** D	**0.49** D	**0.22** D
d_b	1.99, 2.14	0.89–1.02	0.10–0.21	0–0.10	2.55–2.68	0.67–0.78	0–0.06	0.30–0.40	2.81–2.95	0.37–0.48	0–0.10	0.70–0.14
Top-loading variable	bio4, wet	bio5, sph	bio3, bio15	bio12, sc	bio3, bio15	wet	bio12	sc	bio3, bio15	bio4	wet	wet, sph, sc
% variance explained	38.89	19.02	14.63	13.46	42.88	15.79	15.26	12.69	48.62	15.32	13.16	8.13
Biological interpretation	Temperature, Moisture	Temperature, Soil PH	Seasonality	Water, Soil C	Seasonality	Moisture	Water	Soil C	Seasonality	Temperature	Moisture	Moisture, soil PH & C
Correlation longitude	−0.36**	−0.62**	−0.26**	−0.17**	−0.91**	−0.23**	−0.11*	−0.05**	−0.85**	0.11**	0.38**	−0.09**
Correlation latitude	−0.72**	0.42**	0.25**	−0.01	−0.12**	0.25**	−0.39**	0.71**	0.09**	0.84**	−0.08**	0.15**
Correlation altitude	−0.52**	0.59**	0.44**	0.27**	0.81**	0.08**	−0.08**	0.39**	0.81**	0.27**	−0.15**	0.38**

Bold values indicate significant niche divergence (D) or conservatism (C) compared to a 95% null distribution (d_b; t-test, ** for $P<0.01$). Significance of correlations between PC axes and geographical variables is indicated by * for $P<0.05$, and ** for $P<0.01$.
d_n, observed niche divergence; d_b, background divergence (95% null distribution).

Table 6. Correlation of genetic distance (F_{ST}) with geographic and ecological distance (controlling for ecological and geographic distance, respectively) as measured by a partial Mantel test.

Spatial scale	No. of populations	Correlation of F_{ST} with geographic distance		Correlation of F_{ST} with ecological distance	
		r	P	r	P
mtDNA					
Species-wide	16	0.22	0.014	0.28	0.038
Group VI	6	−0.12	0.677	0.20	0.306
Py-eco2	11	0.18	0.039	0.03	0.397
cpDNA					
Species-wide	16	0.25	0.007	−0.06	0.629
Py-eco2	11	0.35	0.006	−0.25	0.927

VII were not established by recent colonization from the central area of the *P. yunnanensis* range. In this scenario, long-term isolation of group I and VII could help to reinforce population differentiation after colonization. Other factors, such as local adaptation during the period of isolation, may also have contributed to genetic divergence of groups I and VII; this issue will be discussed in the next section.

The central region of the *P. yunnanensis* range (groups IV–VI, population nos. 6–14) is characterized by the presence of the M19 mitotype at high frequencies. These populations are all located on the eastern side of the paleo-Honghe River but separated by the paleo-Jinsha, Yalong, and Dadu Rivers into parallel zones. This separation, however, does not seem to have impaired migration across the zones. After the reversal of the middle Jinsha River and the capture of its major tributaries by the East China Sea, the paleo-network became separated into disconnected northern and southern sections [22]. The midstream of the Jinsha River is deeply incised (>1000 m) into bedrock gorges, which could present an effective barrier to gene flow, especially that mediated by seed. The sharing of the M19 mitotype between these areas is not consistent with the modern landscape, but rather reflects the historically continuous distribution of populations along the southward paleo-Honghe drainages. The retention of ancient genetic structure in these central populations is probably attributable to the generally continuous distribution of the species, which reduces the effect of genetic drift.

Two populations (Nos. 4 and 5) in the north-west each had a unique mitotype composition and lacked the M19 mitotype which was found in the neighboring region. MtDNA SAMOVA identified each of these two populations as a distinct group (II and III, respectively). They were distributed along the paleo-Jinsha River, but separated from the other populations along this river by a sharp bend in the current river course. Geological analysis carried out by Clark et al. [22] suggests that the localized reversal of river segments at the capture points of the Yalong and Jinsha bends may be related to the effects of geo-activity on local structures rather than to have resulted from large-scale initiation of plateau uplift along the entire south-eastern Tibetan Plateau margin. Thus, the establishment of these two populations is likely to have been linked to the formation of the modern local topography. Because of their limited distribution, distinct mitotypes could have undergone rapid drift in these populations during range expansion that probably radiated from the neighboring group VI population. Upstream of the Jinsha, Mekong and Salween rivers, there is a region containing another closely related

pine, *P. densata* [32]. The population-specific mitotypes (M25, M27) detected in populations Nos. 4 and 5 were not found in *P. densata* [39], a result which refutes the possibility of maternal introgression from *P. densata*, and further confirms that these two populations have been isolated from seed flow from nearby regions.

Taking all these results into consideration, we propose that the distribution of mtDNA variation in *P. yunnanensis* has been shaped by both the paleo-landscape and the formation of the modern regional topography. The connectivity between populations throughout the main range of the species reflects continuous distribution over a low relief paleo-landscape. This finding is similar to that of other phylogeographic studies of the region [25,26,27,28,29,30], in that the modern landscape does not fully reflect the population structure, but it differs from other observations in that the distinct paleo-river shaped genetic structure seen in other river valley-limited taxa is not apparent in *P. yunnanensis*. Rather than showing that historical drainage systems played a major role in determining current intraspecific genetic structure, the observation of continuous genetic differentiation over the main range of *P. yunnanensis*, together with discrete isolated local clusters, suggests an ancient landscape that imposed little constraint on migration, but which was subsequently disrupted due to regional geo-movements. The discrete differentiation observed at the peripheries appears to reflect both geographic isolation and environment (see the following section).

Ecological Patterns of Divergence

Three distinct ecotypic clusters (Py-eco1, Py-eco2 and Py-eco3) were identified in *P. yunnanensis*. This division is broadly congruent with that based on mtDNA SAMOVA. The two periphery groups I and VII each corresponded to Py-eco1 and Py-eco3, respectively, and all the other five groups (II–VI) in the central area belonged to Py-eco2. Multivariate analysis showed that environmental elements associated with availability of heat energy and water were the main factors that differentiated the three ecotypic clusters. Heat and water availability have strong impacts on the natural distribution of plant species, and are major determinants of plant productivity [70,71]. Thus, the niche diversity detected in *P. yunnanensis* could have important consequences for local adaptation and represent an impediment to gene flow. The Py-eco1 region has a humid subtropical climate. It is warmer and wetter than those where the other ecotypes occur, and is similar to the Mio-Pliocene paleo-climate of SW China [72,73]. Fossil records in SW China indicate that the ancestor of *P. yunnanensis* was present in a milder and moister climate during the Late Miocene than that of today [74]. Triggered by the uplift of the Tibetan Plateau and global cooling in the Late Neogene, *P. yunnanensis* adapted to drier climate in its central distribution area, while in the western periphery (Py-eco1 region) it survived in a warmer and more humid region [74]. Thus, Py-eco1 probably represents a relic ecotype of *P. yunnanensis*. The Py-eco3 region represents a much drier climate than Py-eco1. Populations in this region have a distinct morphology characterized by thin and pendulous needles [35], which is considered to be an adaptation to dry and hot environments [75] [76]. In addition, it has been suggested that a foehn wind specific to the Py-eco3 region is critical for pollination and cone splitting in local *P. yunnanensis* populations [75]. Based on their morphological divergence, some authors [35,77] classified the populations in this region as a variety or ecotype of *P. yunnanensis*. The congruence between genetic and ecological divisions suggests that environmental adaptation could have contributed to the genetic divergence of groups I and VII. Py-eco2 covers the major range of the species, including population groups II–VI. In this region/environment, population differentiation is characterized by IBD as shown by a partial Mantel test. Thus, the genetic groups recognized in this region were shaped by historically neutral processes and local barriers to gene flow rather than by ecological selection.

Neutral DNA markers are not expected to reflect the history of natural selection and adaptation. However, maternally inherited mtDNA, which is dispersed through seeds, is often used to track population establishment and migration history [39,78,79]. Population establishment and forest regeneration is brought about via seeds. Local adaptation to a distinct niche would result in ecological selection against immigrants [3]. Given enough time since a population began to diverge, drift and selection could induce fixation of distinct mitotypes in different niches, and thus shape the genetic structure of local populations [3,8,78]. In this context, the mtDNA pattern might indirectly reflect a population's persistence in, and adaptation to, a specific niche. Most boreal and temperate forest trees retreated into refugia during the Last Glacial Maximum and their current distribution ranges are the result of post-glaciation colonization [80,81,82]. In these species, adaptive evolution may have had insufficient time to induce distinct genetic divergence in recently colonized regions, and population structures revealed by neutral markers have been shaped mainly by periodical isolation and range expansion. In contrast, *P. yunnanensis* is distributed in a subtropical region that is recognized as having been a refuge during the last glaciations [17,83], where population demography was less influenced by climate fluctuations. Therefore, ecological divergence could have developed into a barrier preventing immigrants from surviving and reproducing in new habitats [3], and further strengthened genetic differentiation between *P. yunnanensis* niches. The impact of immigrant inferiority (or inviability) is less visible in a genome that is dispersed through pollen (cpDNA in pines) than in the seeds that produce immigrant plant individuals [3,84]. Immigrant inviability commonly exists between populations that exhibit adaptive ecological divergence, and it plays an important role in ecological modes of speciation [3]. Our finding that ecological and geographic distances have had a significant effect on species-wide genetic divergence supports the

hypothesis that both environment and geographic factors contributed to genetic differentiation in *P. yunnanensis*. The selective pressure exerted by niche divergence upon fitness in this species remains to be explicitly tested.

Conclusions

Integrating the results of genetic analysis and ecological niche modeling revealed the occurrence of ecological and phylogeographic processes in *P. yunnanensis* that were different from those seen in other case studies in SW China. In other taxa from the same region, the intraspecific genetic structure reflects the major role played by historic drainage systems. In contrast, in the case of *P. yunnanensis* our observation of continuous genetic differentiation over the majority of its range, together with discrete isolated local clusters, suggests a paleo-landscape that was generally well connected and imposed few migration constraints, but which was subsequently disrupted as a result of geomorphological movements in response to the uplift of the eastern Tibetan Plateau. The finding of discrete differentiation between two genetic groups in the peripheries is consistent with niche divergence and geographical isolation of these groups. In the central area of the species' range, population structure was shaped mainly by neutral processes and local geography rather than by ecological selection. These results show that geographical and environmental factors acting in combination have created stronger and more discrete genetic differentiation than IBD alone, and illustrate the importance of ecological factors in promoting and maintaining genetic divergence across a complex landscape. Our study highlights the heterogeneous contributions made by historic neutral processes and environment to genetic variation among different taxa in SW China. Further research incorporating multiple approaches applied to additional taxa would permit a better understanding of the origin and maintenance of biological diversity in this geologically unique region.

Supporting Information

Figure S1 SAMOVA analysis of mtDNA and cpDNA. X-axis shows different *K* values (number of groups) and Y-axis shows corresponding F_{CT} values.

Table S1 148 occurrence records for *Pinus yunnanensis*, with the corresponding 14 environmental variables. The eight variables used in ecological niche modeling are indicated in bold.

Table S2 The frequencies of mitotypes and chlorotypes in the 16 sampled *Pinus yunnanensis* populations.

Author Contributions

Conceived and designed the experiments: X-RW. Performed the experiments: BW J-FM. Analyzed the data: BW J-FM WZ. Contributed reagents/materials/analysis tools: BW J-FM WZ. Wrote the paper: BW J-FM X-RW.

References

1. Hickerson MJ, Carstens BC, Cavender-Bares J, Crandall KA, Graham CH, et al. (2010) Phylogeography's past, present, and future: 10 years after Avise, 2000. Mol Phylogenet Evol 54: 291–301.

2. Avise JC (2000) Phylogeography: the history and formation of species. massachusetts: Harvard Unversity Press. 447 p.

3. Nosil P, Vines TH, Funk DJ (2005) Reproductive isolation caused by natural selection against immigrants from divergent habitats. Evolution 59: 705–719.

4. Lozier JD, Mills NJ (2009) Ecological niche models and coalescent analysis of gene flow support recent allopatric isolation of parasitoid wasp populations in the Mediterranean. PLoS One 4: e5901.

5. Terai Y, Seehausen O, Sasaki T, Takahashi K, Mizoiri S, et al. (2006) Divergent selection on opsins drives incipient speciation in Lake Victoria cichlids. PLoS Biol 4: 2244–2251.

6. Thibert-Plante X, Hendry AP (2010) When can ecological speciation be detected with neutral loci? Mol Ecol 19: 2301–2314.

7. Sork VL, Nason J, Campbell DR, Fernandez JF (1999) Landscape approaches to historical and contemporary gene flow in plants. Trends Ecol Evol 14: 219–224.

8. Lee CR, Mitchell-Olds T (2011) Quantifying effects of environmental and geographical factors on patterns of genetic differentiation. Mol Ecol 20: 4631–4642.

9. Manel S, Schwartz MK, Luikart G, Taberlet P (2003) Landscape genetics: Combining landscape ecology and population genetics. Trends Ecol Evol 18: 189–197.

10. Storfer A, Murphy MA, Evans JS, Goldberg CS, Robinson S, et al. (2007) Putting the 'landscape' in landscape genetics. Heredity 98: 128–142.

11. Manel S, Joost S, Epperson BK, Holderegger R, Storfer A, et al. (2010) Perspectives on the use of landscape genetics to detect genetic adaptive variation in the field. Mol Ecol 19: 3760–3772.

12. Chen J, Källman T, Ma X, Gyllenstrand N, Zaina G, et al. (2012) Disentangling the roles of history and local selection in shaping clinal variation of allele frequencies and gene expression in Norway spruce (Picea abies). Genetics 191: 865–881.

13. Eckert AJ, Wegrzyn JL, Pande B, Jermstad KD, Lee JM, et al. (2009) Multilocus patterns of nucleotide diversity and divergence reveal positive selection at candidate genes related to cold hardiness in coastal Douglas fir (Pseudotsuga menziesii var. menziesii). Genetics 183: 289–298.

14. Hall D, Luquez V, Garcia VM, St Onge KR, Jansson S, et al. (2007) Adaptive population differentiation in phenology across a latitudinal gradient in European aspen (Populus Tremula, L.): A comparison of neutral markers, candidate genes and phenotypic traits. Evolution 61: 2849–2860.

15. Myers N, Mittermeier RA, Mittermeier CG, da Fonseca GAB, Kent J (2000) Biodiversity hotspots for conservation priorities. Nature 403: 853–858.

16. Huang J, Chen B, Liu C, Lai J, Zhang J, et al. (2012) Identifying hotspots of endemic woody seed plant diversity in China. Divers Distrib 18: 673–688.

17. Frenzel B, Bräuning A, Adamczyk S (2003) On the problem of possible last-glacial forest-refuge-areas within the deep valleys of eastern Tibet. Erdkunde 57: 182–198.

18. Cook CG, Jones RT, Langdon PG, Leng MJ, Zhang E (2011) New insights on Late Quaternary Asian palaeomonsoon variability and the timing of the Last Glacial Maximum in southwestern China. Quat Sci Rev 30: 808–820.

19. Yao YF, Bruch AA, Cheng YM, Mosbrugger V, Wang YF, et al. (2012) Monsoon versus uplift in southwestern China: Late Pliocene climate in Yuanmou basin, Yunnan. PLoS One 7: e37760.

20. Gao LM, Moeller M, Zhang XM, Hollingsworth ML, Liu J, et al. (2007) High variation and strong phylogeographic pattern among cpDNA haplotypes in Taxus wallichiana (Taxaceae) in China and north Vietnam. Mol Ecol 16: 4684–4698.

21. Li Y, Zhai SN, Qiu YX, Guo YP, Ge XJ, et al. (2011) Glacial survival east and west of the 'Mekong-Salween Divide' in the Himalaya-Hengduan Mountains region as revealed by AFLPs and cpDNA sequence variation in Sinopodophyllum hexandrum (Berberidaceae). Mol Phylogenet Evol 59: 412–424.

22. Clark MK, Schoenbohm LM, Royden LH, Whipple KX, Burchfiel BC, et al. (2004) Surface uplift, tectonics, and erosion of eastern Tibet from large-scale drainage patterns. Tectonics 23: TC1006.

23. Wang FY, Xun G, Hu CM, Hao G (2008) Phylogeography of an alpine species Primula secundiflora inferred from the chloroplast DNA sequence variation. J Syst Evol 46: 13–22.

24. Yuan QJ, Zhang ZY, Peng H, Ge S (2008) Chloroplast phylogeography of Dipentodon (Dipentodontaceae) in southwest China and northern Vietnam. Mol Ecol 17: 1054–1065.

25. Guo XG, He SP, Zhang YG (2005) Phylogeny and biogeography of Chinese sisorid catfishes re-examined using mitochondrial cytochrome b and 16S rRNA gene sequences. Mol Phylogenet Evol 35: 344–362.

26. He DK, Chen YF (2006) Biogeography and molecular phylogeny of the genus Schizothorax (Teleostei : Cyprinidae) in China inferred from cytochrome b sequences. J Biogeogr 33: 1448–1460.

27. Peng ZG, Ho SYW, Zhang YG, He SP (2006) Uplift of the Tibetan Plateau: Evidence from divergence times of glyptosternoid catfishes. Mol Phylogenet Evol 39: 568–572.

28. Zhang DR, Chen MY, Murphy RW, Che J, Pang JF, et al. (2010) Genealogy and palaeodrainage basins in Yunnan Province: Phylogeography of the Yunnan spiny frog, Nanorana yunnanensis (Dicroglossidae). Mol Ecol 19: 3406–3420.

29. Zhang TC, Comes HP, Sun H (2011) Chloroplast phylogeography of Terminalia franchetii (Combretaceae) from the eastern Sino-Himalayan region and its correlation with historical river capture events. Mol Phylogenet Evol 60: 1–12.

30. Yue LL, Chen G, Sun WB, Sun H (2012) Phylogeography of Buddleja crispa (Buddlejaceae) and its correlation with drainage system evolution in southwestern China. Am J Bot 99: 1726–1735.

31. Wu CL (1956) The taxonomic revision and phytogeographical study of Chinese pines. Acta Phytotaxonom Sinica 5: 131–163.

32. Mao JF, Wang XR (2011) Distinct niche divergence characterizes the homoploid hybrid speciation of Pinus densata on the Tibetan Plateau. Am Nat 177: 424–439.

33. Yu H, Zheng SH, Huang RF (1998) Polymorphism of male cones in populations of Pinus yunnanensis Franch. Biodiv Sci 6: 267–271.

34. Yu H, Ge S, Huang RF, Jiang HQ (2000) A preliminary study on genetic variation and relationships of Pinus yunnanensis and its closely related species. Acta Bot Sinica 42: 107–110.

35. Fu LK, Li N, Mill R (1999) Pinus. In: Wu ZY, Raven P, editors. Flora of China. Beijing: Sciense Press; and St Louis: Missouri Botanical Garden Press. pp. 11–25.

36. Mao JF, Li Y, Wang XR (2009) Empirical assessment of the reproductive fitness components of the hybrid pine Pinus densata on the Tibetan Plateau. Evol Ecol 23: 447–462.

37. Wang XR, Szmidt AE (1994) Hybridization and chloroplast DNA variation in a Pinus species complex from Asian. Evolution 48: 1020–1031.

38. Wang XR, Szmidt AE, Savolainen O (2001) Genetic composition and diploid hybrid speciation of a high mountain pine, Pinus densata, native to the Tibetan Plateau. Genetics 159: 337–346.

39. Wang B, Mao JF, Gao J, Zhao W, Wang XR (2011) Colonization of the Tibetan Plateau by the homoploid hybrid pine Pinus densata. Mol Ecol 20: 3796–3811.

40. Gao J, Wang B, Mao JF, Ingvarsson P, Zeng QY, et al. (2012) Demography and speciation history of the homoploid hybrid pine Pinus densata on the Tibetan Plateau. Mol Ecol 21: 4811–4827.

41. Song BH, Wang XQ, Wang XR, Ding KY, Hong DY (2003) Cytoplasmic composition in Pinus densata and population establishment of the diploid hybrid pine. Mol Ecol 12: 2995–3001.

42. Vendramin GG, Lelli L, Rossi P, Morgante M (1996) A set of primers for the amplification of 20 chloroplast microsatellites in Pinaceae. Mol Ecol 5: 595–598.

43. Provan J, Soranzo N, Wilson NJ, McNicol JW, Forrest GI, et al. (1998) Gene-pool variation in Caledonian and European Scots pine (Pinus sylvestris L.) revealed by chloroplast simple-sequence repeats. Proc R Soc B 265: 1697–1705.

44. Thompson JD, Gibson TJ, Plewniak F, Jeanmougin F, Higgins DG (1997) The CLUSTAL_X windows interface: Flexible strategies for multiple sequence alignment aided by quality analysis tools. Nucleic Acids Res 25: 4876–4882.

45. Bandelt HJ, Forster P, Rohl A (1999) Median-joining networks for inferring intraspecific phylogenies. Mol Biol Evol 16: 37–48.

46. Excoffier L, Smouse PE, Quattro JM (1992) Analysis of molecular variance inferred from metric distances among DNA haplotypes: Application to human mitochondrial DNA restriction data. Genetics 131: 479–491.

47. Excoffier L, Laval G, Schneider S (2005) Arlequin (version 3.0): An integrated software package for population genetics data analysis. Evol Bioinform 1: 47–50.

48. Jost L (2008) G_{ST} and its relatives do not measure differentiation. Mol Ecol 17: 4015–4026.

49. Pons O, Petit RJ (1996) Measuring and testing genetic differentiation with ordered versus unordered alleles. Genetics 144: 1237–1245.

50. Dupanloup I, Schneider S, Excoffier L (2002) A simulated annealing approach to define the genetic structure of populations. Mol Ecol 11: 2571–2581.

51. Navascués M, Vaxevanidou Z, Gonzalez-Martinez SC, Climent J, Gil L, et al. (2006) Chloroplast microsatellites reveal colonization and metapopulation dynamics in the Canary Island pine. Molecular Ecology 15: 2691–2698.

52. Rogers AR, Harpending H (1992) Population growth makes waves in the distribution of pairwise genetic differences. Mol Biol Evol, 9: 552–569.

53. Romesburg H (1985) Exploring, confirming, and randomization tests. Comput Geosci 11: 19–37.

54. Dray S, Dufour AB (2007) The ade4 package: Implementing the duality diagram for ecologists. J Stat Softw 22: 1–20.

55. Phillips SJ, Anderson RP, Schapire RE (2006) Maximum entropy modeling of species geographic distributions. Ecol Model 190: 231–259.

56. Fielding AH, Bell JF (1997) A review of methods for the assessment of prediction errors in conservation presence/absence models. Environ Conserv 24: 38–49.

57. Swets J (1988) Measuring the accuracy of diagnostic systems. Science 240: 1285.

58. Warren DL, Glor RE, Turelli M (2010) ENMTools: a toolbox for comparative studies of environmental niche models. Ecography 33: 607–611.

59. Schoener TW (1968) The anolis lizards of Bimini: Resource partitioning in a complex fauna. Ecology 49: 704.

60. Warren DL, Glor RE, Turelli M (2008) Environmental niche equivalency versus conservatism quantitative approaches to niche evolution. Evolution 62: 2868–2883.

61. McCormack JE, Zellmer AJ, Knowles LL (2010) Dos niche divergence accompany allopatric divergence in Aphelocoma Jays as predicted under ecological speciation?: Insights from tests with niche models. Evolution 64: 1231–1244.

62. Fay MP, Shaw PA (2010) Exact and asymptotic weighted logrank tests for interval censored data: The interval R package. J Stat Softw 36: 1–34.

63. Neale DB, Sederoff RR (1989) Paternal inheritance of chloroplast DNA and maternal inheritance of mitochondrial DNA in loblolly pine. Theor Appl Genet 77: 212–216.

64. Wang XR, Szmidt AE, Lu MZ (1996) Genetic evidence for the presence of cytoplasmic DNA in pollen and megagametophytes and maternal inheritance of mitochondrial DNA in Pinus. Forest genet 3: 37–44.

65. Clark MK, House MA, Royden LH, Whipple KX, Burchfiel BC, et al. (2005) Late Cenozoic uplift of southeastern Tibet. Geology 33: 525–528.

66. Clark MK, Royden LH, Whipple KX, Burchfiel BC, Zhang X, et al. (2006) Use of a regional, relict landscape to measure vertical deformation of the eastern Tibetan Plateau. J Geophys Res 111: F03002.

67. Ouimet W, Whipple K, Royden L, Reiners P, Hodges K, et al. (2010) Regional incision of the eastern margin of the Tibetan Plateau. Lithosphere 2: 50–63.

68. Klopfstein S, Currat M, Excoffier L (2006) The fate of mutations surfing on the wave of a range expansion. Mol Biol Evol 23: 482–490.

69. Excoffier L, Foll M, Petit RJ (2009) Genetic Consequences of Range Expansions. Ann Rev Ecol Evol Syst 40: 481–501.

70. Kozlowski TT, Pallardy SG (2002) Acclimation and adaptive responses of woody plants to environmental stresses. Bot Rev 68: 270–334.

71. Körner C (2003) Alpine plant life: functional plant ecology of high mountain ecosystems. Berlin: Springer Verlag. 344p.

72. Xu JX, Ferguson DK, Li CS, Wang YF (2008) Late Miocene vegetation and climate of the Luhe region in Yunnan, southwestern China. Rev Palaeobot and Palyno 148: 36–59.

73. Jacques FMB, Guo SX, Su T, Xing YW, Huang YJ, et al. (2011) Quantitative reconstruction of the late Miocene monsoon climates of southwest China: A case study of the Lincang flora from Yunnan Province. Palaeogeogr Palaeoclimatol Palaeoecol 304: 318–327.

74. Xing Y, Liu Y-S, Su T, Jacques FMB, Zhou Z (2010) *Pinus prekesiya* sp. nov. from the upper Miocene of Yunnan, southwestern China and its biogeographical implications. Rev Palaeobot Palyno 160: 1–9.

75. Li ZJ, Wang XP (1981) The Distribution of *Pinus yunnanensis* var. *tenuifolia* in relation to the environmental conditions. J Plant Ecol 5: 28–37.

76. Rundel PW, Yoder B (1998) Ecophysiology. In: Richardson DM, editor. Ecology and Biogeography of *Pinus*. Cambridge: Cambridge University Press. 296–323.

77. Cheng WC, Fu LG, Cheng CY (1975) *Pinus yunnanensis* Franch. *var. tenuifolia*. Acta Phytotaxonom Sinica 13: 85.

78. Petit RJ, Brewer S, Bordacs S, Burg K, Cheddadi R, et al. (2002) Identification of refugia and post-glacial colonisation routes of European white oaks based on chloroplast DNA and fossil pollen evidence. Forest Ecol Manag 156: 49–74.

79. Petit RJ, Duminil J, Fineschi S, Hampe A, Salvini D, et al. (2005) Comparative organization of chloroplast, mitochondrial and nuclear diversity in plant populations. Mol Ecol 14: 689–701.

80. Gugerli F, Sperisen C, Buchler U, Magni F, Geburek T, et al. (2001) Haplotype variation in a mitochondrial tandem repeat of Norway spruce (*Picea abies*) populations suggests a serious founder effect during postglacial re-colonization of the western Alps. Mol Ecol 10: 1255–1263.

81. Petit RJ, Csaikl UM, Bordacs S, Burg K, Coart E, et al. (2002) Chloroplast DNA variation in European white oaks: Phylogeography and patterns of diversity based on data from over 2600 populations. Forest Ecol Manag 156: 5–26.

82. Naydenov K, Senneville S, Beaulieu J, Tremblay F, Bousquet J (2007) Glacial vicariance in Eurasia: Mitochondrial DNA evidence from Scots pine for a complex heritage involving genetically distinct refugia at mid-northern latitudes and in Asia Minor. BMC Evol Biol 7: 233.

83. Lehmkuhl F (1998) Extent and spatial distribution of Pleistocene glaciations in eastern Tibet. Quatern Int 45–6: 123–134.

84. Arnold ML, Buckner CM, Robinson JJ (1991) Pollen-mediated introgression and hybrid speciation in Louisiana irises. Proc Natl Acad Sci USA 88: 1398–1402.

17

Persistence and Change in Community Composition of Reef Corals through Present, Past, and Future Climates

Peter J. Edmunds[1]*[¶], Mehdi Adjeroud[2,3], Marissa L. Baskett[4], Iliana B. Baums[5], Ann F. Budd[6], Robert C. Carpenter[1], Nicholas S. Fabina[7], Tung-Yung Fan[8], Erik C. Franklin[9], Kevin Gross[10], Xueying Han[11,12], Lianne Jacobson[1,13], James S. Klaus[14], Tim R. McClanahan[15], Jennifer K. O'Leary[12], Madeleine J. H. van Oppen[16], Xavier Pochon[17], Hollie M. Putnam[9], Tyler B. Smith[18], Michael Stat[19], Hugh Sweatman[16], Robert van Woesik[20], Ruth D. Gates[9¶]

1 Department of Biology, California State University Northridge, Northridge, California, United States of America, 2 Institut de Recherche pour le Développement, Unité de Recherche CoReUs, Observatoire Océanologique de Banyuls, Banyuls-sur-Mer, France, 3 Laboratoire d'Excellence "CORAIL", Perpignan, France, 4 Department of Environmental Science and Policy, University of California Davis, Davis, California, United States of America, 5 Department of Biology, The Pennsylvania State University, University Park, Pennsylvania, United States of America, 6 Department of Earth and Environmental Sciences, University of Iowa, Iowa City, Iowa, United States of America, 7 Center for Population Biology, University of California Davis, Davis, California, United States of America, 8 National Museum of Marine Biology and Aquarium, Taiwan, Republic of China, 9 Hawaii Institute of Marine Biology, School of Ocean and Earth Science and Technology, University of Hawaii, Kaneohe, Hawaii, United States of America, 10 Biomathematics Program, North Carolina State University, Raleigh, North Carolina, United States of America, 11 Department of Ecology, Evolution and Marine Biology and the Coastal Research Center, Marine Science Institute, University of California Santa Barbara, Santa Barbara, California, United States of America, 12 National Center for Ecological Analysis and Synthesis, Santa Barbara, California, United States of America, 13 Department of Biology, University of Florida, Gainesville, Florida, United States of America, 14 Department of Geological Sciences, University of Miami, Coral Gables, Florida, United States of America, 15 Wildlife Conservation Society, Marine Program, Bronx, New York, United States of America, 16 Australian Institute of Marine Science, Townsville, Queensland, Australia, 17 The Cawthron Institute, Nelson, New Zealand, 18 Center for Marine and Environmental Studies, University of the Virgin Islands, St. Thomas, Virgin Islands, United States of America, 19 The University of Western Australia Oceans Institute and the Centre for Microscopy, Characterisation and Analysis, University of Western Australia, Crawley, Western Australia, Australia, 20 Department of Biological Sciences, Florida Institute of Technology, Melbourne, Florida, United States of America

Abstract

The reduction in coral cover on many contemporary tropical reefs suggests a different set of coral community assemblages will dominate future reefs. To evaluate the capacity of reef corals to persist over various time scales, we examined coral community dynamics in contemporary, fossil, and simulated future coral reef ecosystems. Based on studies between 1987 and 2012 at two locations in the Caribbean, and between 1981 and 2013 at five locations in the Indo-Pacific, we show that many coral genera declined in abundance, some showed no change in abundance, and a few coral genera increased in abundance. Whether the abundance of a genus declined, increased, or was conserved, was independent of coral family. An analysis of fossil-reef communities in the Caribbean revealed changes in numerical dominance and relative abundances of coral genera, and demonstrated that neither dominance nor taxon was associated with persistence. As coral family was a poor predictor of performance on contemporary reefs, a trait-based, dynamic, multi-patch model was developed to explore the phenotypic basis of ecological performance in a warmer future. Sensitivity analyses revealed that upon exposure to thermal stress, thermal tolerance, growth rate, and longevity were the most important predictors of coral persistence. Together, our results underscore the high variation in the rates and direction of change in coral abundances on contemporary and fossil reefs. Given this variation, it remains possible that coral reefs will be populated by a subset of the present coral fauna in a future that is warmer than the recent past.

Editor: Erik Sotka, College of Charleston, United States of America

Funding: This work was conducted as part of the "Tropical coral reefs of the future: modeling ecological outcomes from the analyses of current and historical trends" Working Group (to RDG and PJE), and while RDG was a Center Fellow supported by the National Center for Ecological Analysis and Synthesis, both funded by NSF (Grant #EF-0553768), the University of California, Santa Barbara, and the State of California. The authors acknowledge additional support from NSF (OCE 04-17413 and 10-26851 to PJE and RCC, DEB 03-43570 and 08-51441 to PJE, OCE 07-52604 to RDG, OCE-1041673 for ECF, DEB 01-02544, 97-05199, EAR 92-19138, and 04-45789 to AFB), the US EPA (FP917096 to ECF and FP917199 to HMP), and a postdoctoral fellowship to MS from the UWA-AIMS-CSIRO collaborative agreement. The funders had no role in study design, data collection and analysis, decision to publish, or preparation of the manuscript.

Competing Interests: The authors have declared that no competing interests exist.

* Email: peter.edmunds@csun.edu

¶ First and last author determined by organizational roles in the project, other authors listed alphabetically.

Introduction

Most present-day coral reefs differ from the reefs that were first described by ecologists and explorers [1], and recent evidence suggests that the rate of change in environmental factors affecting coral survival is accelerating as a result of global climate change (GCC) and ocean acidification (OA) [2]. Many coral reefs have changed dramatically in benthic community structure over the last few decades [3], but contemporary research has focused on declining abundances of scleractinian corals rather than on the few cases where reefs have retained coral cover (or recovered following losses), and where some scleractinian corals have maintained or increased in abundance [4–6].

Coral reefs in remote settings provide some of the best examples of reefs with high coral cover and intact trophic structures [6–8], and their distance from localized anthropogenic effects suggest isolation and protection, rather than global climate, are determinants of their present condition. In addition, coral reefs with diverse scleractinian faunas and relatively high coral cover also can be found in marginal locations characterized by high temperature fluctuations [9], thermal extremes [10], and turbidity [11]. Moreover, while many coral genera have declined in abundance, some persist in ecologically dominant roles, which have led to the suggestion that corals on contemporary reefs can be categorized as "losers" or "winners" [12]. Massive *Porites* spp. is an example of a group of corals that is faring better than others and is increasing in abundance in the Pacific and Caribbean [4,13], and also is showing signs of resistance to OA, both in mesocosms [14] and in at least one reef environment where volcanic carbon dioxide seeps into the seawater [15].

Considerable effort is being dedicated to elucidating the processes driving shifts in coral community structure on contemporary reefs, and characterizing the biological and ecological traits of scleractinian corals that are resistant to disturbances [16,17]. Information is still needed to advance this effort, for example, to evaluate whether shifts in coral community structure are a result of reduced coral recruitment, increased mortality of adult corals, or both. These and other processes interact to determine the trajectories of change in the composition of coral communities. For instance, with increased coral mortality driving regional reductions in fecundity and population size, coral recruitment likely will decline and create compensatory density dependence favoring further reductions in coral cover. Such population-level events are also affected by processes such as herbivory, predation, regional oceanography, and climate change, which alter coral reef communities over short periods. Over geologic time, macroprocesses such as ice ages cause changes in the composition of coral reef communities that are captured in fossilized reefs, where the success of coral species may be discovered based on their retention (or loss) from the fossil record [18]. The fossil record therefore provides a tool through which it is possible to analyze how corals responded to environmental or biological changes in the past, and over much longer time frames than is covered by ecological studies.

The goals of this study were to use long-term data from modern and fossil coral reefs to test for variation among coral genera in the rates and directions of change in abundance over time, to use these trends to consider which genera have the potential to persist as seawater warms through climate change, and to evaluate in what form these genera might assemble in the future to form coral communities. To achieve these goals, we synthesized data from extant reefs at seven locations ("case studies") and from fossil coral reef communities in the Caribbean, and developed a mathematical model to evaluate which traits are most likely advantageous in promoting persistence of coral genera in warming oceans. We present our analyses in three parts: first, we describe the events taking place on extant reefs by examining aspects of ecological records from our case studies (i.e., the Present); second, we use the fossil record (i.e., the Past) to gain insight into the temporal novelty of the changes affecting the community ecology of extant reefs, and whether clues to the ultimate outcomes of these changes might be found in the past; finally, we use a mathematical model to offer

insight into the potential ecological fate of coral reefs under increased thermal stress (i.e., the Future).

Present

Recent efforts to describe changes in the composition of coral reef communities have typically focused on scleractinian corals and their performance relative to other functional groups such as macroalgae [19]. These efforts have brought attention to the large losses of coral cover that have taken place since the 1960s [20]. It is uncommon however, for such studies to explicitly focus on the extent to which coral taxa differ in the way they respond to climate change, a characteristic that could play an important role in determining the future community structure of coral reefs [21]. One example of the value of such approaches is the analysis of coral bleaching on the reefs of Okinawa in 1998, the results of which allowed corals to be categorized based on whether they survived (i.e., winners) or died (i.e., losers) on the short-term following the disturbance [12]. Analysis of the same community over 14 years revealed discrepancies between short-term and long-term winners, both in the trajectories of changing abundance as well as in the demographic mechanisms underpinning those trajectories [22]. Nonetheless, understanding of the community dynamics of a reef in Okinawa was well served by considering variability in the response of corals to a disturbance. In the first portion of our analysis, we focused on long-term trajectories of change in cover of scleractinian corals at several well-studied locations that represent case-studies for the present study, and sought to determine the extent to which these trajectories differed among coral genera. Ecological data for coral genera at two Caribbean and five Indo-Pacific locations were used to explore changes in absolute and relative coral cover over time.

To support our analysis, data were gathered from nine projects in seven locations where multiple sites have been censused frequently, and together span up to 33 y (1981–2013, although not all studies were of equal length) (Table S1 in File S1). The present authors either collected these data (for the US Virgin Islands, Belize, Kenya, Moorea [2005–2010], and Taiwan), or were directly associated with the agencies that collected the data. Most data came from shallow reefs (≤10 m depth), with some from 17 m depth (Moorea), 11–25 m depth (parts of the US Virgin Islands), or>25 m depth (northern US Virgin Islands) (Table S1 in File S1). For all locations, except the Great Barrier Reef (GBR, Australia), data were averaged across sites on a scale of ~10 km. Data from the GBR posed special challenges because it encompassed a large number of sites representing an extensive area (>150,000 km^2) that would individually have considerable leverage on the analysis. The GBR data were therefore collapsed into three habitats - inshore (11 reefs), mid shelf (18 reefs), and outer shelf (18 reefs) - and pooled among latitudes. For all sites, data were summarized annually as percentage cover of scleractinian corals by genus, using taxonomy as described in recent papers [23,24].

The rate of change in coral cover by genus over the duration of each study was evaluated using least-squares linear regression, with analyses separated for the Caribbean (two locations) and Indo-Pacific (five locations) (Table S2 in File S1). Changes in abundance by genus were expressed on absolute and relative scales, with relative cover determined by dividing the cover for each genus by the total coral cover at the study location at the same time. Regression slopes for each coral taxon (i.e., change in coral cover over time, % y^{-1}) were used in subsequent analyses, and slopes were used regardless of their statistical significance. While the significance of any one slope can be evaluated with P values based on the ratio of the mean sum of squares explained by the regression and unexplained variance, in data compilations such as used in the present study, the least squares estimate of the slope is an unbiased estimate of the true slope that is preferable to the biased slope estimate derived by assuming non-significant slopes have a value of zero. The frequency distribution of these slopes would then be distributed more uniformly than one in which non-significant slopes were set to zero, which would create an ecologically unrealistic gap between actual zero slopes and the larger slopes that are statistically significant. Having calculated the slopes of the relationships between coral cover and time, the frequency distributions of the slopes were tested for skewness using a g_2 test, and for normality using a Kolmogorov-Smirnov (K-S) test. Our objective was to understand how corals were responding to the combined effects of biotic and abiotic disturbances extending over multiple decades, and we did not intend to partition these changes to the effects of individual pulse or press disturbances. Such disturbances might reflect rapidly-acting events such as severe storms, or a large-scale predator outbreaks (for example, the corallivorous seastar *Acanthaster planci*), or chronic effects such as rising seawater temperature or declining seawater pH. Therefore the rates of change in coral cover we report cannot be attributed to specific causal processes and cannot be used to distinguish between the response of corals to recent dramatic and local events versus long-lasting, chronic, and regional-scale events. Instead, our analyses attempted to 'capture' the culmination of the above-mentioned processes and events as time-averaged rates of change in cover of each coral genus.

The changes in cover over time in genus-level coral abundance on absolute and relative scales were used to test the hypotheses that: (1) changes in abundance were independent of overall dominance in the community, meaning that abundant and rare coral genera were likely to share similar trajectories of change; and (2) the covariance between changes in absolute and relative abundance was random with regard to distributing coral genera on these axes, meaning that coral genera were equally likely to have any fate defined by all possible combinations of changes in absolute and relative abundances. These analyses were conducted to evaluate the effect of abundance on change in relative abundance (i.e., a measure of success), and were designed so that they could be completed for fossil data as well as ecological data from our case-study locations. Dominance was evaluated as the rank abundance by genus across the entire data set, and the relationship between abundance and success was evaluated separately for our case-history sites from the Caribbean and Indo-Pacific.

The associations between change in relative abundance and rank dominance were tested using Pearson correlations, first by genus, and second by family. The two analyses were used to evaluate the extent to which the dominance–success relationships were independent of taxon. The analyses of covariance between absolute and relative abundance were used to identify genera that had increased in absolute and relative cover, and to evaluate regional variation in these characteristics. Based on the ratio of the change in absolute to relative coral cover, coral genera were separated into four domains with differing trajectories of change in cover: (1) S-domain corals showed the strongest ecological performance by increasing cover on both absolute and relative scales; (2) M-domain corals showed moderate ecological performance by increasing in absolute cover but declining in relative cover, because other taxa increased faster still; (3) W-domain corals showed weak ecological performance by declining in absolute cover but increasing in relative cover, because other taxa declined in coral cover at a faster rate; and (4) F-domain corals

showing *failing* ecological performance by decreasing in cover on both absolute and relative scales. The covariance between changes in absolute and relative abundance was analyzed with Pearson correlations to test for a random association between variables. Rejection of the null hypothesis would indicate a positive or negative association. Additionally, the distribution of the coral genera among the S-, M-, W- and F- domains was tested for equality using a χ^2 test.

Our compilation generated 78 trajectories of changing coral cover by genus from the US Virgin Islands and Belize in the Caribbean, and 153 trajectories from Moorea, Hawaii, Taiwan, Kenya, and the GBR in the Indo-Pacific (Table S2 in File S1). The frequency distributions of changes in absolute coral cover were leptokurtic (based on the statistic $g_2 \geq 14.9$ [25]), centered on stable cover (i.e., 0% y^{-1}), and departed significantly from a normal distribution (K-S D statistic ≥ 0.790, $P<0.001$) (Fig. 1). At least 46% of trajectories in each region showed declines in cover.

Overall, coral cover represented 17 coral families (Table S2 in File S1), and when analyzed by location and habitat, the changes in absolute cover (mean ± SD) ranged from $-0.573 \pm 0.502\%$ y^{-1} (*Orbicella annularis* complex, n = 9) to $0.004 \pm 0.005\%$ y^{-1} (*Stephanocoenia*, n = 6) for the Caribbean, and from $-0.439 \pm 1.056\%$ y^{-1} (*Montipora*, n = 11) to $0.638 \pm 0.004\%$ y^{-1} (*Dipsastrea*, n = 3) for the Indo-Pacific (all mean ± SD). On a relative scale, the range of changes were $-0.149 \pm 0.319\%$ y^{-1} (*Orbicella*, n = 9) to $0.172 \pm 0.236\%$ y^{-1} (*Porites*, n = 9) for the Caribbean, and ranged from $-0.869 \pm 1.572\%$ y^{-1} (*Acropora*, n = 10) to $1.274 \pm 2.695\%$ y^{-1} (*Dipsastrea*, n = 3) for the Indo-Pacific. Mean coral abundance by genus, averaged across all study locations and dates within each region, ranged from 0.002% (*Mussa*) to 15.538% (*Orbicella*) for the Caribbean, and from 0.002% (*Stylocoeniella*) to 7.031% (*Acropora*) for the Indo-Pacific. The regions were characterized by 16 and 41 coral genera respectively, with these representing 9 families in the Caribbean and 12 families in the Indo-Pacific. Project-wide mean coral covers were used to establish a ranking scheme for the abundance of coral genera and families in each location (Table S4 in File S1). For both regions, abundance (i.e., high dominance) of coral genera was associated with more extreme trajectories of relative abundance (Fig. S1 in File S1), although neither relationship was significant (r≤ |0.080|, df≤39, P≥0.619). Analysis of these relationships at a high taxonomic level was problematic because of the limited replication of genera within each family, although this was accomplished for Meandrinidae (n = 3 genera) and Mussidae (n = 6 genera) in the Caribbean, and Acroporidae (n = 4 genera), Agariciidae (n = 3 genera), Lobophylliidae (n = 5 genera), Merulinidae (n = 12 genera), Pocilloporidae (n = 5 genera) and Poritidae (n = 4 genera) in the Indo-Pacific. For the Caribbean, the trajectories of change in relative cover did not vary between families (U = 16, P = 0.071), and for the Mussidae, they were unrelated to relative dominance in the community (r = −0.556, df = 4, P>0.050) (Fig. S1 in File S1). For the Indo-Pacific, the trajectories of change in relative cover also did not vary among families (H = 9.238, P = 0.100), and for the Acroporidae, Lobophylliidae, Merulinidae, Pocilloporidae, and Poritidae, were unrelated to relative dominance in the community (r<0.692, 10≥df≥2, P>0.050) (Fig. S1 in File S1).

Analyses of the covariation between relative and absolute cover by genus (Table S3 in File S1) revealed relationships (Fig. 2) that departed significantly from random for the Caribbean (r = 0.575, df = 14, P<0.050) and Indo-Pacific (r = 0.707, df = 39, P<0.010). Overall, there were positive relationships between the rates of change in relative and absolute cover, showing that there were more genera in the S- and F- domains than chance alone would

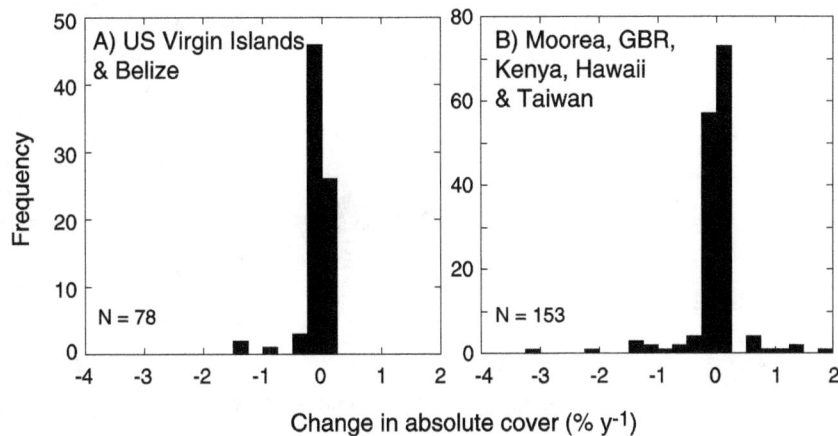

Figure 1. Frequency distributions of changes in absolute coral cover by genus between 1987 and 2012 (two Caribbean sites) and 1987 and 2013 (five Indo-Pacific sites). (A) US Virgin Islands and Belize, and (B) Moorea, Hawaii, Taiwan, the Great Barrier Reef, and Kenya; sample sizes (*n*) shown in all panels. See http://knb.ecoinformatics.org/knb/metacat/nceas.973/knb for raw data, and Table S3 for changes in coral abundances described by these distributions.

predict. This outcome was consistent with the distribution of genera among the F-, M-, S- and W- domains, which departed significantly from an expectation of equal representations for the Caribbean ($\chi^2 = 8.5$, df = 3, $P \leq 0.05$) and Indo-Pacific ($\chi^2 = 20.56$, df = 3, $P \leq 0.001$); in these cases, significance was a result of the relatively large number of genera in F-domain for the Caribbean and in the F- and S- domains for the Indo-Pacific. S- and F-domain corals were found in the Caribbean and in the Indo-Pacific, but ≤ 3 genera were categorized as M- or W-domain corals in either region (Table S3 in File S1). The trajectories of changing coral cover in this two-dimensional space (Fig. 2) were clustered around the origin for the two Caribbean locations – showing that most genera changed little over the study period; trajectories were more divergent for the five Indo-Pacific locations. Additionally, trajectories were consistent in the Caribbean (i.e., SDs based on site replicates for genera were small) and more variable in the Indo-Pacific (i.e., SDs were large), although these differences in SDs were not statistically significant (U = 121, $n_1 = 10$, $n_{2v} = 21$, $P = 0.302$).

Our results from reefs in seven locations revealed stasis in genus-level coral cover (i.e., near-zero absolute change) rather than unequivocal declines [2,3,26], and a high degree of among-genus variation in response to the combined biotic and abiotic drivers of change that have affected these reefs over the last few decades. As has recently been suggested [21], these outcomes are likely to be important in evaluating the composition of coral communities that might persist in these locations in the future. Our analyses have a number of limitations, notably that region-wide inferences are based on only a small number of locations. Further, the temporal trends were constrained by the time period of sampling at each case-study site, which lead to haphazard sampling of biotic and abiotic disturbances affecting coral cover, and to the exclusion of important events that occurred outside of the sampling period. In the Caribbean, our sampling also occurred after large declines in coral cover had already occurred [26], and after the regional near-extirpation of *Acropora* spp. [27].

In summary, our analyses of coral communities at seven locations describe trends that provide higher resolution details of the large-scale declines of coral cover that have been reported elsewhere [3,28], notably demonstrating that absolute and relative cover of many coral genera have at least remained relatively unchanged at least over the last few decades. It is important to

note, however, that our analyses do not imply a 'rosy' future for tropical coral reefs: the future of many coral genera remains uncertain, and relatively few S-domain corals display a strong capacity to increase in cover (Fig. 2). Overall, our genus-level analysis revealed that: (1) many corals have changed little in cover over the last 20–30 y, (2) the absolute and relative cover of a few genera, like *Orbicella* in the US Virgin Islands and Belize and *Pocillopora* and *Acropora* in Moorea, the GBR, Kenya, Hawaii, and Taiwan, have declined rapidly, and (3) the absolute or relative covers of only a few coral genera have increased.

Past

Motivated by the findings from our analysis of extant coral reefs, which revealed evidence of diverse trajectories of change that was dependent on genus (Fig. 2), but that was not related to family-level clades based on molecular trees [24], we asked whether the fossil record contained evidence of similar patterns. To answer this question, we focused on the geological record (6.8 to 0.125 Ma) during the late Miocene to late Pleistocene epochs. During the early Pliocene (5.3 to 3.6 Ma), mean sea-surface temperatures in the Caribbean were elevated 1–2°C — reflecting global mean temperatures 2–4°C higher than present [29] — and atmospheric $p\mathrm{CO}_2$ was as high as 400 ppmv [30]. This was followed by the colder and more thermally variable Pleistocene, which included intermittent glaciation.

To describe the dynamics of coral communities on fossil reefs, we drew on records extending from 0.125 to 6.8 Ma, which encompassed the separation of the Caribbean and Indo-Pacific biogeographic regions. Samples were extracted from outcrop exposures at 70 localities through four Miocene-Pleistocene sequences: Costa Rica [31]; Curacao [32]; Dominican Republic [33]; and Jamaica [34]. The collections comprised ~6,528 specimens and 154 species, deposited at the US National Museum of Natural History (USNM), the University of Iowa (SUI) and the Natural History Museum in Basel, Switzerland (NMB). The specimens were identified to species using a standard set of morphological characters and character states, established in part by comparing morphological and molecular data and as detailed in the Neogene Biota of Tropical America (NMITA) taxonomic database [35]. Localities were grouped into faunules, which are defined as a set of lithologically similar localities from a small

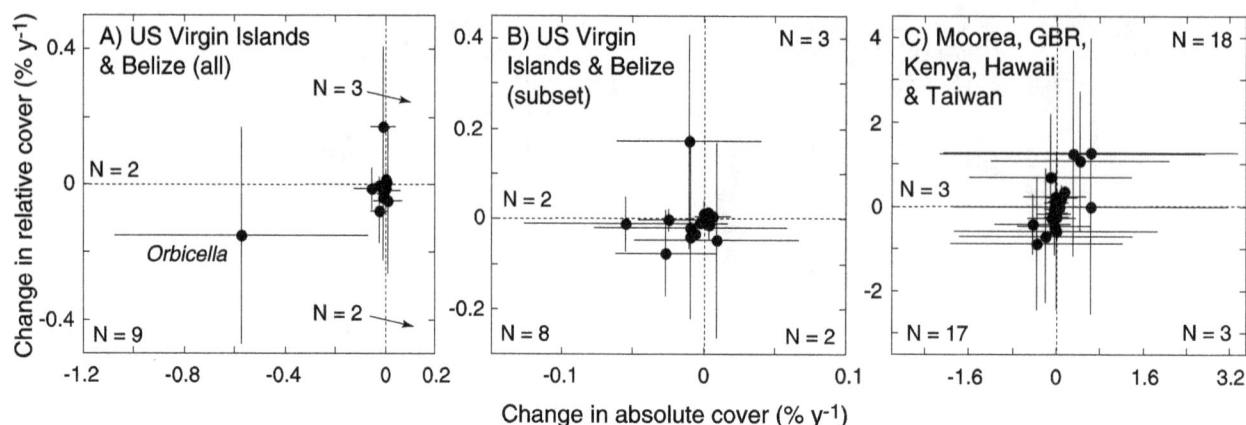

Figure 2. Scatterplots displaying changes in relative abundance (ordinates) and changes in absolute abundance (abscissas) for scleractinian corals in the Caribbean (A and B) and Indo-Pacific (C). Data show mean ± SD (where $n>1$) on both abundance scales for genus [Table S4 in File S1]; N= sample sizes for each quadrant. (A) Caribbean scaled to show *Orbicella*, (B) Caribbean scaled to show taxa other than *Orbicella*, and (C) Indo-Pacific; axes vary among plots. The two axes of these plots define a two-dimensional performance space separated into four quadrats: S-domain corals (top right, increasing on both absolute and relative scales), M-domain corals (top left, increasing in absolute cover, but declining in relative cover), W-domain corals (bottom right, declining in absolute cover, increasing in relative cover), and F-domain corals (bottom left, declining in both absolute and relative cover).

geographical area (usually <1 km) and restricted stratigraphic intervals (usually <20 m). The ages of faunules were assigned by integrating data from high-resolution chronostratigraphic methods that included nanofossil and planktonic foraminiferal biostratigraphy, paleomagnetics, and strontium isotope analyses [35,36] and that generally ranged in accuracy from 0.5–2 Myr. The dataset consisted of counts of specimens belonging to species within each faunule, and is available on the NMITA website (http://nmita. iowa.uiowa.edu/index.htm).

Late Miocene and Pliocene coral reefs supported more diverse coral assemblages than extant coral reefs in the Caribbean, and their taxonomic composition overlapped at the genus level with contemporary Indo-Pacific coral communities. Approximately 80% of the >100 coral species became extinct during the Plio-Pleistocene, and over 60% of the species that now live in the modern Caribbean originated during that period [37]. The relative abundance of fossil genera for each faunule was assessed by dividing the number of specimens of a given genus by the total number of specimens collected within the faunule. The success of the fossil genera was examined by evaluating shifts in relative abundance over time using ordinary linear regression, and the slopes (% Myr^{-1}) were used as a measure of relative success or failure (Table S2 in File S1). In a manner similar to that described above for contemporary reefs, coral genera on fossil reefs were defined as S-corals when their slopes were ≥0% Myr^{-1}, and F-corals when their slopes were <0% Myr^{-1}. Shifts in relative abundances also were scored based on whether genera subsequently became extinct in the Caribbean, and this analysis was used to provide insight into the evolutionary fates of S-corals and F-corals. To evaluate the relationship between overall abundance, trajectories of change, and extinction, the changes in relative abundances (% Myr^{-1}) were plotted against the rank of relative dominance.

Fossil data revealed information on 39 coral genera, which showed normally distributed changes in relative abundances ($D = 0.078$, $n = 39$, $P = 0.971$), ranging from -1.758% Myr^{-1} (*Trachyphyllia*) to 2.650% Myr^{-1} (*Acropora*) (Fig. 3). Of the 39 genera, 15 became extinct and 10 of genera declined in relative abundance over the ~6.7 Myr of the study. The probability of extinction tended to be dependent on the sign (i.e., ≥0 versus <

0% Myr^{-1}) of the slope of abundance on time ($\chi^2 = 3.143$, df = 1, $P = 0.076$). Analysis of the relationship between changes in relative abundance and dominance rank by genus (Fig. S2 in File S1) revealed no significant linear relationship between the two ($r = 0.153$, df = 37, $P = 0.351$), although more abundant taxa displayed larger shifts in relative abundances (both increases and decreases). Interestingly, coral families often were represented by some genera that increased in relative abundance and others that decreased. This suggests that the relative success of coral genera is independent of family affiliation. These results collectively are consistent with other work that focused on the fossil record and past extinction events [38,39]. Extinction rates were higher in species with small colony sizes, but did not differ among species based on colony shape, corallite size, or reproductive mode [38]. However, during the Plio-Pleistocene faunal turnover in response to climate change, ecologically dominant and rare species appear equally susceptible to extinction, making S- and F- corals difficult to predict from these population characteristics [39].

In summary, our analysis of fossil communities demonstrated that coral genera responded differently to Plio-Pleistocene environmental perturbations (Table S3 in File S1). Approximately equal numbers of genera increased (51%) as decreased (49%) in relative abundance, and most extinctions (73%) affected coral genera that decreased in relative abundances. Our results reinforce the point that ecological dominance is not always linked to probability of extinction [39,40]. However, dominance does affect the magnitude of changes in relative abundance, with dominant taxa prone to larger swings in relative abundance. While operating on a much longer time scale, the variable rates at which genera responded to environmental change in the fossil record provides a framework for interpreting differences in response of coral genera that are observed today. Assuming that changes in abundance over ecological time ultimately sum to create similar changes in abundance over geological time, then coral genera that are declining in abundance on contemporary reefs may be destined for extinction, while the relative abundances of genera that are currently dominant in coral communities are likely to change dramatically.

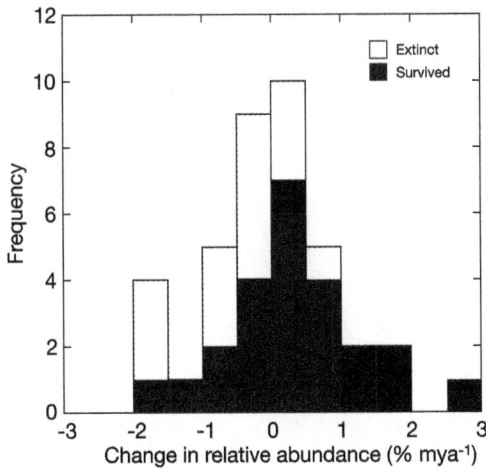

Figure 3. Changes in past coral communities as represented by the fossil record between 6.8 and 0.125 Ma. Changes in relative abundance of genera (% Myr^{-1}) are shown. Histogram shows the changes in relative abundances for tropical taxa that became extinct (open bars) or survived (filled bars) in the Caribbean.

Future

To build on our empirical support for the hypothesis that some corals have the potential to respond favorably to contemporary environmental changes, we developed a model of coral community composition to test the effects of specific coral traits on the trajectories of change in coral cover under increased thermal stress that is expected with future climate change. The traits that we examined included the classic life history traits of reproduction, growth, survivorship (mortality), maturation, and dispersal [41], the traits influencing coral-macroalgal interactions such as overgrowth resistance and herbivore recruitment, and thermal tolerance, which is a composite of physiological and morphological traits as well as symbiont composition. We focus on thermal stress because knowledge of its impact on coral demography allows explicit quantitative modeling. As thermal stress is only one aspect of current and future environmental change, we use a sensitivity analysis to explore the potential impacts of other components of change for which less is known (e.g., ocean acidification). The objective of our model was to evaluate the relative importance of coral life-history traits as predictors of coral cover under future climates [42]. We favored a trait-based model over a taxon-based model because our empirical evidence indicated that ecologically successful genera were scattered across multiple families. Thus, we reasoned that ecological function (i.e., traits) was a better indicator of ecological success than taxonomy, at least when success was gauged by changes in cover occurring over a time interval in which a variety of biotic and abiotic disturbances occurred unpredictably, as in the present analysis. As all of the traits considered affect overall population growth under disturbance conditions, and therefore persistence, there is no *a priori* reason to expect a particular trait to have more influence than another.

Mathematical models are used regularly to explore coral dynamics under general or specific climate change scenarios [e.g., 21,43,44]. Our goal was not to synthesize the existing models or to explicitly forecast the fate of any particular coral taxon. Instead, our intent was to build a generic model of the ecological dynamics of the full suite of different possible corals under expected future environments, where we consider each coral in isolation and use global sensitivity analysis (GSA [45]) to ask which

biological processes and traits are most important in determining coral success. In other words, as a complement to previous analyses that investigate how a specified coral or coral reef community might respond to future disturbance, we sought to understand which coral traits mattered most to that response across reefs. Here, the GSA methodology lets us investigate those responses in the absence of knowledge on appropriate parameter values for the large number of coral species that exist.

Model methods

Our model (Fig. 4A) is a stage-structured, continuous-time compartment model that tracks the proportion cover of coral (recruits and adults treated separately) and macroalgae in multiple patches connected by larval dispersal. In notation, let R_i, A_i, and M_i represent the proportion cover of coral recruits, coral adults, and macroalgae in each patch i of n patches in total. Changes in each of these state variables are given by the differential equations:

$$\frac{dR_i}{dt} = r_A\left((1-\delta)A_i + \frac{\delta}{n}\sum_{j=1}^{n}A_j\right)(1-M_i-A_i-R_i) - aR_i - r_M R_i M_i - d_R R_i$$

$$\frac{dA_i}{dt} = aR_i + gA_i(1-M_i-A_i-R_i) - \beta A_i M_i - d_A A_i$$

$$\frac{dM_i}{dt} = M_i\left(r_M(1-M_i-A_i) + \beta A_i M_i - h_b - h_s\frac{\omega A_i}{1+\omega A_i}\right)$$

where r_A represents coral recruitment (from reproduction by adults), δ is the proportion of dispersing larvae, a quantifies coral maturation, g represents adult coral growth, d_R and d_A represent recruit and adult coral loss of cover due to mortality and/or shrinkage, respectively; r_M is algal growth, h_b is baseline algal mortality (any herbivory and other algal senescence that would occur independent of coral density), h_s is additional algal mortality from recruited herbivores (a rate with the same 1/time units as h_b), ω scales the rate at which adult corals provide habitat for herbivores (units of 1/coral cover, such that the additional herbivory from corals is a saturating function that has a rate of saturation dictated by ω), and $\beta \leq r_M$ is the rate at which algae overgrow adult corals depending on the degree of overgrowth resistance. Note our assumption here that the relative coral versus macroalgal cover affects grazing rate, as suggested in [46,47]. We parameterized the model by using ranges of values based on comparable parameters used in recent models [19,44] that encompassed empirical variation across coral taxa for ecological processes and biological traits (Table S4 in File S1). To focus the analysis on coral characteristics, we fixed parameters related to algal dynamics at single values that were determined similarly.

The model is driven by stochastic thermal anomalies with frequency and intensity drawn from the GFDL 2.1 climate model for 2051–2100 [48]. We compared model output under four IPCC scenarios: "Commit" with zero future greenhouse gas (GHG) emissions; B1 (550 ppm stabilization in 2100) representing large-scale future GHG emission reductions; A1B (700 ppm stabilization in 2100) and A1F1 describing future GHG emissions following "business-as-usual" trajectories, with greater emissions in A1F1. These climate scenarios were chosen to illustrate the range of possible future outcomes rather than to forecast a particular outcome. We applied annual, stochastic disturbances, using random draws from the degree heating months (DHMs) predicted for 2051–2100 [48] with normally distributed spatial variation among patches. Spatially heterogeneous thermal stresses allow for less impacted patches to supply larvae to more impacted patches (the "rescue effect" [49]), and permit the quantitative comparison of larval production and dispersal to reef persistence

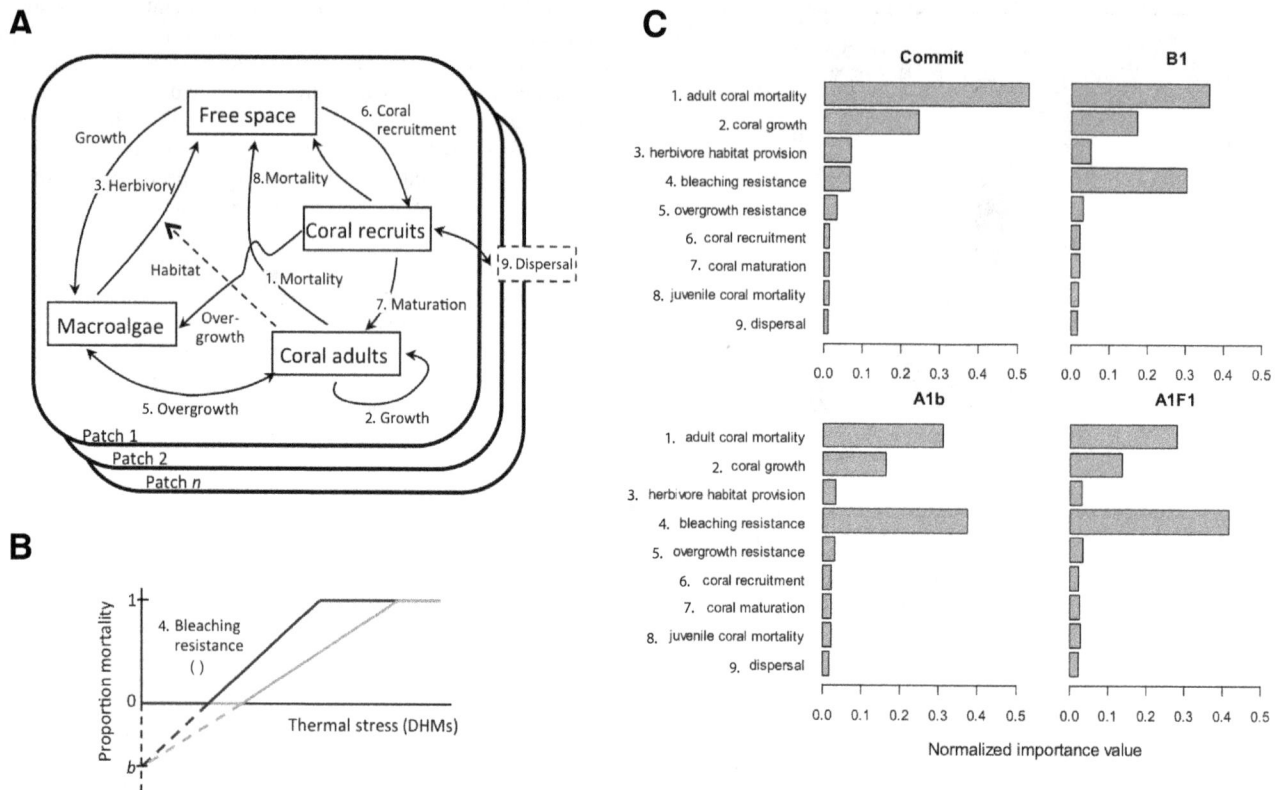

Figure 4. Model design. (A) The model tracks cover for coral recruits, coral adults, macroalgae, and free space across connected patches. Ecological processes (arrows; dashed line = indirect effect; solid lines = flow of energy, matter, organisms) govern changes in cover of state variables, numbered arrows correspond to processes evaluated through sensitivity analyses (C) for impact on community trajectories. Coral mortality subsumes shrinkage and baseline mortality. Parameter values are unboxed except for dispersal. (B) Bleaching resistance as revealed by the translation of degree heating months (DHMs) into mortality as shown by a line with an initial slope η. (C) Normalized importance values [66] for ecological processes for future climate change scenarios [40]. These GSA importance values represent the results of 1,000 simulations for each scenario, each with a unique set of randomly drawn parameter values, analyzed for which parameters were most influential for coral persistence.

relative to other ecological processes. DHMs measure both the magnitude and duration that the temperature is above the average summer maximum, and are a predictor of bleaching events (η) [50]. Here, DHMs are translated into coral (both recruit and adult) mortality using a linear relationship, whose slope and zero-intercept quantify coral resistance to thermal anomalies ("bleaching resistance", Fig. 4B). We ran simulations based on four of the locations from which the empirical data used in this study were obtained (Taiwan, Moorea, St. John, and Belize).

For each location and each climate scenario, we simulated dynamics for 1000 different collections of randomly chosen coral parameters. After discarding 50 y of transients, we recorded the average coral cover across all patches for 20 y. We then pooled simulations across locations and ran a global sensitivity analysis (GSA [45]) for each climate-change scenario to quantify the importance of each process and each trait in determining the coral cover on a multi-decadal scale. In brief, a GSA uses a 'random forest' of regression trees to create a predictive relationship between the randomly drawn coral parameters and the total coral cover under environmental disturbance. An 'importance value' is then calculated for each coral parameter by comparing the prediction accuracy of each tree with the parameter included versus excluded as a predictor. The importance value encompasses all effects of the parameter, including linear and non-linear effects and interactions with other parameters. Here, we used mean-squared error to quantify prediction accuracy, and normalized

importance values to sum to 1 for each climate scenario. Processes and traits with large GSA scores strongly influenced long-term cover, and thus were considered influential in distinguishing coral taxa that function as strong ecological performers (S-corals) under the conditions specified.

We repeated the exercise for different simulation assumptions to determine the robustness of our results. In particular, we varied the number of patches, spatial variance in thermal stress, and the parameter range for coral recruitment. We investigated the recruit and adult cover sensitivity separately, and explored sensitivity with coral dynamics only (no macroalgae). The relative rankings presented are consistent across all of these tests as well as consistent across the four locations (consolidated in the results presented here).

As is inevitable with models, this model excludes more than it includes. For example, the model does not explicitly include other components of predicted future environments (e.g., increasing ocean acidity, increasing frequency and/or intensity hurricanes or typhoons, changing herbivory, nutrient runoff, over fishing, etc.), although we quantify the importance of the demographic processes that other environmental changes are expected to impact. For example, a high importance value for coral growth rates would suggest that total coral cover would be strongly influenced by decreases in growth rates from ocean acidification. The model also does not account for factors such as genetic adaptation of corals or their symbionts, competition or other interspecific interactions among multiple coral species on the same reef, hydrodynamical differences in susceptibility to disturbance based on relative position in the reef, or size-dependent demographics and disturbance susceptibility beyond our two stage classes (recruits and adults). These exclusions are necessary to keep the model tractable and transparent, but provide ample scope for future work.

Model results

Our GSA suggested that the most important ecological processes and biological traits favoring coral persistence (Fig. 4C) were adult coral mortality (i.e., mortality unrelated to bleaching) and adult coral growth (i.e., linear extension), with thermal tolerance becoming increasingly important under severe climate-change scenarios. Further analyses of the model results revealed a strong interaction between adult coral mortality and adult coral growth (Figs. 5, 6). Corals with rapid growth and moderate mortality are likely to persist, as are those with moderate growth and low mortality. Corals with slow growth and high adult mortality are unlikely to persist. The range of growth and mortality that favored persistence depended on the severity of the climate-change scenarios. In the most dire forecast, when temperature anomalies occurred nearly every year, corals needed to have at least two out of the three following traits: low adult mortality, rapid linear extension rates, and high tolerance of upward temperature excursions.

Of the possible two-way combinations between fast growth, low mortality, and high thermal tolerance, a coral having both low adult mortality and high thermal tolerance is most likely given life history trade-offs between growth and longevity, and the tendency for massive coral morphology to be associated with both slow growth and greater thermal tolerance [12]. Thermal tolerance, modeled here in terms of the minimum level of thermal stress before bleaching occurs and the rate of increase in mortality with increasing thermal stress (Fig. 4B), is a property that relates to a variety of characteristics, such as coral morphology [14], identity and flexibility of symbionts [51], genetic variation in thermal tolerance as it relates to adaptive capacity [52], and capacity for coral heterotrophy to supply energy and nutrients during warm thermal anomalies [53]. The importance of coral growth, adult mortality, and thermal tolerance to future coral reef growth parallels the findings from local sensitivity analyses and analogous results in other models that explore coral dynamics under expected future stress [e.g., 44,54–59]. The recurring identification of these three processes across a variety of models with different assumptions reinforces their biological significance.

While the most influential parameters: bleaching resistance, coral growth, and adult coral mortality are intuitively important, what is perhaps more surprising is that these three parameters clearly stand apart from others that relate to processes that have previously been identified as essential to future reef persistence. In particular, our model suggests that larval dispersal and recruitment, juvenile maturation, and competition with macroalgae (i.e., resistance to macroalgal overgrowth and provision of herbivore habitat), while always part of our modeled dynamics, are less influential in comparison with other processes in determining projected coral abundance. These results might seem to contradict those of other models in which recruitment dynamics and coral-macroalgal interactions are important to future coral persistence [19,44,55,57,58]. However, previous studies indicate that the presence versus absence of such dynamics is crucial, whereas our analysis addresses the separate question of whether or not the exact amount of recruitment and coral-macroalgal competition has a major role.

With respect to recruitment, previous studies indicate that larval exchange between multiple patches of coral experiencing different stress levels affects the outcome of coral population projections [56,58]. Therefore, when evaluating coral community dynamics in models such as the one proposed here, it might be more important to incorporate the capacity for dispersal among reefs rather than the exact number of recruits involved in dispersal events. In addition, previous studies highlighting the importance of recruitment to the persistence of coral reefs note its greater influence on short-lived species than on long-lived species [59]. Therefore, if lower mortality and therefore a longer lifespan is central to persistence, then the set of surviving species is exactly the set for which recruitment plays less of a role. Analogously, a study of sea fans that used local sensitivity analysis noted that the observed low sensitivity to recruitment could be due, in part, to longevity [60]. In addition to shorter-lived species, recruitment and relative dispersal might be of greater importance for coral species with pulsed or stochastic recruitment [56], or when stress differs predictably among locations [56,58].

With respect to coral-macroalgal competition, previous studies found coral cover was highly sensitive to competition with macroalgae, particularly where disturbances pushed coral cover close to an unstable threshold between coral-dominated and macroalgal-dominated states [58]. Therefore, parameters related to coral-macroalgal interactions might be particularly important under disturbance scenarios where switches between alternative stable states are likely. However, these parameters might be less influential relative to other processes outside this parameter space, reducing the overall importance, on average, across the broader disturbance regime and across the parameter ranges that drive the location of this threshold. The potential for alternative stable states in coral reefs is strongly debated, e.g., [61,62], and might depend on the degree of anthropogenic degradation as well as on the processes that drive macroalgal growth capacity, which vary substantially within and across the Caribbean and Indo-Pacific [63,64]. Elsewhere we demonstrate that the existence of alternative stable states can have the greatest effect in intermediate bleaching regimes, and change the relative importance of coral life history traits [65]. Lastly, our model makes several (common) simplifying assumptions that may have reduced the apparent importance of algal-coral competition and herbivore habitat provisioning. First, while our model aggregates algal somatic growth and reproduction (similar to, e.g., [19,66]) more detailed simulation models that separate these processes tend to reveal stronger impacts of herbivory on coral-algal competition [67]. Second, skeletons of dead coral may continue to provide habitat for herbivores as they erode [68] and thus promote coral persistence in ways that our model does not capture.

We simulated coral reef community dynamics using climate model output from the fourth IPCC assessment report, released in 2007. Sea surface temperature projections are available from the fifth IPCC assessment report, released in 2013. However, large increases in mean and maximum sea surface temperature in AR5 relative to AR4 caused decreases in simulated coral cover to under 10% in all model run and extirpation in>90% (Fig. S3 in File S1). Despite these major quantitative differences, the AR4 and AR5 model output generate the same qualitative life- history trait importance values (Fig. S4 in File S1). Thus, we chose to illustrate our qualitative results using the AR4 model output because the variance is clearer. The pessimistic nature of the AR5 simulation results should be contextualized in the fact that our model did not consider coral and symbiont acclimation and adaptation, which could increase coral cover or persistence [43,58,69]. Moreover,

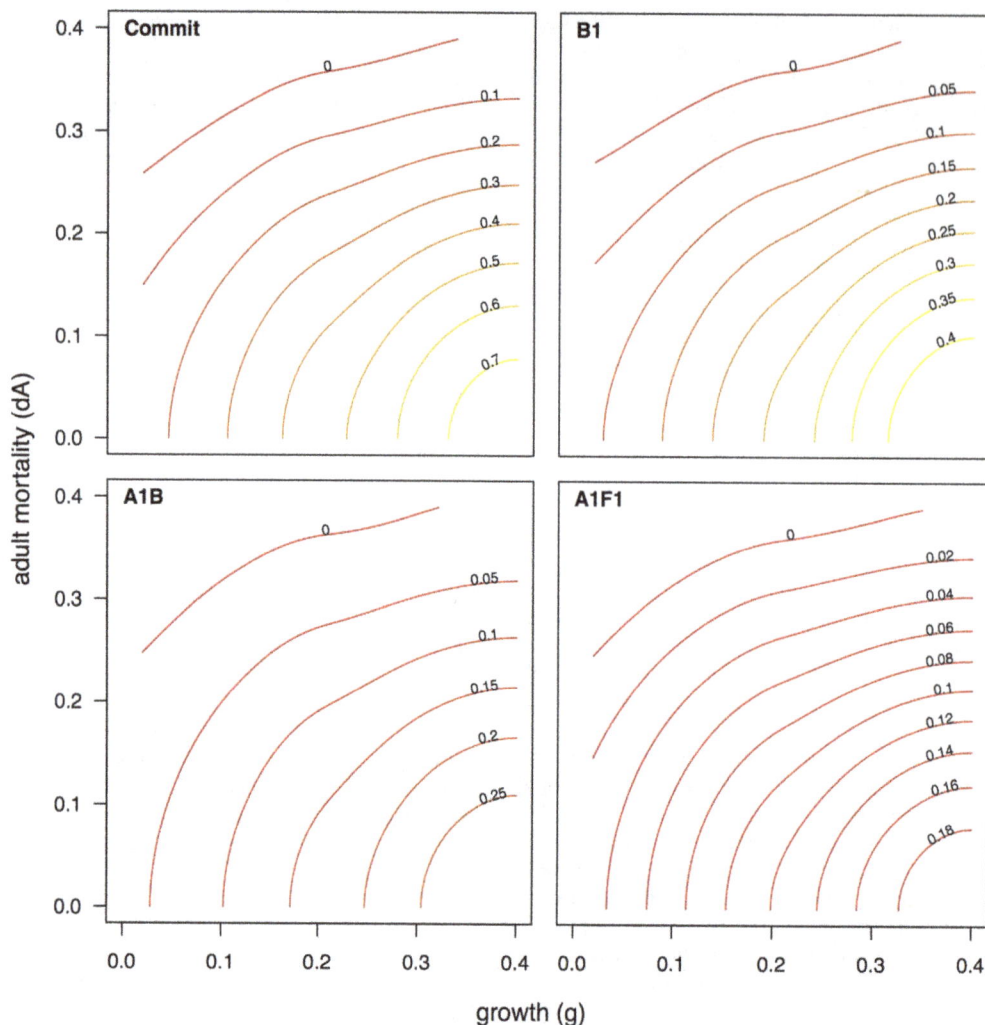

Figure 5. Contour plots for four IPCC climate-change scenarios (Commit, B1, A1b, and A1F1) showing the relationship between adult mortality (dA) versus growth rate (g) with the contours displaying the proportional coral cover. Contour plots were created by using a Loess smoother with average coral cover as the response variable, and g and dA as the predictors. Coral cover is greatest when growth rate is high and mortality is low. Increasingly severe bleaching mortality (i.e., as occurs in IPCC scenario A1F1) results in lower coral cover for a given combination of adult mortality and coral growth.

the AR4 model output may be more accurate than AR5 for specific locations or time periods [70].

Overall, our analysis suggests that future reefs will be populated at low densities by scleractinian corals with at least two of the following traits: high thermal tolerance, fast growth, and low adult mortality (Figs. 4C, 5, 6). The interactive dependence of reef persistence on bleaching resistance, growth, and mortality suggests potential synergistic interaction of future thermal stress with ocean acidification (which affects coral calcification and therefore growth) and temperature-triggered diseases (which affect coral mortality). It is likely that coral species will differ in their susceptibility to the effects of ocean acidification [71] and diseases [72] as much as they will to future bleaching. Therefore, a critical challenge for future research will be to combine the GSA approach used here with models that incorporate multiple interacting impacts (e.g., [55]) to provide an understanding of characteristics most relevant to the persistence of corals under multiple types of disturbances.

Synthesis and Conclusions

We asked whether empirical data from the US Virgin Islands and Belize in the Caribbean, and Moorea, Taiwan, Hawaii, the Great Barrier Reef, and Kenya in the Indo-Pacific, supported a diversity of outcomes for the changing cover of coral genera over the coming century. Our results confirm that coral genera have experienced widespread declines in abundance since the 1980s, with the changes more acute in the Caribbean than in the Indo-Pacific. The results also reveal that some coral genera have maintained or increased their relative and absolute abundance in the study locations over this same period. In our two Caribbean locations, more genera decreased (n = 11) than increased (n = 5) in absolute abundance, but almost equal numbers increased (n = 21) and decreased (n = 20) in the five Indo-Pacific locations. These results suggest that the future of scleractinian corals over the current century will likely include the persistence and perhaps increased abundances of some coral genera, even while others become less common. These results highlight the strong likelihood that changes in the generic-level composition of coral communities will occur in coming decades [73–75]. Changes in generic-level

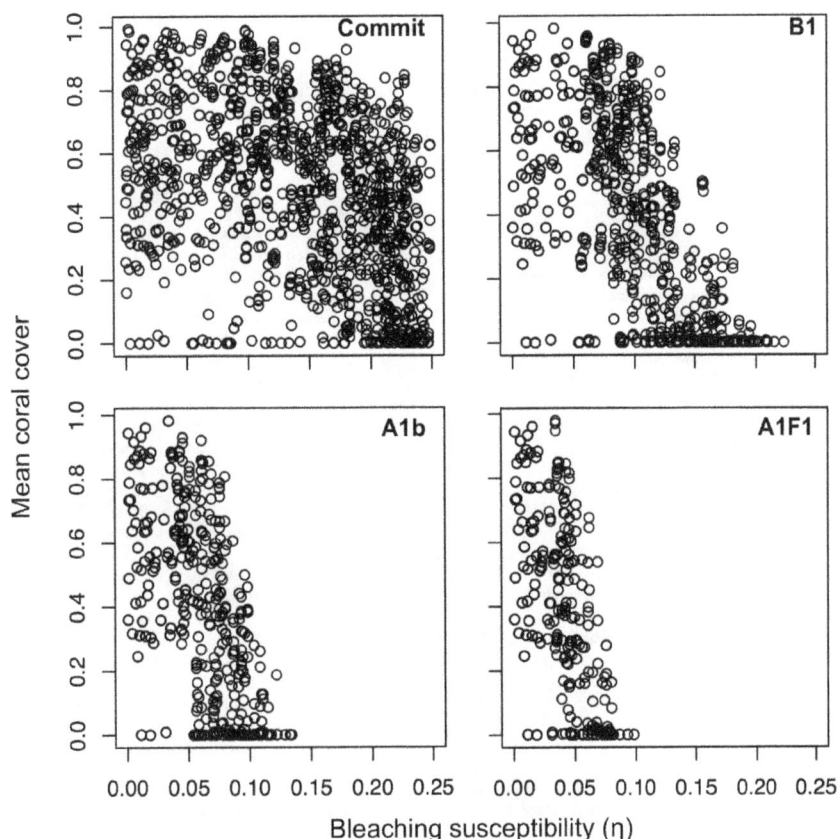

Figure 6. Mean coral cover versus bleaching susceptibility (η) for four IPCC climate-change scenarios (Commit, B1, A1b, and A1F1). Smaller (larger) values of η correspond to decreased (increased) susceptibility to bleaching mortality. In climate scenarios that predict more frequent and more intense thermal anomalies (e.g., A1b and A1F1), corals are only able to persist if their susceptibility to bleaching is below a threshold value. Points with zero coral cover are not shown.

coral-community structure likely will affect the capacity of coral reefs to deliver the goods and services for which they have been well known. The coral genera that will populate future reefs will probably be the strong ecological performers on contemporary reefs, although this conclusion must be tempered by the reality that past performance is an imperfect indicator of future success [76], particularly as the time horizon lengthens. Although the fossil record suggests that community dominance is not a good indicator of long-term persistence, nevertheless long-term increases in relative abundance are generally characteristic of coral genera that persist rather than go extinct, whereas long-term declines tend to culminate in extinction. Translating changes in community structure that have occurred over millennia to implications for coral dynamics over ecological time is problematic, however the fossil record offers a pessimistic prognosis for coral genera that are not currently achieving ecological success (i.e., F-domain corals).

We used modeling to identify traits associated with S-domain corals under a limited subset of future climatic conditions, and this effort highlighted the importance of thermal resistance, rapid growth of reproductively mature corals, and coral longevity. Corals possessing at least two of these three traits are most likely to dominate coral-reef communities in coming decades. It would be desirable to screen extant coral genera for these traits and use the results to couple taxonomic data to the model predictions, but the empirical data to fuel this effort are still sparse [16,77]. While considerable knowledge exists for a small number of coral taxa – like Indo-Pacific *Acropora*, *Stylophora*, and *Pocillopora*, and

Caribbean *Acropora*, *Orbicella*, and *Madracis* – relatively little is known about the functional biology of taxa that are emerging as S-domain corals (Table S3 in File S1). Describing the functional biology of these genera and elucidating how functional traits scale up to determine critical demographic properties, like the intrinsic rate of population increase, must be a research priority for future studies. Such efforts will be critical to guide future empirical work and to predict the functionality of future coral reef ecosystems.

Acknowledgments

This work was conducted as part of the "Tropical coral reefs of the future: modeling ecological outcomes from the analyses of current and historical trends" Working Group (to RDG and PJE), and while RDG was a Center Fellow supported by the National Center for Ecological Analysis and Synthesis. This is a contribution of the Moorea Coral Reef Long Term Ecological Research site, SOEST (number 9192), Hawaii Institute of Marine Biology (number 1599), California State University, Northridge (number 220), the CRIOBE USR 3278 CNRS-EPHE at Moorea, and the Virgin Islands Center for Marine and Environmental Studies (number 118). We thank S. Donner for supplying the DHM data used in the model analysis, and three reviewers that improved an earlier draft of this paper.

Author Contributions

Conceived and designed the experiments: PJE RDG. Performed the experiments: PJE MA MLB IBB AFB RCC NSF TF ECF KG XH LJ JSK TRM JKO MJHO XP HMP TBS MS HS RvW RDG. Analyzed the data: PJE MA MLB IBB AFB RCC NSF TF ECF KG XH LJ JSK TRM JKO MJHO XP HMP TBS MS HS RvW RDG. Contributed reagents/materials/analysis tools: PJE MA MLB IBB AFB RCC NSF TF ECF KG XH LJ JSK TRM JKO MJHO XP HMP TBS MS HS RvW RDG. Wrote the paper: PJE MA MLB IBB AFB RCC NSF TF ECF KG XH LJ JSK TRM JKO MJHO XP HMP TBS MS HS RvW RDG.

References

1. Jackson JBC (1997) Reefs since Columbus. Coral Reefs 16: S23–S32 doi:10.1007/s003380050238
2. Hoegh-Guldberg O, Mumby PJ, Hootem AJ, Steneck RS, Greenfield P, et al. (2007) Coral reefs under rapid climate change and ocean acidification. Science 318: 1737–1742 doi:10.1126/science.1152509
3. Bellwood DR, Hughes TP, Folk C, Nystrom M (2004) Confronting the coral reef crisis. Nature 429: 827–833 doi:10.1038/nature02691
4. Adjeroud M, Michonneau F, Edmunds PJ, Chancerelle Y, Lison de Loma T, et al. (2009) Recurrent disturbances, recovery trajectories, and resilience of coral assemblages on a South Central Pacific reef. Coral Reefs 28: 775–780 doi:10.1007/s00338-009-0515-7
5. Sweatman H, Delean S, Syms C (2011) Assessing loss of coral cover in Australia's Great Barrier Reef over two decades, with implications for longer-term trends. Coral Reefs 30: 521–531 doi:10.1007/s00338-010-0715-1
6. Gilmour JP, Smith LD, Heyward AJ, Baird AH, Pratchett MS (2013) Recovery of an isolated coral reef system following severe disturbance. Science 340: 69–71.
7. Sandin SA, Smith JE, DeMartini EE, Dinsdale EA, Donner SD, et al. (2008) Baselines and degradation of coral reefs in the northern Line Islands. PLoS ONE 3: e1548 doi:10.1371/journal.pone.0001548
8. Adjeroud M, Chancerelle Y, Schrimm M, Perez T, Lecchini D, et al. (2005) Detecting the effects of natural disturbances on coral assemblages in French Polynesia: a decade survey at multiple scales. Aquat Living Resour 18: 111–123 doi:10.1051/alr:2005014
9. McClanahan TR, Ateweberhan M, Muhando CA, Maina J, Mohammed MS (2007) Effects of climate and seawater temperature variation on coral bleaching and mortality. Ecol Monog 77: 503–525 doi: 10.1890/06-1182.1
10. Craig P, Birkeland C, Belliveau S (2001) High temperatures tolerated by a diverse assemblage of shallow-water corals in American Samoa. Coral Reefs 20: 185–189 doi:10.1007/s003380100159
11. Perry CT, Smithers SG, Johnson KG (2009) Long-term coral community records from Lugger Shoal on the terrigenous inner-shelf of the central Great Barrier Reef, Australia. Coral Reefs 28: 941–948 doi:10.1007/s00338-009-0528-2
12. Loya Y, Sakai K, Yamazato K, Nakano Y, Sambali H, et al. (2001) Coral bleaching: the winners and the losers. Ecol Lett 4: 122–131 doi:10.1046/j.1461-0248.2001.00203.x
13. Green DH, Edmunds PJ, Carpenter RC (2008) Increasing relative abundance of *Porites astreoides* on Caribbean reefs mediated by an overall decline in coral cover. Mar Ecol Prog Ser 359: 1–10 doi:10.3354/meps07454
14. Edmunds PJ (2011) Zooplanktivory ameliorates the effects of ocean acidification on the reef coral Porites spp. Limnol Oceangr 56: 2402–2410.
15. Fabricius KE, Langdon C, Uthicke S, Humphrey C, Noonan S, et al. (2011) Losers and winners in coral reefs acclimatized to elevated carbon dioxide concentrations. Nature Clim Change 1: 165–169 doi:10.1038/nclimate1122
16. Darling ES, Alvarez-Filip L, Oliver TA, McClanahan TR, Cote IM (2012) Evaluating life-history strategies of reef corals from species traits. Ecol Lett 15: 1378–1386 doi: 10.1111/j.1461-0248.2012.01861.x
17. Hughes TP, Rodrigues MJ, Bellwood DR, Ceccarelli D, Hoegh-Guldberg O, et al. (2007) Phase shifts, herbivory, and the resilience of coral reefs to climate change. Curr Biol 17: 360–365 doi:10.1016/j.cub.2006.12.049
18. Budd AF, Klaus JS, Johnson KG (2011) Cenozoic diversification and extinction patterns in Caribbean reef corals: A review. Paleontol Soc Papers 17: 79–94.
19. Mumby PJ, Hastings A, Edwards VGW (2007) Thresholds and the resilience of Caribbean coral reefs. Nature 450: 98–101.
20. Bruno JF, Selig ER (2007) Regional decline of coral cover in the Indo-Pacific: timing, extent, and subregional comparisons. PloSONE 2: e711.
21. Pandolfi JM, Connolly SR, Marshall DJ, Cohen AL (2011) Projecting coral reef futures under global warming and ocean acidification. Science 333: 418–422.
22. van Woesik R, Sakai K, Ganase A, Loya Y (2011) Revisiting the winners and the losers a decade after coral bleaching. Mar Ecol Prog Ser 434: 67–76.
23. Fukami H, Chen CA, Budd AF, Collins A, Wallace C, et al. (2008) Mitochondrial and nuclear genes suggest that stony corals are monophyletic but most families of stony corals are not (Order Scleractinia, Class Anthozoa, Phylum Cnidaria). PLoS ONE 3: e3222 doi:10.1371/journal.pone.0003222
24. Budd AF, Fukami H, Smith ND, Knowlton N (2012) Taxonomic classification of the reef coral family Mussidae (Cnidaria: Anthozoa: Scleractinia). Zool J Linn Soc 166: 465–529.
25. Joanes DN, Gill CA (1998) Comparing measures of sample skewness and kurtosis. The Statistician 47: 183–189.
26. Gardner TA, Cote IM, Gill JA, Grant A, Watkins AR (2003) Long-term region-wide declines in Caribbean corals. Science 301: 958–960 doi:10.1126/science.1086050
27. Aronson RB, Precht WF (2001) White-band disease and the changing face of Caribbean coral reefs. Hydrobiologia 460: 25–38.
28. Schutte VGW, Selig ER, Bruno JF (2010) Regional spatio-temporal trends in Caribbean coral reef benthic communities. Mar Ecol Prog Ser 402: 115–122 doi:10.3354/meps08438
29. Dowsett HJ, Robinson MM (2009) Mid-Pliocene equatorial Pacific sea surface temperature reconstruction: a multi-proxy perspective. Phil Trans R Soc A 367: 109–125 doi:10.1098/rsta.2008.0206
30. Pagani M, Zhonghui L, LaRiviere J, Ravelo AC (2010) High Earth-system climate sensitivity determined from Pliocene carbon dioxide concentrations. Nat Geosci 3: 27–30 doi:10.1038/ngeo724
31. Budd AF, Johnson KG, Stemann TA, Tompkins BH (1999) Pliocene to Pleistocene reef coral assemblages in the Limon Group of Costa Rica. In: Collins LS, Coates AG, editors. A paleobiotic survey of the Caribbean faunas from the Neogene of the Isthmus of Panama. Bull Am Paleo 357: 119–158.
32. Budd AF, Petersen RA, McNeill DF (1998) Stepwise faunal change during evolutionary turnover: a case study from the Neogene of Curaçao, Netherlands Antilles. Palaios 13: 167–185.
33. Klaus JS, Budd AF (2003) Comparison of Caribbean coral reef communities before and after Plio-Pleistocene faunal turnover: Analyses of two Dominican Republic reef sequences. Palaios 18: 3–21 doi: 10.1669/0883-1351(2003) 018 <0003:COCCRC> 2.0.CO;2
34. Budd AF, McNeill DF (1998) Zooxanthellate scleractinian corals from the Bowden Shell Bed, SE Jamaica. Contrib Tert Quat Geol 35: 49–65.
35. McNeill DF, Coates AG, Budd AF, Borne PF (2000) Integrated paleontologic and paleomagnetic stratigraphy of the upper Neogene deposits around Limon, Costa Rica: A coastal emergence record of the Central American Isthmus. Geol Soc Am Bull 112: 963–981 doi:10.1130/0016-7606(2000)112<963:IPAPSO> 2.0.CO;2
36. McNeill DF, Klaus JS, Budd AF, Lutz B, Ishman S (2012) Late Neogene chronology and sequence stratigraphy of mixed carbonate-siliciclastic deposits of the Cibao Basin, Dominican Republic. Geol Soc Am Bull 124: 35–58 doi:10.1130/0016-7606(2000)112<963:IPAPSO>2.0.CO;2
37. Budd AF, Johnson KG (1999) Origination preceding extinction during Late Cenozoic turnover of Caribbean reefs. Paleobiology 25: 188–200.
38. Johnson KG, Budd AF, Stemann TA (1995) Extinction selectivity and ecology of Neogene Caribbean reef corals. Paleobiology 21: 52–73.
39. Budd AF, Johnson KG (2001) Contrasting evolutionary patterns in rare and abundant species during Plio-Pleistocene turnover of Caribbean reef corals. In: Jackson JBC, Lidgard S, McKinney FK, editors. Evolutionary patterns: growth, form, and tempo in the fossil record. Chicago, IL: Univ. Chicago Press. pp.295–325.
40. van Woesik R, Franklin EC, O'Leary J, McClanahann TR, Klaus JS, et al. (2012) Hosts of the plio-Pleistocene past reflect modern-day coral vulnerability. Proc Roy Soc 279: 2448–2456
41. Stearns SC (1992) The Evolution of Life Histories. Oxford, UK: Oxford University Press.
42. IPCC (2007) Climate Change 2007: Synthesis Report. Contribution of Working Groups I, II and III to the Fourth Assessment Report of the Intergovernmental Panel on Climate Change [Core Writing Team, Pachauri, R. K and Reisinger, A. (eds.)]. IPCC, Geneva, Switzerland, pp 104.
43. Baskett ML, Gaines SD, Nisbet RM (2009) Symbiont diversity may help coral reefs survive moderate climate change. Ecol Appl 19: 3–17.
44. Fung T, Seymour RM, Johnson CR (2011) Alternative stable states and phase shifts in coral reefs under anthropogenic stress. Ecology, 92: 967–982.
45. Harper EB, Stella JC, Fremier AK (2011) Global sensitivity analysis for complex ecological models: a case study of riparian cottonwood population dynamics. Ecol Appl 21: 1225–1240.
46. Hoey AS, Bellwood DR (2011) Suppression of herbivory by macroalgal density: a critical feedback on coral reefs? Ecol Lett 14: 267–273.
47. Williams I, Polunin N (2001) Large-scale associations between macroalgal cover and grazer biomass on mid-depth reefs in the Caribbean. Coral Reefs 19: 358–366.
48. Donner SD (2009) Coping with commitment: projected thermal stress on coral reefs under different future scenarios. PLoS ONE 4: e5712. Doi: 10.1371/journal. pone.0005712
49. Stephens PA, Sutherland WJ (1999) Consequences of the Allee effect for behavior, ecology and conservation. Trends Ecol Evol 14: 401–405 doi:10.1016/S0169-5347(99)01684-5
50. Donner SD, Skirving WJ, Little CM, Oppenheimer M, Hoegh-Guldberg O (2005) Global assessment of coral bleaching and required rates of adaptation under climate changes. Global Change Biol 11: 2251–2265.
51. Sotka EE, Thacker RW (2005) Do some corals like it hot? Trends Ecol Evol 20: 59–62 doi:10.1016/j.tree.2004.11.015
52. Meyer E, Davies S, Wang S, Willis BL, Abrego D, et al. (2009) Genetic variation in responses to a settlement cue and elevated temperature in the reef-building

coral *Acropora millepora*. Mar Ecol Prog Ser 392: 81–92 doi:10.3354/meps08208

53. Grottoli AG, Rodrigues LJ, Palardy JE (2006) Heterotrophic plasticity and resilience in bleached corals. Nature 440: 1186–1189 doi:10.1038/nature04565

54. Langmead O, Sheppard C (2004) Coral reef community dynamics and disturbance: a simulation model. Ecol Model 175: 271–290.

55. Anthony KRN, Maynard JA, Diaz-Pullido G, Mumby PJ, Marshall PA, et al. (2011) Ocean acidification and warming will lower coral reef resilience. Glob Change Biol 17: 1798–1808.

56. Reigl BM, Purkis SJ (2009) Model of coral population response to accelerated bleaching and mass mortality in a changing climate. Ecol Model 220: 192–208.

57. Fong P, Glynn PW (2000) A regional model to predict coral population dynamics in response to El-Nino southern oscillations. Ecol Appl 10: 842–854.

58. Baskett ML, Nisbet RM, Kappel CV, Mumby PJ, Gaines SD (2010) Conservation management approaches to protecting the capacity for corals to respond to climate change: a theoretical comparison. Global Change Biol 16: 1229–1246 doi: 10.1111/j.1365-2486.2009.02062.x

59. Hughes TP, Tanner JE (2000) Recruitment failure, life histories, and long-term decline of Carribean corals. Ecology 81: 2250–2263.

60. Bruno JF, Ellner SP, Vu I, Kin K, Harvell CD (2011) Impacts of *Aspergillosis* on sea fan coral demography: modeling a moving target. Ecol Monogr 81: 123–139.

61. Zychaluk K., Bruno JF, Clancy D, McClanahan TR, Spencer M (2012) Data-driven models for regional coral-reef dynamics. Ecol Lett 15: 151–158.

62. Mumby PJ, Steneck RS, Hastings A (2013) Evidence for and against the existence of alternate attractors on coral reefs. Oikos 122: 481–491.

63. Dudgeon SR, Aronson RB, Bruno JF, Precht WF (2010) Phase shifts and stable states on coral reefs. Mar Ecol Prog Ser 413: 201–216.

64. Roff G, Mumby PJ (2012) Global disparity in the resilience of coral reefs. Trends Ecol Evol 27: 404–413.

65. Fabina NS, Baskett ML, Gross K (2014) The differential effects of increasing frequency and magnitude of extreme events on coral populations. Ecol Appl (in revision).

66. Blackwood KC, Hastings A, Mumby PJ (2012) The effect of fishing on hysteresis in Caribbean coral reefs. Theoretical Biol 5: 105–114.

67. Mumby PJ (2006) The impacts of exploiting grazers (Scaridae) on the dynamics of Caribbean coral reefs. Ecol Appl 16: 747–769.

68. Bozec YM, Yakob L, Bejarano S, Mumby PJ (2012) Reciprocal facilitation and non-linearity maintain habitat engineering on coral reefs. Oikos 122: 428–440

69. Logan CA, Dunne JP, Eakin CM, Donner SD (2014) Incorporating adaptive responses into future projections of coral bleaching. Global Change Biol 20: 125–139

70. Kumar D, Kodram E, Ganguly AR (2014) Regional and seasonal intercomparison of CMIP3 and CMIP5 climate model ensembles for temperature and precipitation. Climate Dynamics doi 10.1007/s00382-014-2070-3

71. Comeau S, Edmunds PJ, Spindel NB, Carpenter RC (2013) The responses of eight coral reef calcifiers to increasing partial pressure of CO_2 do not exhibit a tipping point. Limnol Oceanogr 58: 388–398.

72. Harvell CD, Kim K, Burkholder JM, Colwell RR, Epstein PR, et al. (1999) Emerging marine diseases – climate links and anthropogenic factors. Science 285: 1505–1510.

73. Burman S, Aronson R, van Woesik R (2013) Homogenization of coral assemblages along the Florida reef tract. Mar Ecol Prog Ser 467: 89–96.

74. Riegl BH, Bruckner AW, Rowlands GP, Purkis SJ, Renaud P (2012) Red sea coral reef trajectories over two decades suggest increasing community homogenization and decline in coral size. PlosONE 7: e38396.

75. Darling ES, McClanahan TR, Cote I (2013) Life histories predict coral community disassembly under multiple stressors. Glob Change Biology 19: 1930–1940.

76. Aronson RB, Macintyre IG, Wapnick CM, O'Neill MW (2004) Phase shifts, alternative states, and the unprecedented convergence of two reef systems. Ecology 85: 1876–1891.

77. Edmunds PJ, Putnam HM, Nisbet RM, Muller EB (2011) Benchmarks in organism performance and their use in comparative analyses. Oecologia 167: 379–390.

18

Prolonged Instability Prior to a Regime Shift

Trisha L. Spanbauer[1]*, **Craig R. Allen[2]**, **David G. Angeler[3]**, **Tarsha Eason[4]**, **Sherilyn C. Fritz[1]**, **Ahjond S. Garmestani[4]**, **Kirsty L. Nash[5]**, **Jeffery R. Stone[6]**

1 Department of Earth and Atmospheric Sciences and School of Biological Sciences, University of Nebraska–Lincoln, Lincoln, Nebraska, United States of America, **2** U.S. Geological Survey, Nebraska Cooperative Fish and Wildlife Research Unit, School of Natural Resources, University of Nebraska–Lincoln, Lincoln, Nebraska, United States of America, **3** Department of Aquatic Sciences and Assessment, Swedish University of Agricultural Sciences, Uppsala, Sweden, **4** Office of Research and Development, National Risk Management Research Laboratory, U.S. Environmental Protection Agency, Cincinnati, Ohio, United States of America, **5** Australian Research Council Centre of Excellence for Coral Reef Studies, James Cook University, Townsville, Queensland, Australia, **6** Department of Earth and Environmental Systems, Indiana State University, Terre Haute, Indiana, United States of America

Abstract

Regime shifts are generally defined as the point of 'abrupt' change in the state of a system. However, a seemingly abrupt transition can be the product of a system reorganization that has been ongoing much longer than is evident in statistical analysis of a single component of the system. Using both univariate and multivariate statistical methods, we tested a long-term high-resolution paleoecological dataset with a known change in species assemblage for a regime shift. Analysis of this dataset with Fisher Information and multivariate time series modeling showed that there was a ~2000 year period of instability prior to the regime shift. This period of instability and the subsequent regime shift coincide with regional climate change, indicating that the system is undergoing extrinsic forcing. Paleoecological records offer a unique opportunity to test tools for the detection of thresholds and stable-states, and thus to examine the long-term stability of ecosystems over periods of multiple millennia.

Editor: John A. D. Aston, University of Cambridge, United Kingdom

Funding: The Nebraska Cooperative Fish and Wildlife Research Unit is jointly supported by a cooperative agreement between the U.S. Geological Survey, the Nebraska Game and Parks Commission, the University of Nebraska–Lincoln, the United States Fish and Wildlife Service, and the Wildlife Management Institute. This work was supported in part by the August T. Larsson Foundation of the Swedish University of Agricultural Sciences and the NSF's Integrative Graduate Education and Research Traineeship (IGERT) program (NSF #0903469) and the Sedimentary Geology & Paleobiology program (NSF #1251678). The funders had no role in study design, data collection and analysis, decision to publish, or preparation of the manuscript.

Competing Interests: The authors have declared that no competing interests exist.

* Email: tspanbauer@unl.edu

Introduction

Ecosystems can undergo regime shifts and reorganize into an alternative state when a critical threshold is exceeded [1–3]. Most quantitative regime shift research has focused on abrupt shifts that have occurred during a period of human observation; this has resulted in a better understanding of how fast variables (e.g. nutrient loading) erode resilience, but it hasn't addressed how slow variables (e.g. long-term changes in climate) can alter ecosystem state. Paleoecological records can provide insight on the frequency and duration of transitions between alternative states in systems that are affected by both fast and slow variables, at timescales not accessible in the observed record.

To test for regime shifts in the paleoecological record, we used a long-term high-resolution sedimentological record from Foy Lake (Montana, USA) that showed abrupt changes in diatom community structure at ~1.3 ka (thousands of years before present, with present defined as AD 1950). Foy Lake (48.1648°N, 1143589°W, 1005 m elevation) is a deep freshwater lake situated in the drought-sensitive Flathead River Basin in the Northern Rocky Mountains [4,5]. Diatom assemblages in this system are sensitive to changes in lake depth driven by changes in effective moisture [6] and represent one metric of ecological resilience. The percent abundances of 109 diatom species were collected from a lake sediment core that was sampled continuously at an interval of every ~5–20 years, yielding a ~7 kyr record of 800 time-steps.

To determine if regimes shifts could be anticipated in this paleoecological data set we (i) plotted several indicators proposed to be early-warning signals of approaching critical thresholds (increasing variance, skewed responses, kurtosis, and the autocorrelation at lag-1) [7] against time, (ii) collapsed the 109 species variables into the system's mean Fisher information (FI) [8], and (iii) used multivariate time series modeling based on canonical ordination [9]. Many of these statistical early-warning signals have been developed based on bifurcation theory, and they have successfully anticipated regime shifts in many [10–13], but not all [14] systems tested. Increasing variance, skewed responses, and kurtosis in time-series data may be indicative of flickering, the rapid alternating between two different states prior to a regime shift [15]. Along with autocorrelation at lag-1, increasing variance in time-series data can be caused by critical slowing down, where a system is slow to recover from minor disturbances as it approaches a critical transition [7]. These univariate metrics can be limited in their utility, because appreciable signals often occur at the onset of the regime shift, which is generally too late to implement effective management actions [16]. Hence, we sought methods (FI and multivariate time series modeling) that more effectively investigate the dynamics of complex multivariate systems. FI, an integrated index based on information theory, declines as it approaches a regime shift, indicating loss of order and increasing variability, and the regime shift is typically identified as a minimum FI value. Afterward, FI will often increase before settling into a new regime [8]. FI has been used to evaluate stability, regime shifts, and resilience in real complex systems, including ecosystems, climate data, urban systems, and nation states [8,17–25]. Multivariate time series modeling, which models the fluctuation of the frequencies of species or groups of species at distinct temporal scales [9], complements the FI approach. Multivariate time series modeling is sensitive to changes in the abundance and occurrence structure of species in the community. It is capable of identifying scale-specific temporal patterns (fluctuations at scales of decades, centuries, and millennia) in the data and therefore permits assessing how transitional and regime dynamics manifest across the modeled time scales. A key advantage of using these two methods with paleoecological data is that neither requires *a priori* knowledge of system structure or dynamics [8–9].

Results and Discussion

Of the indicators used, we found that univariate species-level indicators were weak predictors of regime shifts. Skewness, kurtosis, and critical slowing down showed minor changes in the frequency patterns of some variables. Several species showed increased variance prior to the abrupt change in species composition at ~1.3 ka. However, most of the species provided no warning signal; hence, conclusions about the dynamics of the overall system were unclear (Fig. 1). Since indicators must be computed for each variable (i.e., diatom species) individually, characterization of the overall system is difficult [8]. For example, the variance of two diatom species, *Cymbella cymbiformis* and *Amphora veneta*, showed very different patterns in variance (Fig. 2). The former would be a good candidate for anticipating the transition in community structure in Foy Lake, while variance in the latter species was random in relation to large scale community shifts. While some particular species might serve as a leading indicator of a regime shift in this system, it is impossible, *a priori*, to identify which species might be appropriate to monitor. In addition, an early-warning indicator species that is effective in Foy Lake may not be useful in other systems, because of differences in physical, chemical, or biological variables that affect community interactions. In summary, it was difficult to detect a community-level regime shift from any of the traditional indicators of early-warning signals, because of the multivariate nature of the study system and the univariate capacity of indicators.

Fisher information identified a substantial regime shift in the system prior to the abrupt community change. The mean FI results indicated that the system was in a steady state (regime one) from ~7.0 to ~4.5 ka. This was followed by a ~2 kyr period of instability, before it returned to a steady state (regime two) at ~1.3 ka (Fig. 3). The long period of instability was followed by an abrupt increase in mean FI at ~2 ka denoting a regime shift [23], which preceded the system regaining stability at ~1.3 ka, and, thus, returning to a steady state. Regimes one and two are considered stable states, because there is no overall directional trend in mean FI values during those periods [23]. During the ~2 kyr period of instability, the mean FI decreased steadily, indicating the system was losing dynamic order, and therefore resilience [8]; this slow period of change is a warning of the impending regime shift at ~2.0 ka.

Multivariate time series modeling revealed eight different temporal patterns in the diatom data set that were associated with eight significant canonical axes in the redundancy analysis (RDA) model. Each of these canonical axes reflects a modeled frequency pattern of individual species or groups of species in the diatom data set. The first three canonical axes capture 55% of the variance used to summarize the transitional dynamics and regime shifts (Fig. 4). The first axis explained the most important pattern in the data set (29% of adjusted variance explained); it separated regime two at ~1.7 ka from all prior time points (Fig. 4). Axes two and three, which explain 18% and 8% of the variability, respectively, separated the time series into three periods: the first regime from the beginning of the record to ~4.8 ka, the period of instability that lasts ~2 kyr, and a second regime that begins at ~1.7 ka. The frequency patterns in the three axes generated with RDA showed temporal patterns of change that are not exactly the same as those detected in the FI results, but that are complementary. The areas that differ most are the ages of both the onset of instability and of the regime shift; these differences likely occur because FI is a composite of all species, whereas the multivariate analysis partitions species into groups. RDA axis one is a long time interval that includes both regime one and the subsequent transition period between regimes one and two. The major axis break at the onset of regime two suggests that regime two is the stronger of the two stable-states in the system's history. This interpretation is supported by the higher mean FI and lower standard deviation in FI of the second regime (Fig. 4). This pattern was driven by a sudden shift in the relative abundance of diatoms, marked by the onset in numerical dominance of one species (*Cyclotella bodanica* var. *lemanica*) during the second regime (Fig. 1). The transitional period, delineated by mean FI, is not present in the first axis of the time series analysis (Fig. 4). However, it is evident in subsequent axes and reflects gradual changes in species composition and dominance patterns (Fig. 1, 4).

The regimes, transitional period, and regime shift detected by FI and time series modeling are consistent with ecological and regional climate patterns. Foy Lake was a moderately deep lake with a diverse planktic and benthic flora during regime one. Throughout the period of instability, the lake was much shallower and dominated by a benthic flora, and during the more recent regime two, Foy Lake was a deep lake dominated by *Cyclotella bodanica* var. *lemanica*, a planktic species [26]. It is possible that either intrinsic (e.g. nutrients) or extrinsic (e.g. climate change) drivers, or a combination of both are responsible for the abrupt ecological change [27]. However, synchronous change in multiple climate records from the region suggests that extrinsic drivers are likely the cause of the changes to the diatom community structure at Foy Lake. A pattern of recurrent multi-decadal drought in the Foy Lake region ended abruptly ~4.5 ka [26]; this is at the approximate time that regime one ends and the ~2 kyr period of instability begins. A shift in the dynamics of the climate system is also evident in multiple other mid-continental paleoclimatic records at ~4.2 ka [28]. At ~1.3 ka multiple regional lake records show a synchronous shift in diatom community structure [29], and regular patterns of reoccurring drought returned to the Foy Lake region [30]. Thus, the intervals of recurrent drought on multi-decadal scales coincide with the identified stable regimes in Foy Lake, whereas the onset of the period of instability occurs during a time of persistent severe drought in the mid-continent. There is a lag between the FI identified regime shift and the abrupt change in diatom community structure (from ~2 ka to 1.3 ka). This lag period is coincident with regional synchronous shifts in diatom communities at multiple lakes at ~2.2 ka, ~1.7 ka, and ~1.35 ka. This suggests that emerging from a period of instability may involve several smaller short-lived transitions in ecosystem state before long term stability is achieved.

Paleoenvironmental and paleoecological data provide a vital and fundamental perspective on the long-term functioning of

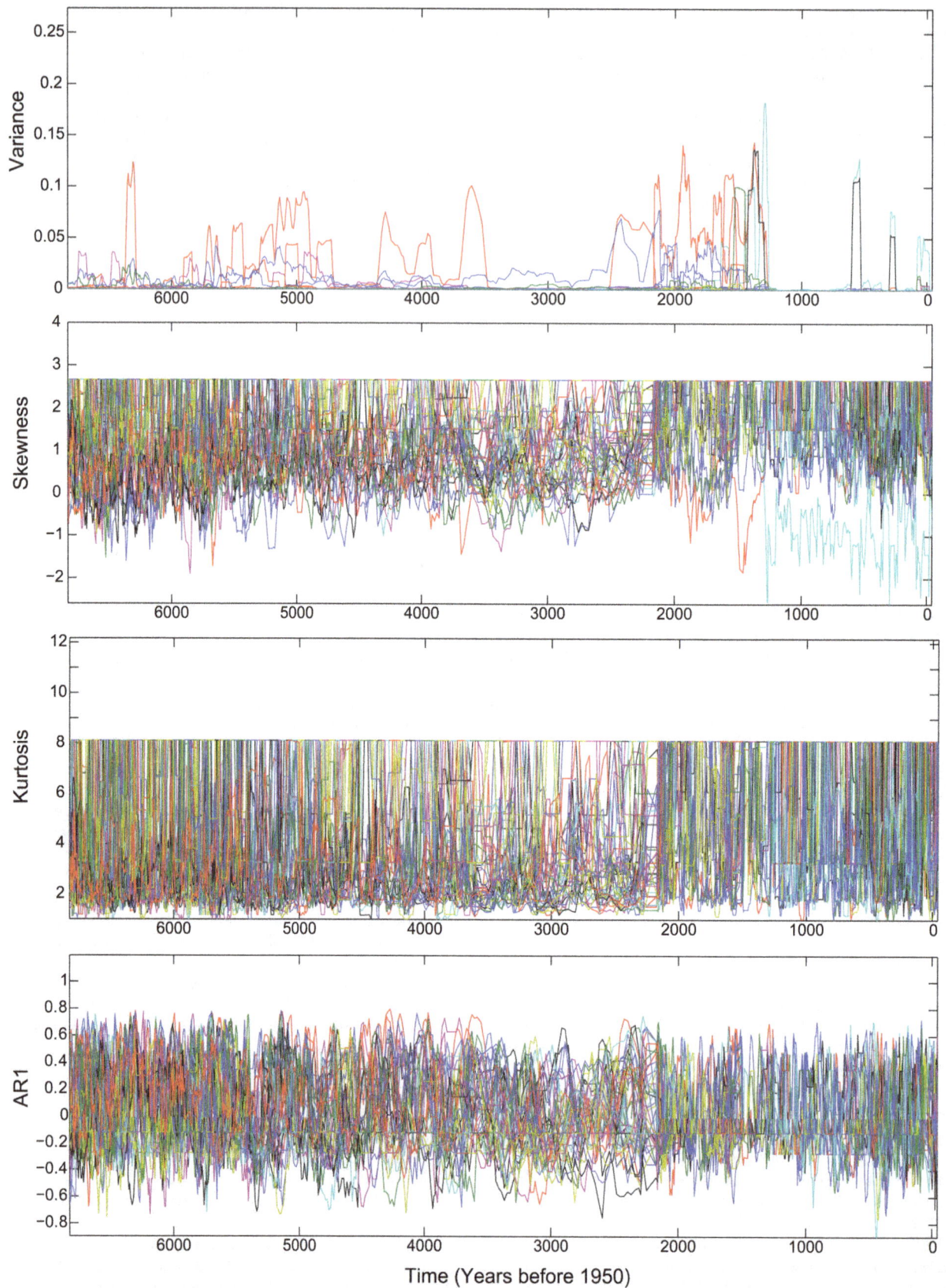

Figure 1. Early warning signals of regime shifts applied to 109 diatom species from Foy Lake. Several populations of species experienced increased variability in the Foy Lake record; this increased variability peaks prior to ~1.3 ka (**A**). Skewness (**B**), kurtosis (**C**), and critical slowing down (**D**) show no clear trends, although, slight frequency changes can be detected at approximately ~4.5 ka and ~2.0 ka.

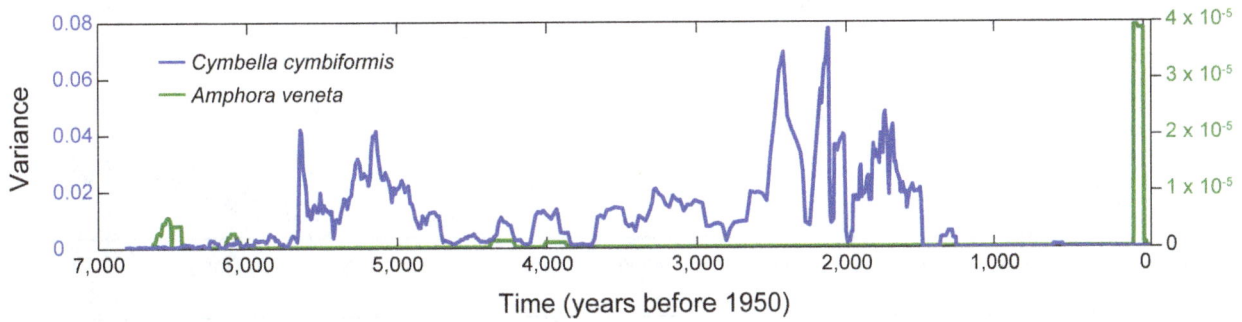

Figure 2. The variance of two diatom species. While *Cymbella cymbiformis* displayed a pattern of increasing variance prior to ~1.3 ka, *Amphora veneta* did not. Conflicting patterns make it difficult to use univariate statistics to characterize the behavior of a complex multivariate system.

complex ecological systems. Here we reveal that climate-driven regime shifts may be infrequent over time in systems not impacted by anthropogenic change, and that transitional periods leading to a regime shift can last a relatively long time (~2.0 kyr). Delayed responses and time lags have been found in other ecosystems [31–33], and these may provide a false sense that the ecosystems are stable, leading to their mismanagement [34]. It is likely that some ecosystems are currently in prolonged periods of instability, whereby they are losing resilience and are exposed to compounding

stresses driven by anthropogenic change. Moreover, when disturbance is large-scale and long-term, some early-warning signals may occur long before the system settles into an alternate stable regime, and the lag between signal and stability may be difficult to predict. Here we suggest effective tools (FI and multivariate time series modeling) to detect and understand changes in those ecosystems that are susceptible to periods of prolonged instability prior to regime shifts.

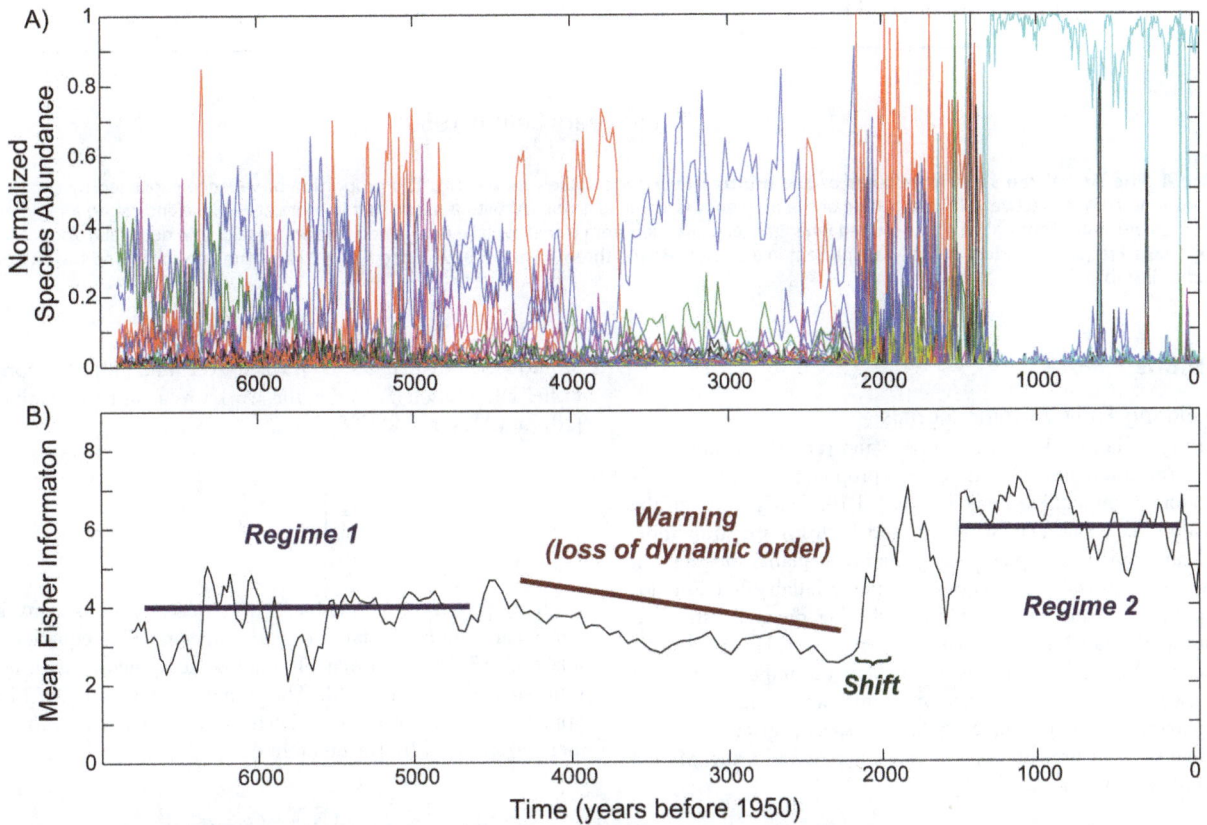

Figure 3. Normalized diatom species abundance for all species (A) and mean Fisher Information (B) for Foy Lake. Prior to ~4.5 ka the system had episodic fluctuations in species composition and mean FI, but the overall mean of the FI is unchanging; this suggests that this period was a stable regime characterized by high variability. At ~4.5 ka species evenness decreases, and the system begins a ~2 kyr gradual decrease in mean FI. Decreases in FI suggest the system is becoming unstable; as instability increases resilience decreases, warning of a possible regime shift. The system was in this unstable transitional period until ~2 ka, but it did not attain a new stable-state until ~1.3 ka.

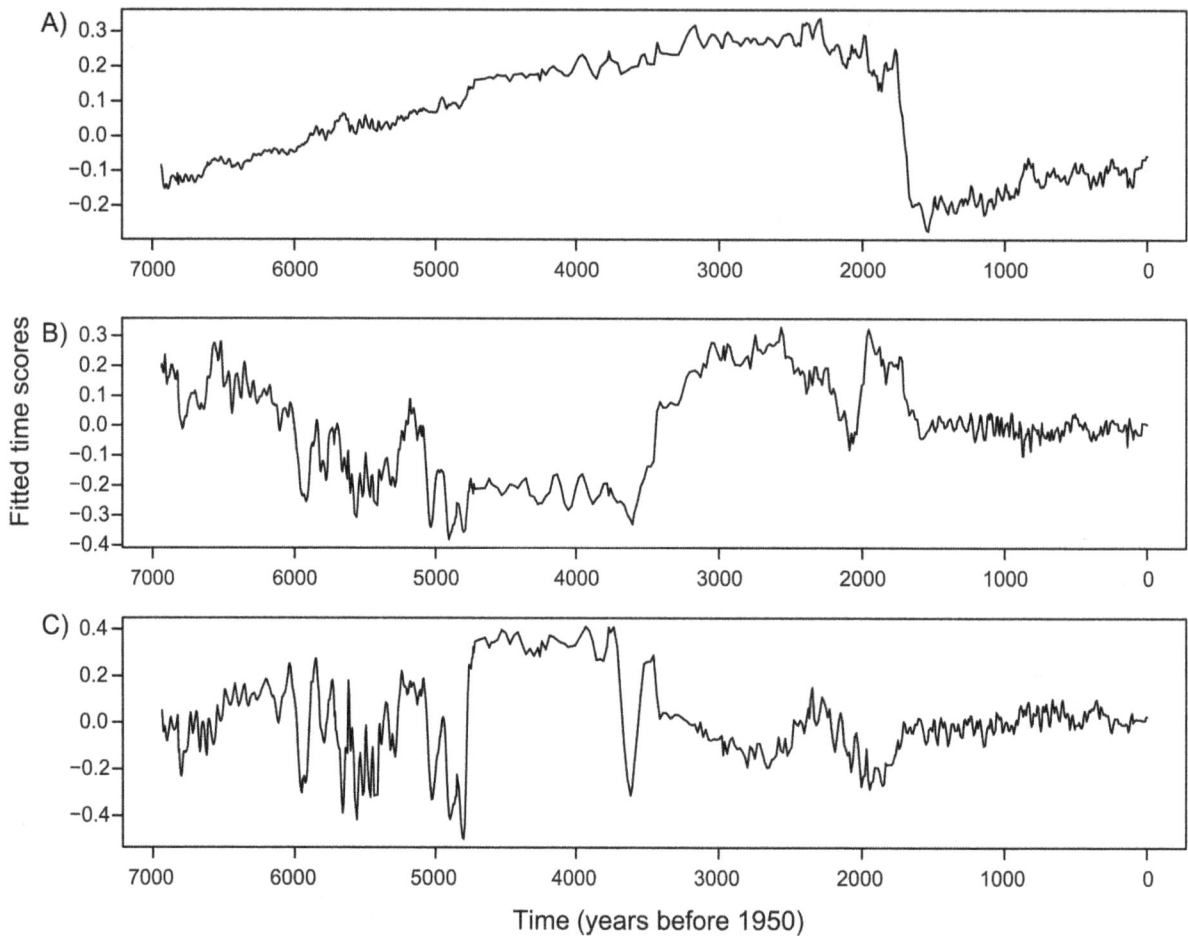

Figure 4. The first three significant axes of the multivariate time-series modeling. The proportion of variance explained by each axis is 29%, 18%, and 8% respectively. The amplitude of the frequency is low in axis one (**A**) with a major shift in score at ~1.7 ka, indicating a regime shift to an alternate state. This regime shift occurred when the lake changed from a shallow lake dominated by benthic taxa to a deep lake dominated by planktic taxa. Frequency pattern changes are present in axes two (**B**) and three (**C**) at ~4.8–5 ka and ~2–1.3 ka, at the beginning and the end of the period of instability.

Methods

Calculating Early Warning Signals

Rising variance, skewness, kurtosis, and critical slowing down are statistical measures that have been proposed and employed as indicators of impending regime shifts [11,16,35–37]. Most of the indicators (i.e., variance, skewness, and kurtosis) are straightforward and can be computed using readily available functions in standard statistical packages (e.g., the Matlab function for computing variance is var). Critical slowing down is estimated by using the lag-1 autocorrelation coefficient [11]. Hence, the autocorrelation function is used to calculate this indicator. For the sake of consistency, all statistical indicators were computed from the percent abundance of each diatom species given the same window size (10 time steps) over the 7 kyr record using Matlab (Release 2012a, Mathworks, Inc.).

Fisher Information

Fisher information (FI) can be used to evaluate the dynamic order of ecosystems, including regimes and regime shifts [23–24]. Unlike early warning signals, FI characterizes changes in complex system dynamics as a function of patterns in underlying variables (e.g., species abundances of diatoms) by collapsing their behavior

of into an index that can be tracked over time [23]. The form of Fisher information (I) used in this work was adapted by Fath et al. [18] and Mayer et al. [38].

$$I = \int \frac{1}{p(s)} \left[\frac{dp(s)}{ds} \right]^2 ds \qquad (1)$$

Here, p(s) is the probability of observing the system in a particular condition (state, s) of the system. This equation was adapted [17–18], such that FI could be computed analytically or estimated numerically [23]. The numerical approach of FI was applied in this work and calculated from the following expression (derived in detail by Karunanithi et al. [23]:

$$FI = 4 \sum_{i=1}^{n} [q_i - q_{i+1}]^2 \qquad (2)$$

where, the probability density p(s) is replaced by its amplitude ($q^2(s) \equiv p(s)$) in order to minimize calculation errors from very small p(s). From Equation 1, note that FI is proportional to the change in the probability of a system being in a particular state ($p(s)$) versus the change in state ds, i.e., $FI \propto dp/ds$ [18].

Calculating FI

Assessing the dynamic changes in system behavior requires gathering information on its condition (state) through time; hence, measurable variables (x_i) are selected such that a time varying system has a trajectory in a phase space defined by the n-dimensions of its system variables and time. Each point in the trajectory is defined by specific values for each of the n variables (i.e., a point at time i is defined as $[x_1(t_i), x_2(t_i) x_3(t_i)...x_n(t_i)]$). Since uncertainty is inherent in any measurement and system variables may fluctuate within a stable state, a state is defined as a region bounded by a level of uncertainty (or size of states for each dimension (i): $sost_i$), such that if $|x_i(t_i) - x_i(t_j)| \le sost_i$ is true for all variables then the two points at times i and j are indistinguishable and are identified as being in the same state of the system. There are a number of methods for defining the sizes of states parameter, but the general idea is to assign a level of uncertainty for each variable based on either knowledge of the system (empirically or theoretically) or estimation [24]. Given this conceptual description of systems and states, the probability $p(s)$ of a system being in a particular state (s) can be estimated by counting the number of observational data points that meet the size of states criteria. Using this approach, it is possible to designate all possible states of the system over time.

The basic steps employed to compute FI for the Foy Lake system were as follows: (1) the diatom time series data (consisting of the relative abundances of all 109 species) were divided into a sequence of overlapping time windows with each window containing 10 time steps. Since the goal is to capture changing patterns, there is no particular window size that must be used to compute FI. The window size is set based on available data and from empirical studies, it is typically at least eight time steps [39]. (2) The level of uncertainty was estimated by searching for the window (i.e., 10 time steps) within the diatom time series with the least amount of variability. The standard deviation for each species was then calculated to establish the size of states criteria and bin points into states. (3) The binned points were then used to generate probability densities, $p(s)$, for each state. (4) Equation 2 was used to compute a unique FI for each window resulting in a sequence of FI values over time. The algorithm for computing FI was coded in Matlab (Release 2012a, Mathworks, Inc). Additional details of the FI derivation, calculation methodology, and computer code can be found in [23,39].

Interpreting FI

Assessing system behavior using FI is based on the fundamental idea that different regimes (set of system conditions) exhibit different degrees of dynamic order [23]. In practical terms, a regime fluctuates within a range of variation, such that the overall condition does not change from one observation to another. Hence, the resulting FI is non-zero and remains relatively stable through time. Steadily decreasing FI signifies loss of dynamic order and resilience of a regime and provides warning of an impending regime shift. A decrease in FI between two stable dynamic regimes denotes a regime shift [8,23]. This shift point is typically identified as a minimum FI value after which FI will often increase. While steadily rising FI is indicative of increasing dynamic order, it denotes a shift to a new regime, only if the increase is followed by a new stable regime (i.e., period in which $d\langle FI \rangle/dt \approx 0$). Note that there is no guarantee that the latter regime is more desirable than the former, i.e., while the condition of the system may be stable, the system could have organized into a less desirable regime (e.g., eutrophic lake). Hence, FI affords the ability to assess the stability of a system, not the quality of its condition [25]. Further evaluation of the underlying variables is required to determine whether the system state is desirable.

Multivariate time series modeling

To assess patterns and scales of diatom fluctuations, we constructed time series models based on redundancy analysis (RDA) [9], and used temporal variables extracted by PCNM (Principal Coordinates of Neighbor Matrices) analysis [40–41]. Briefly, the PCNM analysis converts the linear time vector that comprises the sampling frequency and length of the study period into a set of orthogonal temporal variables. In our study, the time vector consisted of 800 time steps during the 7 kyr study period. The PCNM analysis yielded 517 variables with sine-wave properties from the conversion of the linear time vector. Each PCNM variable corresponds to a specific temporal frequency in the diatom dynamics. That is, the first PCNM variable models the longest temporal frequency while the subsequent variables capture temporal variability from longer to increasingly shorter fluctuation frequencies in the community data over the study period. We constructed a parsimonious RDA model for diatom community dynamics by running a forward selection on the 517 PCNM variables.

The RDA retains significant PCNM variables, and these are linearly combined to extract temporal patterns from the Hellinger-transformed species matrices [42]; that is, the RDA identifies species with similar temporal patterns in the species × time matrix and uses their temporal patterns to calculate a modeled species group trend for these species based on linearly combined PCNMs. The significance of the temporal patterns of all modeled fluctuation patterns of species groups revealed by the RDA is tested by means of permutation tests. The RDA relates each modeled temporal fluctuation pattern with a significant canonical axis. The R software generates linear combination (lc) score plots, which visually present the modeled fit of temporal patterns of species groups that are associated with each canonical axis. Because the canonical axes are orthogonal (independent from each other), one can assess the number of temporal scales at which community dynamics unfold. All relevant steps in the time series analysis are carried out using the "quickPCNM" function in R 2.15.0 (R Development Core Team).

Supporting Information

Dataset S1   Percent abundances of diatom species from Foy Lake calculated relative to the total number of diatom valves counted in each sample. Time steps with no diatom data, due to poor preservation, were removed from the dataset. Time steps 301–312 were averaged for these analyses, because they were assigned the same age, as per the age model.

Acknowledgments

We thank two anonymous reviewers for comments that greatly enhanced this manuscript. Data reported in this paper are archived at the National Climatic Data Center, Data Contribution Series # 2008-070. Any use of trade, firm, or product names is for descriptive purposes only and does not imply endorsement by the U.S. Government. The views expressed in this paper are those of the authors and do not necessarily represent the views or policies of the U.S. Environmental Protection Agency.

Author Contributions

TLS led the writing of the manuscript with contribution from all of the authors. Conceived and designed the experiments: TLS CRA DGA TE. Performed the experiments: JRS. Analyzed the data: DGA TE KLN TLS CRA SCF ASG. Wrote the paper: TLS DGA TE.

References

1. Scheffer M, Carpenter SR, Foley JA, Folke C, Walker B (2001) Catastrophic shifts in ecosystems. Nature 413: 591–596.

2. Scheffer M, Carpenter SR (2003) Catastrophic regime shifts in ecosystems: linking theory to observation. Trends Ecol Evol 18: 648–656.

3. Folke C, Carpenter S, Walker B, Scheffer M, Elmqvist T, et al. (2004) Regime shifts, resilience, and biodiversity in ecosystem management. Annu Rev Ecol Evol Syst. 35: 557–581.

4. McCabe GJ, Palecki MA, Betancourt JL (2004) Pacific and Atlantic Ocean influences on multidecadal drought frequency in the United States. Proc Natl Acad Sci USA 101: 4136–4141.

5. Pederson GT, Fagre DB, Gray ST, Graumlich LJ (2004) Decadal-scale climate drivers for glacial dynamics in Glacier National Park, Montana, USA. Geophys Res Lett 31: doi:10.1029/2004GL019770.

6. Stone JR, Fritz SC (2004) Three-dimensional modeling of lacustrine diatom habitat areas: Improving paleolimnological interpretation of planktic:benthic ratios. Limnol Oceanogr 49: 1540–1548.

7. Scheffer M, Carpenter SR, Lenton TM, Bascompte J, Brock W, et al. (2012) Anticipating critical transitions. Science 338: 344–348.

8. Eason T, Garmestani A, Cabezas H (2014) Managing for resilience: early detection of catastrophic shifts in complex systems. Clean Techn Environl Policy, 16: 773–783.

9. Angeler DG, Viedma O, Moreno J (2009) Statistical performance and information content of time lag analysis and redundancy analysis in time series modeling. Ecology 90: 3245–3257.

10. Carpenter SR, Brock WA (2006) Rising variance: a leading indicator of ecological transition. Ecol Lett 9: 311–318.

11. Dakos V, Scheffer M, Van Nes E, Brovkin V, Petoukhov V, et al. (2008) Slowing down as an early warning signal for abrupt climate change. Proc Natl Acad Sci USA 105: 14308–14312.

12. Drake JM, Griffen BD (2010) Early warning signals of extinction in deteriorating environments. Nature 467: 456–459.

13. Dai L, Vorselen D, Korolev KS, Gore J (2012) Generic indicators for loss of resilience before a tipping point leading to population collapse. Science 336: 1175–1177.

14. Hastings A, Wysham DB (2010) Regime shifts in ecological systems can occur with no warning. Ecol Lett 13: 464–472.

15. Dakos V, Carpenter SR, Brock WA, Ellison AM, Guttal V, et al. (2012) Methods for Detecting Early Warnings of Critical Transitions in Time Series Illustrated Using Simulated Ecological Data. PLoS ONE 7: e41010. doi: 10.1371/journal.pone.0041010.

16. Biggs R, Carpenter SR, Brock WA (2009) Turning back from the brink: detecting an impending regime shift in time to avert it. Proc Natl Acad Sci USA 106: 826–831.

17. Mayer AL, Pawlowski CW, Cabezas H (2006) Fisher information and dynamic regime changes in ecological systems. Ecol Model 195: 72–82.

18. Fath BD, Cabezas H, Pawlowski CW (2003) Regime changes in ecological systems: an information theory approach. J Theor Biol 222: 517–530.

19. Fath BD, Cabezas H (2004) Exergy and Fisher information as ecological indices. Ecol Model 174: 25–35.

20. Cabezas H, Pawlowski CW, Mayer AL, Hoagland NT (2005) Sustainable systems theory: ecological and other aspects. J Clean Prod 13: 455–467.

21. Rico-Ramirez V, Reyes-Mendoza PA, Ortiz-Cruz JA (2010) Fisher Information on the performance of dynamic systems. Ind Eng Chem Res 49: 1812–1821.

22. Shastri Y, Diwekar U, Cabezas H, Williamson J (2008) Is sustainability achievable? Exploring the limits of sustainability with model systems. Environ Sci Technol 42: 6710–6716.

23. Karunanithi AT, Cabezas H, Frieden BR, Pawlowski CW (2008) Detection and Assessment of ecosystem regime shifts from Fisher information. Ecol Soc 13: 22 URL: http://www.ecologyandsociety.org/vol13/iss1/art22/.

24. Eason T, Cabezas H (2012) Evaluating the sustainability of a regional system using Fisher information, San Luis Basin, Colorado. J Environ Manage 94: 41–49.

25. Eason T, Garmestani AS (2012) Cross-scale dynamics of a regional urban system through time. Region et Developpement 36: 55–77.

26. Stone JR, Fritz SC (2006) Multidecadal drought and Holocene climate instability in the Rocky Mountains. Geology 34: 409–412.

27. Williams JW, Blois JL, Shuman BN (2011) Extrinsic and intrinsic forcing of abrupt ecological change: case studies from the late Quaternary. J Ecology 99: 664–667.

28. Booth RK, Jackson ST, Forman SL, Kutzbach JE, Bettis III EA, et al. (2005) A severe centennial-scale drought in mid-continental North America 4200 years ago and apparent global linkages. The Holocene 15: 321–328.

29. Bracht-Flyr B, Fritz SC (2012) Synchronous climatic change inferred from diatom records in four western Montana lakes in the U.S. Rocky Mountains. Quat Res 77: 456–467.

30. Stevens LR, Stone JR, Campbell J, Fritz SC (2006) A 2200-yr record of hydrologic variability from Foy Lake, Montana, USA, inferred from diatom and geochemical data. Quat Res 65: 264–274.

31. Frank KT, Petrie B, Fisher JAD, Leggett WC (2011) Transient dynamics of an altered large marine ecosystem. Nature 477: 86–89.

32. Krauss J, Bommarco R, Guardiola M, Heikkinen RK, Helm A, et al. (2010) Habitat fragmentation causes immediate and time delayed biodiversity loss at different trophic levels. Ecol Lett 13: 597–605.

33. Menéndez R, González Megías A, Hill JK, Braschler B, Willis SC, et al. (2006) Species richness changes lag behind climate change. Proc R Soc London Ser B 273: 1465–1470.

34. Hughes TP, Linares C, Dakos V, van de Leemput IA, van Nes EH (2013) Living dangerously on borrowed time during slow, unrecognized regime shifts. Trends Ecol Evol 28: 149–155.

35. Brock WA, Carpenter SR (2006) Variance as a leading indicator of regime shift in ecosystem services. Ecol Soc 11: 9 URL: http://www.ecologyandsociety.org/vol11/iss2/art9.

36. van Nes EH, Scheffer M (2007) Slow recovery from perturbations as a generic indicator of a nearby catastrophic shift. Am Nat 169: 738–747.

37. Scheffer M, Bascompte J, Brock W, Brovkin V, Carpenter S, et al. (2009) Early-warning signals for critical transitions. Nature 461: 53–59.

38. Mayer AL, Pawlowski CW, Fath BD, Cabezas H (2007) In: Frieden BR, Gatenby RA, editors. Exploratory Data Analysis Using Fisher Information. London: Springer-Verlag. pp. 217–244.

39. Cabezas H, Eason T (2010) Fisher information and order. In: Heberling MT, Hopton ME, editors. San Luis Basin Sustainability Metrics Project: A Methodology for Assessing Regional Sustainability. US EPA Report: EPA/600/R-10/182. pp. 163–222.

40. Borcard D, Legendre P (2002) All-scale spatial analysis of ecological data by means of principal coordinates of neighbour matrices. Ecol Model 153: 51–68.

41. Borcard D, Legendre P, Avois-Jacquet C, Tuomisto H (2004) Dissecting the spatial structure of ecological data at multiple scales. Ecology 85: 1826–1832.

42. Legendre P, Gallagher ED (2001) Ecologically meaningful transformations for ordination of species data. Oecologia 129: 271–280.

19

Dissecting the Space-Time Structure of Tree-Ring Datasets Using the Partial Triadic Analysis

Jean-Pierre Rossi[1]*, Maxime Nardin[2], Martin Godefroid[1], Manuela Ruiz-Diaz[3], Anne-Sophie Sergent[4], Alejandro Martinez-Meier[5], Luc Pâques[2], Philippe Rozenberg[2]

1 Unité Mixte de Recherche 1062 Centre de Biologie pour la Gestion des Populations, Institut National Recherche Agronomique, Centre de Montpellier, France, **2** Unité de Recherche 588 Amélioration Génétique et Physiologie Forestières, Institut National Recherche Agronomique, Centre d'Orléans, France, **3** Parque Tecnológico Misiones, Universidad Nacional de Misiones, Misiones, Argentina, **4** Consejo Nacional de Investigaciones Científicas y Técnicas, Bariloche, Argentina, **5** Grupo de Ecologia Forestal, Instituto Nacional de Tecnología Agropecuaria, Estación Experimental Bariloche (INTA EEA Bariloche), Bariloche, Argentina

Abstract

Tree-ring datasets are used in a variety of circumstances, including archeology, climatology, forest ecology, and wood technology. These data are based on microdensity profiles and consist of a set of tree-ring descriptors, such as ring width or early/latewood density, measured for a set of individual trees. Because successive rings correspond to successive years, the resulting dataset is a *ring variables* × *trees* × *time* datacube. Multivariate statistical analyses, such as principal component analysis, have been widely used for extracting worthwhile information from ring datasets, but they typically address two-way matrices, such as *ring variables* × *trees* or *ring variables* × *time*. Here, we explore the potential of the partial triadic analysis (PTA), a multivariate method dedicated to the analysis of three-way datasets, to apprehend the space-time structure of tree-ring datasets. We analyzed a set of 11 tree-ring descriptors measured in 149 georeferenced individuals of European larch (*Larix decidua* Miller) during the period of 1967–2007. The processing of densitometry profiles led to a set of ring descriptors for each tree and for each year from 1967–2007. The resulting three-way data table was subjected to two distinct analyses in order to explore i) the temporal evolution of spatial structures and ii) the spatial structure of temporal dynamics. We report the presence of a spatial structure common to the different years, highlighting the inter-individual variability of the ring descriptors at the stand scale. We found a temporal trajectory common to the trees that could be separated into a high and low frequency signal, corresponding to inter-annual variations possibly related to defoliation events and a long-term trend possibly related to climate change. We conclude that PTA is a powerful tool to unravel and hierarchize the different sources of variation within tree-ring datasets.

Editor: Jean Thioulouse, CNRS - Université Lyon 1, France

Funding: This research was funded by the French ministère de l'agriculture and the French ministère de l'écologie. The funders had no role in study design, data collection and analysis, decision to publish, or preparation of the manuscript.

Competing Interests: The authors have declared that no competing interests exist.

* Email: rossi@supagro.inra.fr

Introduction

Tree-ring datasets are widely used to reconstruct histories of disturbance events and forest dynamics [1–3], infer large-scale patterns of climate variation (dendrochronology) [4–8], assess trends in tree growth and forest management options [9–11], and regulate wood production and wood quality by controlling site, silviculture, and genetics. Tree-ring data based on microdensity profiles are collected in stems of a set of individual trees, which contains a number of successive annual rings [12] related to the age of the tree, since a new ring is added each year. The most evident structure in a temperate tree ring, especially in conifers, is the earlywood-latewood succession. The light-colored, low-density earlywood is the first part of the ring, formed at the beginning of the growing season (spring and early summer), when temperature is mild, soil water content is high, and the photoperiod is increasing. The darker, higher-density latewood forms during the second part of the growing season (summer and early autumn), when temperature is higher, soil water content is lower, and the photoperiod is decreasing. Earlywood and latewood width and density are variable, and transition from earlywood to latewood is

more or less gradual, affected by species, genetics, tree age, and environment, including climatic variation from the first part to the second part of the growing season. Ring width, earlywood width, latewood width, earlywood density, and latewood density are frequently used to describe a single ring [13]. A basic microdensity table for a single annual ring is a two-way matrix containing as many lines as the number of trees under study and as many columns as the number of variables used to describe each annual ring.

A tree-ring dataset is a three-way dataset of the form *ring variables* × *trees* × *time*. These datasets are often considered two-way matrices, such as *ring variables* × *trees* or *ring variables* × *time*, and subsequently analyzed by multivariate analyses such as principal component analysis (PCA) [14]. Unfortunately, this strategy provides only an incomplete picture of the multivariate space-time variation within the aforementioned datacube [15]. The recent decades have experienced the development of various tools to explore and interpret three- or higher way structure of the data. Popular multiway methods include models from the PARAFAC [16] and Tucker [17] families and alternative models such as the family of STATIS methods [18,19]. The STATIS methods were introduced by L'Hermier des Plantes [20] and Robert and Escoufier [21] and developed by various authors [22,23]. Partial triadic analysis is one of the simplest STATIS methods. It is derived from the triadic analysis [17] and was introduced in ecology under the name of "triadic analysis" by Thioulouse and Chessel [15]. Kroonenberg [24] further renamed the method "partial triadic analysis", emphasizing the difference between the original triadic analysis. The PTA is an exploratory multivariable technique based on PCA. In the case of a spatial structure with repeated measurements, it allows for depicting of temporal variability of the multivariable spatial structure and/or the spatial structure of the temporal trajectories. The PTA has been used in a vast array of situations and disciplines including limnology [25–28], marine ecology [29], soil ecology [30–34], landscape ecology [35], hydrology [36], and food science [37], among others. To our knowledge, however, the potential of PTA has never been assessed in the framework of tree-ring data analysis.

The objective of this study was to show how this technique could be used to explore the variability in a forest tree stand (spatial structure), where each tree is described using several annual tree rings (repeated measurements). This dataset was derived from a set of 149 neighboring, georeferenced trees, constituting a small forest stand. One increment core was collected from each tree. A 41-year, annual-ring time series was obtained from each increment core. Each ring microdensity profile was described by means of 11 ring variables. Although not involved in PTA calculations per se, a large climate dataset was available [38] and could be used to explore possible correlations between average annual minimum and maximum temperatures (1967–2007) and some PTA outputs. We explored our data from two points of view: the analysis depicting the temporal evolution of spatial structures and that portraying the spatial structure of temporal dynamics.

Statistical Background

The theoretical background of PTA is available in various publications [15,24,25,28], and readers are referred to these publications for a formal presentation of the method. We will focus here on an application in the context of tree-ring data analysis and will solely provide an overview of the statistical background. The PTA is designed to analyze the realizations of a set of random variables measured in the same individuals (trees in this case) at different sampling occasions (years of formation of the annual rings). It is based on PCA [14,39] and processes a three-way table consisting of a data matrix with three subscripts (X_{ijt}) that stand for trees, descriptors, and dates (Figure 1). The PTA searches for structures that are stable across a set of two-way tables derived from X_{ijt}. This can be considered in two ways: either the focus is the *trees × ring descriptors* or the *dates × ring descriptors* tables (Figure 1). The first strategy highlights the temporal variability of ring microdensity profile spatial structures (Figure 1A), while the second indicates the spatial structure of temporal trajectories (Figure 1B). This paper reports both of these complementary points of view.

The PTA involves three steps: the interstructure, compromise, and intrastructure analyses [15,23]. Readers are referred to previous work [23] for a formal definition of these terms. The goal of the interstructure is to make a typology of the tables. If we consider the *trees × descriptors* two-way tables, the interstructure yields a typology of the dates (Figure 1A). In that case, the typology is based on the analysis of the *trees × ring descriptors* tables taken as the individuals of PCA [23]. Data preprocessing is an important step that should be considered carefully [40]. The

two-way tables X (either *trees × ring descriptors* or *dates × ring descriptors*) were centered and scaled in order to remove the differences among ring descriptors due to different measurement units or scales without altering the differences between trees or sampling dates. The mean of the correlation coefficients for similar variable *j* between X_k and X_l defines the vectorial correlation coefficient R between these tables. The R coefficient ranges from −1 to +1, since it is the mean of a set of correlation coefficients. The date typology is obtained from the non-centered PCA of the $t × t$ matrix of the inter-date R coefficients [23] (Figure 1A).

The second step of the PTA consists of analyzing the compromise table, which is derived from the positive eigenvectors of the PCA of the interstructure (Figure 1). It contains the factorial coordinates of the trees (dates) for each microdensity profile descriptor (see [25] for a graphical representation). The compromise table is a two-way table summarizing the initial three-way datacube and is analyzed by means of PCA to depict the multivariate structure common to all tables. If we focus on the temporal variability of the *trees × ring descriptors* two-way table, the compromise table will consist of a *trees × ring descriptors* two-way table. In this example, it will encapsulate the multivariable spatial structure common to dates (Figure 1A and Figure 1 in [30]).

The last step of PTA is called the intrastructure [15]. It consists of projecting the initial two-way matrices as complementary tables upon the axis of the PCA of the compromise. This allows assessing which table fits (or does not fit) the structure encapsulated in the compromise. Again, if we consider analyzing the trees × *ring descriptors* two-way tables along dates, the intrastructure provides a picture of the departure of the spatial structure observed at each date from the spatial structure common to all sampling occasions (Figure 1).

Materials and Methods

Ethics statements

This study was approved by the National des Forêts and allowed by the municipality of Villar-Saint-Pancrace (Hautes-Alpes, France). This survey did not involve endangered or protected species.

Site and species

The study site is located close to Villar-Saint-Pancrace (44°52′N, 6°41′E; Hautes-Alpes, France) in the French Alps. The sampling site used in this study is one of four plots of an altitudinal gradient extending from 1350–2300 m above sea level (asl). This experimental site was sampled from 2008–2012 with the goal of studying the adaptation of larch to climate. The altitude of the sloping survey plot ranges from 1640–1683 m, and the vegetation is a continuous natural population of European larch (*Larix decidua* Miller). Increment cores were collected at breast height from 149 trees and used then to study spatial and temporal relationships between annual ring characteristics and climate. Genetic markers were also used to investigate genetic diversity and local adaptations. We focused on the period of 1967–2007, during which the total annual precipitation ranged from 425.5–1078.2 mm in Briançon (44°53′N, 6°38′E), where the nearest "Météo-France" climate station is located. The mean annual temperature, mean annual temperature of the coldest month, mean annual temperature of the warmest month, and mean number of frosty days were 6.32°C, −1.49°C, 15.37°C, and 170, respectively, during the 1967–2007 period. The soil is a colluviosol type. More details are available in [38].

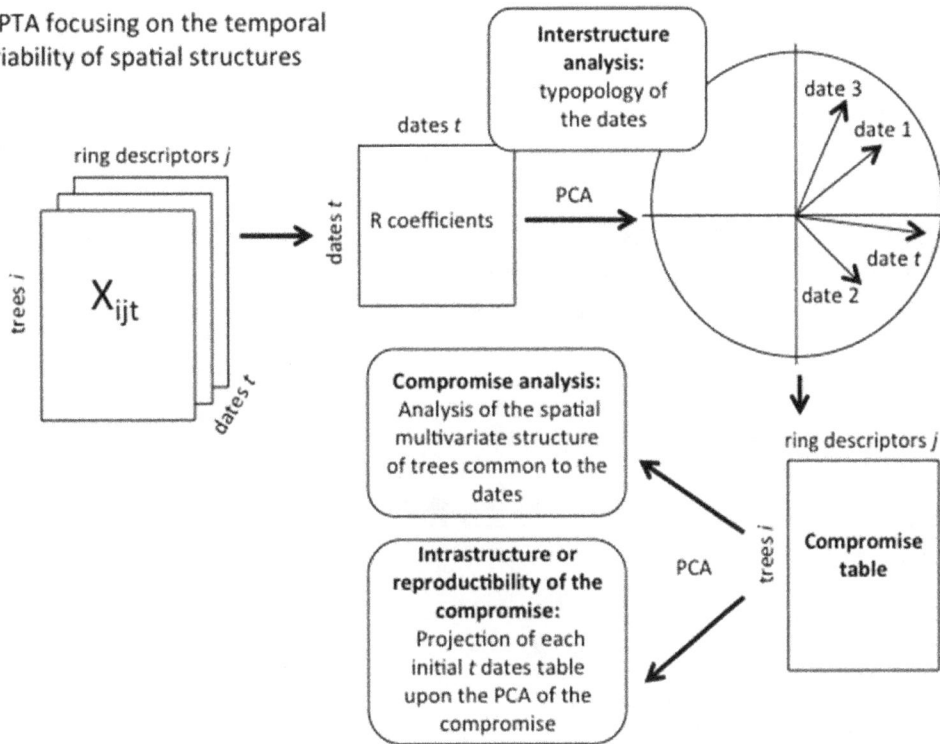

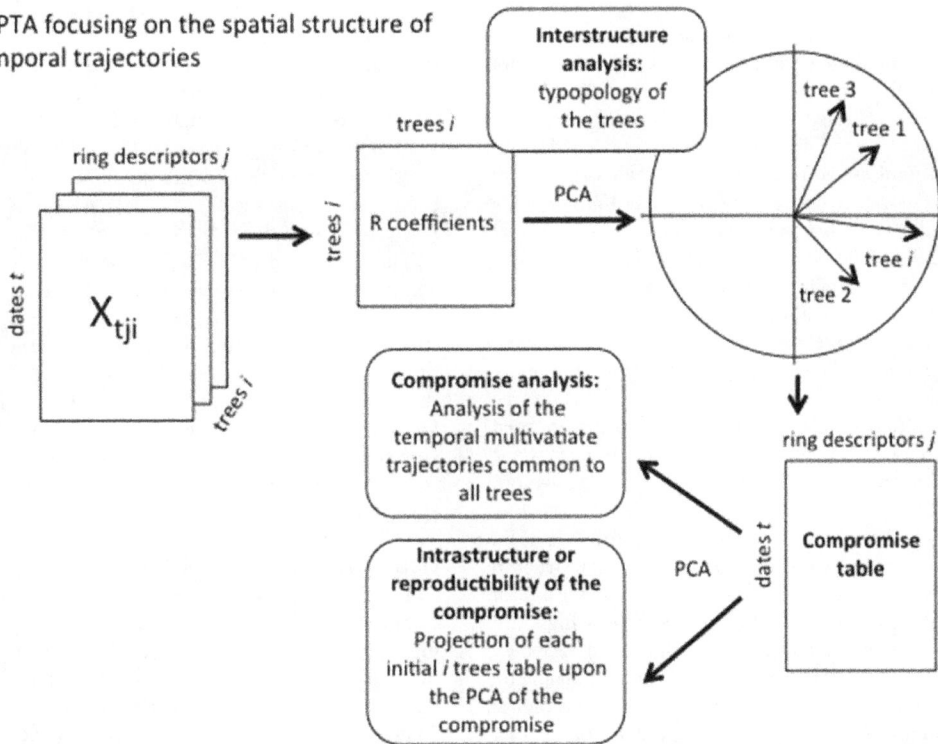

Figure 1. The partial triadic analysis is designed to analyze the realizations of a set of random variables (ring descriptors) measured on a set of points (trees) at different sampling occasions (dates). This corresponds to a three-way table with three subscripts (X_{ijt}) standing for trees, descriptors, and dates, respectively. A given dataset can be analyzed from two complementary viewpoints: seeking for either the temporal evolution of spatial structures (1A) or the spatial structure of temporal dynamics (1B).

Table 1. Ring variables used in the study.

Abbreviation	Definition	Unit
RW	Ring width: the length (or width) of the annual ring along the radius	mm
LW	Latewood percentage: the length of the latewood part of the annual ring along the radius divided by the ring width (RW). Earlywood percentage is 1 - LW.	mm
RD	Mean of the ring microdensity profile	g/dm^3
ED	Mean of the earlywood part of the ring microdensity profile	g/dm^3
LD	Mean of the latewood part of the ring microdensity profile	g/dm^3
MID	Minimum ring density	g/dm^3
MAD	Maximum ring density	g/dm^3
Co	Density contrast: maximum ring density minus minimum ring density	g/dm^3
RSD	Standard deviation of the ring microdensity profile	g/dm^3
ESD	Standard deviation of the earlywood part of the ring microdensity profile	g/dm^3
LSD	Standard deviation of the latewood part of the ring microdensity profile	g/dm^3

Sampling and measurements

The area of the plot was 5815 m^2 and featured 149 trees homogeneously distributed in space with an inter-individual separating distance ranging from 1.1–112.4 m. The stand density was 354 trees/ha, the mean tree age was approximately 150 years, and the average tree height was about 27 m. Increment cores collected at breast height (see details below) provided annual ring data for each tree from 1967–2007 (i.e., 41 successive years [38]).

The numbers and characteristics of the annual rings were estimated using a microdensitometry approach, which also allowed estimating the age of each tree. Pith-to-bark radial increment cores were collected at breast height using a 5.5-mm Pressler increment borer following a constant north-south orientation. The samples were dried to 12% water content and X-rayed [41,42]. The X-ray films were scanned at 4000 dpi. The microdensity profiles were obtained using the software WIND-ENDRO (Windendro 2008e, Regent Instruments Canada, Inc.) [42]. The microdensity profiles were cross-dated (Interdat.exe version 1.1, Dupouey J-L, unpublished work), and the number of rings in each increment core was counted. The ring variables (Table 1) were measured following methods described previously [43]. Most variables are conventional ones, but others rarely employed (i.e., the three standard deviation variables) were used. The majority of these descriptors are based on the earlywood-latewood model, which divides the ring into two successive and contrasting parts [44].

The ring variables describe some aspects of the structure of the whole annual ring. The earlywood variables describe the structure of the wood formed during the first part of the growing season when the temperature is mild, the soil water content is high, and the photoperiod is increasing. The latewood variables describe the structure of the second part of the ring, which is formed during the second part of the growing season when the temperature is higher, the soil water content is lower, and the photoperiod is decreasing. The emphasis given to ring variables differs according to the discipline and the objective of the study. For example, earlywood is known to be by far the most conductive part of the ring; thus, earlywood density is important for sap conduction. Latewood maximum density is related to summer temperature and is used by dendrochronologists to reconstruct past climates. Latewood proportion and latewood density have been found to be strongly related to wood mechanical properties, such as the modulus of elasticity, and can be used to estimate wood value.

The ring data used in this study are provided as supporting information (Dataset S1) and their change from 1967 to 2007 is shown in Figure S1.

Data analysis

The ontological tree-ring age has potentially strong effects on tree-ring width and wood properties [45]. We explored this effect with our time series running from 1967–2007. In these series, tree age ranged from approximately 100–250 yr. As a consequence, the annual ring time series covered a 41-yr period corresponding to cambial ages ranging from approximately 50–250 yr. Because the effect of cambial age upon ring variables is significant for ages <30 yr, we did not observe such an effect in our dataset [38].

All data analyses and graphics preparations were performed using the R statistical software package [46]. The PTA was performed using the R package ade4 [47].

The score of the trees upon the first axis of the PCA of the compromise table was analyzed by means of the Moran's I autocorrelogram. We assessed the presence of a significant spatial structure i.e. a significant departure from randomness, using 1000 random permutations following [48]. The global significance of the correlogram was statistically assessed using Holm's correction for multiple testing as described in Legendre and Legendre [14]. Correlograms were computed using the R package ncf [49].

Results

Depicting the temporal evolution of spatial structures

The first PTA described here aimed at depicting the temporal evolution of spatial structures (Figure 1A). In that case, the interstructure analysis allowed for weighing the dates, thus providing a compromise matrix where more weight is given to dates exhibiting similar spatial structures. As a consequence, the compromise picks up spatial structures (i.e., a spatial typology common to the sampling dates) (Figure 1A). The intrastructure of this PTA will assess the reproducibility of the former compromise, which attempts to assess which dates fit the global spatial structure summarized in the compromise (model) (and which do not) and to identify the ring variables that might explain these patterns (Figure 1A).

Interstructure. The interstructure table was a 41×41 square matrix containing the vectorial correlation (R) between the *trees* × ring *descriptors* sub-matrices (Figure 1A). The PCA of this matrix

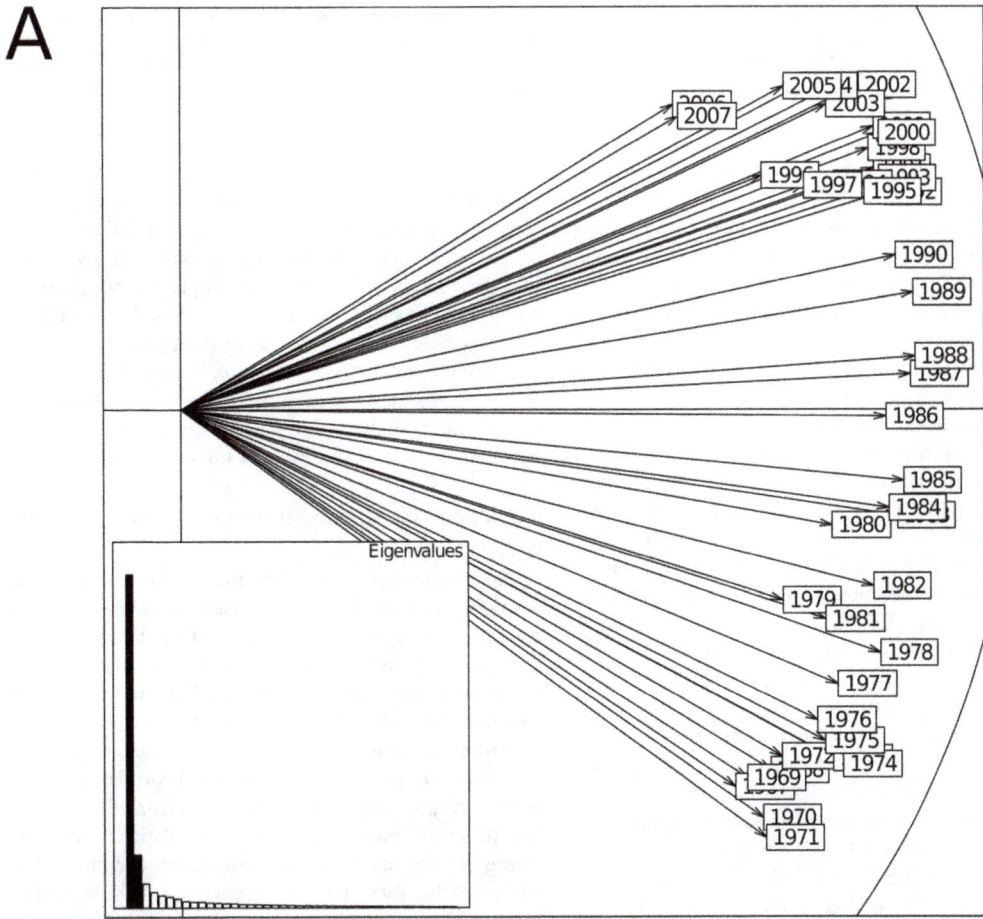

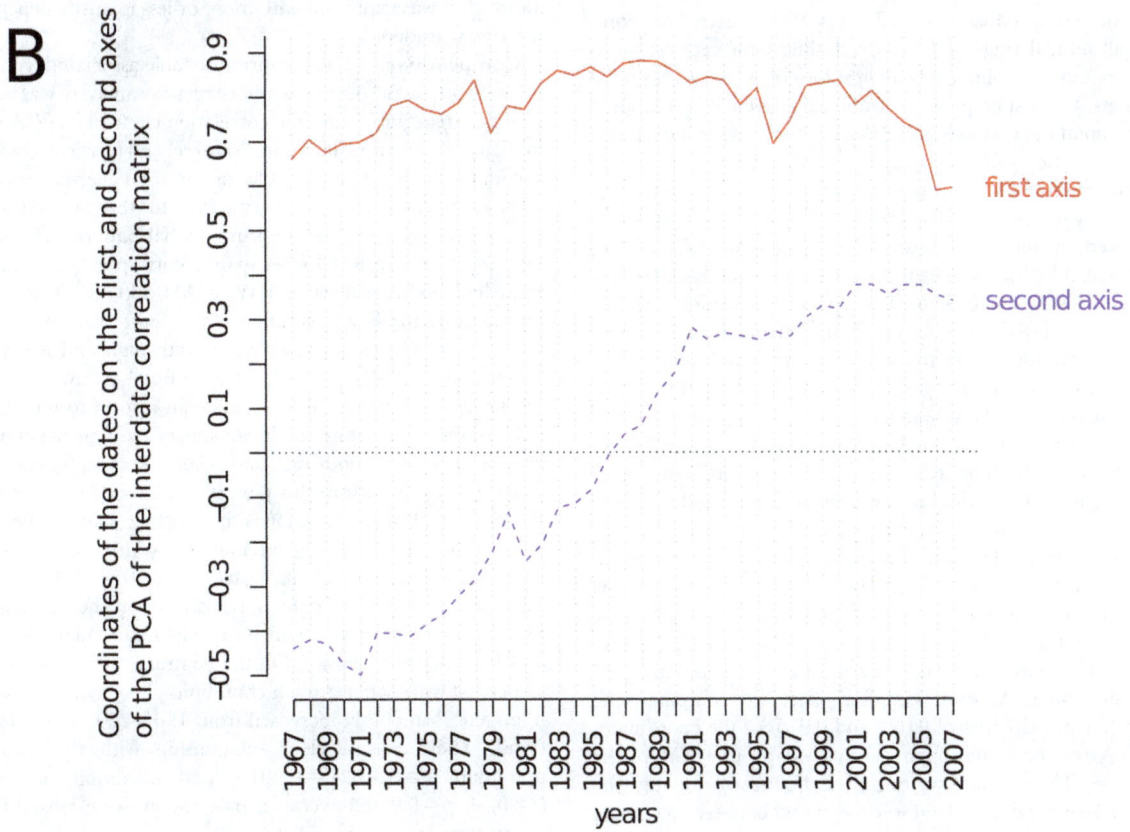

Figure 2. Interstructure analysis of the partial triadic analysis depicting the temporal evolution of spatial structures. A. Scores of the sampling dates upon the principal components of the principal component analysis of the inter-date correlation matrix. The first principal component (horizontal axis) represented 61.2% of the inertia. The second component (vertical axis) accounted for 9.7% of the total inertia. B. Scores of the sampling dates upon both first and second axes as a function of the years.

yielded a clear typology, with the first two principal components accounting for 61.2 and 9.7% of the total inertia, respectively. The corresponding correlation circle is given in Figure 2A. All the dates displayed positive scores upon axis 1, indicating the presence of a structure common to all dates. The case of the second axis was somewhat different, as dates displayed positive and negative values on this axis. Figure 2B shows how coordinates of dates upon both axes changed from 1967–2007. Whereas no clear trend was evident from axis 1, overall, the dates displayed increasing coordinates upon the second axis over the annual ring time series.

Compromise. The compromise table associated with the first component of the PCA of the interstructure table was a *ring variables × tree* two-way table. The correlation circle of the PCA of this compromise is shown in Figure 3A. The first two principal components accounted for 55.4 and 23.9% of the total inertia, respectively. The first axis showed opposite effects between trees having higher latewood density (LD) and maximum ring density (MAD) with trees for which these variables exhibited lower values. These variables were associated with descriptors of intra-ring variation, including the standard deviation of ring microdensity profile (RSD) and density contrast (Co), and, to a lesser extent, the standard deviations of the latewood and earlywood parts of the ring microdensity profiles (LSD and ESD, respectively), which conveyed larger variability in the latewood density compared to earlywood. The second axis separated trees with higher intra-ring density variations (RSD, Co, LSD, and ESD) from trees with higher earlywood density (ED), higher minimum ring density (MID), and, to a lesser extent, higher mean ring density (RD).

Since the compromise encapsulates spatial information common to all annual rings and we are dealing with georeferenced sampling points, we could explicitly test for departure from spatial randomness. We first employed a graphical approach by mapping tree score upon the first axis of the PCA in the geographical space (Figure 3B). The values appeared to be strongly spatially correlated with trees, forming patches of positive values alternating with gaps (negative values). The presence of spatial autocorrelation was assessed by means of Moran's I correlogram (Figure S2), which revealed highly significant departure from randomness ($p <$ 0.05 after Holm's correction). The spatial correlation analysis provided an interesting clue toward the spatial scale of these structures that ranged below roughly 30–40 m, as indicated by the shape of Moran's I correlogram. This means that the spatial structure isolated by PTA, which was common to all sampling occasions, corresponded to patches of trees with higher values of MAD, LD, RW, Co, and RSD (i.e., negative coordinates upon axis 1) and gaps of trees with lower values for these ring descriptors (i.e., positive coordinates upon axis 1).

Intrastructure. The last step of PTA is intended to reveal which original table exhibits departures from (or fits) the model expressed through the compromise. The original tables consisting of 41 *trees × ring* descriptor tables were projected as complementary tables onto the first axis of the PCA of the first compromise table. At each sampling date, we computed the quantiles (for probabilities of 0.025 and 0.975) of the coordinates of the 149 trees projected upon the first axis of the PCA of the compromise. These values were used as bounds to identify the trees that exhibited the largest differences between a model common to all dates (i.e., the compromise) (Figure 4). Fifty-three

trees fell out of that envelope at least on one date, 18 of which fell out only once. A small number of trees differed from the model for the majority of the sampling dates (e.g., 34 and 30 values out of the envelope). Nonetheless, the map shown in Figure 4 revealed that the distribution of these trees displayed no particular spatial pattern at the plot scale. Figure 5 shows the coordinates of each ring descriptor as projected upon the first axis of the PCA of the compromise table for each date. These trajectories fluctuated according to dates, but their position with respect to each other did not change much and followed the pattern displayed in the correlation circle of the PCA of the compromise (Figure 3A).

Depicting the spatial structure of temporal dynamics (trajectories)

The second set of analyses focused on the construction of a temporal typology common to each tree, which is equivalent to extracting the spatially stable part of the temporal structure. For that purpose, we considered the initial tables of *dates × ring* descriptors and performed the PTA on the interstructure table containing inter-tree R coefficients (Figure 1B).

Interstructure. The interstructure table was a $149×149$ square matrix containing the vectorial correlation R between the *dates × ring* descriptor sub-matrices (Figure 1B). The PCA of the interstructure matrix led to the correlation circle (not shown), where the two first principal components accounted for 35.0 and 7.0% of the total inertia, respectively. The correlation circle showed that all trees exhibited a common temporal dynamic, although it was expressed with more or less intensity depending on the tree considered.

Compromise. The compromise table associated with the first component of the previous interstructure analysis was a *date × ring descriptor* table, and the correlation circle of its PCA is shown in Figure 6. The first (horizontal) and second (vertical) components accounted for 63.4 and 30.0% of the total inertia, respectively. This correlation circle is very close to the one displayed in Figure 3A and expresses the same global pattern. The temporal typology common to all trees mainly corresponds to years where trees had higher values of RW, MAD, Co, RSD, and LD, as indicated by the first axis (Figure 6). Another source of variation common to all trees opposed years with high and low values of MID, ED, and RD (axis 2, Figure 6). A convenient way of displaying the patterns of the years with respect to the axis of the PCA of the compromise table consists of plotting the coordinates of each date upon both first and second axes as functions of the years (Figure 6B). From this plot, it can be seen that the first axis showed no clear temporal trend; a zigzag shape reflecting the alternating periods of high or low ring width, ring density, and latewood density is evident. Years 1972, 1979, 1980, 1996, and 2006 corresponded to high coordinates upon the first axis of the PCA of the compromise, which indicated particularly low growth and low latewood density. On the contrary, 1994 and 2000 were associated with better growth conditions. The coordinates of dates upon axis 2 smoothly increased from 1967–2007 in the form of a trend. There was a clear relationship with the mean daily minimum ($r = 0.48$, $p = 0.0013$) and maximum temperatures ($r = 0.54$, $p = 0.0002$) averaged over the growing season (March–September), as shown in Figure S3.

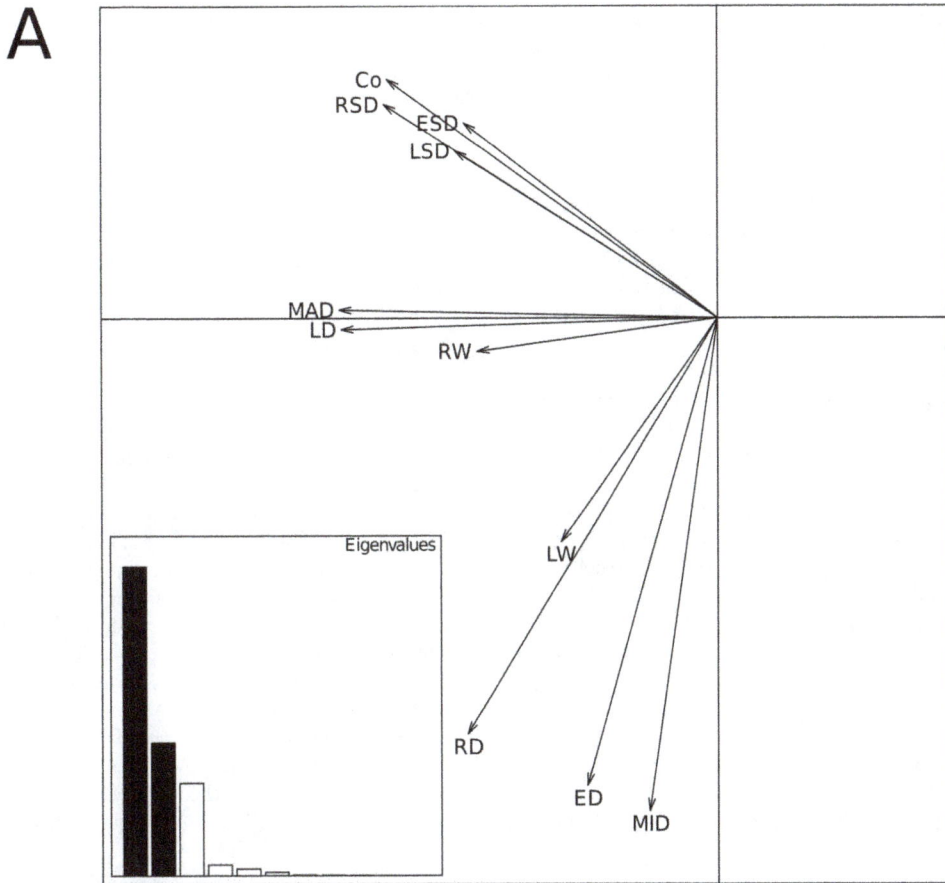

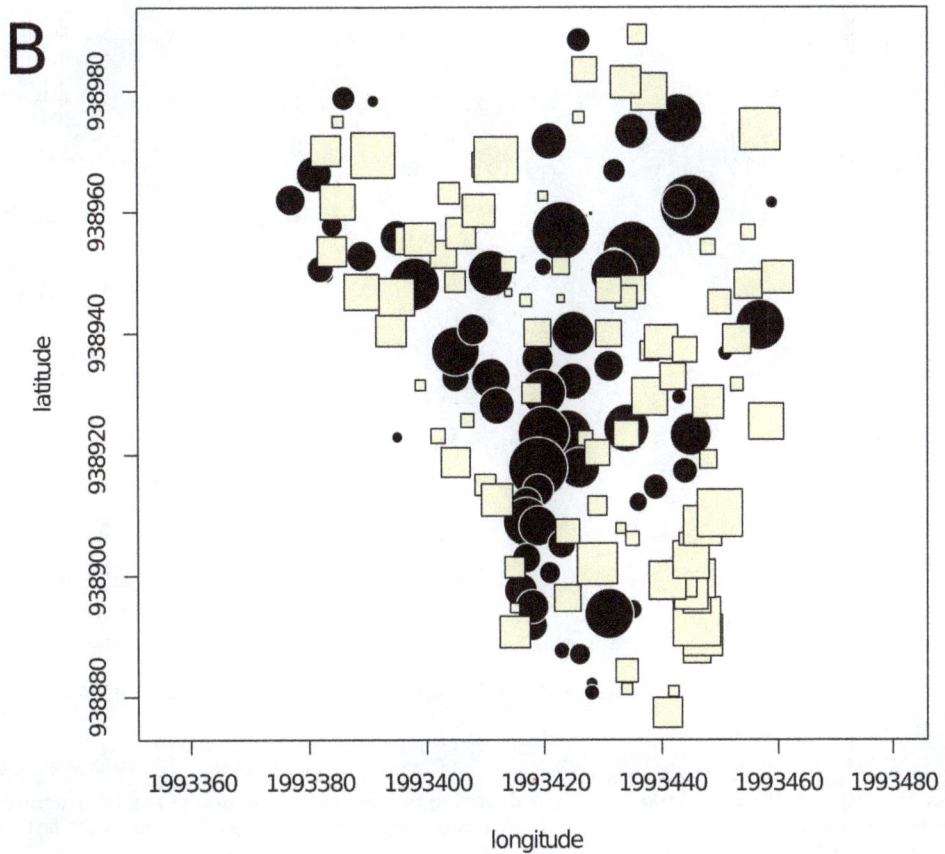

Figure 3. Compromise analysis of the partial triadic analysis depicting the temporal evolution of spatial structures. A. Correlation circle of the principal component analysis of the compromise table (a *tree × ring* descriptors table). The first (horizontal) and second (vertical) principal components accounted for 55.4 and 23.9% of the inertia, respectively. B. Map of the tree score upon the first axis of the PCA of the compromise table showing a strong spatial structure with alternating humps and bumps corresponding to patches of negative and positive scores. The symbol size is proportional to the absolute value of the score. Circles (squares) stand for positive (negative) values.

Intrastructure. The intrastructure of the PTA depicting the spatial structure of temporal dynamics indicates the departure of certain trees from the common model and the ring descriptors that could explain the phenomenon. The 149 original *dates × ring descriptor* data tables were projected upon the first axis of the PCA of the compromise. At each sampling date, we computed the quantiles (for probabilities of 0.025 and 0.975) of the coordinates of the 149 trees projected upon the first axis of the PCA of the compromise. These values were used as bounds to identify the trees that exhibited the largest differences between the model common to all trees (i.e., the compromise). A total of 132 trees (88.6%) fell outside the envelope at least once during the study period. Figure 7 shows the number of occasions that each tree fell outside the envelope. No spatial structure appears, which means that departure from the compromise is not spatially dependent.

Figure S4 shows the coordinates of each ring descriptor projected onto the first axis of the PCA of the compromise at each tree. For most of the ring descriptors, the coordinates upon axis 1 are fairly homogeneous, indicating a lack of spatial variability and limited departure from the structure encapsulated in the compromise. Some limited divergence with respect to the compromise appeared for the variable ED (negative values for some trees scattered across the plot), whereas ED had a positive

coordinate onto the first axis of the PCA of the compromise (Figure 6A). Similarly, several trees displayed positive coordinates for ESD, while that variable had a negative coordinate in Figure 6A.

Discussion

Temporal trajectories

The PTA that focused on the spatial structure of temporal dynamics yielded a temporal typology of wood features that opposed years where climatic conditions allowed growth to years where growth was limited with subsequent alteration of wood characteristics. The PTA showed that the main temporal structure was the separation between the slow growth during years 1972, 1979, 1980, 1996, and 2006 from the rest of the dates (Figure 6B). Some years appeared particularly favorable, such as 1994 and 2000. This pattern is fairly common to all trees because individuals experienced similar climatic conditions due to the relatively small size of the survey area. In the Alps, larch is recurrently defoliated by the larch budmoth (*Zeiraphera diniana*) [50]. Tree defoliation strongly decreases radial growth and produces anomalies in ring density variables during the year of defoliation and/or the year immediately following [51]. A. Roques (personal communication)

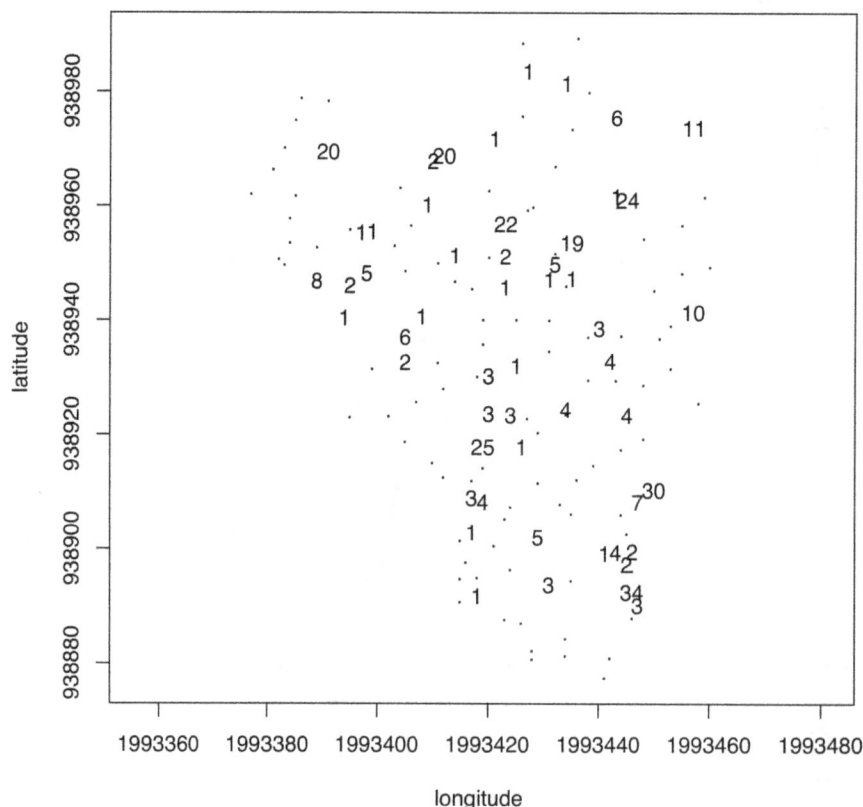

Figure 4. Intrastructure analysis of the partial triadic analysis depicting the temporal evolution of spatial structures. Map of the trees showing the number of times each one fell outside of the 95% envelopes of the tree coordinates projected onto the first axis of the principal component analysis of the compromise for each date.

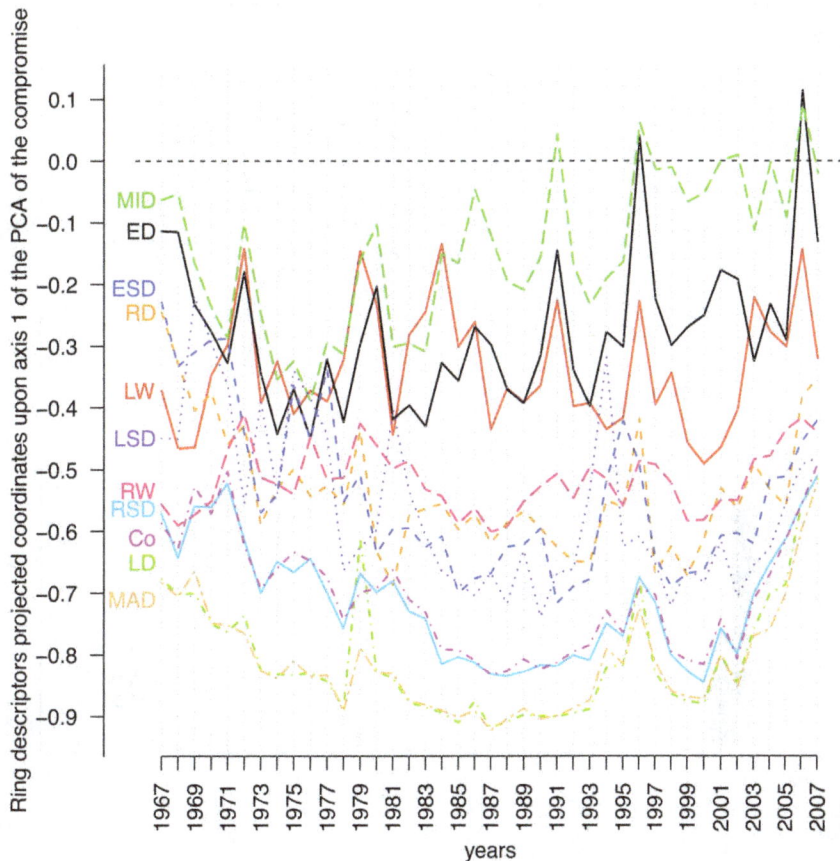

Figure 5. Intrastructure analysis of the partial triadic analysis depicting the temporal evolution of spatial structures. Scatter plot showing the coordinates of the ring variables projected onto the first axis of the principal component analysis of the compromise table across dates.

provided historical records of defoliation by the larch budmoth in the Briançon. These data indicated that defoliation occurred in 1971, 1979, 1996, 1997, and 2006 in an experimental site located 6 km from our study site, at 1800 m asl. There is very close agreement between the direct observations of *Zeiraphera* defoliation in the URZF experimental site and the outputs of the PTA: the 1971 defoliation strongly decreased 1972 annual ring width and latewood formation; the 1979 defoliation affected both 1979 and 1980 rings; the 1996–1997 defoliation affected the 1996 and 1997 rings; and the 2006 defoliation only affected the 2006 annual ring.

Equally interesting is the trend identified in the second axis of the PCA of the compromise (Figure 6B). Such an effect may be related to longer-term climate evolution (e.g., temperature increase, as suggested in this study) (Figure S3). This is consistent with other results indicating that the mean annual temperature at Briançon increased by about 1.5°C from 1967–2007, which may have modified the annual ring structure (Figure S3) [38]. In the present survey, temporal changes of wood quality primarily involved a decrease of earlywood density and an increase of latewood density as well as a decrease of ring homogeneity. This is consistent with the conclusions of a recent survey on Siberian larch (a species closely related to European larch), which showed that warming favors wider earlywood cell lumen (i.e., lower density earlywood), thicker latewood walls (i.e., higher density latewood), denser maximum latewood, and wider rings [52]. Both the *Zeiraphera* defoliation and the temperature trend affected the temporal dynamics of tree-ring characteristics in a similar way for

all trees, and a limited number of individuals displayed departure from that model for some ring descriptors (Figures 7 and S4). This rather homogeneous response of trees to climate could differ in other situations, such as field trials, where various families of genetically selected trees are planted (see below).

It should be noted that some factors may markedly affect tree ring variables, such as cambial age, competition, or heartwood formation. Here, we did not consider these factors because we focused on illustrating the methodology and the use of PTA, for which raw data appeared to be the best option. In some situations however, it may necessary to adjust the annual ring time series according to the objective of the study. While no adjustment is necessary for a wood quality study, it is generally mandatory for climatic or ecological studies.

Spatial variability at the plot scale

The results reported in this paper highlight both the spatial structure of a set of descriptors of wood characteristics of European larch trees and the temporal trajectories of each individual tree across a series of 41 consecutive years. The PTA allowed us to clearly identify the presence of a spatial structure of wood descriptors that was common to all years (i.e., the first PTA presented in the study). This long-term stand structure appeared spatially dependent at a short scale. These results suggest that trees were submitted to contrasting environmental conditions, which affected their growth and the quality of wood in a manner that is constant across years. Radial growth traits, such as ring width

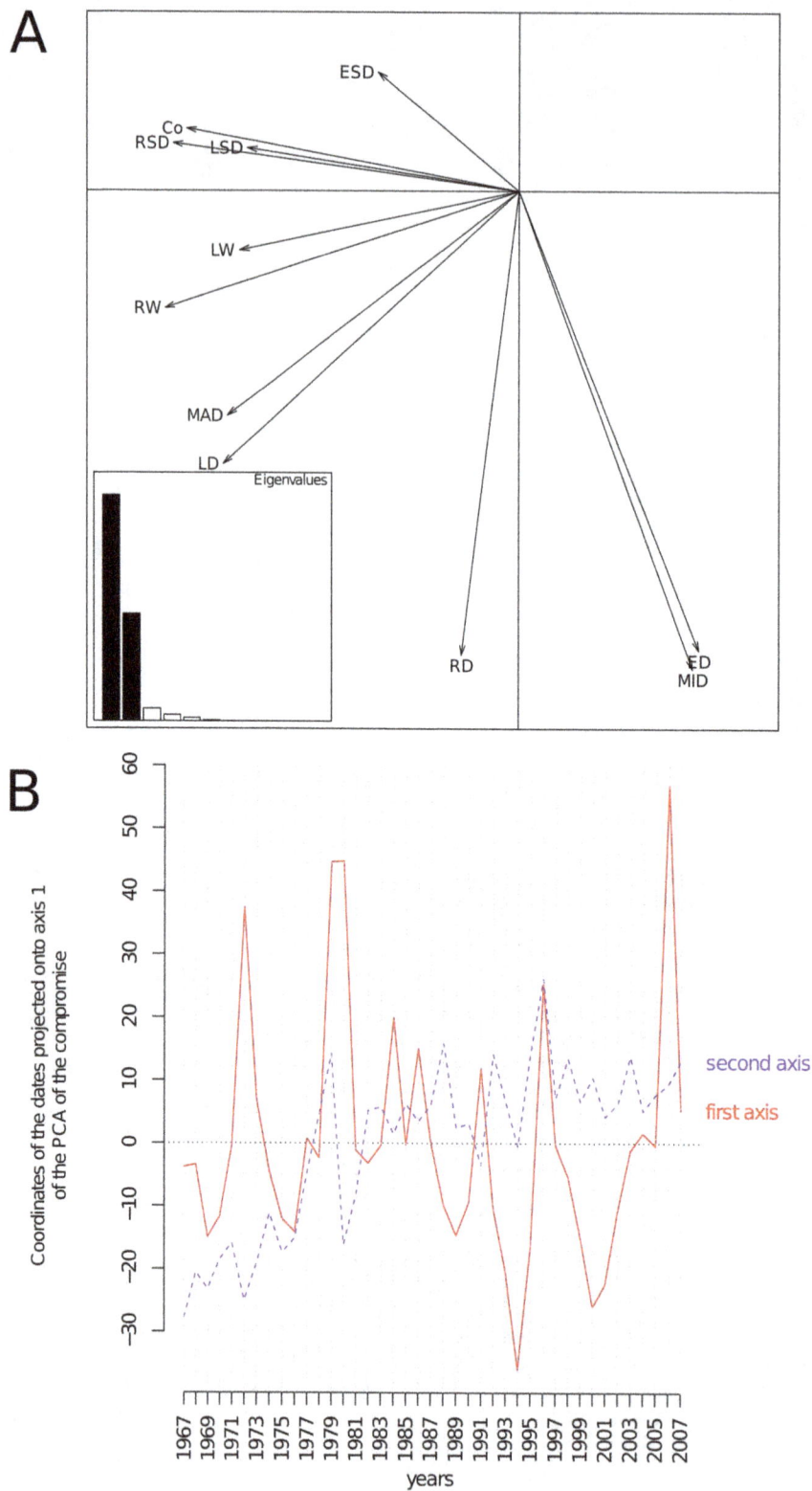

Figure 6. Compromise analysis of the partial triadic analysis depicting the spatial structure of temporal dynamics. Correlation circle of the principal component analysis (PCA) of the compromise table (a *date × descriptor* two-way table). The first principal component (horizontal axis) represented 63.4% of the inertia. The second component (vertical axis) accounted for 30.0% of the total inertia. The first axis expresses the opposition between dates where ring growth and latewood production was high with less favorable periods. B. Changes of dates coordinate upon the first (solid line) and second (dashed line) axes of the PCA of the compromise table. The coordinates of the dates onto the first axis showed no temporal trend but, rather, the presence of punctual events that strongly affected ring descriptors. Dates coordinated onto the second axis revealed a temporal trend over the period of the survey.

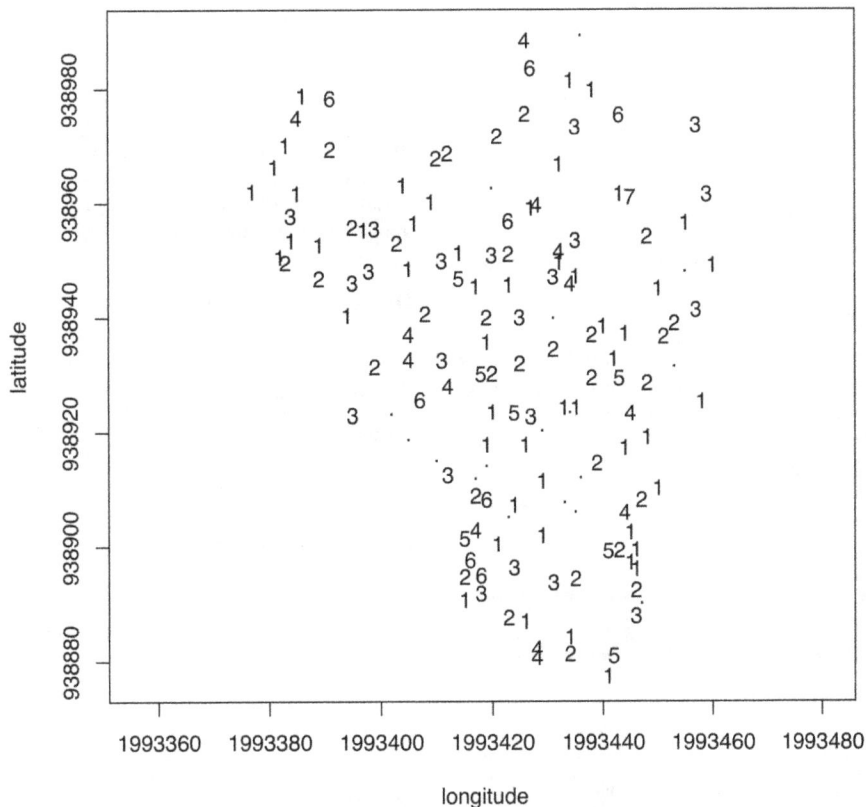

Figure 7. Intrastructure analysis of the partial triadic analysis depicting the spatial structure of temporal dynamics. Map of the trees showing the number of times each one fell outside the 95% envelopes of the tree coordinates projected onto the first axis of the principal component analysis of the compromise for each date.

(RW) and latewood width (LW), are mostly affected by environmental factors, among which the most influential is competition among trees [53]. For the ring density traits, the portion of the genetic variation in the total variance is generally much higher [54].

Soil variability is also a good candidate to explain stand spatial structure, as it is strongly variable in space at very different scales [55] and has substantial impact on plants [56]. The slope of the plot may also constitute an important environmental source of functional heterogeneity, as it affects water flow and often corresponds to spatial gradients of soil texture and overall soil fertility. The present study lacks environmental data allowing to fully explore the environmental drivers of stand spatial variability, but one highlight of PTA is that the compromise can easily be coupled with additional tools, such as spatial statistics. Trees of interest could be identified and monitored, and complementary data could be gathered, thus enabling a better exploration of their relationships with abiotic or biotic environmental factors (i.e., neighboring trees could be incorporated into additional analyses to assess the competition and other micro-environmental effects).

The potential of PTA in tree-ring dataset analyses

The PTA proved to be a precious tool by allowing the proper identification of spatial structures common to all dates and exploration of their potential links with environmental drivers (e.g., soil fertility or water status). An obvious highlight of the approach was that it also allowed a direct access to temporal trajectories common to all trees. In many different cases, this would be helpful for exploring correlations with indices of climate change as well as

identifying which ring descriptors (i.e., wood characteristics) are affected. Multivariate exploration of ring characteristics has been relatively uncommon. The PTA offers a complete space-time framework using a relatively simple mathematical basis, since the central tool is PCA [23] and software resources are available [47]. In our survey, the PTA allowed a proper separation between high and low frequency signals corresponding to inter-annual variations (at least partly linked to forest pest/insect outbreaks in this study) and long-term trends, respectively, possibly related to climate change. Because the PTA allowed for extraction, quantification, and hierarchization of these superimposed sources of variation, it is relevant for climate change research by processing the growing number of databases documenting tree chronologies that are compiled and made available [57].

In common garden experiments with genetic entities, the environmental heterogeneity is largely dictated by the design of the experiment (e.g., blocks of trees of various genetic lineages). The PTA would be helpful to stress the common temporal evolution of all trees, and the intrastructure would discern those lineages that differ from the common model and for which ring characteristics differ. The PTA would thus contribute to identifying lineages with potentially interesting wood characteristics. Considering that the annual ring variables constitute basic wood properties directly linked to the value of wood products, PTA could also be used as an efficient tool to describe variation of overall trunk structure from a wood-quality standpoint.

Finally, PTA can be seen as a pipeline (i.e., a set of connected data-processing elements). The output of some elements, such as the compromise table, can be used in various complementary analyses not strictly related to the PTA. We employed this strategy with the correlogram analysis, but the compromise table may also be taken as a summary of the data and co-analyzed with external data tables [14,23,30,33,34,58]. Examples may also include genetic data describing each tree or data conveying the between-tree competition within the neighboring environment.

Supporting Information

Figure S1   Changes of tree ring variables from 1967–2007.

Figure S2   Spatial analysis of the partial triadic analysis depicting the temporal evolution of spatial structures. Moran's I correlogram of the tree scores upon the first axis of the PCA of the compromise table. Black (open) symbols indicate significant (non-significant) values at $p = 0.05$. The spatial structure proved globally significant ($p < 0.05$) when assessed by means of the Holm's correction test for simultaneous testing.

Figure S3   Average daily minimum and maximum temperatures in Briançon (44°53′N, 6°38′E) averaged over the growing season of the European larch (March–

September) from 1967–2007. Data source: Météo-France.

Figure S4   Intrastructure analysis of the partial triadic analysis depicting the spatial structure of temporal dynamics. Scatter plot showing the coordinates of the ring variables projected onto the first axis of the principal component analysis of the compromise table for each tree.

Dataset S1   Ring dataset.

Acknowledgments

We thank all the people who participated to the fieldwork, including Kevin Ader, Bérangère Bougué, Jean-Paul Charpentier, Philippe Label, Sara Marin, Nathalie Mayeur and Frédéric Millier.

Author Contributions

Conceived and designed the experiments: JPR MN LP PR. Performed the experiments: MN LP PR. Analyzed the data: JPR MG LP PR. Contributed reagents/materials/analysis tools: MRD ASS AMM. Wrote the paper: JPR PR.

References

1. Bergeron Y, Denneler B, Charron D, Girardin M-P (2002) Using dendrochronology to reconstruct disturbance and forest dynamics around Lake Duparquet, northwestern Quebec. Dendrochronologia 20: 175–189.
2. Rubino DL, McCarthyz BC (2004) Comparative analysis of dendroecological methods used to assess disturbance events. Dendrochronologia 21: 97–115.
3. Arseneault D, Payette S (1997) Reconstruction of millennial forest dynamics from tree remains in a subarctic tree line peatland. Ecology 78: 1873–1883.
4. Davi NK, Jacoby GC, Wiles GC (2003) Boreal temperature variability inferred from maximum latewood density and tree-ring width data, Wrangell Mountain region, Alaska. Quaternary Res 60: 252–262.
5. Martinelli N (2004) Climate from dendrochronology: latest developments and results. Global Planet Change 40: 129–139.
6. Koprowski M, Duncker P (2012) Tree-ring width and wood density as the indicators of climatic factors and insect outbreaks affecting spruce growth. Ecol Indicat 23: 332–337.
7. Briffa KR, Osborn TJ, Schweingruber FH (2004) Large-scale temperature inferences from tree rings: a review. Global and Planet Change 40: 11–26.
8. Badeau V, Dupouey J-L, Becker M, Picard J-F (1995) Long-term growth trends of *Fagus sylvatica* L. in northeastern France: a comparison between high- and low-density stands. Acta Oecol 16: 571–583.
9. Bowman DMJS, Brienen RJW, Gloor E, Phillips OL, Prior LD (2013) Detecting trends in tree growth: not so simple. Trends Plant Sci 18: 11–17.
10. Spiecker H (2002) Tree rings and forest management in Europe. Dendrochronologia 20: 191–202.
11. Bräker OU (2002) Measuring and data processing in tree-ring research: a methodological introduction. Dendrochronologia 20: 203–216.
12. Swing RE (1998) An introduction to microdensitometry. Ballingham: SPIE Press. 243 p.
13. Schweingruber FH (1988) Tree-ring basics and applications of dendrochronology: D. Reidel Publishing Company. 276 p.
14. Legendre P, Legendre L (1998) Numerical ecology. Amsterdam: Elsevier. 852 p.
15. Thioulouse J, Chessel D (1987) Analyzing multi-factorial ecology tables. I: From the state to the type of operation by triadic analysis typology. Acta Oecol Oec Gen 8: 463–480.
16. Harshman RA, Lundy ME (1994) PARAFAC: Parallel factor analysis. Comput Stat Data An 18: 39–72.
17. Tucker LR (1966) Some mathematical notes on three-mode factor analysis. Psychometrika 31: 279–311.
18. Acar E, Yener B (2009) Unsupervised multiway data analysis: A literature survey. IEEE Trans Knowl Data Eng 21: 6–20.
19. Stanimirovaa I, Walczaka B, Massarta DL, Simeonovc V, Sabyd CA, et al. (2004) STATIS, a three-way method for data analysis. Application to environmental data. Chemometr Intell Lab 73: 219–233.
20. L'Hermier des Plantes H (1976) Structure of tables with three indices: theory and application of a joint analysis method. Ph.D. dissertation, University of Sciences and Techniques of Languedoc, Montpellier II (in French).
21. Robert P, Escoufier Y (1976) A unifying tool for linear multivariate statistical methods: the RV coefficient. Appl Statist 25: 257–265.
22. Lavit C, Escoufier Y, Sabatier R, Traissac P (1994) The act (STATIS method). Comput Stat Data Anal 18: 97–119.
23. Thioulouse J, Simier M, Chessel D (2004) Simultaneous analysis of a sequence of paired ecological tables. Ecology 85: 272–283.
24. Kroonenberg P (1989) The analysis of multiple tables in factorial ecology. III: Three-mode principal component analysis: "full triadic analysis". Acta Oecol Oec Gen 10: 245–256.
25. Centofanti M, Chessel D, Dolédec S (1989) Stability of spatial structure and statistical analysis of compromise multi-tables: application to the physical chemistry of a lake reservoir. Rev Sci Eau 2: 71–93.
26. Gaertner JC, Chessel D, Bertrand J (1998) Stability of spatial structures of demersal assemblages: a multitable approach. Aquat Living Resour 11: 75–85.
27. Rolland A, Bertrand F, Maumy M, Jacquet S (2009) Assessing phytoplankton structure and spatio-temporal dynamics in a freshwater ecosystem using a powerful multiway statistical analysis. Water Res 43: 3155–3168.
28. Bertrand F, Maumy M (2010) Using partial triadic analysis for depicting the temporal evolution of spatial structures: assessing phytoplankton structure and succession in a water reservoir. CS-BIGS 4: 23–43.
29. Carassou L, Ponton D (2007) Spatio-temporal structure of pelagic larval and juvenile fish assemblages in coastal areas of New Caledonia, southwest Pacific. Mar Biol 150: 697–711.
30. Rossi JP (2003) The spatio-temporal pattern of a tropical earthworm species assemblage and its relationship with soil structure. Pedobiologia 47: 497–503.
31. Jiménez J-J, Decaëns T, Rossi J-P (2006) Stability of the spatio-temporal distribution and niche overlap in neotropical earthworm assemblages. Acta Oecol 30: 299–311.
32. Decaëns T, Jiménez JJ, Rossi J-P (2009) A null-model analysis of the spatio-temporal distribution of earthworm species assemblages in Colombian grasslands. J Trop Ecol 25: 415–427.
33. Decaëns T, Rossi JP (2001) Spatio-temporal structure of earthworm community and soil heterogeneity in a tropical pasture. Ecography 24: 671–682.
34. Godefroid M, Delaville L, Marie-Luce S, Quénéhervé P (2013) Spatial stability of a plant-feeding nematode community in relation to macro-scale soil properties. Soil Biol Biochem 57: 173–181.
35. Ernoult A, Freiré-Diaz S, Langlois E, Alard D (2006) Are similar landscapes the result of similar histories? Landscape Ecol 21: 631–639.
36. Gourdol L, Hissler C, Hoffmann L, Pfister L (2013) On the potential for the partial triadic analysis to grasp the spatio-temporal variability of groundwater hydrochemistry. Appl Geochem 33: 93–107.
37. Martin N, Molimard P, Spinnler HE, Schlich P (2000) Comparison of odor sensory profiles performed by two independently trained panels following the same descriptive analysis procedures. Food Qual Prefer 11: 487–495.
38. Nardin M (2013) Biological adjustment of larch environmental variations along an altitudinal gradient: a microdensitometric approach to climate response. Ph.D. dissertation, University of Orléans (in French). 263 p.
39. Manly BF (1994) Multivariate statistical methods. A primer. Boca raton: Chapman & Hall/CRC. 208 p.
40. Bro R, Smilde AK (2003) Centering and scaling in component analysis. J Chemometrics 17: 16–33.

41. Polge H (1978) Fifteen years of wood radiation densitometry. Wood Sci Tech 12: 187–196.

42. Guay R, Gagnon R, Morin H (1992) A new automatic and interactive tree-ring measurement system based on a line-scan camera. Forest Chron 68: 138–141.

43. Ruiz Diaz Britez M, Sergent A-S, Martinez Meier A, Bréda N, Rozenberg P (in press) Wood density proxies of adaptive traits linked with resistance to drought in Douglas fir (*Pseudotsuga menziesii* [Mirb.] Franco). Trees: DOI 10.1007/s00468-00014-01003-00464.

44. Ivković M, Rozenberg P (2004) A method for describing and modeling of within-ring wood density distribution in clones of three coniferous species. Ann Forest Sci 61: 759–769.

45. Ivković M, Gapare W, Wu H, Espinoza S, Rozenberg P (2013) Influence of cambial age and climate on ring width and wood density in *Pinus radiata* families. Ann Forest Sci 70: 525–534.

46. R Core Team (2014) R: a language and environment for statistical computing. Vienna: R Foundation for Statistical Computing. Available: http://CRAN. R-project.org/. Accessed: 24 June 2014.

47. Dray S, Dufour A-B (2007) The ade4 package: implementing the duality diagram for ecologists. J Stat Soft 22: 1–20.

48. Fortin MJ, Dale MRT (2005) Spatial analysis: a guide for ecologists. Cambridge: Cambridge University Press. 365 p.

49. Bjornstad ON (2012) Ncf: spatial nonparametric covariance functions. R package version 1.1–4. Available: http://CRAN.R-project.org/package=ncf. Accessed: 24 June 2014.

50. Dormont L, Baltensweiler W, Choquet R, Roques A (2006) Larch-and pine-feeding host races of the larch bud moth (*Zeiraphera diniana*) have cyclic and synchronous population fluctuations. Oikos 115: 299–307.

51. Rolland C, Petitcolas V, Michalet R (1998) Changes in radial tree growth for *Picea abies*, *Larix decidua*, *Pinus cembra*, and *Pinus uncinata* near the alpine timberline since 1750. Trees 13: 40–53.

52. Fonti P, Bryukhanova MV, Myglan VS, Kirdyanov AV, Naumova OV, et al. (2013) Temperature-induced responses of xylem structure of *Larix sibirica* (Pinaceae) from the Russian Altay. Am J Bot 100: 1332–1343.

53. Stadt KJ, Huston C, Coates KD, Feng Z, Dale MRT, et al. (2007) Evaluation of competition and light estimation indices for predicting diameter growth in mature boreal mixed forests. Ann For Sci 64: 477–490.

54. Zobel BJ, Jett JB (1995) Genetics of wood production. berlin: Springer-Verlag. 352 p.

55. Goovaerts P (1998) Geostatistical tools for characterizing the spatial variability of microbiological and physico-chemical soil properties. Biol Fertil Soils 27: 315–334.

56. Cambardella CA, Moorman TB, Parkin TB, Karlen DL, Novak JM, et al. (1994) Field-scale variability of soil properties in central Iowa soils. Soil Sci Soc Am J 58: 1501–1511.

57. de Luis M, Čufar K, Di Filippo A, Novak K, Papadopoulos A, et al. (2013) Plasticity in dendroclimatic response across the distribution range of Aleppo pine (*Pinus halepensis*). PLoS ONE 8: e83550.

58. Vivien M, Sabatier R (2003) Generalized orthogonal multiple co-inertia analysis(–PLS): new multiblock component and regression methods. J Chemom 17: 287–301.

20

Wood Anatomy Reveals High Theoretical Hydraulic Conductivity and Low Resistance to Vessel Implosion in a Cretaceous Fossil Forest from Northern Mexico

Hugo I. Martínez-Cabrera[1]*, Emilio Estrada-Ruiz[2]

1 Estación Regional del Noroeste, Instituto de Geología, Universidad Nacional Autónoma de México, Hermosillo, México, **2** Laboratorio de Ecología, Departamento de Zoología, Escuela Nacional de Ciencias Biológicas – Instituto Politécnico Nacional, Ciudad de México, México

Abstract

The Olmos Formation (upper Campanian), with over 60 angiosperm leaf morphotypes, is Mexico's richest Cretaceous flora. Paleoclimate leaf physiognomy estimates indicate that the Olmos paleoforest grew under wet and warm conditions, similar to those present in modern tropical rainforests. Leaf surface area, tree size and climate reconstructions suggest that this was a highly productive system. Efficient carbon fixation requires hydraulic efficiency to meet the evaporative demands of the photosynthetic surface, but it comes at the expense of increased risk of drought-induced cavitation. Here we tested the hypothesis that the Olmos paleoforest had high hydraulic efficiency, but was prone to cavitation. We characterized the hydraulic properties of the Olmos paleoforest using theoretical conductivity (K_s), vessel composition (S) and vessel fraction (F), and measured drought resistance using vessel implosion resistance $(t/b)_h^2$ and the water potential at which there is 50% loss of hydraulic conductivity (P_{50}). We found that the Olmos paleoforest had high hydraulic efficiency, similar to that present in several extant tropical-wet or semi-deciduous forest communities. Remarkably, the fossil flora had the lowest $(t/b)_h^2$, which, together with low median P_{50} (-1.9 MPa), indicate that the Olmos paleoforest species were extremely vulnerable to drought-induced cavitation. Our findings support paleoclimate inferences from leaf physiognomy and paleoclimatic models suggesting it represented a highly productive wet tropical rainforest. Our results also indicate that the Olmos Formation plants had a large range of water conduction strategies, but more restricted variation in cavitation resistance. These straightforward methods for measuring hydraulic properties, used herein for the first time, can provide useful information on the ecological strategies of paleofloras and on temporal shifts in ecological function of fossil forests chronosequences.

Editor: Paul V. A. Fine, Berkeley, United States of America

Funding: This work was supported by funds from "Apoyos Complementarios para la Consolidación Institucional de Grupos de Investigación" CONACYT to HIMC. The funder had no role in study design, data collection and analysis, decision to publish, or preparation of the manuscript.

Competing Interests: The authors have declared that no competing interests exist.

* Email: hugomartinez2w@gmail.com.mx

Introduction

The expression of anatomical and morphological traits is subject to biophysical constraints imposed by environmental demands. Consequently, they can provide important information about ecological strategies of fossil assemblages [1,2] and paleoclimate (e.g. [3,4]). Because a large amount of water is needed to maintain plant growth, water is a major factor limiting plant distribution and trait expression. Water loss and carbon fixation are linked because during CO_2 uptake water is transpired to the atmosphere [5], and because diffusion coefficients for water are larger than for CO_2, efficient carbon fixation and plant growth requires a disproportionately high hydraulic supply to the leaves to meet evaporative demands during photosynthesis. This relationship between stem hydraulic capacity with tree growth rate [6,7] and leaf photosynthetic capacity is well established [8].

Across vegetation types and biomes, at low latitudes and altitudes, there is a positive relationship between water availability and xylem conduit size, such that in xeric environments vessel size is smaller on average than in tropical humid environments [9,10,11]. Plant water transport efficiency is directly related to the hydraulic conductivity of xylem, which is mostly determined by conduit size [12,13,14]. As warm wet environments are also more productive regions, hydraulic efficiency should be directly linked to carbon fixation, as has been empirically confirmed (e.g. [15]). In places with high water availability hydraulic capacity is increased by decreasing resistance to water flow in the xylem (e.g. increasing vessel diameter). In these warm wet environments, plants can maintain high transpiration rates and maximize carbon fixation and growth [16]. However, in dry conditions, wide, hydraulically efficient vessels are also more prone to experience mechanical failure [17,18,19] or drought induced cavitation [14,20,21].

Although drought induced cavitation occurs along the entire water availability continuum [22,23], plants from drier regions with smaller vessel diameters are in general more capable to cope with cavitation because it occurs at higher xylem tensions (lower water potential) than in wet adapted plants with larger vessel diameters. The consequence of vessel cavitation is the disruption of the water column, which reduces xylem hydraulic conductivity and the overall plant water supply to the photosynthetic surface [24]. Cavitation can thus be translated into a decrease in photosynthesis [15] and a drop in stomatal conductance [25], which further increase xylem tension and cavitation [24,26]. Cavitation resistance in extant plants is usually quantified using the water potential at which there is 50% loss of hydraulic conductivity (P_{50}). As P_{50} values are impossible to measure in fossil wood samples, we used a metric developed by Hacke et al. [18], the vessel resistance to implosion metric ($(t/b)_h^2$, defined below) that can be used to approximate cavitation resistance. $(t/b)_h^2$ explains between 80% [18] and 95% [19] of P_{50} variation and is essentially a measure of vessel wall reinforcement. $(t/b)_h^2$ is entirely based on vessel anatomy and therefore can be used to determine drought tolerance thresholds (P_{50}) in fossil woods.

In this paper we used vessel anatomy to determine key functional traits related to drought resistance ($(t/b)_h^2$ and P_{50}) and hydraulic capacity (potential conductivity K_s, vessel fraction F, and vessel size contribution metric S) of a fossil forest from the Olmos Formation (upper Campanian), Coahuila, Mexico. These two sets of functional traits provide information about the hydraulic functional strategies of the fossil forest. Because of their close link to the environment, they also provide hints as to the climate regime in which the fossil plants grew. With more than 80 different leaf morphotypes [27,28,29,30,31,32,33,34], mainly angiosperms (80%), the Olmos Formation is one of the richest Cretaceous paleoforests from Mexico and south-central Western Interior of North America (WINA). According to leaf physiognomy paleoclimate estimates, the Olmos paleoforest grew under a tropical climate (MAT 20–23°C) with high water availability (1.5 to 3 m) [28]. The Olmos paleoforest was probably more mesic-adapted than other floras from the southern part of North America during this period [28]. Based on these paleoclimate reconstructions and leaf physiognomy, we expected that this paleoforest would have had a very efficient hydraulic system to meet high evaporative demands, but we also expected that this hydraulic efficiency should be paired with a high risk of embolism formation. In other words, in the conduction efficiency-cavitation risk trade-off continuum, the Olmos paleoforest should represent a highly hydraulically efficient and embolism-prone ecological strategy only suitable under warm, wet environments. Moreover, large leaf area and tree heights, estimated to be up to 35 m in some species [28], suggest a highly productive environment and efficient carbon fixation, which would require an efficient hydraulic system.

Material and Methods

Geological setting

The outcrops of the Olmos Formation (upper Campanian) are found in the Sabinas Basin in the state of Coahuila in northern Mexico (Fig. 1) [35,36,37,38]. The Olmos Formation represents a fluvial-deltaic system with four main depositional sub-environments [36] that include: 1) swampy areas with restricted circulation, 2) floodplain environments and/or a lagoon system with open circulation, 3) fluvial environments, likely including braided rivers and 4) meandering rivers [36]. The angiosperm woods, along with numerous dinosaur bones, were collected in environments representing meandering rivers [36]. The woods were not found in growth position and despite the large size of some samples, we assumed that some transport occurred.

All necessary permits were obtained for the described study, which complied with all relevant regulations. These permits were issued by the Instituto Nacional de Antropología e Historia (INAH) and the material is housed at the National Collection of Paleontology (Universidad Nacional Autónoma de México).

Anatomical measurements

The anatomical characteristics of the 10 wood xylotypes thus far described for the Olmos Formation [27,39] were measured on transverse sections obtained using a standard thin-section technique. In 11 field trips we have collected nearly 100 samples, from these, we have recognized 10 dicot, 5 palms and 2 gymnosperm (Podocarpaceae and Taxodiaceae) xylotypes. Despite that most of the xylotypes were identified in the first two field visits, and no new xylotypes have been collected in recent visits, we do not discard the possibility of finding new morphospecies given the outstandingly high diversity of the leaf flora. The10 wood xylotypes we studied here represent all the dicot fossil woods collected so far and therefore, despite their relatively low number, they are a good representation of the dicot flora. We measured more than one sample for six of the ten species/xylotypes (i.e. *Coahuiloxylon terrazasiae*, *Javelinoxylon* xylotype 2, *Javelinoxylon weberi*, *Metcalfeoxylon* xylotype 1, *Muzquizoxylon porrasii* and *Wheeleroxylon atascosense*). See Table 1 for species means and standard deviations, and Table S1 and Table S2 for values of each measured vessel and means per sample, respectively. Species means for each of the measured functional traits were calculated based on the mean of the each of the samples per xylotype.

To determine potential conductivity (K_s), vessel resistance to vessel implosion ($t/b)_h^2$, and S and F metrics, we calculated dimensions of 62 to 398 vessels per species (See Table S1 for the number of vessels measured per sample and species/xylotype). The number of vessels measured varied as a function of their density. To calculate vessel dimensions we first randomly selected radial sectors (cross sectional area limited by rays) and measured all the vessels contained in the sector. For species with a limited number of vessels (low vessel density) such as *Quecinium centenoae* and *Metcalfeoxylon* xylotype 1, we measured all the vessels that the preservation allowed. Vessel outlines were drawn using a graphics tablet (Intuous 3, Wacom, Kita Saitama-Gun, Saitama, Japan). Vessel area was calculated using XTools of ArcView (version 3.2, ESRI, Redlands, CA, USA). For details of the measuring protocol, see Martínez-Cabrera et al. [40]. We calculated vessel lumen diameter, including mean vessel diameter (d_{mean}) and hydraulic mean (d_h), using diameters of circles with the same area as the individual vessel lumens, following Kolb and Sperry [22]. d_h was calculated as the sum of the contribution of all conduit diameters ($\sum d^5$) divided by the total number of vessels ($\sum d^4$) [22,41]. d_{mean} was then used, together with other variables (see below), to calculate potential conductivity, while d_h was used for estimating the squared vessel-wall thickness-to-span ratio $(t/b)_h^2$, a proxy for the resistance to vessel implosion.

We calculated potential conductivity per stem cross sectional area (K_s) following Zanne et al. [13]: $K_s \propto F^{1.5} S^{0.5}$, where F is the vessel fraction and S is a vessel size contribution metric. Vessel fraction F is mean vessel area $\bar{A}$ times vessel density (N) ($F = \bar{A} * N$; mm^2·mm^{-2}), and S is the ratio between the same anatomical traits ($S = \bar{A}/N$; mm^4). $\bar{A}$ is the mean individual vessel cross sectional area and N is the vessel number per unit of sapwood area [13]. Vessel fraction F approximates the fraction of cross sectional area occupied by vessel space [13,42] and S measures the variation in vessel size composition. Higher values of

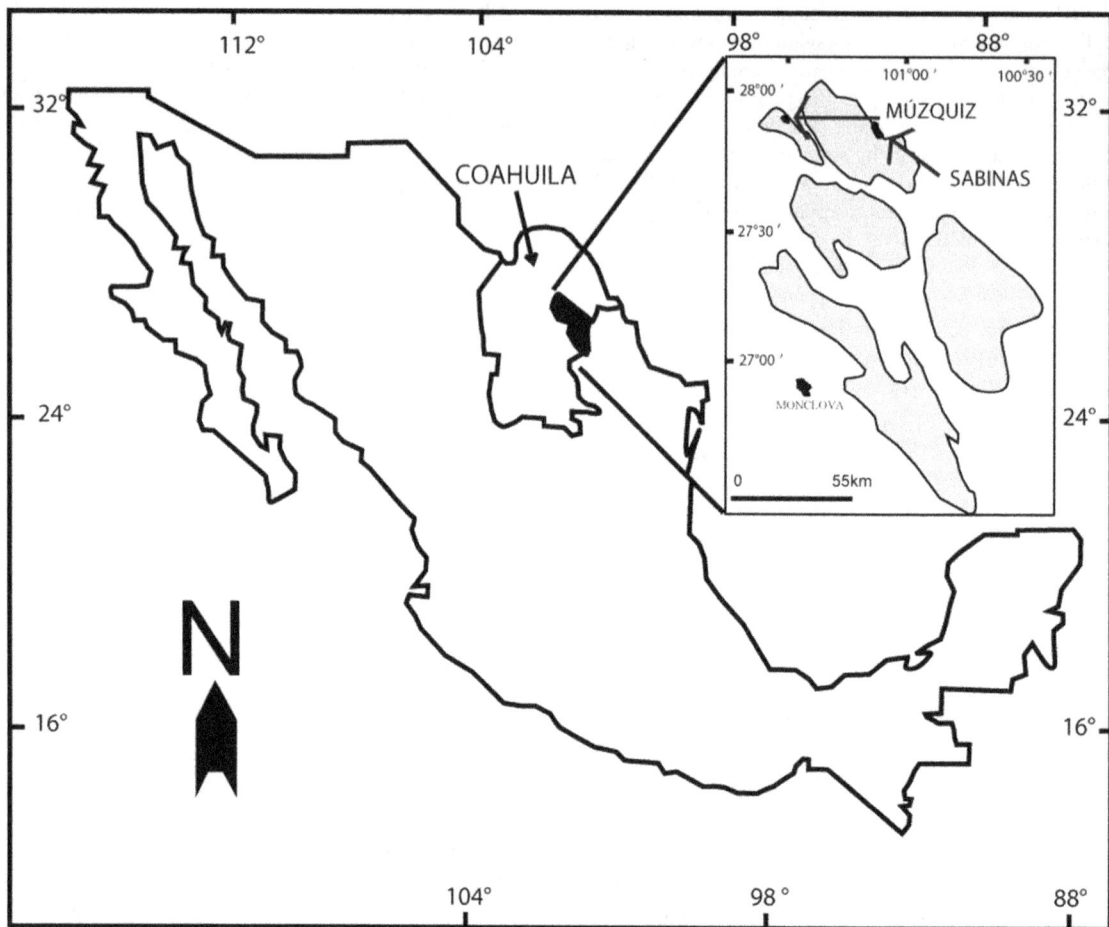

Figure 1. Location map of the Olmos Formation in the Sabinas Basin (in gray). Sampled sites in Sabinas and Múzquiz (Atascoso and Santa Elena Ranchs) are identified by arrows.

S indicate a greater contribution of wide vessels to water conduction in a given area [13].

The safety factor for vessel implosion [18] was calculated using the squared vessel-wall thickness-to-span ratio $(t/b)_h^2$, where the paired vessel wall thickness, t, is the thickness of the wall between two adjacent vessels and b is the diameter of the conduit closest to the hydraulic mean diameter (d_h). The $(t/b)_h^2$ was measured in 16 to 58 vessel pairs per sample, in which at least one of the vessel in a pair was within ± 5 µm of the hydraulic mean diameter (d_h). The number of vessels used for each sample and species/xylotype is presented in Table S2 and Table S3. b has a strong influence on $(t/b)_h^2$; therefore, we do not expect much variation between vessels. The relatively small sample size for some species is because of the insufficient number of vessel pairs within the desired vessel diameter. We also measured implosion resistance in species with solitary vessels that had conducting cells other than vessels (e.g. vascular and vasicentric tracheids) associated with them, as was the case for *Sabinoxylon pasac*, *Quercinium centenoae*, and *Metcalfeoxylon* xylotype 1. Since conduit $(t/b)_h^2$ captures between 80 and 95% of P_{50} variation, we used the relationship between these two variables to estimate cavitation resistance of the fossils ($P_{50} = -0.662 - 154.646\ (t/b)_h^2$; regression equation was kindly provided by Uwe Hacke, University of Alberta).

Comparison with extant communities

The values of the characteristics measured are a good proxy for the hydraulic function of the individual/species studied, in this sense they are by themselves measurements, not predictions of a value (as are the paleoclimate estimates using plant structures, which require a minimum number of species). We compared the calculated K_s, $(t/b)_h^2$, F and S values from the Olmos Formation woods with those from nine extant communities. All the species analyzed in these extant communities were dicots, and the anatomical information was taken from previous studies [1,40,43,44]. These extant communities included a tropical rain forest, a dry deciduous forest and a montane forest from Mexico [1,43] and several communities from sites with a wide range of precipitation in North and South America [40,44]. These last extant communities varied from hardwood forest to desert scrub. For this comparison, K_s, S and F were calculated for the entire set of extant communities. The lack of information necessary to calculate resistance to vessel implosion in the Mexican communities (tropical rain forest, Veracruz; montane forest, Estado de México; dry deciduous forest, Jalisco) restricted our comparisons for this metric to a smaller (6) set of extant localities. Because K_s and S had non-homogeneous error variance term (Levene's test, $P<0.0001$ in both cases) with regard to vegetation type (i.e. variance of the error term is not constant throughout vegetation types) we used Kruskal-Wallis test to detect overall difference

Table 1. Means and standard deviations of the hydraulic and drought resistance traits for the Olmos Formation species/xylotypes.

Species	Affinities	K_s	F	S	$(t/b)_h^2$	P_{50}	N
Coahuiloxylon terrazasiae	Anacardiaceae/Burseraceae	4.28 (2.31)	0.22 (0.55)	0.00168 (0.0005)	0.0038 (0.0014)	−1.25 (0.22)	2
Javelinoxylon xylotype 1	Malvaceae	12.62	0.33	0.0045	0.0103	−2.26	1
Javelinoxylon xylotype 2	Malvaceae	1.63 (0.22)	0.16 (0.011)	0.00069 (0.00004)	0.0346 (0.0065)	−6.01 (1.005)	2
Javelinoxylon weberi	Malvaceae	6.97 (2.24)	0.36 (0.72)	0.0009 (0.00005)	0.0106 (0.0059)	−2.30 (0.091)	2
Metcalfeoxylon xylotype 1	Incertae sedis	5.8 (0.93)	0.15 (0.012)	0.0095 (0.0007)	0.0106 (0.0028)	−2.3 (0.043)	3
Muzquizoxylon porrasii	Cornaceae	0.92 (0.45)	0.17 (0.065)	0.00016 (0.00002)	0.064 (0.0005)	−10.58 (0.084)	2
Olmosoxylon upchurchii	Lauraceae	54.46	0.68	0.0094	0.00159	−0.909	1
Quercinium centenoae	Fagaceae	69.82	0.65	0.0175	0.00101	−0.82	1
Sabinoxylon pasac	Ericales	30.81	0.43	0.0116	0.00303	−1.13	1
Wheeleroxylon atascosense	Malvaceae	38.94 (3.94)	0.66 (0.033)	0.00524 (0.00026)	0.00607 (0.0004)	−1.6 (0.064)	2

K_s = potential conductivity; F = vessel fraction; S = vessel size to number ratio; $(t/b)_h^2$ = vessel implosion resistance; P_{50}= cavitation resistance; N = number of studied samples per xylotype.

among groups. Then, to detect pairwise differences between the Olmos paleoforest and the extant communities, we performed Mann-Whitney U test. The homoscedastic variables (F, implosion resistance and P_{50}) were compared using ANOVA. Similarly, we ran a Levene's test to assess the homoscedasticity of the error term of the functional traits at different MAP levels. The relationship of those traits with homogeneous variance and MAP (F and S) was then analyzed, prior a log transformation, using simple linear regression. For those traits with non-homogeneous variance (implosion resistance and K_s,) we ran weighed least squares regressions using variance at each MAP level as weights to fit the model.

Additionally, we used PCA to visualize 1) hydraulic properties (K_s, S and F) of the Olmos Formation woods with the full extant data set and 2) these same hydraulic properties plus implosion resistance in the smaller subset of communities. To determine whether vegetation type/communities were well discriminated by their hydraulic properties we carried out a between-group PCA and assessed the significance of the results using a Monte-Carlo permutation test on the between group inertia percentage [45,46]. Significance was calculated by comparing the between-group observed differences with the distribution of 999 permutations simulated. We used the R package ade4 [47,48] to perform this analysis.

Results

Potential hydraulic conductivity

There were significant differences of conduction capacities among all communities (Kruskal-Wallis chi squared = 106.3237, p<0.001). Although the Olmos paleoforest had the highest mean potential conductivity per stem cross sectional area ($K_s = 22.6$ g·mm^{-1}·Mpa^{-1}·s^{-1}), it was not significantly different from the hydraulic conductivity of the tropical rain forest (Mann-Whitney Z = 265, p=0.93; $K_s = 17.81$ g·mm^{-1}·Mpa^{-1}·s^{-1}), the montane forest (Mann-Whitney Z = 103, p=0.43; $K_s = 9.41$ g·mm^{-1}·Mpa^{-1}·s^{-1}) and the dry deciduous forest (Mann-Whitney Z = 317, p=0.51; $K_s = 3.06$ g·mm^{-1}·Mpa^{-1}·s^{-1}) (Fig. 2a). The Olmos paleoforest had significantly higher K_s than the dry the North American hardwood forest (Mann-Whitney Z = 14, p=0.005; $K_s = 2.71$ g·mm^{-1}·Mpa^{-1}·s^{-1}) and the remaining extant communities (Fig. 2a). The South American mesquite savanna had the lowest theoretical hydraulic capacity ($K_s = 0.28$ g·mm^{-1}·Mpa^{-1}·s^{-1}). Although the Olmos paleoforest had a high spread in K_s, ranging from 0.9 (g·mm^{-1}·Mpa^{-1}·s^{-1}) in *Muzquizoxylon* to 69.8 (g·mm^{-1}·Mpa^{-1}·s^{-1}) in *Quercinium* (Table 1), the tropical rain forest from Los Tuxtlas, Veracruz had a larger range in values.

Vessel composition and vessel fraction

Patterns of vessel composition S (not shown) were parallel to those for K_s. There was an overall difference in S values across the studied communities (Kruskal-Wallis chi squared = 124.5, p< 0.001). S values of the tropical rainforest and the Olmos paleoforest were not significantly different (Mann-Whitney Z = 302, p=0.62; 0.006 mm^4 vs. 0.0054 mm^4). In these two communities, higher S indicates that fewer larger vessels have greater hydraulic contribution. The S metric was significantly lower in the dry-deciduous forest (Mann-Whitney Z = 435, p = 0.008; 0.0034 mm^4) and the in remaining of the extant communities. The montane forest (0.0014 mm^4), the juniper/mesquite savanna (0.00013 mm^4), North American hardwood forest (0.000119 mm^4), palm forest (0.000114 mm^4), desert (7.6e-05 mm^4), mesquite savanna (5.3e-05 mm^4) and sage scrub

(4.9e-05 mm^4) had S values several orders of magnitude lower than the Olmos paleoforest, indicating that numerous small vessels comprise the conducting area in the woods from these communities.

We detected an overall difference among vegetation types in F values (ANOVA, $F_{9, 198} = 4.5$, p<0.001). Vessel lumen fraction (F) was largest in the montane forest, where more than 50% (0.5 mm^2·mm^{-2}) of the cross sectional area is occupied by vessels, followed by the dry-deciduous forest (0.46 mm^2·mm^{-2}), hardwood forest (0.44 mm^2·mm^{-2}), and tropical rainforest (0.38 mm^2·mm^{-2}). The proportion of cross sectional area occupied by vessels in the Olmos paleoforest is around 38%, while drier communities had lower F values, ranging from 0.31 in the desert vegetation to 0.23 in palm forest, mesquite savanna and juniper mesquite savanna (Fig. 2b). The Olmos paleoforest was only significantly different from the montane forest ($F_{1, 34} = 5.88$, p = 0.02).

The safety factor for vessel implosion

There was a significant difference in resistance to implosion $(t/b)_h^2$ among the compared vegetation types (ANOVA, $F_{6,65} = 3.59$, p = 0.0038). Vessel implosion resistance was the lowest in the Olmos paleoforest (conduit $(t/b)_h^2 = 0.0145$) and ranged from 0.001 in *Quercinium* to 0.064 in *Muzquizoxylon* (Table 1). *Muzquizoxylon* drove mean paleocommunity conduit $(t/b)_h^2$ to higher values (Fig. 2c). As a result, the Olmos paleoforest was not significantly different ($F_{1,19} = 0.0027$, p = 0.95) from the North American hardwood forest and Argentinean palm forest ($F_{1,19} = 0.95$, p = 0.34) with a conduit $(t/b)_h^2 = 0.0139$ and 0.021, respectively. If we compare the medians, instead of the means, the vessel implosion thresholds of the fossil assemblage (0.0082) are lower to those of the hard wood (0.0145) and palm forests (0.0149). The desert ($F_{1,26} = 11.6$, p = 0.002) and the remaining vegetation types had significantly higher vessel implosion resistance than the Olmos paleoforest.

The estimated P_{50} values in the fossil assemblage, calculated with the regression equation describing its relationship with $(t/b)_h^2$, ranged from -0.082 MPa in *Quercinium* to -6 and -10.58 Mpa in *Javelinoxylon* xylotype 2 and *Muzquizoxylon*, respectively (Table 1). The mean paleoforest P_{50} (-2.9 Mpa, 95% CI = $-4.82, -1.01$) was driven to higher values by these two species. The median P_{50} of the fossil assemblage was -1.9 Mpa. We found a significant trade-off between K_s and resistance to vessel implosion ($R^2 = 0.14$, p<0.001; y = 0.014–0.008x), indicating that the very high hydraulic capacity of the Olmos fossils comes at the expense of high susceptibility to vessel implosion/cavitation (Fig. 3a). Low P_{50} values indicate a high risk of cavitation at relatively low water stress (Table 1).

In the extant species' data set, mean annual precipitation (MAP) was significantly related with resistance to vessel implosion ($R^2 = 0.32$, p<0.001; y = 3.8–1.72x, Fig. 3b). None of the vessel conduction metrics F ($R^2 = 0.02$, p = 0.27; y = 0.025+2.2e–5x), S ($R^2 = 0.003$, p = 0.67; y = 1.19–5.5e–5) or K_s ($R^2 = 0.01$, p = 0.78; y = 5.6–4.19x) was significantly associated with MAP, suggesting that resistance to implosion has a greater potential to infer precipitation than any other metric describing water conduction properties. Assuming that resistance to vessel implosion can provide at least a rough approximation of water availability and based on the lines in Figure 3b representing median and the first and third quartiles, we suggest that precipitation in the Olmos paleoforest was comparable to that of our wettest extant sites.

Between-PCA analysis

Species from the Olmos Formation occupied a different region of the functional space in the PCA analysis including hydraulic properties (K_s, S and F) and implosion resistance (Fig. 4). The three first PCA axes explained 99% of the variation. In this analysis, hydraulic capacity and implosion resistance are orthogonal (PC loadings for both PCA analysis are presented in Table S4). In the first PCA axis, K_s had the highest loading, while implosion resistance was highest in the second, and S and F in the third. For instance, in the PCA plot the Olmos paleoforest represents the hydraulically efficient highly prone to vessel implosion extreme. The PCA analysis also revealed that the Olmos paleoforest had greater variation along the hydraulic efficiency axis relative to the implosion resistance axis (the ellipse of the Olmos flora is elongated parallel to the efficient conduction axis). This pattern suggests that the Olmos paleoforest had a large variety of hydraulic strategies but a more constrained number of strategies in implosion/cavitation resistance. The drier communities showed the opposite pattern, with a large breadth in cavitation resistance variation (and also higher values) but highly constrained hydraulic capacity (with low absolute values) (see positive association between axes in Fig. 4). The Monte Carlo permutation test of the ratio of between-class and total inertia in the between PCA analysis supports the discrimination of these groups (ratio = 0.78, P<0.001; Fig. 4). In the second PCA analysis (only K_s, S and F; for all 9 extant communities and the Olmos forest), where the first two principal components explained 93% of the variation, significant differences in functional space among communities are also supported (ratio = 0.25, P<0.001; Fig. 5). In this analysis it is clear that the tropical rain forest exhibits a larger variation in hydraulic strategies than the rest of the communities, including the Olmos paleoforest. In addition, this second analysis reveals a positive relationship between the first two first axes in the tropical rain forest (high K_s is correlated with high S: large conduction capacity is reached with large vessels, see the ellipse orientation in Figure 5) while in drier communities the relationship between these two is negative (large K_s, although overall low in most of them, is reached by having many small vessels; low S). The Olmos paleoforest did not have any strong pattern in this regard.

Discussion

In this paper we set out to determine key properties of the hydraulic system of the Olmos Formation woods. Traits related to hydraulic capacity and vulnerability are important because, as biophysical constraints link them to environmental variation, they provide direct information on ecological strategies of fossil assemblages. For instance, given the warm, wet environment estimated for the Olmos Formation from leaves [28], we expected a highly efficient, highly prone to cavitation hydraulic system. Indeed, our analyses suggest that hydraulic capacity and vulnerability to cavitation at low water stress were high in the Olmos Formation fossil woods.

Hydraulic capacity of The Olmos Formation

The hydraulic capacity of the Olmos paleoforest was similar to several extant communities including wet tropical, dry deciduous, or montane forest. This suggests that the use of K_s in fossil woods provides information allowing characterization of conduction efficiency differences among communities, but that similar conduction efficiencies can be found under relatively different precipitation regimes. For instance, K_s, has enough power to detect differences in conduction capability between dry and wet floras and therefore may provide only rough estimates of water

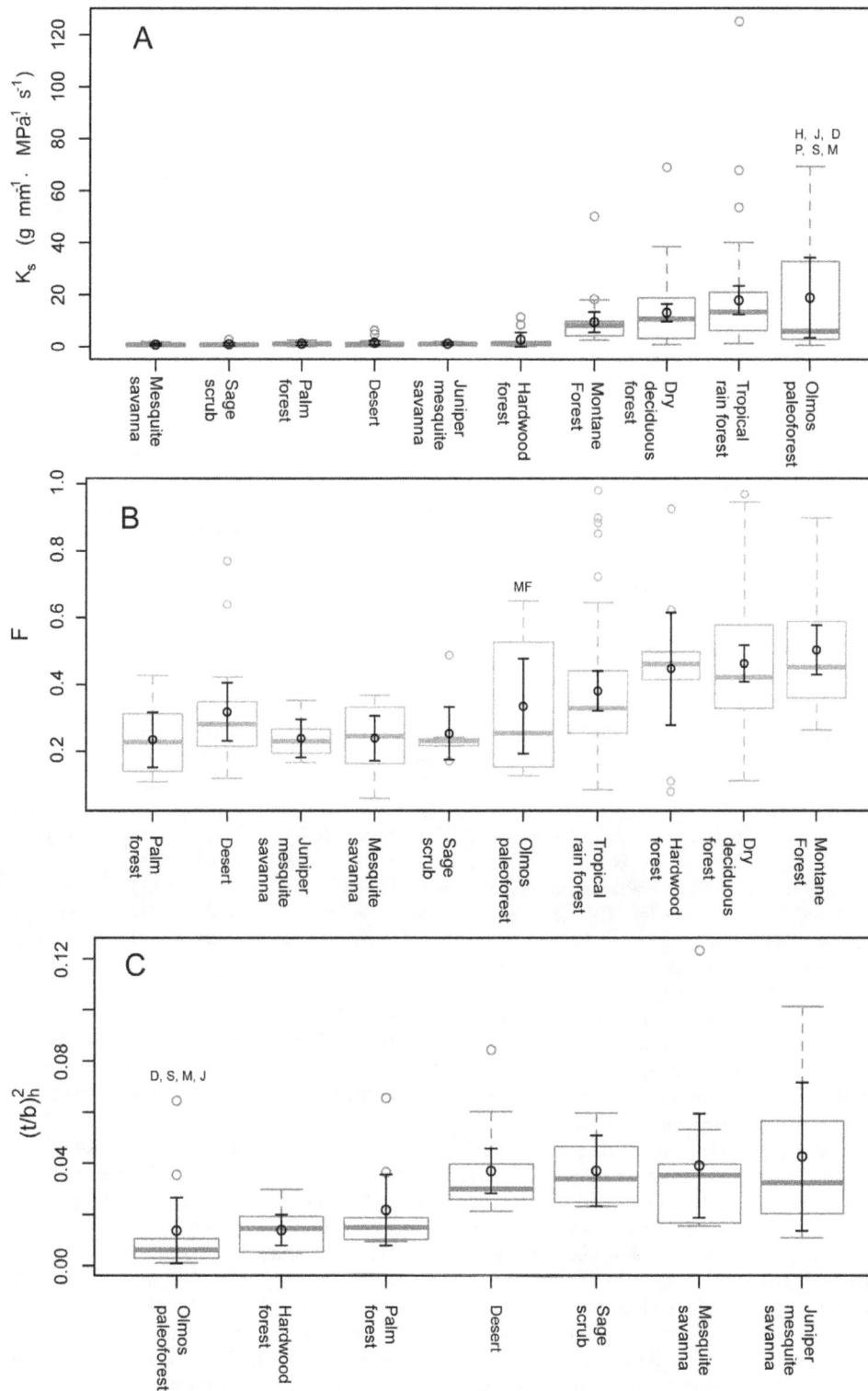

Figure 2. Comparison of a) potential conductivity, b) vessel fraction and c) implosion resistance between the Olmos Formation fossil woods and extant communities. Box plots in grey show the median and interquartile distance of each one of the variables. The black circle and error bars represent the mean and ±95% confidence intervals. Box plots showing implosion resistance only includes the drier extant communities and the Olmos formation flora. Sample sizes: mesquite savanna (M) = 11, sage scrub (S) = 8, palm forest (P) = 9, desert (D) = 17, juniper-mesquite savanna (J) = 7, hardwood forest (H) = 10, montane forest (MF) = 25, dry deciduous forest = 56, tropical rain forest = 54, Olmos flora = 10. The two different desert communities were pooled for this analysis. Significant difference between the Olmos pleoflora and extant vegetation types were determined using Mann-Whitney test for potential conductivity, and ANOVA for *F* and implosion resistance.

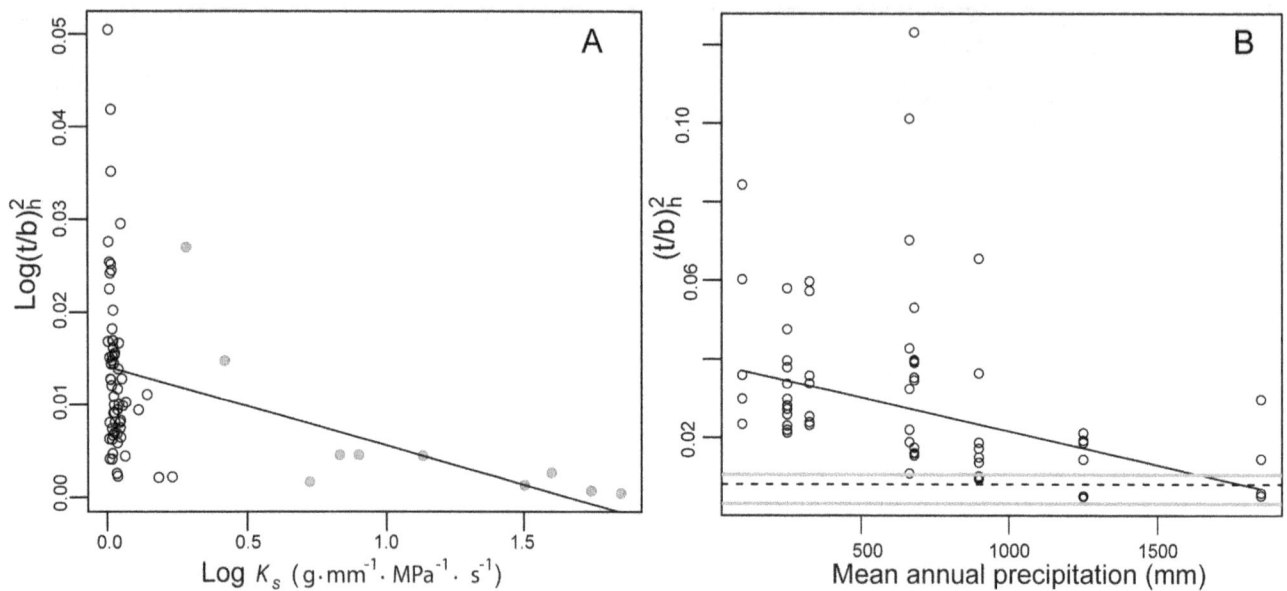

Figure 3. Implosion resistance as a function of a) potential conductivity, b) MAP. In figure 3a the Olmos formation xylotypes are in grey; black circles are the extant communities presented in Figure 2c. In figure 3b the horizontal lines show the median (dashed) and first and third quartiles (grey lines) of implosion resistance for the Olmos Formation, the circles are the species of all extant communities in Figure 2c. Regression line in 3a was fitted using simple linear regression on the log-transformed variables, while in 3b using weighed least square regression.

availability in paleoforests. This is further supported by the absence of a significant relationship between K_s and MAP in our analysis of extant communities. These two variables have been found to be independent in evergreen angiosperms [49]. There is, however, a significant inverse relationship (instead of positive as would be expected) between these two variables in deciduous angiosperms that has been interpreted as the adaptation to water limitation in plant with this ecological strategy [49]. The inverse relationship observed in that study is because most of the deciduous species analyzed [49] are winter deciduous, and thus, a product of the decrease in vessel size associated with higher thresholds to freezing induced cavitation of small vessels (e.g., [50,51]). It is also possible that K_s is maximized in seasonally dry habitats (e.g., tropical deciduous forest) because of the need for fast growth and carbon fixation during restricted periods of water availability [49,52]. In the context of our study, the absence of ring porous species and the faint growth rings in the Olmos Formation species indicate that the flora was not subject to sharp seasonal fluctuations of water transport and growth. Instead, high hydraulic efficiency of the woods analyzed, along with constancy of vessel diameter across growth rings, suggests environmental conditions allowing constant hydraulic capacity throughout the year.

As expected, variation in S and F yielded similar results. However, S and F describe different functional aspects of water conduction [13]. Low S indicates the presence of many small vessels, which are selective under conditions promoting freezing-induced embolism [13,16]. Thus the large S value of the Olmos woods, which was similar to the S value of the tropical rainforest, suggests the absence of freezing temperatures. F, on the other hand, is related to potential conductivity, but also describes the amount of cross sectional space occupied by lumen and, by extension, non-lumen (cell wall) area. This metric is bounded at higher values by mechanical support needs and at lower values by hydraulic requirements [13,42]. All else being equal, high F would indicate low construction cost because growth is achieved by greater gas fraction represented by vessel space [12,13,53]. The F

value of the Olmos paleoforest, which is not the highest in our study (around 0.38 $mm^2 \cdot mm^{-2}$ compared to up to 0.5 and 0.46 $mm^2 \cdot mm^{-2}$ in the montane and dry deciduous forest), suggests that construction cost was high relative to hydraulic efficiency. It is important to keep in mind, however, that the amount of variation of wood density (a proxy for construction cost) explained by vessel fraction or other vessel traits is low [13,19,40] and that density is mainly driven by fiber traits (e.g. [40]). High wood fraction is positively related with survival and slow growth rate in shade tolerant species, while increased gas fraction is related to high light requirements and adult stature in fast growing species [12]. In sum, our analysis indicate that the Olmos paleoforest had high hydraulic capacity (high K_s) carried out by few very efficient vessels (high S), but the amount of wood committed to water conduction was not very large (F).

Resistance to implosion and cavitation

Given the low estimated cavitation resistance (mean $P_{50} = -2.9$ MPa) and vessel wall reinforcement metric (($t/b)^2_h = 0.0145$) of the Olmos Formation plants, it seems that, on average, they were at high risk of cavitation even at high water potential (low water stress). This pattern is even more evident if the median P_{50} (-1.9 MPa) or the mean without outliers (-1.57 MPa) is considered. In a large study considering 230 observations belonging to 167 species from several vegetation types, Maherali et al. [49] showed that the median P_{50} for extant tropical rain forest was around -1 MPa (n = 41), and ranged from very close to 0 MPa to a little over -6 MPa. Therefore, despite most of species from tropical rain forests being very prone to experience cavitation at high water potential, this biome also has some species that can be quite resistant to drought-induced cavitation. Maherali et al. [49] also found that despite its adaptive significance, P_{50} exhibited large variation both within and across climates. In some species from the Olmos paleoforest such as *Quercinium* and *Olmosoxylon*, with $(t/b)^2_h$ values in the order of 0.001 and estimated P_{50} of -0.82 and -0.91 MPa, respectively, it is likely that mild water

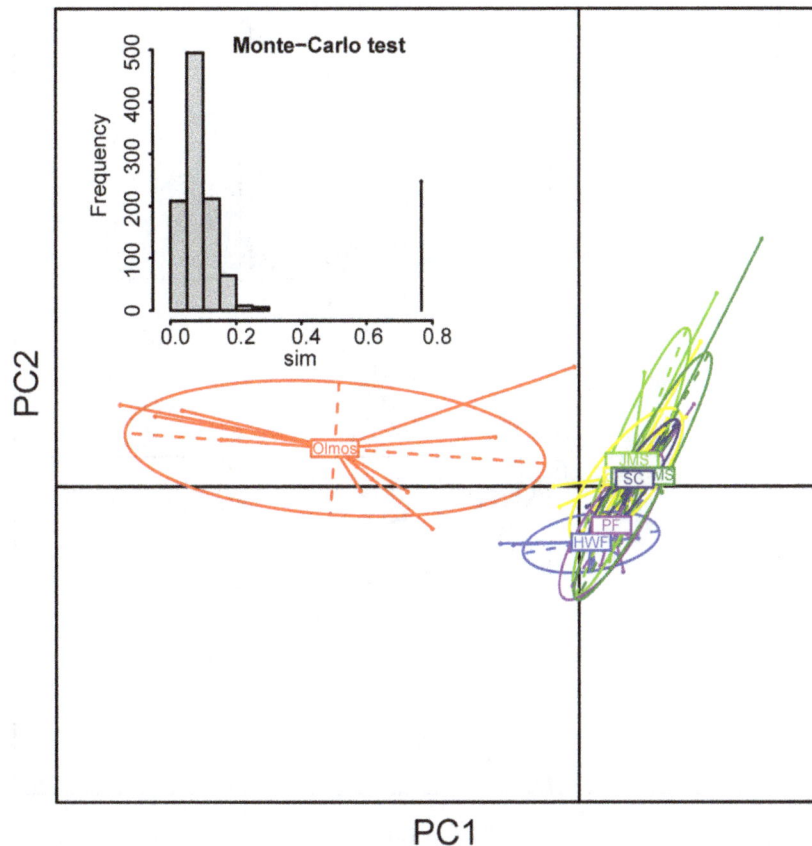

Figure 4. PCA plot portraying hydraulic strategies among communities. Between-Class principal component analysis showing ellipses and gravity centers for extant communities and the Olmos paleoforest. Wet tropical, dry deciduous and montane forest were not included in this analysis. The insert in the upper left corner is the histogram of 1000 simulated values for the between-groups PCA and the observed value (vertical line at sim = 0.78). Sim = ratio of between class and total inertia. HWF = hardwood forest, JMS = juniper-mesquite savanna, MS = mesquite savanna, PF = palm forest, SC = sage scrub.

stress would have driven embolism formation and/or vessel collapse. Median P_{50} for tropical dry forest in the study of Maherali et al. [49] was almost 1 MPa lower than our median (around -2.5 MPa, n = 19), but some species reached over -10 MPa. Of the Olmos Formation woods, only *Muzquizoxylon* reached such low values. It could be argued that the reason behind the extremely high $(t/b)_h^2$ and P_{50} values in *Muzquizoxylon* is deficiency in preservation, but our observations do not support this conclusion. As relatively high values of cavitation resistance are not absent from extant tropical communities, the high inferred cavitation resistance of *Muzquizoxylon* could also be attributed to intra-community variation.

We showed that there is a significant relationship between $(t/b)_h^2$ and MAP in extant communities and that the wall reinforcement parameter of the Olmos species were in general within the ranges of extant vegetations with higher precipitation. It has been found that P_{50} decreases (becomes more negative) with decreasing precipitation in extant evergreen angiosperms and conifers [49]. High resistance to embolism (more negative P_{50}) has therefore evolved as a strategy to cope with drought. It seems that the utility of $(t/b)_h^2$ and P_{50} in detecting water stress in fossil communities is higher than that for K_s, and the calculated values of these metrics for the Olmos paleoforest indicate high water availability.

Paleoclimate and vegetation

The high hydraulic capacity and low resistance to drought we calculated here reinforces evidence from foliar physiognomy indicating a wet warm climate for the Olmos paleoforest. Around 72% of the species are entire-margined, and 50% of the species with preserved leaf apices have drip tips [28]. Foliar physiognomy estimates a MAT of 20–23°C and a MAP of 1.5–3 m and growing season precipitation of 2 m. In addition, the occurrence of palms and the absence of growth rings in the dicot woods indicate cold month mean temperature >5°C [54]. Our results also agree with the current understanding of the climate regime of the southern WINA during the Late Cretaceous as having tropical temperatures (MAT 18–25°C) and above-freezing annual minimum temperatures [28,55,56]. Tropical rainforest is defined by a combination of climatic parameters and plant physiognomical features [57]. These include a MAP of over 1.8 m year^{-1}, high MAT (>18°C), a seasonal variation in temperature of less than 7°C and high percent of species with large, entire margined leaves and dip tips [57]. The climate calculated for the Olmos Formation and the physiognomic characteristics of the leaves suggest it was a tropical rainforest. Paleoclimate models show that the frost line during the Maastrichtian was well to the north [58], and based on these simulations [58], Lomax et al. [59] predicted high net primary productivity and leaf area index for the Olmos Formation region. Significantly, the leaf flora of the Olmos Formation also indicates greater precipitation levels than other North American and

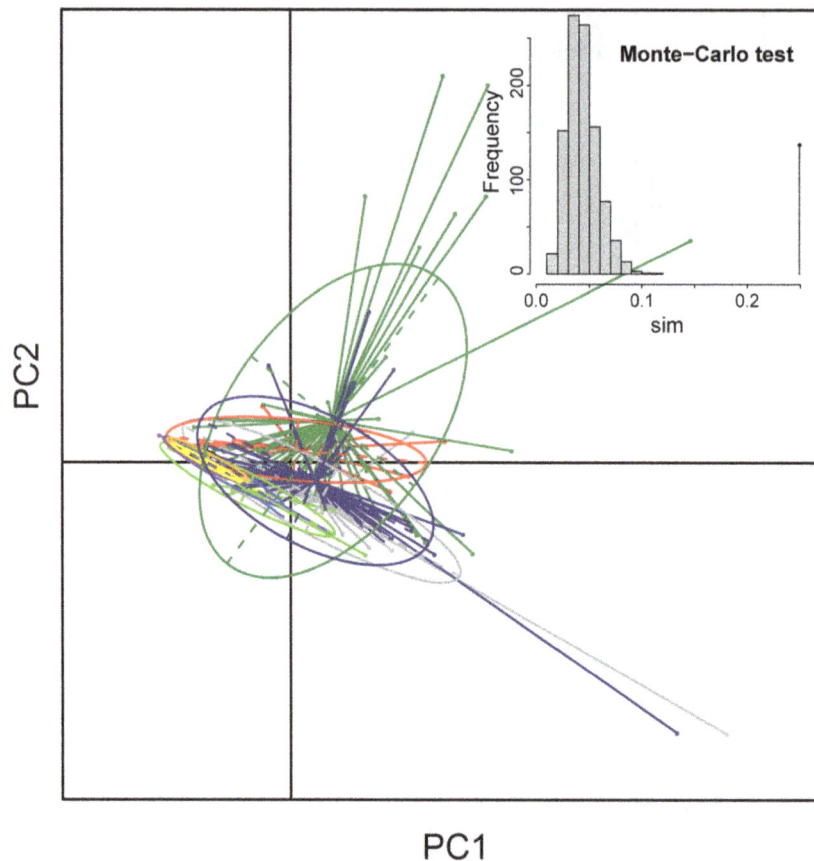

Figure 5. PCA plot portraying hydraulic strategies among all the studied communities. Between-Class principal component analysis shows the ellipses and gravity centers. This plot is based on only hydraulic traits (F, S and K_s). The insert in the upper right corner is the histogram of 1000 simulated values for the between-groups PCA and the observed value (vertical line at sim = 0.25). Sim = ratio of between class and total inertia. Red = Olmos paleoforest, dark green = tropical rain forest, light green = hardwood forest, light yellow = sage scrub, dark yellow = palm forest, light blue = desert, dark blue = dry deciduous forest, grey = montane forest, black = juniper mesquite savanna, purple = mesquite savanna.

Western Interior floras during the Late Cretaceous. Based on the small leaf size and low prevalence of drip tips in the paleofloras of the region, Wolfe and Upchurch [56] suggest a subhumid climate throughout the late Cretaceous. Other calculations, based on Climate Multivariate Leaf Analysis Program (CLAMP), support those estimates and propose that in the southern Western Interior precipitation reached up to 1.5 m but it could have been less than 1 m [60,61]. These estimates contrast with the more mesic conditions of the Olmos Formation flora, with up to 3 m of precipitation, and high prevalence of drip tips. In this sense, the Olmos paleoforest, is physiognomically closer to younger (Paleocene) southern Western Interior floras, where larger leaves, with up to 50% of them having drip tips, are present [56,28,36].

The high hydraulic capacity and low resistance to embolism agrees with the high water availability inferred by the leaf physiognomy [28]. Since the fossil woods we studied here were collected in the meandering rivers facies and some of them have pelecypod perforations, it possible that, given their very low cavitation resistance values, the assemblage could be riparian. However, at this point, this is uncertain since the fossil woods were not collected in growth position, despite that several of them reached over 40 m (*Metcalfeoxylon* and *Javelinoxylon* xylotype 1). Whether the woods represent a riparian environment or not, extreme humidity of the Olmos Formation is independently confirmed by the leaf flora as it was collected in the flood plain-lagoon lithofacies [28].

Globally, it has been suggested that an increase in xylem and leaf hydraulic capacity during the mid to late Cretaceous [62,63] likely influenced a significant amplification of angiosperm's forest biomass [64], and contributed to maximize carbon fixation [62] and expansion of tree size [64]. Indeed, our results indicate that the high conduction capacity of the woods from the Olmos paleoforest was necessary to sustain high leaf area [28] and high productivity predicted by paleoclimatic models [58,59] for the region. High hydraulic capacity in the Olmos Formation paleoforest supports the current understanding of late cretaceous angiosperm hydraulic function, which proposes an increased conduction efficiency compared to early cretaceous short statured trees/shrubs [62,63].

We suggest that the climate for the Olmos paleoforest, and likely other floras of the WINA, selected for an ecological strategy that maximized conductance and efficient carbon gain, and penalized high cavitation resistance because of its associated cost in hydraulic efficiency. As the probability of finding large pores in the pit membrane increases with vessel size [14,20,21], resistance to cavitation comes at expense of vessel size and conduction capacity. We suggest that the probability of drought in the Olmos paleoforest was low, otherwise the high vulnerability to cavitation of most species, together with a narrow range of ecological strategies along this functional aspect, would be an extremely risky strategy.

Supporting Information

Table S1 Vessel dimensions data for each of the samples/morphotypes.

Table S2 Sample means for each functional trait.

Table S3 Individual measurements of implosion resistance and estimated cavitation resistance.

Table S4 Trait loadings in the three first PC axes.

Acknowledgments

We thank Uwe Hacke for providing the regression equation to calculate P_{50} values, Jochen Schenk and Cynthia Jones for providing data for some of the extant communities, and Deborah Woodcock and Garland Upchurch Jr. for their comments on a previous draft. We thank Roberto Pujana, Paul Fine and an anonymous reviewer for their constructive criticism.

Author Contributions

Conceived and designed the experiments: HIMC EER. Performed the experiments: HIMC EER. Analyzed the data: HIMC EER. Contributed reagents/materials/analysis tools: HIMC EER. Wrote the paper: HIMC EER.

References

1. Martínez-Cabrera HI, Estrada-Ruiz E, Castañeda-Posadas C, Woodcock D (2012) Wood specific gravity estimation based on wood anatomical traits: Inference of key ecological characteristics in fossil assemblages. Review of Palaeobotany and Palynology 187: 1–10.
2. Royer DL, Sack L, Wilf P, Lusk CH, Jordan GJ, et al. (2007) Fossil leaf economics quantified: calibration, Eocene case study, and implications. Paleobiology 33: 574–589.
3. Wilf P, Wing SL, Greenwood DR, Greenwood CL (1998) Using fossil leaves as paleoprecipitation indicators. An Eocene example. Geology 26: 203–206.
4. Wolfe JA (1971) Tertiary climatic fluctuations and methods of analysis of Tertiary floras. Palaeogeography, Palaeoclimatology, Palaeoecology 9: 27–57.
5. Lambers H, Chapin FS, Pons TL (1998) Plant physiological ecology. New York: Springer-Verlag. 540 p.
6. Machado J-L, Tyree MT (1994) Patterns of hydraulic architecture and water relations of two tropical canopy trees with contrasting leaf phonologies: Ochroma pyramidale and Pseudobombax septenatum. Tree Physiology 14: 219–240.
7. Tyree MT, Sneidermann DA, Wilmot TR, Machado JL (1991) Water relations and hydraulic architecture of a tropical tree (Scheflera morototoni): data, models and a comparison to two temperate species (Acer saccharum and Thuja occidentalis). Plant Physiology 96: 1105–1113.
8. Santiago LS, Goldstein G, Meinzer FC, Fisher JB, Machado K, et al. (2004) Leaf photosynthetic traits scale with hydraulic conductivity and wood density in Panamanian forest canopy trees. Oecologia 140: 543–550.
9. Carlquist S (1975) Ecological strategies of xylem evolution. Berkeley: Univ. Calif. Press. 243 p.
10. Carlquist S (1988) Comparative wood anatomy. Berlin: Springer-Verlag. 436 p.
11. Wheeler EA, Baas P, Rodgers S (2007) Variations in dicot wood anatomy. A global analysis. IAWA Journal 28: 229–258.
12. Poorter L (2008) The relationships of wood-, gas-, and water fractions of tree stems to performance and life history variation in tropical trees. Annals of Botany 102: 367–375.
13. Zanne AE, Westoby M, Falster DS, Ackerly DD, Loarie SR, et al. (2010) Angiosperm wood structure: Global patterns in vessel anatomy and their relation to wood density and potential conductivity. American Journal of Botany 97: 207–215.
14. Zimmerman MH (1983) Xylem structure and ascent of the sap. Berlin: Springer-Verlag. 143 p.
15. Brodribb TJ, Feild TS (2000) Stem hydraulic supply is linked to leaf photosynthetic capacity: evidence from new Caledonian and Tasmanian rainforest. Plan Cell and Environment 23: 1381–1388.
16. Tyree MT (2003) Hydraulic limits on tree performance: transpiration, carbon gain and growth of trees. Trees 17: 95–100.
17. Hacke UG, Sperry JS (2001) Functional and ecological xylem anatomy. Perspectives in Plant Ecology, Evolution and Systematics 4: 97–115.
18. Hacke UG, Sperry JS, Pockman WT, Davis SD, Mcculloh KA (2001) Trends in wood density and structure are linked to prevention of xylem implosion by negative pressure. Oecologia 126: 457–461.
19. Jacobsen AL, Ewers FW, Pratt RB, Paddock WA, Davis SD (2005) Do xylem fibers affect vessel cavitation resistance? Plant Physiology 139: 546–556.
20. Jarbeau JA, Ewers FW, Davis SD (1995) The mechanism of water stress induced embolism in two species of chaparral shrubs. Plant Cell and Environment 18: 189–196.
21. Wheeler JK, Sperry JS, Hacke UG, Hoang N (2005) Inter-vessel pitting and cavitationin woody Rosaceae and other vesselled plants: a basis for a safety versus efficiency trade-off in xylem transport. Plant, Cell and Environment 28: 800–812.
22. Kolb KJ, Sperry JS (1999) Differences in drought adaptation between subspecies of sagebrush (Artemisia tridentata). Ecology 80: 2373–2384.
23. Sperry JS, Pockman WT (1993) Limitation of transpiration by hydraulic conductance and xylem cavitation in Betula occidentalis. Plant Cell and Environment 16: 279–288.
24. Meinzer FC, Clearwater MJM, Goldstein G (2001) Water transport in trees: current perspectives, new insights and some controversies. Environmental and Experimental Botany 45: 239–262.
25. Pratt RB, Ewers FW, Lawson MC, Jacobsen AL, Brediger M, et al. (2005) Mechanism for toleratin freeze-thaw stress of two evergreen chaparral species: Rhus ovata and Malosma laurina (Anacardiaceae). American Journal of Botany 92: 1102–1113.
26. Tyree MT, Sperry JS (1989) Vulnerability of xylem to cavitation and embolism. Annual Review of Plant Physiology and Plant Molecular Biology 40: 19–38.
27. Estrada-Ruiz E, Martínez-Cabrera HI, Cevallos-Ferriz SRS (2010) Fossil woods from the Olmos Formation (late Campanian-early Maastrichtian), Coahuila, Mexico. 97: 1179–1194.
28. Estrada-Ruiz E, Upchurch Jr GR, Cevallos-Ferriz SRS (2008) Flora and climate of the Olmos Formation (upper Campanian-lower Maastrichtian), Coahuila, Mexico: A preliminary report. Gulf Coast Association of Geological Societies Transactions 58: 273–283.
29. Estrada-Ruiz E, Upchurch Jr GR, Wolfe JA, Cevallos-Ferriz SRS (2011) Comparative morphology of fossil and extant leaves of Nelumbonaceae, including a new genus from the Late Cretaceous of Western North America. Systematic Botany 32: 337–351.
30. Serlin B, Delevoryas TH, Weber R (1980) A new conifer pollen cone from the Upper Cretaceous of Coahuila, Mexico. Review of Palaeobotany and Palynology 31: 241–248.
31. Weber R (1972) La vegetación maestrichtiana de la Formación Olmos de Coahuila, México. Boletín de la Sociedad Geológica Mexicana 33: 5–19.
32. Weber R (1973) Salvinia coahuilensis nov. sp. del Cretácico Superior de México. Ameghiniana 10: 173–190.
33. Weber R (1975) Aachenia knoblochii n. sp. an interesting conifer of the Upper Cretaceous Olmos Formation of Northeastern Mexico. Palaeontographica 152B: 76–83.
34. Weber R (1978) Some aspects of the Upper Cretaceous angiosperm flora of Coahuila, Mexico. Courier Forschungsinstitut Senckenberg 30: 38–46.
35. Eguiluz de Antuñano S (2001) Geologic evolution and gas resources of the Sabinas Basin in northeastern Mexico. In: Bartolini C, Buffler RT, Cantú-Chapa A, editors. The western Gulf of Mexico basin: Tectonics, sedimentary basins, and petroleum systems: American Association of Petroleum Geologists Memoir. pp. 241–270.
36. Estrada-Ruiz E (2009) Reconstrucción de los ambientes de depósito y paleoclima de la región de Sabinas-Saltillo, estado de Coahuila, con base en plantas fósiles del Cretácico Superior. Doctoral Thesis, Universidad Nacional Autónoma de México.
37. Flores Espinoza E (1989) Stratigraphy and sedimentology of the Upper Cretaceous terrigenous rocks and coal of the Sabinas-Monclova area, northern Mexico. Austin: University of Texas at Austin. 315 p.
38. Robeck RC, Pesquera VR, Ulloa AS (1956) Geología y depósitos de carbón de la región de Sabinas, Estado de Coahuila. XX Congreso Geológico Internacional, México. pp. 109.
39. Estrada-Ruiz E, Martínez-Cabrera HI, Cevallos-Ferriz SRS (2007) Fossil wood from the late Campanian-early Maastrichtian Olmos Formation, Coahuila, Mexico. Review of Palaeobotany and Palynology 145: 123–133.
40. Martínez-Cabrera HI, Jones CS, Espino S, Schenk HJ (2009) Wood anatomy and wood density in shrubs: Responses to varying aridity along transcontinental transects. American Journal of Botany 96: 1388–1398.
41. Davis SD, Sperry JS, Hacke UG (1999) The relationship between xylem conduit diameter and cavitation caused by freezing. American Journal of Botany 86: 1367–1372.
42. Preston KA, Cornwell WK, DeNoyer JL (2006) Wood density and vessel traits as distinct correlates of ecological strategy in 51 California coast range angiosperms. New Phytologist 170: 807–818.
43. Martínez-Cabrera HI, Cevallos-Ferriz SRS (2008) Palaeoecology of the Miocene El Cien Formation (Mexico) as determined from wood anatomical characters. Review of Palaeobotany and Palynology 150: 154–167.

44. Schenk HJ, Espino S, Goedhart CM, Nordenstahl M, Martínez-Cabrera HI, et al. (2008) Hydraulic integration and shrub growth form linked across continental aridity gradients. Proceedings of the National Academy of Sciences 105: 11248–11253.

45. Baty F, Facompré M, Wiegand J, Schwager J, Brutsche MH (2006) Analysis with respect to instrumental variables for the exploration of microarray data structures. BMC Bioinformatics 7: 422.

46. Dolédec S, Chessel D (1987) Rythmes saisonniers et composantes stationnelles en milieu aquatique I: Description d'un plan d'observations complet par projection de variables. Acta Oecologica, Oecologia Generalis 8: 403–426

47. Chessel D, Dufour AB, Thioulouse J (2004) The ade4 package-I: One-table methods. R News 4: 5–10.

48. Thioulouse J, Chessel D, Dolédec S, Olivier JM (1996) ADE-4: A multivariate analysis and graphical display software. Statistics and Computing 7: 75–83.

49. Maherali H, Pockman WT, Jackson RB (2004) Adaptive variation in the vulnerability of woody plants to xylem cavitation. Ecology 85: 2184–2199.

50. Sperry JS, Sullivan JEM (1992) Xylem embolism in response to freeze-thaw cycles and water stress in ring-porous, diffuse-porous, and conifer species. Plant Physiology 100: 605–613.

51. Wang J, Ives N, Lechowicz MJ (1992) The relation of foliar phenology to xylem embolism in trees. Functional Ecology 6: 469–475.

52. Reich PB, Ellsworth DS, Walters MB, Vose J, Gresham C, et al. (1999) Generality of leaf traits relationships: a test across six biomes. Ecology 80: 1955–1969.

53. Gartner BL, Moore JR, Gardiner BA (2004) Gas in stems: abundance and potential consequences for tree biomechanics. Tree Physiology 24: 1239–1250.

54. Greenwood MC, Wing SL (1995) Eocene continental climates and latitude temperature gradients. Geology 23: 1044–1048.

55. Spicer RA, Parrish JT (1986) Paleobotanical evidence for cool North Polar climates in middle Cretaceous (Albian-Cenomanian) time. Geology 14: 703–706.

56. Wolfe JA, Upchurch Jr GR (1987) North American nonmarine climates and vegetation during the Late Cretaceous. Palaeogeography, Palaeoclimatology, Palaeoecology 61: 33–77.

57. Jaramillo C, Cárdenas A (2013) Global warming and neotropical rainforests: A historical perspective. Annual Review of Earth and Planetary Sciences 41: 741–766.

58. Upchurch Jr GR, Otto-Bliesner BL, Scotese C (1998) Vegetation-atmosphere interactions and their role in global warming during the latest Cretaceous. Phil Trans R Soc Lond B 353: 97–112.

59. Lomax BH, Beerling DJ, Upchurch Jr GR, Otto-Bliesner BL (2000) Terrestrial ecosystem responses to global environmental change across the Cretaceous-Tertiary boundary. Geophysical Research Letters 27: 2149–2152.

60. Wolfe JA (1990) Paleobotanical evidence for a marked temperature increase following the Cretaceous-Tertiary boundary. Nature 343: 153–156.

61. Johnson KR, Reynolds ML, Wert KW, Thomasson JR (2003) Overvew of the Late Cretaceous, early Paleocene, and Early Eocene megafloras of the Denver Basin, Colorado. Rocky Mountain Geology 38: 101–120.

62. Feild TS, Brodribb TJ, Iglesias A, Chatelet DS, Baresch A, et al. (2011) Fossil evidence for Cretaceous escalation in angiosperm leaf vein evolution. Proceedings of the National Academy of Science 108: 8363–8366.

63. Feild TS, Wilson JP (2012) Evolutionary Voyage of Angiosperm Vessel Structure-Function and Its Significance for Early Angiosperm Success. International Journal of Plant Sciences 173: 596–609.

64. Upchurch Jr GR, Wolfe JA (1993) Cretaceous vegetation of the Western Interior and adjacent regions of North America. In: Kauffman EG, Caldwell WGE, editors. Cretaceous evolution of the Western Interior Basin. Geological Association of Canada Special Paper 39. pp. 243–281.

Permissions

All chapters in this book were first published in PLOS ONE, by The Public Library of Science; hereby published with permission under the Creative Commons Attribution License or equivalent. Every chapter published in this book has been scrutinized by our experts. Their significance has been extensively debated. The topics covered herein carry significant findings which will fuel the growth of the discipline. They may even be implemented as practical applications or may be referred to as a beginning point for another development.

The contributors of this book come from diverse backgrounds, making this book a truly international effort. This book will bring forth new frontiers with its revolutionizing research information and detailed analysis of the nascent developments around the world.

We would like to thank all the contributing authors for lending their expertise to make the book truly unique. They have played a crucial role in the development of this book. Without their invaluable contributions this book wouldn't have been possible. They have made vital efforts to compile up to date information on the varied aspects of this subject to make this book a valuable addition to the collection of many professionals and students.

This book was conceptualized with the vision of imparting up-to-date information and advanced data in this field. To ensure the same, a matchless editorial board was set up. Every individual on the board went through rigorous rounds of assessment to prove their worth. After which they invested a large part of their time researching and compiling the most relevant data for our readers.

The editorial board has been involved in producing this book since its inception. They have spent rigorous hours researching and exploring the diverse topics which have resulted in the successful publishing of this book. They have passed on their knowledge of decades through this book. To expedite this challenging task, the publisher supported the team at every step. A small team of assistant editors was also appointed to further simplify the editing procedure and attain best results for the readers.

Apart from the editorial board, the designing team has also invested a significant amount of their time in understanding the subject and creating the most relevant covers. They scrutinized every image to scout for the most suitable representation of the subject and create an appropriate cover for the book.

The publishing team has been an ardent support to the editorial, designing and production team. Their endless efforts to recruit the best for this project, has resulted in the accomplishment of this book. They are a veteran in the field of academics and their pool of knowledge is as vast as their experience in printing. Their expertise and guidance has proved useful at every step. Their uncompromising quality standards have made this book an exceptional effort. Their encouragement from time to time has been an inspiration for everyone.

The publisher and the editorial board hope that this book will prove to be a valuable piece of knowledge for researchers, students, practitioners and scholars across the globe.

List of Contributors

Natasha L. Vokhshoori and Matthew D. McCarthy
Ocean Sciences Department, University of California Santa Cruz, Santa Cruz, California, United States of America

Sahra Talamo and Jean-Jacques Hublin
Department of Human Evolution, Max Planck Institute for Evolutionary Anthropology, Leipzig, Germany

Marco Peresani, Matteo Romandini, Nicola Nannini and Camille Jéquier
Universitá di Ferrara, Dipartimento di Studi Umanistici, Ferrara, Italy

Andreas Pastoors and Gerd-Christian Weniger
Neanderthal Museum, Mettmann, Germany

Andrea Picin
Neanderthal Museum, Mettmann, Germany
Institut Catalá de Paleoecologia Humana i Evolució Social (IPHES), Tarragona, Spain
Universitat Rovira I Virgili, Area de Prehistória, Tarragona, Spain

Rossella Duches
Universitá di Ferrara, Dipartimento di Studi Umanistici, Ferrara, Italy
Museo delle Scienze, Trento, Italy

Manuel Vaquero
Institut Catalá de Paleoecologia Humana i Evolució Social (IPHES), Tarragona, Spain
Universitat Rovira I Virgili, Area de Prehistória, Tarragona, Spain

Guo Liu, Yi Yin, Hongyan Liu and Qian Hao
College of Urban and Environmental Sciences, Peking University, Beijing, China

Ana Rosa Gómez Cano
Departamento de Paleontología, Facultad de Ciencias Geológicas, Universidad Complutense de Madrid, Madrid, Spain

Manuel Hernández Fernández and M. Ángeles Álvarez-Sierra
Departamento de Paleontología, Facultad de Ciencias Geológicas, Universidad Complutense de Madrid, Madrid, Spain
Departamento de Geología Sedimentaria y Cambio Medioambiental, Instituto de Geociencias (UCM, CSIC), Madrid, Spain

Gangsheng Wang, Melanie A. Mayes and Lianhong Gu
Climate Change Science Institute, Oak Ridge National Laboratory, Oak Ridge, Tennessee, United States of America
Environmental Sciences Division, Oak Ridge National Laboratory, Oak Ridge, Tennessee, United States of America

Christopher W. Schadt
Climate Change Science Institute, Oak Ridge National Laboratory, Oak Ridge, Tennessee, United States of America
Biosciences Division, Oak Ridge National Laboratory, Oak Ridge, Tennessee, United States of America

Rafael César Lima Pedroso de Andrade
Graduate Student/Programa de Pós-Graduação (CTG), Universidade Federal de Pernambuco, Recife, Brazil

Juliana Manso Sayão
Centro Acadêmico de Vitória, Universidade Federal de Pernambuco, Bela Vista, Vitória de Santo Antão, Pernambuco, Brazil

David Coty, Romain Garrouste, Frédéric Legendre and AndréNel
Muséum National d'Histoire Naturelle, Institut de Systématique, Evolution, Biodiversité, ISYEB, UMR 7205 CNRS UPMC EPHE, Paris, France

Cédric Aria
Department of NaturalHistory-Palaeobiology, Royal Ontario Museum, Toronto, Ontario, Canada
Department of Ecology & Evolutionary Biology, University of Toronto, Toronto, Ontario, Canada

Patricia Wils
CNRS UMS 2700, Muséum National d'Histoire Naturelle, Paris, France

Valentin Fischer
Department of Geology, University of Liége, Liége, Belgium, Operational Directory 'Earth and History of Life', Royal Belgian Institute of Natural Sciences, Brussels, Belgium

Pascal Godefroit
Operational Directory 'Earth and History of Life', Royal Belgian Institute of Natural Sciences, Brussels, Belgium

Nathalie Bardet
CNRS UMR 7207, Département Histoire de la Terre, Muséum National d'Histoire Naturelle, Paris, France

Myette Guiomar
Réserve naturelle géologique de Haute Provence, Digne-les-bains, France

Javier Fernández-López de Pablo
Institut Cataláde Paleoecologia Humana i Evolució Social, Zona Educacional 4 Campus Sescelades (Edifici W3), Tarragona, Spain
Área de Prehistória, Universitat Rovira i Virgili (URV), Tarragona, Spain

Ernestina Badal
Departament de Prehistória i Arqueologia, Facultat de Geografia i História, Universitat de Valéncia, Valéncia, Spain

Alfred Sanchis Serra and Carlos Ferrer García
Museu de Prehistória de Valéncia, SIP (Servei d'Investigació Prehistórica), Diputacióde Valéncia, Valéncia, Spain

Alberto Martínez-Ortí
Museu Valenciá d'História Natural & i\ Biotaxa, Valencia, Spain

Angelica Feurdean
Senckenberg Research Institute and Natural History Museum and Biodiversity and Climate Research Center, Frankfurt am Main, Germany
Romanian Academy "Emil Racovită" Institute of Speleology, Cluj Napoca, Romania

Shonil A. Bhagwat
Department of Geography, The Open University, Milton Keynes, United Kingdom
School of Geography and the Environment, University of Oxford, Oxford, United Kingdom
Long-Term Ecology Laboratory, Biodiversity Institute, Department of Zoology, University of Oxford, Oxford, United Kingdom
Biodiversity Institute, Oxford Martin Institute, Department of Zoology, University of Oxford, Oxford, United Kingdom

Katherine J. Willis
Long-Term Ecology Laboratory, Biodiversity Institute, Department of Zoology, University of Oxford, Oxford, United Kingdom
Biodiversity Institute, Oxford Martin Institute, Department of Zoology, University of Oxford, Oxford, United Kingdom
Department of Biology, University of Bergen, Bergen, Norway

H. John B Birks
Department of Geography, The Open University, Milton Keynes, United Kingdom
Department of Biology, University of Bergen, Bergen, Norway
Environmental Change Research Centre, University College London, London, United Kingdom

Heike Lischke
Dynamic Macroecology, Landscape Dynamics, Swiss Federal Institute for Forest, Snow and Landscape Research WSL, Birmensdorf, Switzerland

Thomas Hickler
Senckenberg Research Institute and Natural History Museum and Biodiversity and Climate Research Center, Frankfurt am Main, Germany
Biodiversity and Climate Research Centre and Senckenberg Gesellschaft für Naturforschung and Goethe University, Frankfurt am Main, Germany

Olev Vinn
Department of Geology, University of Tartu, Tartu, Estonia

Mark A. Wilson
Department of Geology, The College of Wooster, Wooster, Ohio, United States of America

Mari-Ann Mõtus
Institute of Geology, Tallinn University of Technology, Tallinn, Estonia

Danielle Fraser and Natalia Rybczynski
Department of Biology, Carleton University, Ottawa, Ontario, Canada
Palaeobiology, Canadian Museum of Nature, Ottawa, Ontario, Canada

Christopher Hassall
Department of Biology, Carleton University, Ottawa, Ontario, Canada
School of Biology, University of Leeds, Leeds, United Kingdom

Root Gorelick
Department of Biology, Carleton University, Ottawa, Ontario, Canada
Department of Mathematics and Statistics, Carleton University, Ottawa, Ontario, Canada
Institute of Interdisciplinary Studies, Carleton University, Ottawa, Ontario Canada

Thomas Tütken
Steinmann-Institut für Geologie, Mineralogie und Paläontologie, University of Bonn, Bonn, Germany

Thomas M. Kaiser
University Hamburg, Biozentrum Grindel, Hamburg, Germany

Torsten Vennemann
Institut de Géochimie, Université de Lausanne, Lausanne, Switzerland

Gildas Merceron
iPHEP UMR 7262 CNRS, Université de Poitiers, Poitiers, France

Jordan C. Mallon
Department of Biological Sciences, University of Calgary, Calgary, Alberta, Canada

Jason S. Anderson
Department of Comparative Biology & Experimental Medicine, University of Calgary, Calgary, Alberta, Canada

William C. H. Parr, Laura A. B. Wilson and Tracey L. Rogers
Evolution and Ecology Research Centre, School of Biological, Earth and Environmental Sciences, University of New South Wales, Sydney, New South Wales, Australia

Stephen Wroe
Function, Evolution and Anatomy Research laboratory, Zoology, School of Environmental and Rural Sciences, University of New England, New South Wales, Australia

Marie R. G. Attard
Evolution and Ecology Research Centre, School of Biological, Earth and Environmental Sciences, University of New South Wales, Sydney, New South Wales, Australia
Function, Evolution and Anatomy Research laboratory, Zoology, School of Environmental and Rural Sciences, University of New England, New South Wales, Australia

Michael Archer and Suzanne J. Hand
Evolution of Earth and Life Sciences Research Centre, School of Biological, Earth and Environmental Sciences, University of New South Wales, Sydney, New South Wales Australia

Baosheng Wang and Xiao-Ru Wang
Department of Ecology and Environmental Science, Umeå University, Umeå, Sweden

Jian-Feng Mao
National Engineering Laboratory for Forest Tree Breeding, Key Laboratory for Genetics and Breeding of Forest Trees and Ornamental Plants of Ministry of Education, Beijing Forestry University, Beijing, People's Republic of China

Wei Zhao
State Key Laboratory of Systematic and Evolutionary Botany, Institute of Botany, Chinese Academy of Sciences, Beijing, People's Republic of China

Peter J. Edmunds and Robert C. Carpenter
Department of Biology, California State University Northridge, Northridge, California, United States of America

Mehdi Adjeroud
Institut de Recherche pour le Développement, Unité de Recherche CoReUs, Observatoire Océanologique de Banyuls, Banyuls-sur-Mer, France
Laboratoire d'Excellence "CORAIL", Perpignan, France

Marissa L. Baskett
Department of Environmental Science and Policy, University of California Davis, Davis, California, United States of America

Iliana B. Baums
Department of Biology, The Pennsylvania State University, University Park, Pennsylvania, United States of America

Ann F. Budd
Department of Earth and Environmental Sciences, University of Iowa, Iowa City, Iowa, United States of America

Nicholas S. Fabina
Center for Population Biology, University of California Davis, Davis, California, United States of America

Tung-Yung Fan
National Museum of Marine Biology and Aquarium, Taiwan, Republic of China

Erik C. Franklin, Hollie M. Putnam and Ruth D. Gates
Hawaii Institute of Marine Biology, School of Ocean and Earth Science and Technology, University of Hawaii, Kaneohe, Hawaii, United States of America

Kevin Gross
Biomathematics Program, North Carolina State University, Raleigh, North Carolina, United States of America

Xueying Han
Department of Ecology, Evolution and Marine Biology and the Coastal Research Center, Marine Science Institute, University of California Santa Barbara, Santa Barbara, California, United States of America
National Center for Ecological Analysis and Synthesis, Santa Barbara, California, United States of America

Jennifer K. O'Leary
National Center for Ecological Analysis and Synthesis, Santa Barbara, California, United States of America

Lianne Jacobson
Department of Biology, California State University Northridge, Northridge, California, United States of America
Department of Biology, University of Florida, Gainesville, Florida, United States of America

James S. Klaus
Department of Geological Sciences, University of Miami, Coral Gables, Florida, United States of America

Tim R. McClanahan
Wildlife Conservation Society, Marine Program, Bronx, New York, United States of America

Madeleine J. H. van Oppen and Hugh Sweatman
Australian Institute of Marine Science, Townsville, Queensland, Australia

Xavier Pochon
The Cawthron Institute, Nelson, New Zealand

Tyler B. Smith
Center for Marine and Environmental Studies, University of the Virgin Islands, St. Thomas, Virgin Islands, United States of America

Michael Stat
The University of Western Australia Oceans Institute and the Centre for Microscopy, Characterisation and Analysis, University of Western Australia, Crawley, Western Australia, Australia

Robert van Woesik
Department of Biological Sciences, Florida Institute of Technology, Melbourne, Florida, United States of America

Trisha L. Spanbauer and Sherilyn C. Fritz
Department of Earth and Atmospheric Sciences and School of Biological Sciences, University of Nebraska–Lincoln, Lincoln, Nebraska, United States of America

Craig R. Allen
U.S. Geological Survey, Nebraska Cooperative Fish and Wildlife Research Unit, School of Natural Resources, University of Nebraska–Lincoln, Lincoln, Nebraska, United States of America

David G. Angeler
Department of Aquatic Sciences and Assessment, Swedish University of Agricultural Sciences, Uppsala, Sweden

Ahjond S. Garmestani and Tarsha Eason
Office of Research and Development, National Risk Management Research Laboratory, U.S. Environmental Protection Agency, Cincinnati, Ohio, United States of America

Kirsty L. Nash
Australian Research Council Centre of Excellence for Coral Reef Studies, James Cook University, Townsville, Queensland, Australia

Jeffery R. Stone
Department of Earth and Environmental Systems, Indiana State University, Terre Haute, Indiana, United States of America

Jean-Pierre Rossi and Martin Godefroid
Unité Mixte de Recherche 1062 Centre de Biologie pour la Gestion des Populations, Institut National Recherche Agronomique, Centre de Montpellier, France

Maxime Nardin, Luc Pâques and Philippe Rozenberg
Unité de Recherche 588 Amélioration Génétique et Physiologie Forestiéres, Institut National Recherche Agronomique, Centre d'Orléans, France

Manuela Ruiz-Diaz
Parque Tecnoló gico Misiones, Universidad Nacional de Misiones, Misiones, Argentina

Anne-Sophie Sergent
Consejo Nacional de Investigaciones Científicas y Técnicas, Bariloche, Argentina

Alejandro Martinez-Meier
Grupo de Ecologia Forestal, Instituto Nacional de Tecnología Agropecuaria, Estación Experimental Bariloche (INTA EEA Bariloche), Bariloche, Argentina

Hugo I. Martínez-Cabrera
Estación Regional del Noroeste, Instituto de Geología, Universidad Nacional Autónoma de México, Hermosillo, México

Emilio Estrada-Ruiz
Laboratorio de Ecología, Departamento de Zoología, Escuela Nacional de Ciencias Biológicas – Instituto Politécnico Nacional, Ciudad de México, México

Index

www.ingramcontent.com/pod-product-compliance
Lightning Source LLC
Chambersburg PA
CBHW080509200326
41458CB00012B/4138